Control Applications for Biomedical Engineering Systems

Control Applications for Biomedical Engineering Systems

Edited by

Ahmad Taher Azar

Robotics and Internet-of-Things Lab (RIOTU),
Prince Sultan University, Riyadh,
Saudi Arabia; Faculty of Computers and
Artificial Intelligence, Benha University,
Benha, Egypt

ACADEMIC PRESS

An imprint of Elsevier

ELSEVIER

Academic Press is an imprint of Elsevier
125 London Wall, London EC2Y 5AS, United Kingdom
525 B Street, Suite 1650, San Diego, CA 92101, United States
50 Hampshire Street, 5th Floor, Cambridge, MA 02139, United States
The Boulevard, Langford Lane, Kidlington, Oxford OX5 1GB, United Kingdom

Notices
Knowledge and best practice in this field are constantly changing. As new research and experience
broaden our understanding, changes in research methods, professional practices, or medical treatment
may become necessary.

Practitioners and researchers must always rely on their own experience and knowledge in evaluating
and using any information, methods, compounds, or experiments described herein. In using such
information or methods they should be mindful of their own safety and the safety of others, including
parties for whom they have a professional responsibility.

To the fullest extent of the law, neither the Publisher nor the authors, contributors, or editors, assume
any liability for any injury and/or damage to persons or property as a matter of products liability,
negligence or otherwise, or from any use or operation of any methods, products, instructions, or ideas
contained in the material herein.

Library of Congress Cataloging-in-Publication Data
A catalog record for this book is available from the Library of Congress

British Library Cataloguing-in-Publication Data
A catalogue record for this book is available from the British Library

ISBN 978-0-12-817461-6

For information on all Academic Press publications
visit our website at https://www.elsevier.com/books-and-journals

Publisher: Mara Conner
Acquisition Editor: Sonnini R. Yura
Editorial Project Manager: Emma Hayes
Production Project Manager: Nirmala Arumugam
Cover Designer: Greg Harris

Typeset by SPi Global, India

Working together
to grow libraries in
developing countries
www.elsevier.com • www.bookaid.org

Contents

Contributors

P. Abuin Institute of Technological Development for the Chemical Industry (INTEC), CONICET—Universidad Nacional del Litoral, Santa Fe, Argentina

Omid Aghajanzadeh Department of Mechanical Engineering, Sharif University of Technology, Tehran, Iran

Alma Y. Alanis Electronics and Computing Division, CUCEI, Universidad de Guadalajara, Guadalajara, Mexico

Mumtaz Ali Neurosurgery Department LRH, Peshawar, Pakistan

Vibha Bhalerao Jijamata Hospital and IVF Center, Nanded, India

Mathias Blandeau Université Polytechnique Hauts-de-France CNRS, Valenciennes, France

Manuchi Dansa Department of Electronics and Telecommunication Engineering (DETEL), State University of Rio de Janeiro (UERJ), Rio de Janeiro, Brazil

Paolo Di Giamberardino Department of Computer, Control and Management Engineering, Sapienza University of Rome, Rome, Italy

Urmila Diwekar Vishwamitra Research Institute, Stochastic Research Technologies LLC, The University of Illinois at Chicago, Chicago, IL, United States

Ali Falsafi Department of Mechanical Engineering, Swiss Federal Institute of Technology Lausanne, Lausanne, Switzerland

A. Ferramosca Facultad Regional de Reconquista, CONICET, UTN, Santa Fe, Argentina

Aldo Pardo Garcia A&C, Automation and Control Group, Universidad de Pamplona, Pamplona, Colombia

J.A. García-Rodríguez Electronics and Computing Division, CUCEI, Universidad de Guadalajara, Guadalajara, Mexico

J.L. Godoy Institute of Technological Development for the Chemical Industry (INTEC), CONICET—Universidad Nacional del Litoral, Santa Fe, Argentina

A.H. González Institute of Technological Development for the Chemical Industry (INTEC), CONICET—Universidad Nacional del Litoral, Santa Fe, Argentina

Thierry-Marie Guerra Université Polytechnique Hauts-de-France CNRS, Valenciennes, France

Wassim M. Haddad School of Aerospace Engineering, Georgia Institute of Technology, Atlanta, GA, United States

Jorge Herrera Departamento de Ingeniería, Universidad de Bogotá Jorge Tadeo Lozano, Bogotá, Colombia

Mehdi Hosseinzadeh Department of Control Engineering and System Analysis, The Free University of Brussels, Brussels, Belgium

Daniela Iacoviello Department of Computer, Control and Management Engineering, Sapienza University of Rome, Rome, Italy

Asier Ibeas Escola d'Enginyeria, Autonomous University of Barcelona, Barcelona, Spain

M. Ishfaq South Al Qunfudhah General Hospital, Al Qunfudhah, Saudi Arabia

Davide Manca PSE-Lab, Process Systems Engineering Laboratory, Dipartimento di Chimica, Materiali e Ingegneria Chimica "Giulio Natta," Politecnico di Milano, Milano, Italy

Eduardo Márquez-Martín Medical-Surgical Unit of Respiratory Diseases, Instituto de Biomedicina de Sevilla (IBiS), University Hospital Virgen del Rocío, Seville, Spain

Nader Meskin Department of Electrical Engineering, Qatar University, Doha, Qatar

NasimUllah Department of Electrical Engineering, College of Engineering, Taif University, Taif, Kingdom of Saudi Arabia

Apoorva Nisal The University of Illinois, Chicago, IL, United States

Tiago Roux Oliveira Department of Electronics and Telecommunication Engineering (DETEL), State University of Rio de Janeiro (UERJ), Rio de Janeiro, Brazil

Francisco Ortega-Ruiz Medical-Surgical Unit of Respiratory Diseases, Instituto de Biomedicina de Sevilla (IBiS), University Hospital Virgen del Rocío, Seville, Spain

Regina Padmanabhan Department of Electrical Engineering, Qatar University, Doha, Qatar

Victor Hugo Pereira Rodrigues Department of Electronics and Telecommunication Engineering (DETEL), State University of Rio de Janeiro (UERJ), Rio de Janeiro, Brazil

Philippe Pudlo Université Polytechnique Hauts-de-France CNRS, Valenciennes, France

Muhammad Mohsin Rafiq Aprus Technologies Pvt Ltd, Peshawar, Pakistan

Javier Reina-Tosina Biomedical Engineering Group, Universidad de Sevilla, Seville, Spain

Y.Yuliana Rios GAICO, Grupo de Automatización y Control, Universidad Tecnológica de Bolívar, Cartagena, Bolívar, Colombia

Pablo S. Rivadeneira Universidad Nacional de Colombia, Facultad de Minas, Grupo GITA, Medellin, Colombia; Institute of Technological Development for the Chemical Industry (INTEC), CONICET—Universidad Nacional del Litoral, Santa Fe, Argentina

Laura María Roa-Romero Biomedical Engineering Group, Universidad de Sevilla, Seville, Spain

E. Ruiz-Velázquez Electronics and Computing Division, CUCEI, Universidad de Guadalajara, Guadalajara, Mexico

Edgar N. Sanchez Electrical Engineering Department, CINVESTAV, Zapopan, Mexico

Adriana Savoca PSE-Lab, Process Systems Engineering Laboratory, Dipartimento di Chimica, Materiali e Ingegneria Chimica "Giulio Natta," Politecnico di Milano, Milano, Italy

J.E. Sereno Universidad Nacional de Colombia, Facultad de Minas, Grupo GITA, Medellin, Colombia; Institute of Technological Development for the Chemical Industry (INTEC), CONICET—Universidad Nacional del Litoral, Santa Fe, Argentina

Mojtaba Sharifi Department of Mechanical Engineering, Sharif University of Technology, Tehran, Iran

Alejandro Talaminos-Barroso Biomedical Engineering Group, Universidad de Sevilla, Seville, Spain

Fleur T. Tehrani California State University, Fullerton, CA, United States

Kirti Yenkie Chemical Engineering, Rowan University, Glassboro, NJ, United States

Foreword

During the past one hundred years, medicine and healthcare have been revolutionized due to numerous innovations and staggering developments in many areas of science and technology. Just gazing around any hospital setting shows the extent that healthcare has been improved due to technological breakthroughs. From the laboratories where various devices such as sophisticated microscopes and gas or blood analyzers are routinely used, to the imaging departments equipped with ultrasound, computerized axial tomography, magnetic resonance imaging, and positron emission tomography scanners, to the intensive care and the operating rooms where numerous medical devices such as ventilators, heart-lung machines, and robotic surgical equipment are utilized, one can see the extent that healthcare has become dependent upon technological advancements.

Biomedical engineering, which is basically the application of science and technology in medicine, has grown to a vast and rapidly developing field of engineering. Biomedical engineers combine the knowledge in physiology, biology, physics, chemistry, and the electrical, mechanical, and chemical engineering fields to design and develop biomaterials, sensors and physiological monitors, prostheses devices and artificial organs, imaging equipment, artificial intelligence (AI) systems, and many automatic controllers in various fields of medicine, to name a few.

Some of the rapidly advancing areas of biomedical engineering at the present time are in the applications of AI and expert systems as well as development of closed-loop controllers for various types of diagnosis and patient treatment. A biomedical engineer involved in the development of an AI or expert system needs to acquire detailed knowledge about the physiology of the part of the human body targeted by the treatment. If the system is developed for drug delivery, the pharmacokinetics and pharmacodynamics of the drug need to be thoroughly studied and included in the system. If the system is designed for diagnosis and/or treatment purposes, the biomedical engineer needs to study and consult various medical protocols and may have to design a simulation model of the part of the body targeted by the treatment and include that model in the system. In order to move away from black-box representation, the simulation model may need to be isomorphic and fairly detailed mathematically. Such models can be quite useful in the diagnosis as well as treatment of patients. The resulting AI or expert systems can provide the foundations for closed-loop controllers for anesthesia and drug delivery, to ventilators, cardiac pacemakers, and artificial organs, as few examples.

Applications of control techniques in biomedical engineering are diverse, numerous, and expanding. Various technologies from deterministic and classical methods to neurofuzzy technologies are applied in control of medical systems and devices.

Biomedical engineers normally go through several steps to develop control systems and devices. The first step is to acquire in-depth knowledge about the physiological/biological characteristics of the part of the patient's body that will be subjected to analysis/treatment. The second step may involve the design, development, testing, and verifying the effectiveness of a model of the patient's organ. The next step is to design and develop a system that can process monitored and/or simulated patient data and can be used for diagnosis or the intended treatment. The biomedical engineer should also develop user-friendly interfaces to communicate data with the medical personnel.

If the system is intended for use as a closed-loop controller for any medical procedure, the next step is to use proper negative feedback techniques and develop a robust and stable system. This system needs to go through thorough clinical testing to assess its effectiveness and safety. If the tests are successful, the final step is to review the details of the requirements of the regulatory organizations (e.g., the FDA) and to make sure that the system's compliance with those requirements is properly documented to obtain the necessary regulatory approvals.

The aforementioned steps in the design, development, testing, and final application and implementation of medical control systems require multiple specialties and may not all be taken by the same biomedical engineer. In fact, these tasks are normally performed by a team of biomedical engineers, scientists, and physicians who have the required specialized knowledge and expertise at different levels of product development.

The end products as medical control systems are normally quite impactful. The successful controllers help provide optimal treatment to patients, significantly reduce medical errors that cause countless morbidities and mortalities all over the world annually, and considerably reduce the healthcare costs. In brief, an effective medical controller can save lives and reduce healthcare costs at the same time.

This book that has been carefully arranged and edited by Prof. Ahmad Taher Azar, is prepared to provide the readers with valuable information about the applications of control techniques in various biomedical engineering systems. The intended readers of the book are researchers in the field of biomedicine. The book provides the descriptions of some of the advanced control systems that are currently in practical use in healthcare worldwide. The subject of the book is timely and the applications of control technologies in medicine are vast and are bound to increase in the years to come.

Fleur T. Tehrani
Electrical Engineering, California State University,
Fullerton, CA, United States

Preface

The field of control and systems has been connected to biological systems and biotechnology for many decades, going back to the work of Norbert Wiener on cybernetics in 1965, the work of Walter Cannon on homeostasis in 1929, and the early work of Claude Bernard on the milieu interieur in 1865. Nonetheless, the impact of control and systems on devices and applications in the field of biology has only emerged in recent years.

There is increasing number of researchers, who fuse biomedical control engineering principles with the knowledge and tools of molecular life sciences in order to solve contemporary problems through the measurement, modeling, and rational manipulation of biological systems. These researchers are more rapidly creating biology-based technologies to benefit a range of diverse areas, including human and environmental health, agriculture, manufacturing, and defense. Recently, many researchers in biomedical control engineering have explored complicated problems arising from societal needs and concerns and directed leading-edge research teams to address those challenge problems, in which design and control of biofunctionalized systems based on measured biological responses are the central issue. Through this book, we wish to deliver essential and advanced bioengineering information in applications of control technologies in life science. In the next few years, there will surely be much more exciting developments in this area. Judging by what we have witnessed so far, this exciting field of control systems in bioengineering is likely to produce revolutionary breakthroughs over the next decade. In this book, we aim at (i) pointing out theoretical and practical issues to biomedical control systems, (ii) bringing together solutions developed under different settings with specific attention to the validation of these tools in biomedical settings using real-life datasets and experiments, and (iii) introducing significant case studies.

About the book

The new Elsevier book, *Control Applications for Biomedical Engineering Systems*, consists of 14 contributed chapters by subject experts who are specialized in the various topics addressed in this book. The special chapters have been brought out in this book after a rigorous review process in the broad areas of control engineering and biomedical systems. Special importance was given to chapters offering practical solutions and novel methods for the recent research problems in the mathematical modeling and control applications of biomedical systems. This book presents some of the latest innovative approaches to medical diagnostics and procedures as well as clinical rehabilitation from a point of view of dynamic modeling, system analysis, and control.

Objectives of the book

The goal of this book is to provide a forum for latest research in biomedical signal measurement and processing, dynamic modeling, analysis, and control for clinical diagnosis, patient health monitoring, drug administration, and biosignal-assisted rehabilitation. There has been a significant increase in research activities in these areas within diverse specialties, including mechanical, electrical, and biomedical engineering. Developing sensors to produce appropriate biosignals, developing dynamic models of biosystems and biosignals for diagnostics, and using biosignals as feedback in controlled processes such as drug delivery and rehabilitation are some of the biggest challenges encountered in these engineering fields.

Organization of the book

This well-structured book consists of 14 full chapters.

Book features

- The book chapters deal with the recent research problems in the areas of biomedical control engineering.
- The book chapters present various mathematical techniques for biomedical systems.
- The book chapters contain a good literature survey with a long list of references.
- The book chapters are well written with a good exposition of the research problem, methodology, block diagrams, and mathematical techniques.
- The book chapters are lucidly illustrated with simulations.
- The book chapters discuss details of engineering applications and future research areas.

Audience

The book is primarily meant for researchers from academia and industry, who are working in the research areas—control engineering, biomedical engineering, electrical engineering, and computer engineering. The book can also be used at the graduate or advanced undergraduate level as a textbook or major reference for courses such as biomedical control systems, modeling of dynamical systems, selected topics in biomedical engineering, numerical simulation, and many others.

Acknowledgments

As editor, I hope that the chapters in this well-structured book will stimulate further research in mathematical modeling and control of biomedical systems, and utilize them in real-world applications.

I hope that this book, covering so many different topics, will be very useful for all readers.

I would like to thank all the reviewers for their diligence in reviewing the chapters.

Special thanks go to Elsevier, especially the book Editorial Project Manager Emma Hayes and Production Project Manager Nirmala Arumugam.

No words can express my gratitude to the Acquisitions Editor, Sonnini Ruiz Yura, for her great effort and support during the publication process.

Special acknowledgment to Prince Sultan University and Robotics and Internet-of-Things Lab (RIOTU), Riyadh, Saudi Arabia for giving me the opportunity to finalize this book.

Ahmad Taher Azar
Robotics and Internet-of-Things Lab (RIOTU),
Prince Sultan University, Riyadh, Saudi Arabia,
Faculty of Computers and Artificial Intelligence,
Benha University, Benha, Egypt
http://www.bu.edu.eg/staff/ahmadazar14
https://sites.google.com/site/drahmadtaherazar/

Neuro-fuzzy inverse optimal control incorporating a multistep predictor as applied to T1DM patients

1

Alma Y. Alanis[a], Y. Yuliana Rios[d], J.A. García-Rodríguez[a], Edgar N. Sanchez[b], E. Ruiz-Velázquez[a], Aldo Pardo Garcia[c]
[a]Electronics and Computing Division, CUCEI, Universidad de Guadalajara, Guadalajara, Mexico, [b]Electrical Engineering Department, CINVESTAV, Zapopan, Mexico, [c]A&C, Automation and Control Group, Universidad de Pamplona, Pamplona, Colombia, [d]GAICO, Grupo de Automatización y Control, Universidad Tecnológica de Bolívar, Cartagena, Bolívar, Colombia

1 Introduction

Diabetes mellitus (DM) is a chronic disorder of carbohydrate, fat, and protein metabolism. Currently, this disease is affecting more than 422 million adults worldwide. There were 1.5 million deaths in 2012 directly related to DM, according to the most recent data reported by the World Health Organization (WHO, 2016). Diabetes is classified into type 1 diabetes mellitus (T1DM) and type 2 diabetes mellitus (T2DM). In T1DM, also known as *autoimmune diabetes*, the pancreas reacts in an autoimmune way and destroys its β-cells which are responsible for insulin production. In the absence of insulin, glucose is not properly metabolized to be used as energy by the body. In fact, patients completely depend on an external infusion of insulin to survive. This type of diabetes is present since birth and usually diagnosed during childhood and adolescence (Katsarou et al., 2017). T2DM is a gradual disease which develops frequently in adulthood; it is more prevalent than T1DM and related to overweight and a sedentary lifestyle. In T2DM the pancreas is able to produce insulin, but the body tissues exhibit *insulin resistance* (International Diabetes Federation, 2009); therefore, insulin does not carry out its metabolizing task and glucose remains in the bloodstream. For both types, glucose levels increase abnormally to more than 300 mg/dL after food intake (postprandial), a condition called *hyperglycemia*. In addition, patients under insulin injections treatment may experience *hypoglycemia* events, with glucose levels below 70 mg/dL, if the dose is larger than required. These events where glucose control is deficient are risky if they last for a long time. Moreover, hyperglycemia and hypoglycemia are strongly associated with macrovascular damage, such as coronary, cerebral, and peripheral vascular failure, as well as, microvascular diseases like retinopathy, nephropathy, and neuropathy. These long-term conditions result in increasing disability, reduced life expectancy, and enormous health costs for families and government (International Diabetes Federation, 2009). Currently, there is no cure

Control Applications for Biomedical Engineering Systems. https://doi.org/10.1016/B978-0-12-817461-6.00001-9

for diabetes. The current and traditional treatment for people with T1DM is based on discrete blood glucose (BG) measurements and insulin injections. Emerging technologies have focused on the development of more accurate sensors for continuous glucose monitoring and also continuous subcutaneous insulin infusion devices. The scientific community is making big efforts to develop the artificial pancreas (AP), a completely autonomous biomedical device to carry out glucose level control for T1DM patients, with the main objective of delivering the required amount of insulin. The AP offers a promising approach: having a strict glycemic control would reduce severe complications linked to the disease and thus achieve a better quality of life. Numerous proposals of control algorithms aimed at the development of the AP have been reported. Laguna Sanz et al. (2017) proposed an optimal insulin infusion algorithm based on a zone model predictive control; information is taken from the so-called prediction residuals and the control algorithm is successfully adapted, with a significant hypoglycemia reduction. Bamgbose et al. (2017) have used the model of a virtual patient from the AIDA diabetes simulator, a time-shifted neural network predictor, and a PI controller to compute the correct insulin infusion; they suggested a semistatic patient data protocol and a meal ingestion pattern for validation. One of the most important contributions is made by Messori et al. (2018) where they have proposed individualized MPC algorithms based on the idea that patient's glucose-insulin dynamics is quite different; the results were completely outstanding when tested in the Uva/Padova simulator research version. In an innovative application, Romero-Aragon et al. (2015) have reported an inverse optimal controller using neural network techniques, which is implemented in a Field Programmable Gate Array, with good performance. Ruiz-Velázquez et al. (2004) developed a proposal based on $\mathcal{H}_\infty$ techniques to obtain a robust controller for BG regulation, with good results for reference tracking. A diversity of relevant works have been done with insulin feedback for BG control (Garcia-Gabin and Jacobsen, 2013; Hashimoto et al., 2014; Liu and Ying, 2015) such as model predictive control (Batora et al., 2015; Messori et al., 2015; Wang et al., 2014), robust control schemes (Colmegna et al., 2014; Femat et al., 2009; Morales-Contreras et al., 2017), and neural networks strategies (Leon et al., 2014; Rios et al., 2018).

In this chapter, a novel neuro-fuzzy control scheme of BG regulation for virtual T1DM patients is proposed. A neural multistep predictor is incorporated in order to estimate glucose dynamics within a 15-min horizon. Thus, the prediction of future data allows determining the convenient basal infusion rate that is defined by fuzzy membership functions. Implementation using the well-known Uva/Padova simulator illustrates that the proposed neuro-fuzzy controller exhibits a good performance to maintain *normoglycemia* for virtual populations of adults, adolescents, and children when compared with two other similar control approaches.

The chapter is outlined as follows: Section 2 reveals the most recent related works. Section 3 presents a recurrent high-order neural network (RHONN) scheme to identify nonlinear systems dynamics using extended Kalman filter (EKF) as the training algorithm, and describes trajectory tracking in positive systems using inverse optimal control (IOC) strategy. In Section 4, the Uva/Padova simulator is briefly described. Section 5 includes the proposed feedback control scheme for BG regulation in

T1DM patients. In Section 6, this scheme is tested using the Uva/Padova simulator; furthermore, three control schemes are compared. In Section 7, a brief results discussion is contained to highlight the proposal relevance and its advantages. Finally, conclusions are stated in Section 8.

2   Related work

The continuous research on diseases treatment has provided the scientific community with very interesting findings. Definitely, the advance in T1DM knowledge brings new important challenges, which must be considered. The most recent works have proposed innovative control proposals, which increase the application landscape of an AP device, as summarized by Cinar (2018). T1DM condition has been treated from cellular levels, that is, glucose sensing. Glucose monitoring is a determining factor for a closed-loop system to be effective. Huyett et al. (2018) have clearly analyzed the effect of the glucose sensor lag on the overall control performance. A relevant improvement in postprandial hyperglycemia could be achieved by reducing sensor lag and intraperitoneal insulin delivery. Many researchers agree that intelligent control and model-based predictive control strategies should be explored to develop the AP. A predictive insulin suspension scheme has been proposed by Bequette et al. (2018); this prediction is based on the Kalman filter and they have reported an interesting risk reduction to overnight hypoglycemia mitigation. Another extensive review was presented by Bondia et al. (2018) from the nonlinear systems point of view. This paper reveals the complex mathematical variability for control strategies based on physiological models. Enhanced glycemic control for the postprandial events can be achieved if the meal absorption dynamics and insulin sensitivity are integrated into the controller algorithm design, as stated by El Fathi et al. (2018). Furthermore, Turksoy et al. (2018) have presented a promising contribution using a multivariable algorithm based on adaptive control techniques. The use and incorporation of different strategies for T1DM treatment are truly interesting. Undoubtedly there are still many open challenges which the science community must deal with T1DM such as treatment during physical activity, carbohydrate and protein content in meal, glucose variability under stress conditions, and sleeping, among others. Thus, it is expected that some of these issues can be covered with the proposals addressed in this chapter.

3   Fundamentals

3.1   Online discrete-time neural network

RHONN for modeling of nonlinear systems was used by Sanchez et al. (2008), Romero-Aragon et al. (2015), and Quintero-Manriquez et al. (2017). In order to obtain such model, a discrete-time nonlinear system with a disturbance is considered,

$$x_{k+1} = f(x_k) + g(u_k) + d(x_k),\tag{1}$$

where $f(\bullet) \in \mathfrak{R}^n \to \mathfrak{R}^n$ and $g(\bullet) \in \mathfrak{R}^n \to \mathfrak{R}^{n \times m}$ are smooth mappings, $d(\cdot)$ is a bounded unknown term which represents internal and external disturbances, and uncertain parameters. $x_k \in \mathfrak{R}^n$ is the state of system at time $k \in \mathbb{N} \cup \{0\}$, $u \in \mathfrak{R}^m$ is the control input vector. On the other hand, the neural identifier model is proposed as follows:

$$\chi_{i,k+1} = w_i^T \varphi_i(x_k, u_k), \quad i = 1, \dots n,\tag{2}$$

where $\chi_i \in \mathfrak{R}^L$ is the ith identifier neuron for the $x_{i,k}$ state, $w_i \in \mathfrak{R}^{L_i}$ is a vector of adapted weights, $u \in \mathfrak{R}^{L_i}$ is the input vector to the RHONN model, and φ_i is defined as

$$\varphi_i(x_k, u_k) = \begin{bmatrix} \varphi_{i_1} \\ \varphi_{i_2} \\ \vdots \\ \varphi_{i_{L_i}} \end{bmatrix} = \begin{bmatrix} \prod_{j \in I_1} \xi_{ij}^{dij(1)} \\ \prod_{j \in I_2} \xi_{ij}^{dij(2)} \\ \vdots \\ \prod_{j \in I_{L_i}} \xi_{ij}^{dij(L_i)} \end{bmatrix},\tag{3}$$

where $\{I_1, I_2, \dots, I_{L_i}\}$ is a collection of not ordered subsets of $\{1, 2, \dots, n + m\}$, n is the dimension of the state, m is the number of external inputs, $d_{ij,k}$ are nonnegative integers, L_i is the number high-order connections. ξ_{ij} can be expressed as

$$\xi_i = \begin{bmatrix} \xi_{i_1} \\ \vdots \\ \xi_{i_n} \\ \xi_{i_{n+1}} \\ \vdots \\ \xi_{i_{n+m}} \end{bmatrix} = \begin{bmatrix} S(x_1) \\ \vdots \\ S(x_n) \\ u_1 \\ \vdots \\ u_m \end{bmatrix},\tag{4}$$

the hyperbolic tangent function $S(\cdot)$ is given as

$$S(\varsigma) = \mu_i \tanh(\beta_i \varsigma),\tag{5}$$

where ς is a real variable, μ and β are positive constants. An RHONN scheme is displayed in Fig. 1.

For controlling nonlinear system (1), its controllability must be guaranteed; due to this reason a modification to Eq. (2) by Rovithakis and Christodoulou (2000) is proposed as follows:

$$\chi_{i,k+1} = w_i^T \varphi_i(x_k) + w_i'^T \psi_i(x_k, u_k),\tag{6}$$

where w_i is an adapted weights vector and w'_i is a fixed weights vector which ensures controllability, ψ_i is a linear function of the dynamical model or network external inputs.

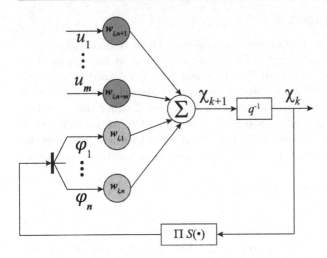

Fig. 1 Discrete-time
RHONN scheme.

The EKF algorithm as a learning method for RHONN training is used. The EFK determines the optimal values of RHONN weights for minimizing the prediction error. This algorithm is defined by

$$K_{i,k} = P_{i,k}H_{i,k}[R_{i,k} + H_{i,k}^T P_{i,k}H_{i,k}]^{-1},$$

$$w_{i,k+1} = w_{i,k} + \eta K_{i,k}e_{i,k}, \tag{7}$$

$$P_{i,k+1} = P_{i,k} - K_{i,k}H_{i,k}^T P_{i,k} + Q_{i,k},$$

with

$$H_{ij,k} = \left[\frac{\partial x_{i,k}}{\partial w_{ij,k}}\right]^T_{w_{i,k}=w_{i,k+1}}, \tag{8}$$

$$e_{i,k} = x_{i,k} - \chi_{i,k}, \tag{9}$$

where $K_i \in \mathfrak{R}^{L_i \times m}$ is the Kalman gain matrix, $P_i \in \mathfrak{R}^{L_i \times L_i}$ is the prediction error-associated covariance matrix, $H_i \in \mathfrak{R}^{L_i \times m}$ is a matrix where each entry $(H_{i,\ j})$ is the derivative of one neural network state x_i respecting the neural network weights vector $wi, j, i = 1, ..., n, j = 1, ..., L_i, R_i \in \mathfrak{R}^{m \times m}$ is the measurement noise-associated covariance matrix, $w_i \in \mathfrak{R}^{L_i}$ is the weight vector, η_i is a design parameter, $e_i \in \mathfrak{R}$ is the respective identification error, $Q_i \in \mathfrak{R}^{L_i \times L_i}$ is the noise-associated covariance matrix of the state, and L_i is the number of neural network weights. $P_i, Q_i,$ and R_i as a diagonal matrix are initialized, the EFK method has been detailed by Song and Grizzle (1992) and an analysis of the learning convergence and robustness was done by Alanis et al. (2007).

3.2 Inverse optimal control

For optimal control of nonlinear systems, a Hamilton-Jacobi-Bellman (HJB) partial differential equation is required to be solved; such solution may not exist or may be extremely difficult to obtain Freeman and Kokotović (2009). The IOC approach for different applications with an effective control law was detailed by Almobaied et al. (2018), Ornelas et al. (2011), and Sanchez and Ornelas-Tellez (2013). The IOC technique avoids to solve the HJB partial differential equation using a quadratic control Lyapunov function (CLF), which guarantees system stability. This function is used to define a cost functional, which is minimized.

A nonlinear affine system can be represented as

$$x_{k+1} = f(x_k) + g(x_k) u_k \quad x_0 = x(0), \tag{10}$$

where $x \in \mathfrak{R}^n$ is the system state at time $k \in \mathbb{N} \cup \{0\}$, $u \in \mathfrak{R}^m$ is the control law, $f: \mathfrak{R}^n \to \mathfrak{R}^n$ and $g: \mathfrak{R}^n \to \mathfrak{R}^{n \times m}$ are smooth and bounded mappings, such as $f(0) = 0$, and $\mathbb{N}$ denotes the nonnegative integers set. The tracking error z_k is defined as follows:

$$z_k = x_k - x_{\delta,k}, \tag{11}$$

where $z_k \in \mathfrak{R}^n$, $x_{\delta,k}$ is a desired trajectory. Then, the error dynamics at $(k + 1)$ is given as

$$\begin{aligned}
z_{k+1} &= x_{k+1} - x_{\delta,k+1}, \quad z(0) = z_0, \\
z_{k+1} &= f(x_k) + g(x_k)u_k - x_{\delta,k+1}.
\end{aligned} \tag{12}$$

To achieve trajectory tracking for system (12), an optimal control law u_k is proposed, using the following cost function:

$$J(z_k) = \sum_{n=k}^{\infty} \left(l(z_n) + u_n^T R(z_n) u_n \right) \quad J(0) = 0, \tag{13}$$

with $J(z_k): \mathfrak{R}^n \to \mathfrak{R}^+$; $l(z_k): \mathfrak{R}^n \to \mathfrak{R}^+$ is a positive semidefinite function, and $R(z_k): \mathfrak{R}^n \to \mathfrak{R}^{m \times m}$, $R(z_k) = R^T(z_k) > 0$ is a matrix-valued function for all x_k, defined by Kirk (1970). Then,

$$\begin{aligned}
J(z_k) &= l(z_k) + u_k^T R(z_k) u_k \\
&\quad + \sum_{n=k+1}^{\infty} \left(l(z_n) + u_n^T R(z_n) u_n \right) \\
&= l(z_k) + u_k^T R(z_k) u_k + J(z_{k+1}), \\
J^*(z_k) &= \min_{u_k} \left(l(z_k) + u_k^T R(z_k) u_k + J^*(z_{k+1}) \right).
\end{aligned} \tag{14}$$

A Lyapunov function $V(x_k)$, using the optimal function $J^*(x_k)$ in Eq. (14), can be established as

$$V(z_k) = \min_{u_k} \left\{ l(z_k) + u_k^T R(z_k) u_k + V(z_{k+1}) \right\}. \tag{15}$$

The boundary condition $V(0) = 0$ should be fulfilled. The value at $(k + 1)$ of this function depends on both z_k and u_k by means of z_{k+1}. According to Basar and Olsder (1999), Ohsawa et al. (2010), and Al-Tamimi et al. (2008), $V(z_k)$ satisfies the discrete-time Bellman equation.

In order to establish conditions for the optimal control law, the discrete-time Hamiltonian function $H(z_k, u_k)$ is introduced as follows.

$$H(z_k, u_k) = l(z_k) + u_k^T R(z_k) u_k + V(z_{k+1}) - V(z_k), \tag{16}$$

with

$$\frac{\partial H(z_k, u_k)}{\partial u_k} = 0. \tag{17}$$

Then,

$$0 = 2R(z_k) u_k + \frac{\partial V(z_{k+1})}{\partial u_k} \tag{18}$$

$$= 2R(z_k) u_k + \frac{\partial z_{k+1}}{\partial u_k} \frac{\partial V(z_{k+1})}{\partial u_k} \tag{19}$$

$$= 2R(z_k) u_k + g^T(x_k) \frac{\partial V(z_{k+1})}{\partial u_k}. \tag{20}$$

Hence, trajectory tracking using optimal control law is defined as

$$u_k^* = -\frac{1}{2} R^{-1}(z_k) g^T(x_k) \frac{\partial V(z_{k+1})}{\partial z_{k+1}}, \quad V(0) = 0. \tag{21}$$

In order to track the desired trajectory for system (12) using u_k^* (Eq. 21), a definition is proposed as follows:

Definition 1. Consider the tracking error as Eq. (11). The control law defined in Eq. (21) will be inverse optimal stabilizing along the desired trajectory $x_{\delta,k}$ if:

(i) for system (10) achieves (global) asymptotic stability of $x_k = 0$, along reference $x_{\delta,k}$; and

(ii) $V(z_k)$ is (radially unbounded) positive definite function such that inequality

$$\overline{V} := V(z_{k+1}) - V(z_k) + u_k^{*T} R(z_k) u_k^* \leq 0$$

is satisfied.

Selecting $l(z_k) := -\overline{V}$, the cost functional (13) is minimized; moreover, V is solution for Eq. (16). In order to satisfy conditions (i) and (ii) in Definition 1, a quadratic candidate CLF $V(z_k)$ is proposed as

$$V(z_k) = \frac{1}{2} z_k^T P z_k \quad P = P^T > 0, \tag{22}$$

with optimal control law u_k^* (Eq. 21), using $V(z_k)$ (Eq. 22), the tracking error stability z_k in Eq. (11) is ensured, and u_k^* is given as

$$u_k^* = -\frac{1}{4} R^{-1}(z_k) g^T(x_k) \frac{\partial z_{k+1}^T P z_{k+1}}{\partial z_{k+1}}. \tag{23}$$

Biological systems are positive, that is, states and outputs are nonnegative; moreover, the initial conditions and the inputs are also nonnegative (Sanchez and Ornelas-Tellez, 2013), the optimal control law is rewritten as

$$u_k^* = \left| -\frac{1}{2} (R(z_k) + P_2(x_k))^{-1} P_1(x_k, x_{\delta,k}) \right|, \tag{24}$$

with

$$P_1(x_k, x_{\delta,k}) = \begin{cases} g^T(x_k) P(f(x_k) - x_{\delta,k+1}) \\ \quad \text{for } f(x_k) \geq x_{\delta,k+1} \\ \\ g^T(x_k) P(x_{\delta,k+1} - f(x_k)) \\ \quad \text{for } f(x_k) \leq x_{\delta,k+1} \end{cases}, \tag{25}$$

$$P_2(x_k) = \frac{1}{2} g^T(x_k) P g(x_k), \tag{26}$$

and

$$R(z_k) = \frac{x_k^T r x_k}{\| x_{\delta,k+1} \|}, \tag{27}$$

where $P_1(\bullet)$, $P_2(x_k)$, and r are positive definite symmetric matrices, $(R(z_k) + P_2(x_k)) > 0$. These conditions ensure the existence of the inverse in Eq. (24). $V(z_k)$ (Eq. 22) is a CLF, then the control law is inverse optimal using the cost functional (13).

4 The Uva/Padova T1DM simulator

The Uva/Padova simulator scheme for the glucose-insulin system is illustrated in Fig. 2. In this scheme, the gastrointestinal tract represents glucose ingestion and absorption. Glucose appearance rate, that is, the glucose transit through the stomach and intestine crosses three compartments (two for stomach and one for gut), is described. The glucose system is represented by two compartments (glucose mass in plasma and rapid equilibrating tissues, and slowly equilibrating tissues). For the insulin system, two compartments (liver and plasma) are considered. These systems have been detailed by Man et al. (2007). In 2013, new features were added to the

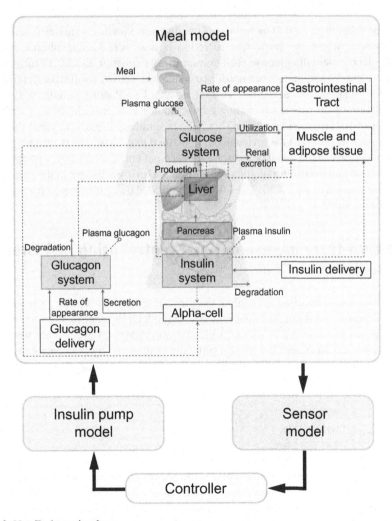

Fig. 2 Uva/Padova simulator.

Table 1 CVGA boundary zones (Magni et al., 2008).

Zones	Control assessment	Limits
A-zone	Accurate control	X: 110–90 and Y: 110–180 mg/dL
B-zone	Benign deviations in hypo/ hyperglycemia	X: 90–70 and Y: 180–300 mg/dL
C-zone	Over-correction in hypo/ hyperglycemia	X: 70–50; 110–90 and Y: 300–400; 110–180 mg/dL
D-zone	Failure to deal with hypo/ hyperglycemia	X: 70–50; 90–70 and Y: 180–300; 300–400 mg/dL
E-zone	Erroneous control	X: 70–50 and Y: 300–400 mg/dL

simulator (Man et al., 2014) as follows. The glucagon kinetics is modeled using one compartment, where the respective secretion is determined using plasma insulin, plasma glucose, and the glucose change rate; for the dynamic model, 18 differential equations and 39 parameters are used. See Man et al. (2014) for further details.

In this chapter, the academic version of the Uva/Padova simulator is used. A population of 30 patients including adults, adolescents, and children (distributed in groups of 10) are considered. The Uva/Padova simulator uses the Control Variability Grid Analysis (CVGA) as a visualization method to evaluate the controller performance according to the localization zone for each patient. This CVGA is plotted with the minimum and maximum of glucose level for each virtual patient in the simulation time lapse (Magni et al., 2008). The CVGA are gridded using different color zones, as described in Table 1.

5 Neuro-fuzzy inverse optimal control using multistep prediction

The control action required to regulate the glucose level of the virtual patients, from the Uva/Padova simulator, is based on a neural model. For identification, the virtual patient inputs are selected as the total glucose absorbed with every meal (carbohydrates intake amount), and the insulin (mU/min) calculated by the control law. The virtual patient output is the BG level (mg/dL); for this identification, an RHONN structure is used as follows:

$$\begin{aligned} \chi_{1,k+1} &= w_1' x_2, \\ \chi_{2,k+1} &= w_1 S(x_1) S(x_2) + w_2 S(x_1) + w_3 S(x_2) \\ &\quad + w_2' x_2 + w_3' u, \end{aligned} \tag{28}$$

where w and w' are vectors, the first one is formed by adjustable weights and the second vector is a set of fixed parameters used to ensure the RHONN controllability as explained in Eq. (6). The RHONN scheme is illustrated in Fig. 3, with the glucose level $x_{1,k} = G_{k-1}$ and $x_{2,k} = G_k$ as state variables, and u_k as the insulin dose.

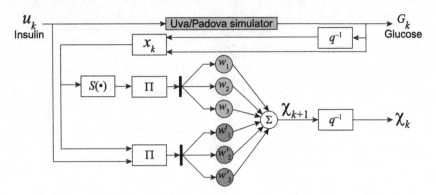

Fig. 3 RHONN scheme for glucose identification.

The main goal of T1DM treatment is to reduce hyperglycemic and hypoglycemic events. In order to achieve such reduction with the methodology proposed by Chen et al. (2013), a multistep prediction (MSP) model based on neural networks is proposed. The neural MSP scheme is the RHONN serial connection with the same structure of the identifier to generate a glucose prediction within 15-min (t) horizon; the corresponding scheme is presented in Fig. 4.

The whole closed-loop control structure for T1DM treatment is presented in Fig. 5. For this structure, in the neuro-fuzzy IOC (NF-IOC) block, the insulin total amount u_k is regulated as follows:

Step 1 The insulin amount u_k^*, using the neural inverse optimal control (NIOC) is calculated using Eq. (24), with

$$
f(x_k) = \begin{bmatrix} w_1' x_{2,k} \\ w_1 S(x_{1,k})S(x_{2,k}) + w_2 S(x_{1,k}) + w_3 S(x_{2,k}) + w_2' x_{2,k} \end{bmatrix},
$$
$$
g(x_k) = \begin{bmatrix} 0 \\ w_3' \end{bmatrix}.
\tag{29}
$$

Step 2 The Takagi-Sugeno (T-S) approach is used to obtain a fuzzy smooth signal (Takagi and Sugeno, 1985). This approach allows preventing hyperglycemia and hypoglycemia zones using T-S inferences with a relationship between variables according to the following propositional sentence.

If *Premise* **then** *Consequence*

where the membership of input variables is represented through *Premise* and the inferred value of output variable through *Consequence*. Using the propositional sentence, a T-S inference is proposed as

R_1: **If** $\widetilde{GL}$ **is** $\widetilde{LG}$ **then** $u_1 = 0$

R_2: **If** $\widetilde{GL}$ **is** $\widetilde{NG}$ **then** $u_2 = u_k^*$

R_3: **If** $\widetilde{GL}$ **is** $\widetilde{HG}$ **then** $u_3 = u_k^* + basal$

R_4: **If** $\widetilde{GL}$ **is** $\widetilde{VHG}$ **then** $u_4 = u_k^* + Hbasal$

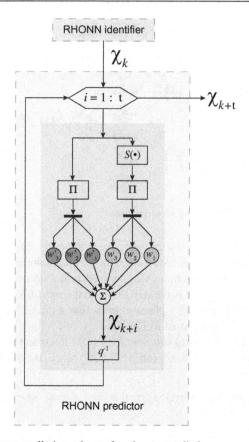

Fig. 4 Neural multistep prediction scheme for glucose prediction.

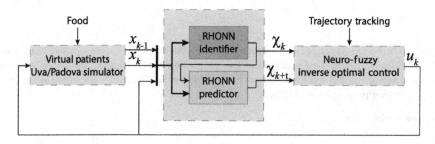

Fig. 5 Block diagram of closed-loop control system with prediction.

where R_i are the implication rules with $i = \{1, ..., 4\}$, $\widetilde{GL}$ (predicted glucose level) is the conditioned variable, fuzzy sets of linear membership functions are $\widetilde{LG}$ (low glucose), $\widetilde{NG}$ (normal glucose), $\widetilde{HG}$ (high glucose), $\widetilde{VHG}$ (very high glucose), and u_i represents the inferred variables according to the implication rules. The fuzzy sets are displayed in Fig. 6.

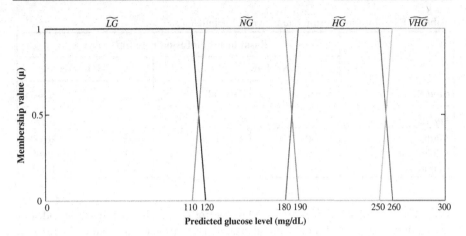

Fig. 6 Fuzzy sets of membership linear functions.

Step 3 The NF-IOC using T-S inferences is given as

$$u_k = \mu_{\widetilde{LG}}u_1 + \mu_{\widetilde{NG}}u_2 + \mu_{\widetilde{HG}}u_3 + \mu_{\widetilde{VHG}}u_4, \tag{30}$$

where $\mu_{(\bullet)}$ is membership value of each fuzzy set with the input variable $\widetilde{GL}$, this value is characterized using two parameters giving the greatest value 1 and the least value 0.

6 Simulation results

In this section, the Uva/Padova simulator is used to test the proposed closed-loop control scheme. It is worth noting that a diabetic patient needs a strictly balanced diet as an amount in grams of carbohydrate (gCH) per day; the total carbohydrates are distributed on the meal protocol which is presented in Table 2. For comparison, three control schemes are considered: The proposed NF-IOC; the second one is the Neuro Inverse Optimal Control with Switching (NIOC-SW), which is a controller with switched approach and MSP, which was detailed by Rios et al. (2018); and the last one is the NIOC, which is an optimal controller with an RHONN to identify the system

Table 2 Meal protocol for virtual populations.

	Meal protocol (gCH)				
Time (h)	**Breakfast (6:00)**	**Snack (12:00)**	**Meal (15:00)**	**Dinner (19:00)**	**Snack (23:00)**
Population					
Adults	35	20	60	35	20
Adolescents	30	15	50	30	15
Children	20	5	35	20	5

Table 3 Basal insulin infusion rate for virtual populations.

	Basal insulin infusion rate (mU/min)			
	NF-IOC		NIOC-SW	
Control	Basal	Hbasal	Basal	Hbasal
Population				
Adults	26.00	28.00	26.00	30.00
Adolescents	22.00	25.00	22.00	25.00
Children	6.00	16.00	6.00	10.00

states. These three algorithms are tested to control the 30 virtual patients included in the simulator; basal insulin infusion rates are defined for each population in Table 3. These insulin amounts are used to help the controller for normoglycemia through *basal* rate and *Hbasal* for hyperglycemia prevention.

The major concern for T1DM patients is to keep BG level within a safe range. Usually, they are prone to go through unnoticed hypoglycemic when sleeping or taking a rest. The BG level value estimated by RHONN identification for the adults average is presented in Fig. 7, including the identification root mean square (RMS) error which is below 2.5 mg/dL for all time simulation lapses. Likewise, a similar performance for adolescents' average and children average is obtained; these performances are presented in Table 4. It can be seen that the RHONN identification one is adequate.

The main objective of the proposed closed-loop control scheme is to determine the adequate insulin rate for the virtual patient, in order to avoid hypoglycemic and hyperglycemic events. In fact, the predicted glucose level within a 15-min horizon allows

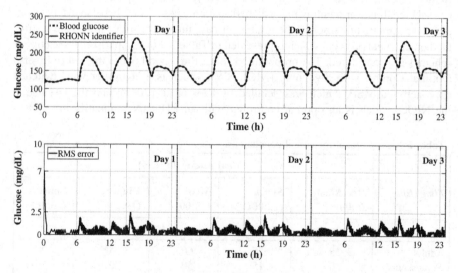

Fig. 7 Blood glucose and RHONN identifier comparison for the adults average.

Table 4 Root mean square error for identification performance in virtual average population.

	Identification performance					
	NF-IOC		NIOC-SW		NIOC	
Control	RMS	Std	RMS	Std	RMS	Std
Population						
Adults	2.5	0.67	2.3	0.67	2.7	0.65
Adolescents	2.5	0.67	2.6	0.67	2.8	0.65
Children	3.0	0.86	3.2	0.89	3.0	0.78

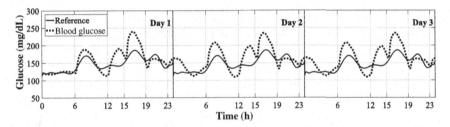

Fig. 8 Blood glucose reference tracking for adults average with NF-IOC.

the optimal control law to improve its performance. In order to carry out the experimentation with the proposed controller for a 3-day simulation environment, the glucose curve of a healthy patient has been used as a reference. The BG dynamics of the adults' average and glucose reference curve using the meal protocol (see Table 2) are presented in Fig. 8. The 15-min prediction illustrated in Fig. 9 allows NF-IOC to change insulin infusion rate for a potential hypoglycemic event before 6:00 hours of Day 2, as well as, a hyperglycemia one just after the meal at 15:00 hours. The postprandial glucose remains below 240 mg/dL for the entire simulation lapse.

The adolescents' average presents a similar behavior as the adults one. The closed-loop control response is displayed in Fig. 10 with the minimum glucose level value of 90 mg/dL and the maximum of 240 mg/dL; these values are not considered severe because they disappear promptly. Fig. 11 clearly shows that glucose prediction allows determining the adequate amount of insulin to be delivered; that is, the NF-IOC controller receives glucose level values *in prediction advance*, which results in a *preventive action*.

For children population, meal intake and *basal* insulin amounts are carefully adapted to their highly complex glucose-insulin behavior (see Tables 2 and 3). In Fig. 12, trajectory tracking for children average is illustrated. The postprandial glucose remains in a safe zone below 250 mg/dL. The released insulin is presented in Fig. 13.

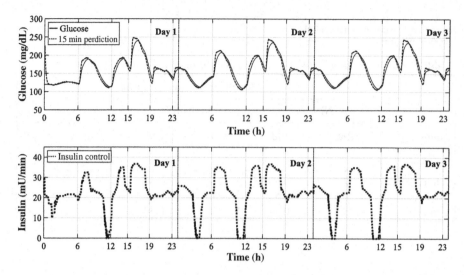

Fig. 9 Closed-loop control for adults average. *Top*: Blood glucose versus 15-min prediction signal. *Bottom*: Insulin rate calculated by the NF-IOC algorithm.

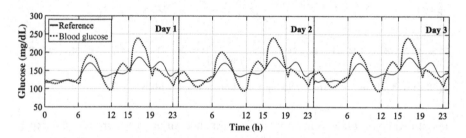

Fig. 10 Blood glucose reference tracking for adolescents average with NF-IOC.

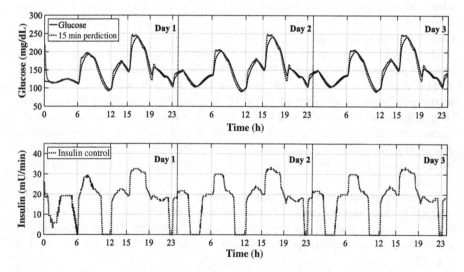

Fig. 11 Closed-loop control for adolescents average. *Top*: Blood glucose versus 15-min prediction signal. *Bottom*: Insulin rate calculated by the NF-IOC algorithm.

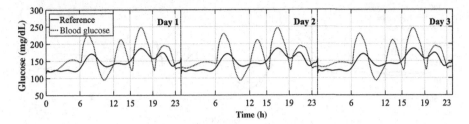

Fig. 12 Blood glucose reference tracking for children average with NF-IOC.

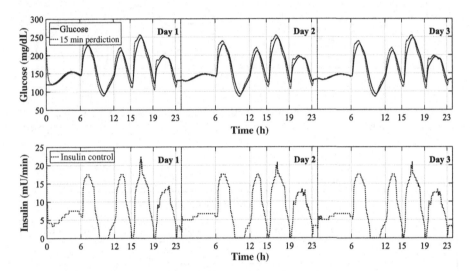

Fig. 13 Closed-loop control for children average. *Top*: Blood glucose versus 15-min prediction signal. *Bottom*: Insulin rate calculated by the NF-IOC algorithm.

For comparison, considering only the children average, the glucose amount trajectory tracking and the insulin delivered by the three different optimal control approaches (NIOC, NIOC-SW, and NF-IOC) are presented in Fig. 14, where it is clear that the best performance is the one corresponding to the NF-IOC controller. Additionally, Table 5 summarizes the total daily delivered insulin (in U/day) for each population and for each control strategy.

The CVGA included in the simulator allows visualizing the minimum and maximum of glucose level for each patient. To illustrate the glucose-insulin dynamical behavior with the control laws, three cases are included for the 30-patient population. The first one in Fig. 15 corresponds to closed-loop control for the whole 30 patients with NF-IOC resulting in 100% of normoglycemia. Similarly, the second scheme corresponds to the NIOC-SW control application law, which results in 93% of normoglycemia, is presented in Fig. 16. Finally, for the last case, Fig. 17 presents performance corresponding to the NIOC control law, whose effectiveness is 63%.

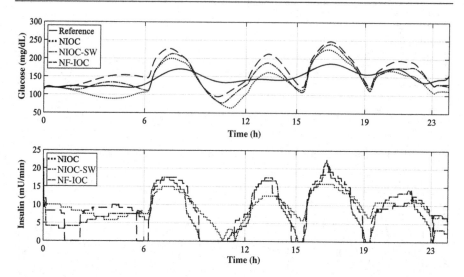

Fig. 14 Closed-loop control for children average. *Top*: Reference versus glucose dynamics under NIOC, NIOC-SW, and NF-IOC. *Bottom*: Insulin rate calculated by the three proposed strategies.

Table 5 Total daily insulin (TDI) delivered per day.

	Total daily insulin (U/day)			
Control	**NF-IOC**	**NIOC-SW**	**NIOC**	**Uva/Padova simulator**
Population				
Adults	35.33	33.73	35.57	39.47
Adolescents	25.9	26.06	26.69	28.05
Children	11.9	12.81	13.47	13.35

7 Discussion

The more important concern about T1DM disease is the mortality rate, which increases considerably worldwide year after year. Therefore, the main goal of scientific research in this field is to develop achievable strategies to reduce the fatal effects of this disease, including the development of the AP. Published results are diverse and have allowed important advances; however, the fully functional AP remains a very important objective, which is not completely functional and available. For this reason, a neuro-fuzzy glucose control scheme with applicability to the AP system is discussed in this chapter; in fact, the proposed control strategy clearly represents a novel application with significant differences regarding existing results, such as Ruiz-Velázquez et al. (2004), Romero-Aragon et al. (2015), Leon et al. (2014), Colmegna et al. (2014), and Messori et al. (2018). The capability of artificial neural networks and their

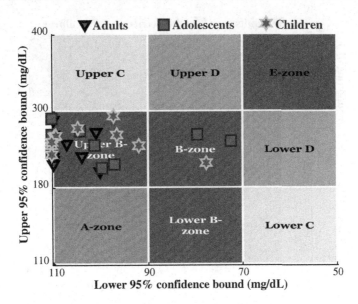

Fig. 15 CVGA chart for NF-IOC closed-loop control in 30 virtual subjects. *Triangles*: 10 of 10 adults are in upper B-zone. *Squares*: 8 of 10 adolescents are in upper B-zone and 2 of 10 fall in B-zone. *Stars*: 9 of 10 children fall in upper B-zone of and 1 child is located in B-zone of benign deviations in hypo/hyperglycemia (see Table 1).

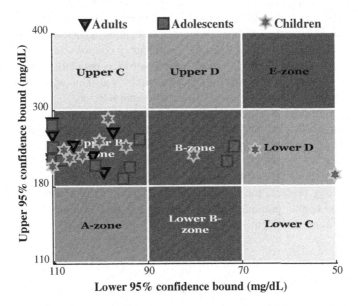

Fig. 16 CVGA chart for NIOC-SW closed-loop control in 30 virtual subjects. *Triangles*: 10 of 10 adults are in upper B-zone. *Squares*: 8 of 10 adolescents are in upper B-zone and 2 of 10 fall in B-zone. *Stars*: 8 of 10 children fall in B-zone and 2 drop in lower D-zone, which means a failure to deal with hypoglycemia (see Table 1).

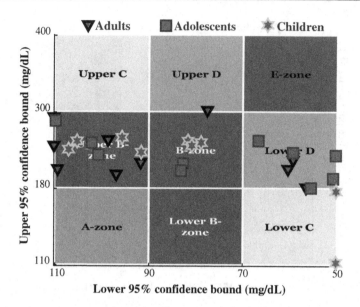

Fig. 17 CVGA chart for NIOC closed-loop control in 30 virtual subjects. *Triangles*: 6 of 10 adults are B-zone and 4 of 10 fall in D-zone. *Squares*: 5 of 10 adolescents are in B-zone and 5 of 10 fall in lower D-zone. *Stars*: 8 of 10 children fall in B-zone and 2 drop in lower C-zone, which means an over correction in hypoglycemia is needed (see Table 1).

combination with other control techniques have provided good results for engineering applications (Quintero-Manriquez et al., 2017; Rios et al., 2018; Sanchez et al., 2008). Digressing from published results, the most outstanding advance discussed in this chapter are the results obtained when the neuro-fuzzy control scheme is applied on the Uva/Padova simulation environment, which considers populations of adults, adolescents, and children. A neuronal multistep predictor is also incorporated in order to know the future dynamics of glucose within a 15-min horizon.

Using the proposed new control scheme, incorporating prediction, glycemic control by insulin infusion is found most appropriate for all considered populations, without events of severe hyperglycemia or hypoglycemia during the 3-day treatment as shown in Figs. 8, 10, and 12. The three control strategies NIOC, NIOC-SW, and NF-IOC are successfully implemented; they are all based on the NIOC strategy. NF-IOC strategy presents the best performance in comparison with the other two (Fig. 14). The results presented in this chapter are very encouraging to continue research for future development of the AP.

8 Conclusions

In recent years, the scientific community has witnessed a great breakthrough in the development of artificial systems. Different algorithms of this type are biologically inspired and have the "learning" capacity. The application of computational intelligence techniques focused on glycemic control for T1DM patients has been discussed in this chapter. The first RHONN configuration helps to define the IOC law, and the second one is used for MSP of glucose, thereby the control action is smoothly determined by fuzzy insulin infusions. A remarkable fact is the implementation of the proposed scheme in the Uva/Padova simulator approved by the FDA to test control strategies for T1DM patients. A comparison of three neural control approaches is also included. The results for glycemic regulation are very encouraging, since the insulin infusion, which is computed by the proposed neuro-fuzzy controller, keeps the glucose level within safe levels for a 3-day simulation lapse. The CVGA evaluation illustrates the benefits of the proposed control scheme. The 30 patients stay on the upper B-zone, which corresponds to an adequate behavior.

Acknowledgments

The authors thank the support of CONACYT México, through Projects CB256769, CB256880, and CB257200 (projects supported by Fondo Sectorial de Investigación para la Educación).

References

Al-Tamimi, A., Lewis, F.L., Abu-Khalaf, M., 2008. Discrete-time nonlinear HJB solution using approximate dynamic programming: convergence proof. IEEE Trans. Syst. Man Cybern. B Cybern. 38 (4), 943–949. https://doi.org/10.1109/TSMCB.2008.926614.

Alanis, A.Y., Sanchez, E.N., Loukianov, A.G., 2007. Discrete-time adaptive backstepping nonlinear control via high-order neural networks. IEEE Trans. Neural Netw. 18 (4), 1185–1195. https://doi.org/10.1109/TNN.2007.899170.

Almobaied, M., Eksin, I., Guzelkaya, M., 2018. Inverse optimal controller based on extended Kalman filter for discrete-time nonlinear systems. Optimal Control Appl. Methods 39 (1), 19–34. https://doi.org/10.1002/oca.2331.

Bamgbose, S.O., Li, X., Qian, L., 2017. Closed loop control of blood glucose level with neural network predictor for diabetic patients. In: 2017 IEEE 19th International Conference on e-Health Networking, Applications and Services, Healthcom 2017, IEEE, Dalian, pp. 1–6.

Basar, T., Olsder, G.J., 1999. Dynamic Noncooperative Game Theory, second ed. Society for Industrial and Applied Mathematics, New Brunswick, NJ, ISBN: 9780898714296, p. 526 10.1137/1.9781611971132.

Batora, V., Tarnik, M., Murgas, J., Schmidt, S., Norgaard, K., Poulsen, N.K., Madsen, H., 2015. Bihormonal control of blood glucose in people with type 1 diabetes. In: 2015 European Control Conference (ECC), IEEE, Linz, pp. 25–30.

Bequette, B.W., Cameron, F., Buckingham, B.A., Maahs, D.M., Lum, J., 2018. Overnight hypoglycemia and hyperglycemia mitigation for individuals with type 1 diabetes: how risks can be reduced. IEEE Control Syst. 38 (1), 125–134. https://doi.org/10.1109/MCS.2017.2767119.

Bondia, J., Romero-Vivo, S., Ricarte, B., Diez, J.L., 2018. Insulin estimation and prediction: a review of the estimation and prediction of subcutaneous insulin pharmacokinetics in closed-loop glucose control. IEEE Control Syst. 38 (1), 47–66. https://doi.org/10.1109/MCS.2017.2766312.

Chen, P.A., Chang, L.C., Chang, F.J., 2013. Reinforced recurrent neural networks for multi-step-ahead flood forecasts. J. Hydrol. 497 (2013), 71–79. https://doi.org/10.1016/j.jhydrol.2013.05.038.

Cinar, A., 2018. Artificial pancreas systems: an introduction to the special issue. IEEE Control Syst. 38 (1), 26–29. https://doi.org/10.1109/MCS.2017.2766321.

Colmegna, P., Sánchez Peña, R.S., Gondhalekar, R., Dassau, E., Doyle, F.J., 2014. Reducing risks in type 1 diabetes using $\mathcal{H}_\infty$ control. IEEE Trans. Biomed. Eng. 61 (12), 2939–2947. https://doi.org/10.1109/TBME.2014.2336772.

El Fathi, A., Raef Smaou, M., Gingras, V., Boule, B., Haidar, A., 2018. The artificial pancreas and meal control: an overview of postprandial glucose regulation in type 1 diabetes. IEEE Control Syst. 38 (1), 67–85. https://doi.org/10.1109/MCS.2017.2766323.

Femat, R., Ruiz-Velazquez, E., Quiroz, G., 2009. Weighting restriction for intravenous insulin delivery on T1DM patient via $\mathcal{H}_\infty$ control. IEEE Trans. Autom. Sci. Eng. 6 (2), 239–247. https://doi.org/10.1109/TASE.2008.2009089.

Freeman, R.A., Kokotović, P., 2009. Robust Nonlinear Control Design. Birkhäuser, Boston, ISBN: 978-0-8176-4759-9, p. 255. 10.1007/978-0-8176-4759-9.

Garcia-Gabin, W., Jacobsen, E.W., 2013. Multilevel model based glucose control for type-1 diabetes patients. In: Proceedings of the Annual International Conference of the IEEE Engineering in Medicine and Biology Society, EMBS, IEEE, Osaka, pp. 3917–3920.

Hashimoto, S., Noguchi, C.C.Y., Furutani, E., 2014. Postprandial blood glucose control in type 1 diabetes for carbohydrates with varying glycemic index foods. In: 2014 36th Annual International Conference of the IEEE Engineering in Medicine and Biology Society, EMBC 2014, IEEE, Chicago, IL, pp. 4835–4838.

Huyett, L.M., Dassau, E., Zisser, H.C., Doyle III, F.J., 2018. Glucose sensor dynamics and the artificial pancreas: the impact of lag on sensor measurement and controller performance. IEEE Control Syst. 38 (1), 30–46. 10.1109/MCS.2017.2766322.

International Diabetes Federation, 2009. Diabetes Atlas, fourth ed. International Diabetes Federation, Brussels, pp. 1–527. 2-930229-80-2.

Katsarou, A., Gudbjörnsdottir, S., Rawshani, A., Dabelea, D., Bonifacio, E., Anderson, B.J., Jacobsen, L.M., Schatz, D.A., Lernmark, Å., 2017. Type 1 diabetes mellitus. Nat. Rev. Dis. Primers 3 (17016), 1–17. https://doi.org/10.1038/nrdp.2017.16.

Kirk, D.E., 1970. Optimal Control Theory: An Introduction. Springer-Verlag, Austin, TX, ISBN: 9783319982366, p. 564. 10.1007/978-3-319-98237-3.

Laguna Sanz, A.J., Doyle, F.J., Dassau, E., 2017. An enhanced model predictive control for the artificial pancreas using a confidence index based on residual analysis of past predictions. J. Diabetes Sci. Technol. 11 (3), 537–544. https://doi.org/10.1177/1932296816680632.

Leon, B.S., Alanis, A.Y., Sanchez, E.N., Ornelas-Tellez, F., Ruiz-Velazquez, E., 2014. Neural inverse optimal control via passivity for subcutaneous blood glucose regulation in type 1 diabetes mellitus patients. Intell. Autom. Soft Comput. 20 (2), 279–295. https://doi.org/10.1080/10798587.2014.891307.

Liu, B., Ying, H., 2015. Analysis of the islets-based glucose control system involving the nonlinear glucose-insulin metabolism model. In: 2015 IEEE International Conference on Information and Automation, ICIA 2015—In Conjunction With 2015 IEEE International Conference on Automation and Logistics, IEEE, Lijiang, pp. 2373–2378.

Magni, L., Raimondo, D.M., Man, C.D., Breton, M., Patek, S., De Nicolao, G., Cobelli, C., Kovatchev, B.P., 2008. Evaluating the efficacy of closed-loop glucose regulation via control-variability grid analysis. J. Diabetes Sci. Technol. 2 (4), 630–635. https://doi.org/10.1177/193229680800200414.

Man, C.D., Rizza, R.A., Cobelli, C., 2007. Meal simulation model of the glucose-insulin system. IEEE Trans. Biomed. Eng. 54 (10), 1740–1749. https://doi.org/10.1109/TBME.2007.893506.

Man, C.D., Micheletto, F., Lv, D., Breton, M., Kovatchev, B., Cobelli, C., 2014. The Uva/ Padova type 1 diabetes simulator. J. Diabetes Sci. Technol. 8 (1), 26–34. https://doi. org/10.1177/1932296813514502.

Messori, M., Ellis, M., Cobelli, C., Christofides, P.D., Magni, L., 2015. Improved postprandial glucose control with a customized model predictive controller. In: Proceedings of the American Control ConferenceIEEE, Chicago, IL, pp. 5108–5115.

Messori, M., Incremona, G.P., Cobelli, C., Magni, L., 2018. Individualized model predictive control for the artificial pancreas: in silico evaluation of closed-loop glucose control. IEEE Control Syst. 38 (1), 86–104. https://doi.org/10.1109/MCS.2017.2766314.

Morales-Contreras, J., Ruiz-Velazquez, E., Garcia-Rodriguez, J., 2017. Robust glucose control via μ-synthesis in type 1 diabetes mellitus. In: 2017 IEEE International Autumn Meeting on Power, Electronics and Computing (ROPEC), IEEE, Ixtapa, pp. 1–6.

Ohsawa, T., Bloch, A.M., Leok, M., 2010. Discrete Hamilton-Jacobi theory and discrete optimal control. In: Proceedings of the IEEE Conference on Decision and Control, IEEE, Atlanta, pp. 5438–5443.

Ornelas, F., Sanchez, E.N., Loukianov, A.G., 2011. Discrete-time nonlinear systems inverse optimal control: a control Lyapunov function approach. In: Proceedings of the IEEE International Conference on Control Applications, IEEE, Denver, pp. 1431–1436.

Quintero-Manriquez, E., Sanchez, E.N., Harley, R.G., Li, S., Felix, R.A., 2017. Neural sliding mode control for induction motors using rapid control prototyping. IFAC-PapersOnLine 50 (1), 9625–9630. https://doi.org/10.1016/j.ifacol.2017.08.1711.

Rios, Y.Y., García-Rodríguez, J.A., Sánchez, O.D., Sanchez, E.N., Alanis, A.Y., Ruiz-Velázquez, E., Arana-Daniel, N., 2018. Inverse optimal control using a neural multi-step predictor for T1DM treatment. In: Proceedings of the International Joint Conference on Neural Networks, IEEE, Rio de Janeiro, pp. 1–8.

Romero-Aragon, J.C., Sanchez, E.N., Alanis, A.Y., 2015. Glucose level regulation for diabetes mellitus type 1 patients using FPGA neural inverse optimal control. In: IEEE SSCI 2014— 2014 IEEE Symposium Series on Computational Intelligence—CICA 2014: 2014 IEEE Symposium on Computational Intelligence in Control and Automation, Proceedings, IEEE, Orlando, pp. 1–7.

Rovithakis, G.A., Christodoulou, M.A., 2000. Adaptive Control With Recurrent High-Order Neural Networks: Theory and Industrial Applications. Springer, London, ISBN: 9781447107859, p. 196.

Ruiz-Velázquez, E., Femat, R., Campos-Delgado, D.U., 2004. Blood glucose control for type I diabetes mellitus: a robust tracking H_∞ problem. Control Eng. Pract. 12 (9), 1179–1195. https://doi.org/10.1016/j.conengprac.2003.12.004.

Sanchez, E.N., Ornelas-Tellez, F., 2013. Discrete-Time Inverse Optimal Control for Nonlinear Systems. CRC Press, Boca Raton, FL, ISBN: 9781466580886p. 268 10.1201/b14779.

Sanchez, E.N., Alanis, A.Y., Loukianov, A.G., 2008. Discrete-Time High Order Neural Control: Trained With Kalman Filtering. vol. 112. Springer Science & Business Media, Berlin, ISBN: 978-3-540-78288-9, p. 116.

Song, Y., Grizzle, J.W., 1992. The extended Kalman filter as a local asymptotic observer for nonlinear discrete-time systems. In: 1992 American Control Conference, IEEE, Chicago, IL, pp. 3365–3369.

Takagi, T., Sugeno, M., 1985. Fuzzy identification of systems and its applications to modeling and control. IEEE Trans. Syst. Man Cybern. SMC-15 (1), 116–132. https://doi.org/ 10.1109/TSMC.1985.6313399.

Turksoy, K., Littlejohn, E., Cinar, A., 2018. Multimodule, multivariable artificial pancreas for patients with type 1 diabetes: regulating glucose concentration under challenging conditions. IEEE Control Syst. 38 (1), 105–124. https://doi.org/10.1109/MCS.2017.2766326.

Wang, Y., Xie, H., Jiang, X., Liu, B., 2014. Intelligent closed-loop insulin delivery systems for ICU patients. IEEE J. Biomed. Health Inform. 18 (1), 290–299. https://doi.org/10.1109/JBHI.2013.2269699.

WHO, 2016. Global Report on Diabetes. World Health Organization, Geneva, pp. 1–88. 10.1002/psp.

Blood glucose regulation in patients with type 1 diabetes by means of output-feedback sliding mode control

2

Manuchi Dansa, Victor Hugo Pereira Rodrigues, Tiago Roux Oliveira
Department of Electronics and Telecommunication Engineering (DETEL), State University of
Rio de Janeiro (UERJ), Rio de Janeiro, Brazil

1 Introduction

The pancreas, in addition to its digestive functions, secretes two important hormones, *insulin* and *glucagon*, that are crucial for normal regulation of blood glucose concentration (Guyton and Hall, 2016). When the glucose concentration rises above a certain level, insulin is secreted; the insulin in turn causes the blood glucose concentration to drop toward normal. Conversely, a decrease in blood glucose stimulates glucagon secretion; the glucagon then functions in the opposite direction to increase and then steer the glucose back to its normal level.

However, in patients with type 1 diabetes mellitus (T1DM), pancreatic insulin production is impaired, which entails a plenty of risks to these patients health. Tight glucose control reduces the risk of long-term diabetes-related complications, such as kidney disease, heart disease, blindness, and peripheral vascular and nerve damage. Moreover, conventional methods of self-monitoring of blood glucose and multiple daily injections are challenging for patients and families (Bakhtiani et al., 2013).

Against such a background, an automated closed-loop glucose control system stands out as a target that has been pursued by researchers for the past five decades (Kadish, 1964). Such devices, known as artificial pancreas (Bequette, 2012; Cobelli et al., 2011), consist of a glucose sensor, from which data are collected and entered into an algorithm, which in turn compute the amount of insulin to be delivered through a pump. Fig. 1 depicts how such a device could be.

1.1 Motivation

In modeling drug delivery to human body, certain requirements like finite reaching time and robustness to uncertainties should be satisfied (Kaveh and Shtessel, 2008). In addition, the glucose-insulin dynamics is a complex physiologic system that encompass a number of nonlinear process and therefore, cannot be accurately described through linear models (Audoly et al., 2001).

Control Applications for Biomedical Engineering Systems. https://doi.org/10.1016/B978-0-12-817461-6.00002-0

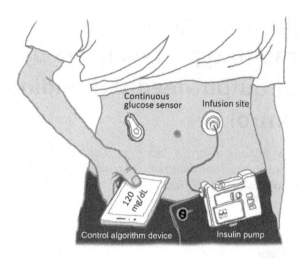

Fig. 1 Mono-hormonal artificial pancreas. A bihormonal artificial pancreas should yet add the hormone glucagon in the pump system. The communication between the glucose sensor and the device running the control algorithm (as well as the latter with the injection pump) is wireless. Adapted from Insulin Pump Awareness Group—IPAG Scotland. Available from: https://www.ipag.co.uk/study-insulin-pump-better-than-injections/ (Accessed April 16, 2019).

As discussed in the following, it is noteworthy that a plenty of different control strategies could have been used in order to assure the maintenance of normal glucose concentrations (euglycemia).

1.2   State of the art of the control algorithms for blood glucose regulation

The most commonly used controllers are based on proportional-integral-derivative control (Huyett et al., 2018; Jacobs et al., 2014; Marchetti et al., 2008), model predictive control (MPC) (Bátora et al., 2015b; Boiroux et al., 2016; Favero et al., 2015; Messori et al., 2018), adaptive control (Bellazzi et al., 1995; El-Khatib et al., 2007; Turksoy et al., 2014), and fuzzy logic and neural networks (Atlas et al., 2010).

Sliding mode control has been proved an efficient technique to provide high-fidelity performance in different control problems for nonlinear systems with uncertainties in system parameters and external disturbances (Shtessel et al., 2003).

In particular, one alternative approach and control technique that has been tested for closed-loop control of blood glucose is the fast terminal sliding mode control (FTSMC) (Jajarm et al., 2012). Although this controller seems to be more robust to uncertainties in the model and presents faster responses when compared to the references above, the full-state measurement is required and no stability proof was carried out for the closed-loop system.

1.3 Dual-hormone strategy

Concerning the pump actuator for drug delivery, there is a question that has generated much controversy in the artificial pancreas community: to utilize single-hormone systems employing only insulin or to adopt a dual-hormone strategy adding glucagon administration to insulin delivery. Although insulin-only closed-loop systems achieve a much better glycemic control than standard open-loop insulin therapy, it does not completely eliminate the risk of hypoglycemia (Bakhtiani et al., 2013; Bequette et al., 2018; Kumareswaran et al., 2009).

On the other hand, a closed-loop system that employs subcutaneous infusion of both insulin and glucagon has proven its efficacy in preventing and treating hypoglycemia (Castle et al., 2010; El-Khatib et al., 2010). Moreover, such bihormonal systems would better emulate the function of the endocrine pancreas (El-Khatib et al., 2009) and, for this sake, will be considered in this chapter.

Pumps used for diabetic patients infuse hormone subcutaneously. In this sense, in silico testing of a dual-hormone control algorithm requires a suitable model that is able to simulate the effects of subcutaneously administered insulin and glucagon (Bátora et al., 2015a). In this essay, we implement the model proposed by Bergman et al. (2009) extended with glucagon action as proposed by Herrero et al. (2013).

1.4 Contributions

The purpose of this contribution is to introduce closed-loop sliding mode control algorithms capable of regulating the blood glucose concentration in subjects with T1DM through a bihormonal pump. A rigorous stability analysis is carried out by means of Lyapunov's theory taking into account parametric uncertainties in the biological model and unmatched disturbances due to food intake. Two different control strategies are proposed: first-order sliding mode control (FOSMC) and nonsingular terminal sliding mode controller (NSTSMC). Continuous versions of the algorithms are also addressed.

Numerical simulations illustrate the efficiency of both control strategies. The output-feedback version of the proposed algorithms using higher-order sliding mode (HOSM) differentiators is also discussed and evaluated.

1.5 Chapter outline

All things considered, it should be said that this chapter is organized as follows. In Section 2, the mathematical model used in this manuscript is exposed. The methodology and the control objectives are the themes of Section 3. In Section 4, we discuss how food intake is handled as an unmatched disturbance of the system dynamics. Section 5 presents the bihormonal actuator description and the physical interpretation of the two control actions of the controller. The subsequent two sections deal with the sliding mode control algorithms applied to achieve the control goals. Section 6 shows the FOSMC; while Section 7 introduces the fast terminal sliding mode control (FTSMC). Next, an exact differentiator based on HOSMs is dully discussed in Section 8. At last, Section 9 presents some numerical examples.

2 Mathematical model

Several authors (Bátora et al., 2015c; El-Khatib et al., 2010; Patti, 2018; Reiter et al., 2016; Russell et al., 2012) believe a bihormonal closed-loop algorithm could provide a safe blood glucose regulation and reduce significantly the risk and time spent in hypoglycemic episodes (Bequette et al., 2018) compared to usual insulin therapy.

The availability of a model that incorporates glucagon as a counter-regulatory hormone to insulin would allow more efficient design of bihormonal glucose controllers (El-Khatib et al., 2010). In this sense, an extended minimal model was proposed (Herrero et al., 2013) to incorporate the glucagon effect.

For simplicity's sake, we consider in controller design only the reduced model (Herrero et al., 2013). Such assumption is reasonable since some dynamics among all presented in the extended model are, in general, faster than the dynamics of plasma glucose concentration, insulin action on glucose production and glucagon action on glucose production. Therefore, the dynamics presented by Herrero et al. (2013) is now represented by

$$\dot{x}_1(t) = -(S_G + x_2(t) - x_3(t))x_1(t) + S_G G_B + \frac{1}{t_{maxG}V}d(t), \tag{1}$$

$$\dot{x}_2(t) = -p_2(t)x_2(t) + p_2(t)S_I(t)(u^+ - I_B), \tag{2}$$

$$\dot{x}_3(t) = -p_3(t)x_3(t) + p_3(t)S_N(t)(u^- - N_B), \tag{3}$$

$$y(t) = x_1(t), \tag{4}$$

where $y(t)$ (mg dL^{-1}) is the output variable, $x_1(t)$ (mg dL^{-1}) is the plasma glucose concentration, $x_2(t)$ (min^{-1}) is the insulin action on glucose production, and $x_3(t)$ (min^{-1}) is the glucagon action on glucose production. $V = 1.7$ dL kg^{-1} is the glucose distribution volume. $u^+ \in \mathbb{R}^+$ (μU dL^{-1}) is the control action and represents the plasma insulin concentration, $u^- \in \mathbb{R}^+$ (pg dL^{-1}) is the plasma glucagon concentration, $d(t)$ (mg kg^{-1}) is the glucose concentration resulting from ingested meals, and $S_G = 0.014$ (min^{-1}) is the glucose effectiveness per unit distribution volume. For simplicity's sake, the parameter $t_{maxG} = 69.6$ (min) was assumed constant, since its fluctuations are too slow, in general (Herrero et al., 2013). All other variables and parameters are described in Table 1, followed by its respective units and descriptions. Finally, we remark that if $u^+ \in \mathbb{R}^+$ and $u^- \in \mathbb{R}^+$, it means that the control signals are positive.

The mathematical description of the time-varying parameter is given below:

$$p_2(t) = 0.012 \; \mathbb{1}(t) - 0.0081 \; \mathbb{1}(t - 300) + 0.0171 \; \mathbb{1}(t - 720) - 0.009 \; \mathbb{1}(t - 1080), \tag{5}$$

$$p_3(t) = 0.017 \; \mathbb{1}(t) - 0.001 \; \mathbb{1}(t - 300) + 0.123 \; \mathbb{1}(t - 720) - 0.122 \; \mathbb{1}(t - 1080), \tag{6}$$

Table 1 Description of extended minimal model parameters

Parameters	Description
S_G (min^{-1})	Glucose effectiveness per unit distribution volume
G_B (mg dL^{-1})	Basal plasma glucose concentration
$t_{maxG}(t)$ (min)	Time-to-maximum glucose absorption
V (dL kg^{-1})	Glucose distribution volume
$p_2(t)$ (min^{-1})	Rate of disappearance of the interstitial insulin effect
$S_I(t)$ (min^{-1} per μU mL^{-1})	Insulin sensitivity
$I_B(t)$ (mU dL^{-1})	Basal plasma insulin concentration
$p_3(t)$ (min^{-1})	Rate constant describing the dynamics of glucagon action
$S_N(t)$ (min^{-1} per pg mL^{-1})	Glucagon sensitivity
$N_B(t)$ (pg mL^{-1})	Basal plasma glucagon concentration

$$S_I(t) = [7.73 \ \mathbb{1}(t) + 0.82 \ \mathbb{1}(t-300) - 1.73 \ \mathbb{1}(t-720) + 0.91 \ \mathbb{1}(t-1080)] \times 10^{-4}, \tag{7}$$

$$S_N(t) = [1.38 \ \mathbb{1}(t) + 0.58 \ \mathbb{1}(t-300) - 1.15 \ \mathbb{1}(t-720) + 0.57 \ \mathbb{1}(t-1080)] \times 10^{-4}, \tag{8}$$

$$I_B(t) = 11.01 \ \mathbb{1}(t) + 8.75 \ \mathbb{1}(t-300) - 9.73 \ \mathbb{1}(t-720) + 0.98 \ \mathbb{1}(t-1080), \tag{9}$$

$$N_B(t) = 46.30 \ \mathbb{1}(t) + 1.83 \ \mathbb{1}(t-300) + 11.10 \ \mathbb{1}(t-720) - 12.93 \ \mathbb{1}(t-1080), \tag{10}$$

where $\mathbb{1}(t)$ represents the unit step function. The delay in the unit step function is measured in minutes. Therefore, the functions $\mathbb{1}(t - 300)$, $\mathbb{1}(t - 720)$, and $\mathbb{1}(t - 1080)$ denote changes in parameters (5)–(10) at 5:00, 12:00, and 18:00, respectively.

The maximum and minimum parameters values (5)–(10) are

$$0.0039 \le p_2(t) \le 0.021, \tag{11}$$

$$0.016 \le p_3(t) \le 0.139, \tag{12}$$

$$6.82 \times 10^{-4} \le S_I(t) \le 8.55 \times 10^{-4}, \tag{13}$$

$$0.81 \times 10^{-4} \le S_N(t) \le 1.96 \times 10^{-4}, \tag{14}$$

$$10.03 \le I_B(t) \le 19.76, \tag{15}$$

$$46.30 \le N_B(t) \le 59.23. \tag{16}$$

During the development of the stability analysis, the system parameters will be considered uncertain in such a way that its bounds are known and described by

$$\underline{p}_2 < p_2(t) < \overline{p}_2, \quad \underline{p}_3 < p_3(t) < \overline{p}_3, \tag{17}$$

$$\underline{S}_I < S_I(t) < \overline{S}_I, \quad \underline{S}_N < S_N(t) < \overline{S}_N, \tag{18}$$

$$\underline{I}_B < I_B(t) < \overline{I}_B, \quad \underline{N}_B < N_B(t) < \overline{N}_B, \tag{19}$$

$$\underline{t}_{\text{maxG}} < t_{\text{maxG}}, \quad G_B < \overline{G}_B, \quad \underline{V} < V, \tag{20}$$

$$\underline{S_G} < S_G < \overline{S_G}. \tag{21}$$

Furthermore, we assume that

$$|d(t)| < \overline{d}, \quad |\dot{d}(t)| < \dot{\overline{d}}, \tag{22}$$

where $\overline{d}$ and $\dot{\overline{d}}$ are positive known constants for which Eq. (22) is satisfied, except for a zero measure set in the sense of Lebesgue.

At last, a couple of remarks concerning the hormones units are presented. Glucagon levels are reported as picogram per milliliter (pg mL^{-1}). Insulin is administrated in units, abbreviated U (international units). One unit of insulin is defined as the amount of insulin that will lower the blood glucose of a healthy 2 kg rabbit that has fasted for 24 h to 45 mg dL^{-1} within 5 h (Hanas, 1998).

Proceeding forward, we present the control objective and the methodology utilized in order to reach it. The tools explored for this purpose will be duly discussed.

3   Methodology and control objectives

The goal of this project is to ensure the regulation of glycemia in a patient with T1DM. Mathematically, it can be represented through output error stabilization:

$$e(t) = G_B - x_1(t), \tag{23}$$

where $G_B = 90$ mg dL^{-1} represents the desired setpoint. This value was chosen so that the glycemia of the patient with T1DM remained within the limits considered safe by the medical community: 80 and 100 mg dL^{-1} (Guyton and Hall, 2016).

Among all the nonlinear control strategies available, we decided to work with sliding modes controllers (SMC). The main advantage of an SMC relies on the fact that its design does not demand the precise knowledge of the model to be controlled. For controller design, it is sufficient that the upper and lower bounds of the plant parameters are known. The knowledge of upper bounds for disturbances is also required. In this sense, the proposed algorithm is said to be robust with respect to parametric uncertainties and exogenous disturbances. In this chapter, the food intake is considered an exogenous disturbance.

3.1 Glycemic curve

An illustrative record of glycemia of a healthy subject in the course of 24 h is depicted in Fig. 2. It is about a glycemic curve of a healthy subject, whose organism is able to safely regulate the blood glucose, avoiding hypoglycemic and hyperglycemic cases (Bequette et al., 2018).

Fig. 3, in turn, illustrates the glycemic curve of a type 1 diabetic patient, who strives to regulate his blood glucose by regular practice of exercises, a balanced diet, and multiple daily injections.

This figure was obtained with the help a *FreeStyle Libre* sensor (Abbott Laboratories Ltd, Alameda, CA, USA). Through this graphic, one can remark that the standard open-loop insulin pump therapy is unable to maintain blood glucose regulated within safe limits during the 24 h of the day.

The contrast between this type of therapy and closed-loop control strategy will become more evident in Section 9, when the simulation results of the control algorithms employed here are presented. The comparison between Figs. 2 and 3 makes evident the undesirable effect that diabetes mellitus has on glycemic regulation.

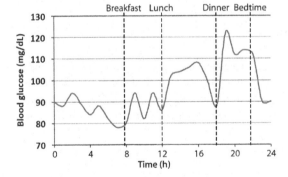

Fig. 2 Glycemia of a health subject.
Adapted from Albertos, P., Mareels, I., 2010. Feedback and Control for Everyone. Springer Science & Business Media, New York, NY.

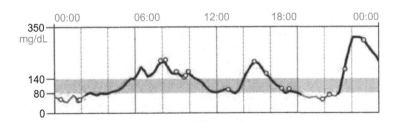

Fig. 3 Blood glucose regulation of a T1DM subject by means of a medical diet. The manufacturer of this sensor considers the safe blood glucose range to be between 80 and 140 mg dL^{-1}, as indicated in the vertical axis.

4 Food ingestion as input disturbances

In this research, it is desired that the patient's blood glucose be regulated over a period of 24 h, which is equivalent to 1440 min. In this context, it is required for the patient to maintain a regular diet consisting of three meals a day, scheduled for the following times: 5:00, 12:00, and 18:00. The food intake modeling was proposed by Kaveh and Shtessel (2008) and Hernández et al. (2013) as

$$
d(t) = \; 80e^{-0.5(t-300)} \, \mathbb{1}(t-300) + 100e^{-0.5(t-720)} \, \mathbb{1}(t-720) \\
+ 70e^{-0.5(t-1080)} \, \mathbb{1}(t-1080),
\tag{24}
$$

where $d(t)$ can be understood as an exogenous disturbance of the system. Therefore, it is assumed that each meal may represent a different rate of appearance of blood glucose. Fig. 4 depicts the effect of meal intake over the patient glycemia.

Hereupon, observe systems (1)–(3). Note the disturbance appears in Eq. (1), while the positive and negative control actions, u^+ and u^-, respectively, appear in Eqs. (2), (3).

Thus, the disturbances are said to be *unmatched* (Edwards and Spurgeon, 1998). It generates an additional challenge in the controller design, since the control gain will have to upper bound the perturbation and its derivatives in amplitude.

5 Bihormonal actuator

Sliding mode controllers have two control actions: a positive control action, u^+, and a negative control action, u^-. In this essay, the positive control action is represented by the amount of insulin administered in the bloodstream, whilst the negative control action is represented by the amount of glucagon administered in the bloodstream.

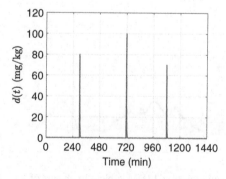

Fig. 4 Glucose concentration resulting from ingested meals along the day.

The algorithm decides which hormone should be injected into the patient at each moment, and this decision satisfies the following rule:

$$u^+ = \begin{cases} \varrho, & \text{if } u > 0 \, (\text{sgn}(\sigma) < 0) \\ 0, & \text{otherwise} \end{cases}, \qquad (25)$$

$$u^- = \begin{cases} \varrho, & \text{if } u < 0 \, (\text{sgn}(\sigma) > 0) \\ 0, & \text{otherwise} \end{cases}, \qquad (26)$$

where ϱ represents the controller modulation function and σ represents the sliding variable. With rules (25)–(26) in mind, one can conclude that insulin and glucagon are never simultaneously administered.

6 FOSMC: Design and stability analysis

Sliding mode control is a well-documented control technique, and its fundamentals can be found in Edwards and Spurgeon (1998) and Utkin (1993).

In what follows, we present the local analysis of the first-order sliding mode controller, by means of the ideal sliding variable. Thus, it is assumed, initially, that the error derivatives are available.

Theorem 1. *Consider the system described by Eqs. (1)–(4) and the bounds (17)–(20). Thus, it is possible to find a sliding mode control law, u, given by*

$$u = -\varrho \, \text{sgn}(\sigma(t)), \quad \varrho > 0, \qquad (27)$$

$$\sigma(t) = \dot{e}(t) + l_0 e(t), \quad l_0 > 0, \qquad (28)$$

with a constant and sufficiently large modulation function, ϱ, and with such a sliding variable, $\sigma(t)$, that the ideal sliding mode, $\sigma(t) = 0$, occurs in finite time $t_s > 0$. Besides, under the sliding regime, the error convergence is exponential $(e(t) = e^{-l_0(t-t_s)} e(0), \forall t \geq t_s)$.

Proof. Consider the following candidate Lyapunov function:

$$V(t) = \sigma^2(t), \qquad (29)$$

where the time derivative of V is $\dot{V}(t) = 2\sigma(t)\dot{\sigma}(t)$. Thus,

$$\dot{V}(t) = 2\sigma(t)(\ddot{e}(t) + l_0 \dot{e}(t)). \qquad (30)$$

Differentiating Eq. (23) yields

$$\dot{e}(t) = -\dot{y}(t) = S_G x_1(t) + x_1(t)x_2(t) - x_1(t)x_3(t) - S_G G_B - \frac{1}{t_{maxG}V}d(t),$$

$$\ddot{e}(t) = (-S_G^2 - p_2(t)S_I(t)I_B(t) + p_3(t)S_N(t)N_B(t))x_1(t) + S_G G_B x_2(t)$$

$$- (S_G G_B x_3(t)) - (2S_G + p_2(t))x_1(t)x_2(t) + (2S_G + p_3(t))x_1(t)x_3(t)$$

$$- x_1(t)x_2^2(t) - x_1(t)x_3^2(t) + 2x_1(t)x_2(t)x_3(t) + \frac{1}{t_{maxG}V}S_G d(t)$$

$$+ \frac{1}{t_{maxG}V}x_2(t)d(t) - \frac{1}{t_{maxG}V}x_3(t)d(t) - \frac{1}{t_{maxG}V}\dot{d}(t) + S_G^2 G_B$$

$$+ p_2(t)S_I(t)x_1(t)u^+ - p_3(t)S_N(t)x_1(t)u^-.$$

$$(31)$$

The time derivative of Eq. (28) is given by

$$\dot{\sigma}(t) = \ddot{e}(t) + l_0\dot{e}(t)$$

$$= (-S_G^2 - p_2(t)S_I(t)I_B(t) + p_3(t)S_N(t)N_B(t))x_1(t) + S_G G_B x_2(t)$$

$$- S_G G_B x_3(t) - (2S_G + p_2(t))x_1(t)x_2(t) + (2S_G + p_3(t))x_1(t)x_3(t)$$

$$- x_1(t)x_2^2(t) - x_1(t)x_3^2(t) + S_G^2 G_B + l_0 S_G x_1(t) + l_0 x_1(t)x_2(t)$$

$$- l_0 x_1(t)x_3(t) l_0 S_G G_B + 2x_1(t)x_2(t)x_3(t) + \frac{1}{t_{maxG}V}S_G d(t)$$

$$+ \frac{1}{t_{maxG}V}x_2(t)d(t) - \frac{1}{t_{maxG}V}x_3(t)d(t) - \frac{1}{t_{maxG}V}\dot{d}(t) - \frac{l_0}{t_{maxG}V}d(t)$$

$$+ p_2(t)S_I(t)x_1(t)u^+ - p_3(t)S_N(t)x_1(t)u^-.$$

$$(32)$$

Therefore, the first derivative of the candidate Lyapunov function is given by

$$\dot{V}(t) = 2\{(-S_G^2 - p_2(t)S_I(t)I_B(t) + (p_3(t)S_N(t)N_B(t))x_1(t) + S_G G_B x_2(t))\sigma(t)$$

$$- S_G G_B x_3(t)\sigma(t) - (2S_G + p_2(t))x_1(t)x_2(t)\sigma(t) + 2S_G x_1(t)x_3(t)\sigma(t)$$

$$+ p_3(t)x_1(t)x_3(t)\sigma(t) + (2x_1(t)x_2(t)x_3(t) - x_1(t)x_2^2(t) + S_G^2 G_B)\sigma(t)$$

$$- x_1(t)x_3^2(t)\sigma(t) + \frac{1}{t_{maxG}V}([S_G + x_2(t) - x_3(t)]d(t) - \dot{d}(t))\sigma(t)$$

$$+ l_0 S_G x_1(t)\sigma(t) + l_0 x_1(t)x_2(t)\sigma(t) - l_0 x_1(t)x_3(t)\sigma(t) - l_0 S_G G_B \sigma(t)$$

$$- \frac{l_0}{t_{maxG}V}d(t)\sigma(t) + p_2(t)S_I(t)x_1(t)u^+\sigma(t) - p_3(t)S_N(t)x_1(t)u^-\sigma(t)\}.$$

$$(33)$$

Eq. (33) can be upper bounded by

$$
\begin{aligned}
\dot{V}(t) \leq 2\{ &(S_G^2 + p_2(t)S_I(t)I_B(t) + p_3(t)S_N(t)N_B(t))|x_1(t)|\,|\sigma(t)| \\
&+ S_G G_B(|x_2(t)| + |x_3(t)|)|\sigma(t)| + (2S_G + p_2(t))|x_1(t)|\,|x_2(t)|\,|\sigma(t)| \\
&+ (2S_G + p_3(t))|x_1(t)|\,|x_3(t)|\,|\sigma(t)| + (2S_G + p_3(t))|x_1(t)|\,|x_3(t)|\,|\sigma(t)| \\
&+ |x_1(t)|\,|x_2^2(t)|\,|\sigma(t)| + |x_1(t)|\,|x_3^2(t)|\,|\sigma(t)| + 2|x_1(t)|\,|x_2(t)|\,|x_3(t)|\,|\sigma(t)| \\
&+ \frac{1}{t_{maxG}V}(S_G + |x_2(t)| + |x_3(t)|)|d(t)|\,|\sigma(t)| + \frac{1}{t_{maxG}V}|\dot{d}(t)|\,|\sigma(t)| \\
&+ S_G^2 G_B|\sigma(t)| + l_0 S_G|x_1(t)|\,|\sigma(t)| + l_0|x_1(t)|\,|x_2(t)|\,|\sigma(t)| \\
&+ l_0|x_1(t)|\,|x_3(t)|\,|\sigma(t)| + l_0 S_G G_B|\sigma(t)|\sigma(t) + \frac{l_0}{t_{maxG}V}|d(t)|\,|\sigma(t)| \\
&+ p_2(t)S_I(t)x_1(t)u^+\sigma(t) - p_3(t)S_N(t)x_1(t)u^- \}.
\end{aligned}
\tag{34}
$$

Since the system parameters are uncertain, by using bounds (17)–(21), inequality (34) can be upper bounded by

$$
\begin{aligned}
\dot{V}(t) \leq 2\Big\{ &\Big[(\overline{S}_G^2 + \overline{p}_2\overline{S}_I\overline{I}_B + \overline{p}_3\overline{S}_N\overline{N}_B)|x_1(t)| + \overline{S}_G\overline{G}_B(|x_2(t)| + |x_3(t)|) \\
&+ (2\overline{S}_G\overline{p}_2)|x_1(t)|\,|x_2(t)| + (2\overline{S}_G + \overline{p}_3)|x_1(t)|\,|x_3(t)| + |x_1(t)|x_2^2(t) \\
&+ |x_1(t)|(x_3^2(t) + 2|x_2(t)|\,|x_3(t)|) + \frac{1}{t_{maxG}\ \underline{V}}(\overline{S}_G + |x_2(t)| + |x_3(t)|)|d(t)| \\
&+ \frac{1}{t_{maxG}\ \underline{V}}|\dot{d}(t)| + \overline{S}_G^2\overline{G}_B + l_0\overline{S}_G|x_1(t)| \\
&+ l_0|x_1(t)|\,|x_2(t)| + l_0|x_1(t)|\,|x_3(t)| + l_0\overline{S}_G\overline{G}_B + \frac{l_0}{t_{maxG}\ \underline{V}}|d(t)|\Big]|\sigma(t)| \\
&+ p_2(t)S_I(t)x_1(t)u^+\sigma(t) - p_3(t)S_N(t)x_1(t)u^-\sigma(t)\Big\}.
\end{aligned}
\tag{35}
$$

The patient receives insulin ($u^+ = \varrho$) if $u > 0$. In this case, sgn(σ) $< 0 \rightarrow \sigma = -|\sigma|$ and $u^- = 0$. So, inequality (34) can be rewritten as

$$
\begin{aligned}
\dot{V}(t) \leq 2\Big\{ &(\overline{S}_G^2 + \overline{p}_2\overline{S}_I\overline{I}_B + \overline{p}_3\overline{S}_N\overline{N}_B)|x_1(t)| + \overline{S}_G\overline{G}_B(|x_2(t)| + |x_3(t)|) \\
&+ (2\overline{S}_G + \overline{p}_2)|x_1(t)|\,|x_2(t)| + (2\overline{S}_G + \overline{p}_3)|x_1(t)|\,|x_3(t)| + |x_1(t)|x_2^2(t) \\
&+ |x_1(t)|x_3^2(t) + 2|x_1(t)|\,|x_2(t)|\,|x_3(t)| + \frac{1}{t_{maxG}\ \underline{V}}(|x_2(t)| + |x_3(t)|)|d(t)| \\
&+ \frac{1}{t_{maxG}\ \underline{V}}\overline{S}_G + \frac{1}{t_{maxG}\ \underline{V}}|\dot{d}(t)| + \overline{S}_G^2\overline{G}_B + l_0\overline{S}_G|x_1(t)| + l_0|x_1(t)|\,|x_2(t)| \\
&+ l_0|x_1(t)|\,|x_3(t)| + l_0\overline{S}_G\overline{G}_B + \frac{l_0}{t_{maxG}\ \underline{V}}|d(t)| - p_2(t)S_I(t)x_1(t)\varrho \Big\}|\sigma(t)|.
\end{aligned}
\tag{36}
$$

The patient receives glucagon ($u^- = \varrho$) if $u < 0$. In this case, $\mathrm{sgn}(\sigma) > 0 \to \sigma = |\sigma|$ and $u^+ = 0$. So, inequality (34) can be rewritten as

$$
\begin{aligned}
\dot{V}(t) \leq 2\Bigg\{ &(\overline{S}_G^2 + \overline{p}_2 \overline{S}_I \overline{I}_B + \overline{p}_3 \overline{S}_N \overline{N}_B)|x_1(t)| + \overline{S}_G \overline{G}_B(|x_2(t)| + |x_3(t)|) \\
&+ (2\overline{S}_G + \overline{p}_2)|x_1(t)|\,|x_2(t)| + (2\overline{S}_G + \overline{p}_3)|x_1(t)|\,|x_3(t)| + |x_1(t)|x_2^2(t) \\
&+ |x_1(t)|x_3^2(t) + 2|x_1(t)|\,|x_2(t)|\,|x_3(t)| + \frac{1}{t_{maxG}\,\underline{V}}(|x_2(t)| + |x_3(t)|)|d(t)| \\
&+ \frac{1}{t_{maxG}\,\underline{V}}\overline{S}_G + \frac{1}{t_{maxG}\,\underline{V}}|\dot{d}(t)| + \overline{S}_G^2 \overline{G}_B + l_0 \overline{S}_G|x_1(t)| + l_0|x_1(t)|\,|x_2(t)| \\
&+ l_0|x_1(t)|\,|x_3(t)| + l_0 \overline{S}_G \overline{G}_B + \frac{l_0}{t_{maxG}\,\underline{V}}|d(t)| - p_3(t)S_N(t)x_1(t)\varrho \Bigg\}|\sigma(t)|.
\end{aligned}
$$

$$(37)$$

Since x_1 stands for glucose, it is straightforward to conclude that at least $x_1 > 1$, $\forall t$. Therefore, inequalities (36), (37) can be upper bounded by

$$
\begin{aligned}
\dot{V}(t) \leq 2\Bigg\{ &(\overline{S}_G^2 + \overline{p}_2 \overline{S}_I \overline{I}_B + \overline{p}_3 \overline{S}_N \overline{N}_B)|x_1(t)| + \overline{S}_G \overline{G}_B(|x_2(t)| + |x_3(t)| + l_0) \\
&+ (2\overline{S}_G + \overline{p}_2)|x_1(t)|\,|x_2(t)| + (2\overline{S}_G + \overline{p}_3)|x_1(t)|\,|x_3(t)| + |x_1(t)|x_2^2(t) \\
&+ |x_1(t)|x_3^2(t) + 2|x_1(t)|\,|x_2(t)|\,|x_3(t)| + \frac{1}{t_{maxG}\,\underline{V}}(|x_2(t)| + |x_3(t)|)|d(t)| \\
&+ \frac{1}{t_{maxG}\,\underline{V}}\overline{S}_G + \frac{1}{t_{maxG}\,\underline{V}}|\dot{d}(t)| + \overline{S}_G^2 \overline{G}_B + l_0 \overline{S}_G|x_1(t)| + l_0|x_1(t)|\,|x_2(t)| \\
&+ l_0|x_1(t)|\,|x_3(t)| + \frac{l_0}{t_{maxG}\,\underline{V}}|d(t)| - \min\{\underline{p}_2,\underline{p}_3\}\min\{\underline{S}_I,\underline{S}_N\}\varrho \Bigg\}|\sigma(t)|.
\end{aligned}
$$

$$(38)$$

From Eq. (38), an upper bound to time-derivative Lyapunov function can be described by

$$
\begin{aligned}
\dot{V}(t) \leq 2\Bigg\{ &\left(\overline{S}_G^2 + \overline{p}_2 \overline{S}_I \overline{I}_B + \overline{p}_3 \overline{S}_N \overline{N}_B + 2\overline{S}_G \overline{G}_B + l_0 \overline{S}_G + \frac{2}{t_{maxG}\,\underline{V}}|d(t)| \right)|x(t)| \\
&+ (4\overline{S}_G + \overline{p}_2 + \overline{p}_3 + 2l_0)|x(t)|^2 + 4|x(t)|^3 + \frac{\overline{S}_G + l_0}{t_{maxG}\,\underline{V}}|d(t)| + \frac{1}{t_{maxG}\,\underline{V}}|\dot{d}(t)| \\
&+ \overline{S}_G^2 \overline{G}_B + l_0 \overline{S}_G \overline{G}_B - \min\{\underline{p}_2,\underline{p}_3\}\min\{\underline{S}_I,\underline{S}_N\}\varrho \Bigg\}|\sigma(t)|.
\end{aligned}
$$

$$(39)$$

The meal disturbance, $d(t)$, and its first time derivative, $\dot{d}(t)$, are bounded by some real positive number described in Eq. (22), such that

$$\dot{V}(t) \leq 2\left\{\left(\overline{S}_G^2 + \overline{p}_2\overline{S}_I\overline{I}_B + \overline{p}_3\overline{S}_N\overline{N}_B + 2\overline{S}_G\overline{G}_B + l_0\overline{S}_G + \frac{2}{t_{maxG}}\underline{V}\overline{d}\right)|x(t)| \right.$$
$$+ (4\overline{S}_G + \overline{p}_2 + \overline{p}_3 + 2l_0)|x(t)|^2 + 4|x(t)|^3 + \frac{\overline{S}_G + l_0}{t_{maxG}}\underline{V}\overline{d} + \frac{1}{t_{maxG}}\underline{V}\overline{\dot{d}}$$
$$\left. + \overline{S}_G^2\overline{G}_B + l_0\overline{S}_G\overline{G}_B - \min\{\underline{p}_2,\underline{p}_3\}\min\{\underline{S}_I,\underline{S}_N\}\varrho\right\}|\sigma(t)|.$$

(40)

Let us define $k_1 = \overline{S}_G^2 + \overline{p}_2\overline{S}_I\overline{I}_B + \overline{p}_3\overline{S}_N\overline{N}_B + 2\overline{S}_G\overline{G}_B + l_0\overline{S}_G + \frac{2}{t_{maxG}}\underline{V}\overline{d}$, $k_2 = 4\overline{S}_G +$

$\overline{p}_2 + \overline{p}_3 + 2l_0$, $k_3 = 4$, and $k_4 = \frac{\overline{S}_G + l_0}{t_{maxG}}\underline{V}\overline{d} + \frac{1}{t_{maxG}}\underline{V}\overline{\dot{d}} + \overline{S}_G^2\overline{G}_B + l_0\overline{S}_G\overline{G}_B$. Thus,

Eq. (40) becomes

$$\dot{V}(t) \leq 2\left\{k_1|x(t)| + k_2|x(t)|^2 + k_3|x(t)|^3 + k_4\right.$$
$$\left. - \min\{\underline{p}_2,\underline{p}_3\}\min\{\underline{S}_I,\underline{S}_N\}\varrho\right\}|\sigma(t)|.$$

(41)

It is worth mentioning that the state is bounded, since we are dealing with a biological system. Thus, we suppose there exists a known upper bound such that the inequality $|x(t)| < \chi$ is satisfied. Therefore, Eq. (41) can be rewritten as

$$\dot{V}(t) \leq 2\{k_1\chi + k_2\chi^2 + k_3\chi^3 + k_4 - \min\{\underline{p}_2,\underline{p}_3\}\min\{\underline{S}_I,\underline{S}_N\}\varrho\}|\sigma(t)|.$$

(42)

The modulation function is defined as

$$\varrho = \frac{k_1\chi + k_2\chi^2 + k_3\chi^3 + k_4 + \delta}{\min\{\underline{p}_2,\underline{p}_3\}\min\{\underline{S}_I,\underline{S}_N\}}, \quad \delta > 0.$$

(43)

If Eq. (43) is replaced in Eq. (42), then we can readily obtain

$$\dot{V}(t) \leq -2\delta|\sigma(t)|, \quad t \geq 0.$$

(44)

Let $\tilde{\sigma}(t) := |\sigma(t)| = \sqrt{V(t)}$ be the auxiliary variable. Thus, we have $\dot{\tilde{\sigma}}(t) = \frac{\dot{V}(t)}{2\sqrt{V(t)}}$.

Proceeding forward, we divide both sides of Eq. (44) by $2\sqrt{V(t)}$, which implies that

$$\frac{\dot{V}(t)}{2\sqrt{V(t)}} \leq -2\delta\frac{|\sigma(t)|}{2\sqrt{V(t)}},$$

(45)

which is identical to

$$\dot{\tilde{\sigma}}(t) \leq -\delta, \quad t \geq 0. \tag{46}$$

By using the comparison lemma, there exists an upper bound $\bar{\sigma}(t)$ of $\tilde{\sigma}(t)$ that satisfies the differential equation

$$\dot{\bar{\sigma}}(t) = -\delta, \quad \bar{\sigma}(0) = \tilde{\sigma}(0) \geq 0, \quad t \geq 0. \tag{47}$$

Integrating both sides of Eq. (47) yields

$$\bar{\sigma}(t) - \bar{\sigma}(0) = -\delta t, \quad t \geq 0. \tag{48}$$

Therefore, the following inequality is valid:

$$\bar{\sigma}(t) = -\delta t + \bar{\sigma}(0), \quad t \geq 0. \tag{49}$$

Since $\bar{\sigma} \geq 0$ is continuous, $\sigma(t)$ becomes identically null $\forall t \geq t_1 = \delta^{-1}\bar{\sigma}(0)$. Proceeding forward, we conclude that there exists a finite time $0 < t_s \leq t_1$, where the sliding mode starts such that $\sigma(t) = 0$, $\forall t \geq t_s$. From Eq. (28), one can conclude that $\dot{e} = -l_0 e$ and then the error $e(t)$ converges exponentially to zero. □

6.1 Boundary layer for chattering alleviation

There are some problems that are intrinsic to the traditional sliding mode controller, such as discontinuous control action and the so-called *chattering* effect. In order to mitigate these effects, we utilize the boundary layer technique (Slotine and Li, 1991).

The boundary layer implementation occurs through the control law design. In this sense, instead of using the relay for implementing sgn(σ) in (27), we use a control action given by the following law:

$$u = -\varrho \frac{\sigma}{|\sigma| + \delta}, \tag{50}$$

where $0 < \delta < 1$. Proceeding this way, attenuation of chattering effect is expected to be achieved.

7 Terminal sliding mode control: Design and stability analysis

The nonsingular terminal SMC differs from the first-order SMC primarily for two reasons. First, its sliding surface is a nonlinear function of the error state variables. Second, while exponential convergence is guaranteed for the first-order SMC when the closed-loop system remains on the sliding surface, the nonsingular terminal algorithm is able to bring the state to the origin in finite time $(e \equiv \dot{e} \equiv 0)$ (Feng et al., 2002).

Theorem 2. *Consider the system described by Eqs. (1)–(4) and the lower-upper bounds in Eqs. (17)–(20). If the constant modulation function ϱ is chosen sufficiently large, then, it is possible to find a nonsingular terminal control law $u(t)$ given by*

$$u = -\varrho \operatorname{sgn}(\sigma), \quad \varrho > 0, \tag{51}$$

$$\sigma(t) = e(t) + \frac{1}{\beta}\dot{e}^{p/q}(t), \tag{52}$$

where $\beta = 2$, $p = 5$, and $q = 3$, such that the ideal sliding mode in $\sigma(t) = 0$ occurs in finite time. Besides, under the sliding regime, the error convergence occurs in finite time.

Proof. Consider the following Lyapunov candidate function:

$$V(t) = \sigma^2(t), \tag{53}$$

with $\sigma(t)$ given in Eq. (52). The derivative of Eq. (53) is

$$
\begin{aligned}
\dot{V}(t) = 2\Bigg\{ & \bigg[S_G x_1(t) + x_1(t)x_2(t) - x_1(t)x_3(t) - S_G G_B - \frac{1}{t_{\max G}V}d(t) \\
& + \frac{1}{\beta}\frac{p}{q}\bigg(S_G x_1(t) + x_1(t)x_2(t) - x_1(t)x_3(t) - S_G G_B - \frac{1}{t_{\max G}V}d(t) \bigg)^{\frac{p-q}{q}} \\
& \times \big[(-S_G^2 - p_2 S_I I_B + p_3 S_N N_B)x_1(t) + S_G G_B(x_2(t) - x_3(t)) - x_1(t)x_3^2(t) \\
& - (2S_G + p_2)x_1(t)x_2(t) + (2S_G + p_3)x_1(t)x_3(t) - x_1(t)x_2^2(t) - x_3(t) \big] d(t) \\
& + 2x_1(t)x_2(t)x_3(t) + \frac{1}{t_{\max G}V}\bigg(S_G + x_2(t) - \frac{1}{t_{\max G}V}\dot{d}(t) \bigg) \bigg] \sigma(t) \\
& + \frac{1}{\beta}\frac{p}{q}\bigg(S_G x_1(t) + x_1(t)x_2(t) - x_1(t)x_3(t) - S_G G_B - \frac{1}{t_{\max G}V}d(t) \bigg)^{\frac{p-q}{q}} \\
& \times \big(p_2 S_I x_1(t)u^+ - p_3 S_N x_1(t)u^- \big)\sigma(t) \Bigg\}.
\end{aligned}
\tag{54}
$$

Insulin will be delivered to the patient $(u^+ = \varrho)$ if $u > 0$. In this case, $\operatorname{sgn}(\sigma) < 0 \to \sigma = -|\sigma|$ and $u^- = 0$. Glucagon will be delivered to the patient $(u^- = \varrho)$ if $u < 0$. In this case, $\operatorname{sgn}(\sigma) > 0 \to \sigma = |\sigma|$ and $u^+ = 0$. Hence, for simplicity's sake, we assume that: (i) $x_1(t) > 1$; (ii) it is always possible to find k_G sufficiently small; and

(iii) $\left(S_G x_1(t) + x_1(t)x_2(t) - x_1(t)x_3(t) - S_G G_B - \dfrac{1}{t_{\max G}V}d(t) \right)^{\frac{p-q}{q}} \geq k_G (S_G G_B)^{\frac{p-q}{q}}$,

equality (54) can be upper bounded by

$$\dot{V}(t) \leq 2\left\{\left[S_G|x_1(t)| + |x_1(t)|\,|x_2(t)| + |x_1(t)|\,|x_3(t)| + S_G G_B\right.\right.$$

$$+\frac{1}{t_{maxG}V}|d(t)| + \frac{1}{\beta}\frac{p}{q}(S_G|x_1(t)| + |x_1(t)|\,|x_2(t)| + |x_1(t)|\,|x_3(t)|$$

$$+S_G G_B + \frac{1}{t_{maxG}V}|d(t)|)^{\frac{p-q}{q}}\left((S_G^2 + p_2 S_I I_B + p_3 S_N N_B)|x_1(t)|\right.$$

$$+S_G G_B(|x_2(t)| + |x_3(t)|) + (2S_G + p_2)|x_1(t)|\,|x_2(t)|$$

$$+(2S_G + p_3)|x_1(t)|\,|x_3(t)| + |x_1(t)|(|x_2(t)|^2 + |x_3(t)|^2)$$

$$+2|x_1(t)|\,|x_2(t)|\,|x_3(t)| + \frac{1}{t_{maxG}V}(S_G + |x_2(t)| + |x_3(t)|)|d(t)|$$

$$\left.\left.+\frac{1}{t_{maxG}V}|\dot{d}(t)|\right) - \frac{1}{\beta}\frac{p}{q}(S_G G_B)^{\frac{p-q}{q}}\min\{\underline{p}_2,\underline{p}_3\}\min\{\underline{S}_I,\underline{S}_N\}\varrho\right]\right\}. \tag{55}$$

By using bounds (17)–(20), $|x(t)| < \chi$, $|d(t)| < \overline{d}$, and $|\dot{d}(t)| < \overline{\dot{d}}$ in Eq. (55), it is possible to find

$$\dot{V}(t) \leq 2\left\{\left[\overline{S}_G\chi + 2\chi^2 + \overline{S}_G\overline{G}_B + \frac{1}{t_{maxG}}\frac{1}{\underline{V}}\overline{d}\right.\right.$$

$$+\frac{1}{\beta}\frac{p}{q}\left(\overline{S}_G\chi + 2\chi^2 + \overline{S}_G\overline{G}_B + \frac{1}{t_{maxG}}\frac{1}{\underline{V}}\overline{d}\right)^{\frac{p-q}{q}}$$

$$\times\left(\left(\overline{S}_G^2 + 2\overline{p}_2\overline{S}_I\overline{I}_B + \overline{p}_3\overline{S}_N\overline{N}_B + 2\overline{S}_G\overline{G}_B + \frac{1}{t_{maxG}}\frac{1}{\underline{V}}\right)\chi\right.$$

$$\left.+(4\overline{S}_G + \overline{p}_2 + \overline{p}_3)\chi^2 + 4\chi^3 + \frac{\overline{S}_G\overline{d} + \overline{\dot{d}}}{t_{maxG}}\frac{1}{\underline{V}}\right) \tag{56}$$

$$\left.\left.-\frac{k_G p}{\beta q}(\underline{S}_G\underline{G}_B)^{\frac{p-q}{q}}\min\{\underline{p}_2,\underline{p}_3\}\min\{\underline{S}_I,\underline{S}_N\}\varrho\right]\right\}.$$

Defining $\overline{\chi}$ as

$$\overline{\chi} = \overline{S}_G\chi + 2\chi^2 + \overline{S}_G\overline{G}_B + \frac{1}{t_{maxG}}\frac{1}{\underline{V}}\overline{d}$$

$$+\frac{1}{\beta}\frac{p}{q}\left(\overline{S}_G\chi + 2\chi^2 + \overline{S}_G\overline{G}_B + \frac{1}{t_{maxG}}\frac{1}{\underline{V}}\overline{d}\right)^{\frac{p-q}{q}}$$

$$\times\left(\left(\overline{S}_G^2 + 2\overline{p}_2\overline{S}_I\overline{I}_B + \overline{p}_3\overline{S}_N\overline{N}_B + 2\overline{S}_G\overline{G}_B + \frac{1}{t_{maxG}}\frac{1}{\underline{V}}\right)\chi\right.$$

$$\left.+(4\overline{S}_G + \overline{p}_2 + \overline{p}_3)\chi^2 + 4\chi^3 + \frac{\overline{S}_G\overline{d} + \overline{\dot{d}}}{t_{maxG}}\frac{1}{\underline{V}}\right), \tag{57}$$

yields

$$\dot{V}(t) \leq 2\left\{\left[\overline{\chi} - \frac{k_G p}{\beta q}(S_G G_B)^{\frac{p-q}{q}} \min\{\underline{p}_2, \underline{p}_3\} \min\{\underline{S}_I, \underline{S}_N\} \varrho\right]\right\}. \tag{58}$$

Then, describing ϱ conveniently as

$$\varrho = \frac{\overline{\chi} + \delta}{\frac{k_G p}{\beta q}(S_G G_B)^{\frac{p-q}{q}} \min\{\underline{p}_2, \underline{p}_3\} \min\{\underline{S}_I, \underline{S}_N\}}, \quad \delta > 0, \tag{59}$$

one can guarantee that the derivative of Eq. (53) satisfies inequality

$$\dot{V}(t) \leq -2\delta|\sigma(t)|. \tag{60}$$

Defining the auxiliary variable $\tilde{\sigma}(t) := |\sigma(t)| = \sqrt{V(t)}$, we can readily obtain $\dot{\tilde{\sigma}}(t) = \frac{\dot{V}(t)}{2\sqrt{V(t)}}$. Next, dividing both sides of Eq. (60) by $2\sqrt{V}(t)$, yields

$$\frac{\dot{V}(t)}{2\sqrt{V(t)}} \leq -2\delta \frac{|\sigma(t)|}{2\sqrt{V(t)}}, \tag{61}$$

that is identically equal to

$$\dot{\tilde{\sigma}}(t) \leq -\delta, \quad t \geq 0. \tag{62}$$

Using the comparison theorem, there is an upper bound $\overline{\sigma}(t)$ of $\tilde{\sigma}(t)$ satisfying the differential equation

$$\dot{\overline{\sigma}}(t) = -\delta, \quad \overline{\sigma}(0) = \tilde{\sigma}(0) \geq 0, \quad t \geq 0. \tag{63}$$

Integrating both sides of Eq. (63) yields

$$\overline{\sigma}(t) - \overline{\sigma}(0) = -\delta t, \quad t \geq 0. \tag{64}$$

Thus, the following equality holds:

$$\overline{\sigma}(t) = -\delta t + \overline{\sigma}(0), \quad t \geq 0. \tag{65}$$

Since $\overline{\sigma} \geq 0$ is a continuous signal, the variable $\sigma(t)$ becomes identically null $\forall t \geq t_1 = \delta^{-1}\overline{\sigma}(0)$. Thereby, one can conclude that there exists a finite time $0 < t_s \leq t_1$, where the sliding mode starts, such that $\sigma(t) = 0$ and $\dot{\sigma}(t) = 0$, $\forall t \geq t_s$. From Eq. (52), one can conclude that

$$|e(t)| = \frac{1}{\beta} \sqrt[p-q]{\left[\sqrt[p]{(\beta|e_0|)^{(p-q)}} - \beta \frac{(p-q)}{p}(t-t_0)\right]^p}, \quad \forall t \geq t_f, \tag{66}$$

where $t_f = t_s + \frac{p\sqrt[p]{(\beta|e_0|)^{(p-q)}}}{\beta(p-q)}$ and then the error $e(t)$ converges to zero in finite time, since the sliding variable σ becomes identically zero. $\qquad\square$

7.1 Continuous nonsingular terminal sliding mode control for chattering alleviation

The extension of the output feedback to the continuous nonsingular terminal sliding mode algorithm (CNTSMA) can be obtained by following the steps introduced by Fridman et al. (2015). It can be understood as a combination of a super-twisting algorithm (Oliveira et al., 2016) with a nonsingular terminal SMC and written as

$$\begin{aligned}
u = & -k_1 \left| \dot{e} + k_2 |e|^{2/3} \operatorname{sgn}(e) \right|^{1/2} \operatorname{sgn}\left(\dot{e} + k_2 |e|^{2/3} \operatorname{sgn}(e) \right) \\
& -k_3 \int_0^t \operatorname{sgn}\left(\dot{e} + k_2 |e|^{2/3} \operatorname{sgn}(e) \right) d\tau,
\end{aligned} \tag{67}$$

where $k_1 > 0$, $k_2 > 0$, and $k_3 > 0$ are appropriate constants.

8 HOSM exact differentiators for output feedback

In this chapter, it was chosen to work with two different control strategies called FOSMC and nonsingular terminal sliding modes control. An issue that arises in this context is the necessity of error derivative signal. Nevertheless, this signal is not available directly. In this section, an exact differentiator based on HOSMs is applied to overcome this drawback. Besides its capability to provide the exact derivative of the output error $e \in \mathbb{R}$, it ensures the attenuation of small high-frequency noises (Levant, 2003). Its structure is given by

$$\begin{aligned}
\dot{\zeta}_0 &= v_0 = -\lambda_0 C^{\frac{1}{p+1}} |\zeta_0 - e(t)|^{\frac{p}{p+1}} \operatorname{sgn}(\zeta_0 - e(t)) + \zeta_1, \\
&\vdots \\
\dot{\zeta}_i &= v_i = -\lambda_i C^{\frac{1}{p-i+1}} |\zeta_i - v_{i-1}|^{\frac{p-i}{p-i+1}} \operatorname{sgn}(\zeta_i - v_{i-1}) + \zeta_{i+1}, \\
&\vdots \\
\dot{\zeta}_p &= -\lambda_p C \operatorname{sgn}(\zeta_p - v_{p-1}),
\end{aligned} \tag{68}$$

where λ_i are appropriate constants, chosen recursively; C is an appropriate constant, such that $C \geq |e^{(\rho)}(t)|$; the state is described by $\zeta = [\zeta_0, ..., \zeta_{\rho-1}]^T$; and $p = \rho - 1$ represents the order of the differentiator. Therefore, the following equalities

$$\zeta_0(t) = e(t), \quad \zeta_i(t) = e^{(i)}(t), \quad i = 1, ..., p, \tag{69}$$

are established in finite time (Levant, 2003), provided that the signal, $e^{(\rho)}(t)$, be uniformly bounded, as assumed in the HOSM differentiator.

As $\rho = 2$, knowing that the plant has dynamics given by Eqs. (1)–(4), from the regulation error (23), it is possible to show that the following inequality is satisfied:

$$
\begin{aligned}
|\ddot{e}(t)| &= \Big| (-S_G^2 - p_2 S_I I_B + p_3 S_N N_B) x_1(t) \\
&\quad + S_G G_B (x_2(t) - x_3(t)) - (2S_G + p_2) x_1(t) x_2(t) \\
&\quad + (2S_G + p_3) x_1(t) x_3(t) - x_1(t) x_2^2(t) - x_1(t) x_3^2(t) \\
&\quad + 2 x_1(t) x_2(t) x_3(t) + \frac{d(t)}{t_{maxG} V} (S_G + x_2(t) - x_3(t)) \\
&\quad - \frac{\dot{d}(t)}{t_{maxG} V} + S_G^2 G_B + p_2 S_I x_1(t) u^+ - p_3 S_N x_1(t) u^- \Big| \\
&\leq \Big(\overline{S}_G^2 + \overline{p}_2 \overline{S}_I \overline{I}_B + \overline{p}_3 \overline{S}_N \overline{N}_B + 2 \overline{S}_G \overline{G}_B + \frac{2 \overline{d}}{t_{maxG}} \underline{V} \\
&\quad + \max\{\overline{p}_2, \overline{p}_3\} + \max\{\overline{S}_I, \overline{S}_N\} \varrho \Big) \chi \\
&\quad + (4 \overline{S}_G + \overline{p}_2 + \overline{p}_3) \chi^2 + 4 \chi^3 + \frac{\overline{S}_G \overline{d} + \overline{\dot{d}}}{t_{maxG}} \underline{V} + \overline{S}_G^2 \overline{G}_B,
\end{aligned}
\tag{70}
$$

where the parameters are upper bounded by Eqs. (17)–(20), the disturbance, $d(t)$, and its derivative, $\dot{d}(t)$, satisfy Eq. (22) and the state norm is upper bounded at least locally, by $|x(t)| < \chi$. Assuming that these upper bounds are constant, and are available for designing the controller, an upper bound for the absolute value of the second time derivative of $e(t)$ can be obtained through

$$
\begin{aligned}
C &= \Big(\overline{S}_G^2 + \overline{p}_2 \overline{S}_I \overline{I}_B + \overline{p}_3 \overline{S}_N \overline{N}_B + 2 \overline{S}_G \overline{G}_B + \frac{2 \overline{d}}{t_{maxG}} \underline{V} \\
&\quad + \max\{\overline{p}_2, \overline{p}_3\} \max\{\overline{S}_I, \overline{S}_N\} \varrho \Big) \chi + (4 \overline{S}_G + \overline{p}_2 + \overline{p}_3) \chi^2 \\
&\quad + 4 \chi^3 + \frac{\overline{S}_G \overline{d} + \overline{\dot{d}}}{t_{maxG}} \underline{V} + \overline{S}_G^2 \overline{G}_B.
\end{aligned}
\tag{71}
$$

Throughout this chapter, the following exact differentiator will be used:

$$
\dot{\zeta}_0 = v_0 = -\lambda_0 C^{1/2} |\zeta_0 - e(t)|^{1/2} \operatorname{sgn}(\zeta_0 - e(t)) + \zeta_1,
\tag{72}
$$

$$
\dot{\zeta}_1 = -\lambda_1 C \operatorname{sgn}(\zeta_1 - v_0),
\tag{73}
$$

with $\lambda_0 = 5$ and $\lambda_1 = 3$ and gain C given in Eq. (71).

The values that have been used in implementing the control system and its parameters are given by: $p_2 = 0.003$; $\overline{p}_2 = 0.03$; $p_3 = 0.01$; $\overline{p}_3 = 0.14$; $S_N = 0.8 \times 10^{-4}$;

$\overline{S}_N = 2 \times 10^{-4}; S_I = 6.8 \times 10^{-4}; \overline{S}_I = 8.6 \times 10^{-4}; \overline{S}_G = 0.015; \overline{N}_B = 60; \overline{I}_B = 20; \underline{V} = 1.5;$
$\underline{t}_{\text{max}G} = 65; \overline{d} = 110; \underline{d} = 60;$ and $\varrho = 100$. These values were chosen from the numeric values of the parameters (5)–(10) which, in turn, can be consulted in Herrero et al. (2013).

9 Numerical examples

In this section, the four strategies presented so far have their performances evaluated by simulation results: the traditional sliding mode controller (27)–(28) and the non-singular terminal sliding mode controller (51)–(52), which have a discontinuous control action; the sliding mode controller with *boundary layer* (50); and the continuous nonsingular terminal sliding mode control (67) whose control action are smooth.

Concerning the plant, the parameter values were chosen according to Herrero et al. (2013) and are described as follows: $S_G = 0.014 \text{ min}^{-1}$; $t_{\text{max}G} = 69.6 \text{ min}$; $V = 1.7 \text{ dL kg}^{-1}$; $p_2 = 0.012 \text{ min}^{-1}$; $p_3 = 0.017 \text{ min}^{-1}$; $S_I = 7.73 \times 10^{-4} \text{ min}^{-1}$ per $\mu\text{U mL}^{-1}$; $S_N = 1.38 \times 10^{-4} \text{ min}^{-1}$ per pg mL^{-1}; $I_B = 11.01 \text{ mU dL}^{-1}$; and $N_B = 46.30 \text{ pg mL}^{-1}$. The desired set point was defined as $G_B = 90 \text{ mg dL}^{-1}$, as discussed in Section 2, and the initial condition of the glycemia was arbitrarily chosen as $G_0 = 120 \text{ mg dL}^{-1}$. The major concern was to define an initial condition that was different from the set point so as to attest the controllers' performance. The initial condition of variables $x_2(t) \text{ (min}^{-1})$ and $x_3(t) \text{ (min}^{-1})$ was set as $x_2(0) = x_3(0) = 0$, as recommended by Herrero et al. (2013).

In what concerns the exact differentiator, the gain was defined as $C = 100$; the parameters as $\lambda_0 = 5$, $\lambda_1 = 3$; and the initial conditions as $\zeta_0(0) = 0$ and $\zeta_1(0) = 0$. The simulation time was set as 1440, since one desires to safely regulate the patient's blood glucose over a period of 24 h, which is equivalent to 1440 min. At last, the sampling step was made equal to 10^{-3}.

9.1 Discontinuous FOSMC with estimate of sliding variable using exact differentiator

Although Theorem 1 shows that it is possible to find a constant ϱ that locally guarantees the sliding mode, the control law (27), (28) cannot be implemented, since signal $\dot{e}$ is not available. Thus, recalling Eq. (69), the control law is redesigned and becomes

$$u = -\varrho \, \text{sgn}(\hat{\sigma}(t)), \quad \varrho > 0, \tag{74}$$

$$\hat{\sigma}(t) = \zeta_1(t) + l_0 \zeta_0(t), \tag{75}$$

where $\hat{\sigma}(t)$ is an estimate for $\sigma(t)$, using Eqs. (72), (73).

In the presented simulations, controller (74), (75) was implemented with $\varrho = 100$, $l_0 = 1 \text{ min}^{-1}$ and with estimated variables ζ_0 and ζ_1 given by Eqs. (72), (73). As for the unit of ϱ, it should be mentioned that this constant is measured in $(\mu\text{U min mg}^{-1})$ during the positive control action; and it is measured in (pg dL^{-1}) during the negative control action.

Next, the simulation results are presented for the first-order sliding mode controller. Fig. 5A illustrates the ability of the controller to safely regulate the blood glucose and avoid hypoglycemic and hyperglycemic episodes over the period assessed. Fig. 5B will be discussed later on.

Fig. 6A and B demonstrates the role of bihormonal actuator, detaching the positive control action, u^+, and the negative control action, u^-, in two isolated and independent actions. The positive control action stands for insulin injection in the organism of the patient with T1DM, while the negative control action represents glucagon injection.

Since the control action can be understood as the union of the negative and the positive control actions, one may conclude that Fig. 6 depicts the control action, which is notably discontinuous and counts with the presence of chattering. The high amplitude of the control action should be highlighted. This amplitude is due to the fact that the controller gains in Eq. (74) are extremely elevated. The consequence of it is the use of a large amount of hormones being injected in the patient every time. This effect, obviously, is undesirable, since it may cause harm to the patient's health.

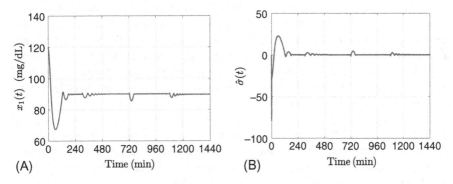

Fig. 5 Sliding surface and glycemia. (A) Blood glucose concentration. (B) Sliding variable estimate $\hat{\sigma}$.

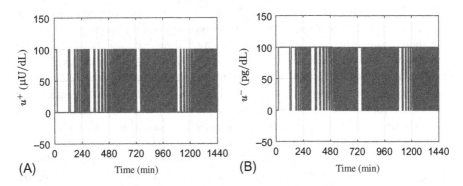

Fig. 6 Insulin and glucagon. (A) Plasma insulin concentration. (B) Plasma glucagon concentration.

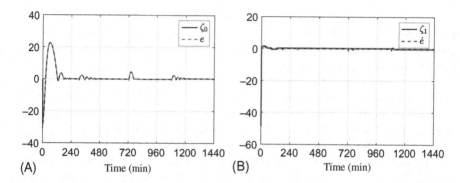

Fig. 7 Exact differentiator. (A) Comparison between ζ_0 and e. (B) Comparison between ζ_1 and $\dot{e}$.

Fig. 7A and B attest in favor of the transient briefness of variables ζ_0 and ζ_1. In another words, within a short period of time, the estimates of the error, e, and its first derivative, $\dot{e}$, are perfectly obtained. Therefore, an estimate of the ideal sliding variable σ can be constructed from $\hat{\sigma}$, which in turn guarantees that the ideal sliding mode $\sigma = 0$ is reached in finite time. Once again, it is noteworthy that the signals illustrated in Fig. 7A are measured in (mg dL^{-1}), while the signals represented in Fig. 7B are measured in (mg dL^{-1} min^{-1}).

As was said in Section 4, the meals made by the patient with T1DM are scheduled for 5:00, 12:00, and 18:00. As it is clear for Fig. 5A, it is noted that in the instants when the patient makes his meals, oscillations appear in the glycemic curve. Indeed, there is a forward relation between feeding, $d(t)$, and the glycemic curve, $x_1(t)$. Next, this relationship will be elucidated in the light of control theory.

Note that during the controller design, the variable $d(t)$ was assumed to be uniformly upper bounded except for a zero measure set in the sense of Lebesgue. It was also discussed the double interpretation that could be made from this variable: from the medical point of view, it represents the effects that feeding has over the patient's glycemic curve; from the control point of view, it represents an unmatched disturbance.

Although any demonstration had been presented, it was said that, in face of a case of unmatched disturbance, it was necessary that both the differentiator gain, C, and the modulation function of the controller, ϱ, must be capable of upper bound the disturbance, $d(t)$, and its first derivative, $\dot{d}(t)$. This is a condition that must necessarily be met in order for the differentiator to be able to provide ζ_1, where ζ_1 is the estimate of the derivative of the tracking error. It was said that the knowledge of this variable was necessary to the construction of the sliding surface $\sigma = 0$.

Notice, however, that if the disturbance $d(t)$ is mathematically represented by Eq. (24), it is easy to demonstrate that $\dot{d}(t)$ is described by a series of three impulses positioned in the same instants in which the meals are made. As known, the amplitude of an impulse is infinite. In this sense, it is important to highlight that neither the differentiator gain, either the modulation function of the controller are capable of upper bound the amplitude of $\dot{d}(t)$ in the instants that occur right after each one of the daily meals.

For this reason, precisely in these moments, it is evident that the construction of the sliding surface, $\sigma = 0$, is not possible, as depicted in Fig. 5B. Finally, a last comment should be made about this figure: the variable $\hat{\sigma} = 0$ is measured in (mg dL^{-1} min^{-1}).

The simulation results concerning to the rest of the controllers will present a great deal of similarities with the results that were discussed in this section. For this reason, in order to avoid the repetition of content, the following sections will provide shorter comments, always made in reference of the aforementioned ones.

9.2 Continuous FOSMC with estimate of sliding variable using exact differentiator and boundary layer

The major disadvantage of a classic sliding mode control strategy is the so-called chattering effect. This phenomenon is characterized by small oscillations in the system output that may impair the performance of the control system. The chattering effect may arise due to unmodeled high-frequency dynamics, or due to digital implementation problems. However, the chattering effects can be smoothed by the implementation of the technique described in Section 6.1: the boundary layer (Utkin, 1993).

As it is noticed from Fig. 8B, the chattering effect has been alleviated from the control action $u(t)$. In other words, the control action now is described by a soft function, different from the discontinuous control action evidenced in Section 9.1. This is due to the use of *boundary layer*. In the other aspects, the same one that was commented on in Section 9.1 stands out (Figs. 9 and 10).

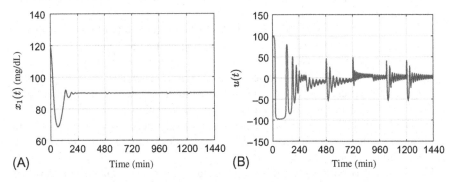

Fig. 8 Glycemia and control action. (A) Blood glucose concentration. (B) Controller action.

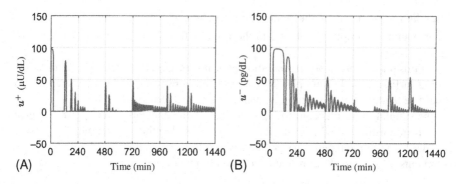

Fig. 9 Insulin and glucagon. (A) Plasma insulin concentration. (B) Plasma glucagon concentration.

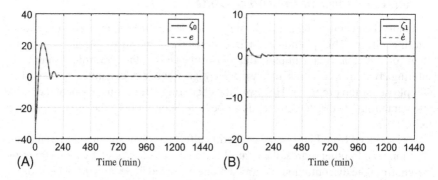

Fig. 10 Exact differentiator. (A) Comparison between ζ_0 and e. (B) Comparison between ζ_1 and $\dot{e}$.

9.3 Discontinuous nonsingular terminal sliding mode control

Figs. 11–13 depict the simulation results obtained through discontinuous nonsingular terminal sliding mode control.

In what concerns the NSTSMC, Fig. 11A illustrates the ability of the controller to safely regulate the blood glucose and avoid hypoglycemic and hyperglycemic episodes over the period assessed.

It is evident from Fig. 11B, the discontinuous control action is affected by constant and high amplitude chattering. Once again, such amplitude is due to the high controller gains presented in Eq. (51). As a consequence, the blood glucose regulation is afforded, provided a large amount of hormone is injected in the patients' bloodstream in the course of the day. Such effect, obviously, is not desirable, since it entails a plenty of risks to these patients health.

Fig. 12A and B portrays the role of bihormonal actuator, detaching the positive control action, u^+, and the negative control action, u^-, in two independent and isolated

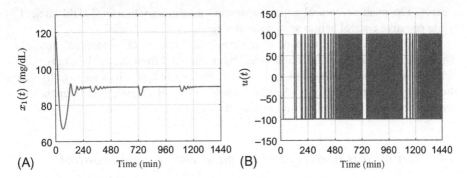

Fig. 11 Curves related to discontinuous nonsingular terminal sliding mode control. (A) Blood glucose concentration. (B) Controller action.

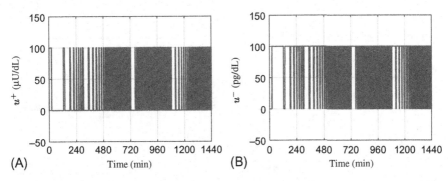

Fig. 12 Curves related to discontinuous nonsingular terminal sliding mode control. (A) Plasma insulin concentration. (B) Plasma glucagon concentration.

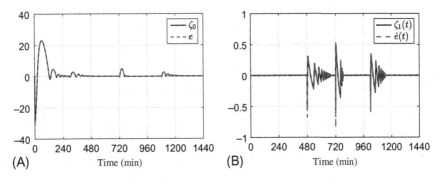

Fig. 13 Curves related to discontinuous nonsingular terminal sliding mode control. (A) Comparison between ζ_0 and e. (B) Comparison between ζ_1 and $\dot{e}$.

signals. The positive control action represents insulin injection in T1DM subject, while negative control action stands for glucagon.

Fig. 13A and B attest in favor of the transient briefness of variables ζ_0 and ζ_1. Therefore, the ideal sliding variable σ can be constructed from $\hat{\sigma}$, which in turn guarantees that the ideal sliding mode $\sigma = 0$ is reached. Once again, it is noteworthy that the signals illustrated in Fig. 13A are measured in (mg dL^{-1}), while the signals represented in Fig. 13B are measured in (mg dL^{-1} min^{-1}).

9.4 Continuous nonsingular terminal sliding mode control

In order to mitigate the chattering effect and reduce the amount of hormone injected into the patients' bloodstream, one possibility is to apply the continuous version of NSTSMC, the so-called CNTSMA (Fridman et al., 2015).

In Figs. 14–16, the simulation results are presented for the continuous NSTSMC.

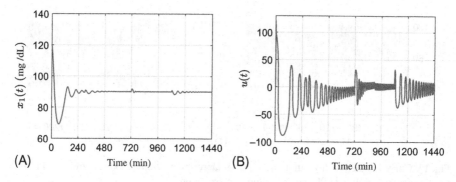

Fig. 14 Control action and glycemia. (A) Blood glucose concentration. (B) Controller action.

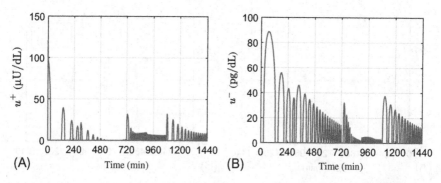

Fig. 15 Insulin and glucagon. (A) Plasma insulin concentration. (B) Plasma glucagon concentration.

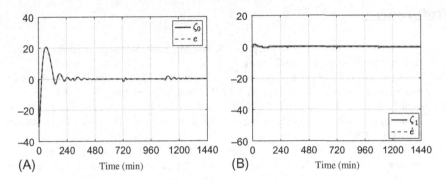

Fig. 16 Exact differentiator. (A) Comparison between ζ_0 and e. (B) Comparison between ζ_1 and $\dot{e}$.

As can be seen from Fig. 14B, the chattering effect was alleviated from the control action $u(t)$. In other words, the control action is now described by a smooth function, while the control action provided in Section 9.3 was discontinuous. Moreover, Fig. 15 depicts also that the control action has a variable amplitude. In the remaining aspects, the issues that are to be underscored here are not different from those presented in Section 9.1.

10 Conclusions

In this chapter, two different control strategies based on sliding mode control were used in order to safely regulate the blood glucose concentration of a T1DM patient. From a control point of view, this is a challenging problem since the process is represented by a nonlinear and time-varying model, where the parameters were considered uncertain. Meals of varying carbohydrate content are understood as unmatched disturbances. The output has relative degree two with regard to control signal, and since the model adopted describes a biological process, the control problem must be treated as a strictly positive system.

Such challenges were overcome by using some upper bounds for the plant parameters and disturbances as well as a bihormonal actuator formulation with the sliding mode control theory. A rigorous stability analysis through Lyapunov functions and numerical simulations shows that the proposed control strategy was efficient in the glycemic regulation task.

As future works, one proposes the experimental validation of the presented controllers. Moreover, in an eventual application, one could replace the fixed-gain differentiator presented here by a variable-gain differentiator (Oliveira et al., 2017) in order to mitigate the sensitivity of the control system in closed loop with respect to the measurement error/noise. At last, one suggests the assessment of different sliding mode control techniques aiming to achieve the euglycemia for patients with T1DM.

References

Atlas, E., Nimri, R., Miller, S., Grunberg, E., Phillip, M., 2010. MD-logic artificial pancreas system. Diabetes Care 33, 1072–1076.

Audoly, S., Bellu, G., D'Angio, L., Saccomani, M., Cobelli, C., 2001. Global identifiability of nonlinear models of biological systems. IEEE Trans. Biomed. Eng. 48, 55–65.

Bakhtiani, P., Zhao, L., El Youssef, J., Castle, J., Ward, W., 2013. A review of artificial pancreas technologies with an emphasis on bi-hormonal therapy. Diabetes Obes. Metab 15, 1065–1070.

Bátora, V., Boiroux, D., Hagdrup, M., Tárník, M., Murgaš, J., Schmidt, S., Nøgaard, K., Poulsen, N.K., Madsen, H., Jøgensen, J.B., 2015. Glucagon administration strategies for a dual-hormone artificial pancreas. IEEE Trans. Biomed. Eng.

Bátora, V., Tárník, M., Murgaš, J., Schmidt, S., Nørgaard, K., Poulsen, N., Madsen, H., Jørgensen, J., 2015. Bihormonal control of blood glucose in people with type 1 diabetes. In: 2015 European Control Conference (ECC). Linz, Austria, pp. 25–30.

Bátora, V., Tárník, M., Murgaš, J., Schmidt, S., Nørgaard, K., Poulsen, N., Madsen, H., Boiroux, D., Jørgensen, J., 2015. The contribution of glucagon in an artificial pancreas for people with type 1 diabetes. In: 2015 American Control Conference pp. 5097–5102 Chicago, IL.

Bellazzi, R., Siviero, C., Stefanelli, M., De Nicolao, G., 1995. Adaptive controllers for intelligent monitoring. Artif. Intell. Med. 7 (6), 515–540.

Bequette, B.W., 2012. Challenges and recent progress in the development of a closed-loop artificial pancreas. Annu. Rev. Control. 36, 255–266.

Bequette, B.W., Cameron, F., Buckingham, B.A., Maahs, D.M., Lum, J., 2018. Overnight hypoglycemia and hyperglycemia mitigation for individuals with type 1 diabetes: how risks can be reduced. IEEE Control. Syst. Mag. 38 (1), 125–134.

Bergman, R., Phillips, L., Cobelli, C., 2009. Physiologic evaluation of factors controlling glucose tolerance in man-measurement of insulin sensitivity and β-cell glucose sensitivity from the response to intravenous glucose. J. Clin. Investig. 68, 1456–1467.

Boiroux, D., Duun-Henriksen, A., Schmidt, S., Nørgaard, K., Poulsen, N., Madsen, H., Jørgensen, J., 2016. Adaptive control in an artificial pancreas for people with type 1 diabetes. Control Eng. Pract. 58, 332–342.

Castle, J., Engle, J., El Youssef, J., Massoud, R., Yuen, K., Kagan, R., Ward, W., 2010. Novel use of glucagon in a closed-loop system for prevention of hypoglycemia in type 1 diabetes. Diabetes Care 33, 1282–1287.

Cobelli, C., Renard, E., Kovatchev, B., 2011. Artificial pancreas: past, present, future. Diabetes 60, 2672–2682.

Edwards, C., Spurgeon, S., 1998. Sliding Mode Control: Theory and Applications. CRC Press, Boca Raton, Florida.

El-Khatib, F., Jiang, B., Damiano, E., 2007. Adaptive closed-loop control provides blood-glucose regulation using dual subcutaneous insulin and glucagon infusion in diabetic swine. J. Diabetes Sci. Technol. 1 (2), 181–192.

El-Khatib, F.H., Jiang, J., Damiano, E.R., 2009. A feasibility study of bihormonal closed-loop blood glucose control using dual subcutaneous infusion of insulin and glucagon in ambulatory diabetic swine. J. Diabetes Sci. Technol. 3, 789–803.

El-Khatib, F., Russell, S., Nathan, D., Sutherlin, R., Damiano, E., 2010. A bihormonal closed-loop artificial pancreas for type 1 diabetes. Sci. Transl. Med. 2 (27), 1–12.

Favero, S.D., Place, J., Kropff, J., Messori, M., Keith-Hynes, P., Visentin, R., Monaro, M., Galasso, S., Boscari, F., Toffanin, C., Palma, F.D., Lanzola, G., Scarpellini, S.,

Farret, A., Kovatchev, B., Avogaro, A., Bruttomesso, D., Magni, L., DeVries, J., Cobelli, C., Renard, E., 2015. Multicenter outpatient dinner/overnight reduction of hypoglycemia and increased time of glucose in target with a wearable artificial pancreas using modular model predictive control in adults with type 1 diabetes. Diabetes Obes. Metab. 17 (5), 468–476.

Feng, Y., Yu, X., Man, Z., 2002. Non-singular terminal sliding mode control of rigid manipulators. Automatica 38, 2159–2167.

Fridman, L., Moreno, J.A., Bandyopadhyay, B., Kamal, S., Chalanga, A., 2015. Continuous nested algorithms: the fifth generation of sliding mode controllers. In: Recent Advances in Sliding Modes: From Control to Intelligent Mechatronics, vol. 24. Springer, New York City, United States of America, pp. 5–36.

Guyton, A., Hall, J., 2016. Textbook of Medical Physiology. Elsevier, Amsterdam, Netherlands.

Hanas, R., 1998. Insulin-Dependent Diabetes in Children, Adolescents and Adults—How to Become an Expert on Your Own Diabetes. Class Publishing, Bristol, United Kingdom.

Hernández, A., Fridman, L., Levant, A., Shtessel, Y., Leder, R., Monsalve, C., Andrade, S., 2013. High-order sliding-mode control for blood glucose: practical relative degree approach. Control Eng. Pract. 21, 747–758.

Herrero, P., Georgiou, P., Oliver, N., Reddy, M., Johnston, D., Toumazou, C., 2013. A composite model of glucagon-glucose dynamics for in silico testing of bihormonal glucose controllers. J. Diabetes Sci. Technol. 7, 941–951.

Huyett, L.M., Dassau, E., Zisser, H.C., Doyle III, F.J., 2018. Glucose sensor dynamics and the artificial pancreas: the impact of lag on sensor measurement and controller performance. IEEE Control Syst. Mag. 38 (1), 30–46.

Jacobs, P., El Youssef, J., Castle, J., Bakhtiani, P., Branigan, D., 2014. Automated control of an adaptive bihormonal, dual-sensor artificial pancreas and evaluation during inpatient studies. IEEE Trans. Biomed. Eng., 61, 2569–2581.

Jajarm, A., Ozgoli, S., Momeni, H., 2012. Blood glucose regulation using fuzzy recursive fast terminal sliding mode control. In: 4th International Conference on Intelligent and Advanced Systems (ICIAS2012) pp. 393–397 Kuala Lumpur, Malaysia.

Kadish, A., 1964. Automation control of blood sugar. Servomechanism for glucose monitoring and control. Am. J. Med. Electron. 9, 82–86.

Kaveh, P., Shtessel, Y., 2008. Blood glucose regulation using higher-order sliding mode control. Int. J. Robust Nonlinear Control 18, 557–569.

Kumareswaran, K., Evans, M., Hovorka, R., 2009. Artificial pancreas: an emerging approach to treat type 1 diabetes. Expert Rev. Med. Devices. 6, 401–410.

Levant, A., 2003. Higher-order sliding modes, differentiation and output-feedback control. Int. J. Control 76, 924–941.

Marchetti, G., Barolo, M., Jovanovic, L., Zisser, H., Seborg, D., 2008. An improved PID switching control strategy for type 1 diabetes. IEEE Trans. Biomed. Eng. 55 (3), 857–865.

Messori, M., Incremona, G.P., Cobelli, C., Magni, L., 2018. Individualized model predictive control for the artificial pancreas: in silico evaluation of closed-loop glucose control. IEEE Control. Syst. Mag. 38 (1), 86–104.

Oliveira, T.R., Estrada, A., Fridman, L.M., 2016. Output-feedback generalization of variable gain super-twisting sliding mode control via global HOSM differentiators. In: 14th International Workshop on Variable Structure Systems. IEEE, pp. 257–262.

Oliveira, T.R., Estrada, A., Fridman, L.M., 2017. Global and exact HOSM differentiator with dynamic gains for output-feedback sliding mode control. Automatica 81, 156–163.

Patti, M., 2018. New Glucagon Delivery System Reduces Episodes of Post-Bariatric Surgery Hypoglycemia. Medical XPress, Douglas, Isle of Man.

Reiter, M., Reiterer, F., del Re, L., 2016. Bihormonal glucose control using a continuous insulin pump and a glucagon-pen. In: European Control Conference (ECC). IEEE, pp. 2435–2440.

Russell, S., El-Khatib, F., Nathan, D., Magyar, K., Jiang, J., Damianod, E., 2012. Blood glucose control in type 1 diabetes with a bihormonal bionic endocrine pancreas. Diabetes Care 35, 2148–2155.

Shtessel, Y., Shkolnikov, I., Brown, M., 2003. An asymptotic second-order smooth sliding mode control. Asian J. Control 5, 498–504.

Slotine, J.J.E., Li, W., 1991. Applied Nonlinear Control. Prentice-Hall, Upper Saddle River, United States of America.

Turksoy, K., Quinn, L., Littlejohn, E., Cina, A., 2014. Multivariable adaptive identification and control for artificial pancreas systems. IEEE Trans. Biomed. Eng. 61 (3), 883–891.

Utkin, V.I., 1993. Sliding mode control design principles and applications to electric drives. IEEE Trans. Ind. Electron. 40, 23–36.

Further readings

Albertos, P., Mareels, I., 2010. Feedback and Control for Everyone. Springer Science & Business Media, New York, NY.

Impulsive MPC schemes for biomedical processes: Application to type 1 diabetes

Pablo S. Rivadeneira[a,b], J.L. Godoy[b], J.E. Sereno[a,b], P. Abuin[b],
A. Ferramosca[c], A.H. González[b]
[a]Universidad Nacional de Colombia, Facultad de Minas, Grupo GITA, Medellin, Colombia,
[b]Institute of Technological Development for the Chemical Industry (INTEC), CONICET—
Universidad Nacional del Litoral, Santa Fe, Argentina, [c]Facultad Regional de Reconquista,
CONICET, UTN, Santa Fe, Argentina

1 Introduction

Impulsive control systems (ICS)—those which show discontinuities in the variables time evolution at certain times and free responses between these times—have received a great attention in the last decade, especially in the field of biomedical research. One central problem has been the scheduling of drug administration in the treatment of several human diseases, as it was stated in the Bellman's seminal work (Bellman, 1971).

Despite its potential use in these important field of application, the regulation to nonzero set points—which is the case in most applications—has received little attention in the literature, since the only one formal equilibrium of an impulsive system is the origin. Recently, by Rivadeneira and Gonzalez (2018) and Rivadeneira et al. (2018), a full description of the generalized equilibrium out of the origin for ICS was developed including its dynamical characterization. It is shown that the state equilibria in this kind of systems can be described by bounded orbits, each of one corresponding to an impulsive control equilibrium sequence. Besides, two underlying discrete-time systems can be obtained if the states are evaluated just before and after the impulsive times, which allows a simplified design of control laws for ICS, as state feedback or model predictive control (MPC).

The first MPC formulation for ICS is presented by Sopasakis et al. (2015). The strategy covers the problem to steer a linear ICS to a zone defined by a "therapeutic window" that does not includes the origin. Furthermore, it accounts for feasibility at both impulsive and continuous time. However, the formulation is based on polytopic invariant target sets, whose calculation is not a trivial task and could be difficult to characterize in many applications. The second MPC formulation for ICS—called impulsive zone model predictive control (iZMPC)—is introduced by

Control Applications for Biomedical Engineering Systems. https://doi.org/10.1016/B978-0-12-817461-6.00003-2

Rivadeneira et al. (2018). This is a zone MPC (Ferramosca et al., 2010; González and Odloak, 2009) which is less general than the one having invariant sets as target sets (González et al., 2014; Sopasakis et al., 2015), but more general than the typical one having equilibrium points as set points (Rawlings and Mayne, 2009). In this framework, the state is steered to an equilibrium set—instead to an equilibrium point—making no differences between any point inside the set. Furthermore, this control strategy is formulated in a tracking scenario where the equilibrium set can be far from the origin. By means of the use of artificial/intermediary variables that are only forced to lie in the equilibrium space, these kind of controllers ensure feasibility for any change of the target set. It also provides an enlarged domain of attraction, given by the controllable set to the entire equilibrium space, instead of the controllable set to a given point or invariant terminal set (as it is described by Rawlings and Mayne, 2009). In spite of being an interesting strategy for ICS, it has not been exploited in the context of biomedical applications such as influenza, HIV, malaria, or type 1 diabetes mellitus (T1DM).

In this chapter, the selected case study is T1DM and its study with impulsive MPC is pursued. T1DM is a illness caused by insulin deficiency due to the destruction of the pancreatic beta cells leading to an alteration in the natural glucose regulation system. People affected by T1DM are prone to have acute complications such as diabetic ketoacidosis (due to an augmented lipolysis in response to a decrease in glucose uptake, alternative metabolic pathway) and chronic complications such as retinopathy, diabetic nephropathy, neuropathy, and cardio-cerebral diseases (due to a persistent high blood glucose [BG] level).

The American Diabetes Association (ADA) recommends the functional insulin therapy (FIT) consisting in a continuous subcutaneous insulin infusion (CSII) and insulin boluses matching carbohydrate (CHO) intake, preprandial BG and activity level (Jeandidier et al., 2008). Many T1DM patients use manually controlled CSII pumps with continuous glucose monitoring (CGM) sensor associated with administer meal-time insulin boluses and correction boluses, and also to program the permanent insulin infusion rate (or basal insulin rate [BIR]). The benefits of using a sensor-augmented insulin pump include more flexible timing of meals and more predictable insulin absorption, thus reducing the variability of daily glucose profiles (Sindelka et al., 1994). Adding a continuous glucose monitor had also shown to further improve glucose control and reduce nocturnal hypoglycemia (Bergenstal et al., 2010, 2013).

An artificial pancreas (AP) is a closed-loop control system that automatically modulates patient's insulin infusion rate in order to maintain their BG within safe limits. The AP is composed by a CSII pump (manipulated variable), a CGM sensor (controlled variable), a meal announcement based on carbohydrate counting (CC) (Schmidt and Nørgaard, 2014) (measured disturbance variable), and a control system (MPC, proportional-integral-derivative [PID], linear-quadratic-Gaussian [LQG], fuzzy, etc.). This automated device significantly reduces the risks of hypoglycemia and hyperglycemia, as it has been reported in the literature (Dassau et al., 2017). The first AP system (approved commercialization in September 2016; Food and

Administration, 2016) can automatically adjust the BIR according to the values of CGM (different to its manual mode in which the BIR is constant); however, the user must manually administer insulin bolus during meals, which could produce a lot of performance of the AP systems, particularly when large meals are badly informed or unannounced (Lee et al., 2009; Ramkisson et al., 2018). So, there are still several issues to be solved to move toward a unified insertion platform for a complete AP.

The core of the AP is given by a model-based controller, which requires an estimation of the actual states and a reliable model of glucose, insulin, and meal absorption dynamics, to perform the predictions and compute the proper (optimal) control actions. Here, a new compartmental model is proposed, which is composed by three subsystems, describing the glucose, the insulin, and the digestion dynamics, respectively. This model permits better long-term prediction than others in the literature (Kirchsteiger et al., 2014; Ruan et al., 2017), and so it is suitable for avoiding undesirable episodes (hypoglycemia and hyperglycemia) in the context of predictive-control strategies (Godoy et al., 2018). Furthermore, to properly exploit the benefits of the latter model and ensure an adequately behavior of the whole MPC control loop, a suitable state estimation accounting for inherent disturbances (as meals intake or exercise) and sensor noise is developed.

Particularly, this work is focused on the case when meals are badly informed or, even, unannounced. In spite of the fact that MPC-based AP is naturally robust to bad announced meals, recent studies have revealed that its performance is degraded when the size of meals increases (Ramkisson et al., 2018). Therefore, meal detection systems and strategies to compensate postprandial hyperglycemia episodes are necessary. In this context, several authors have proposed meal detection systems, which can be classified as follows: (i) Meal detection: the algorithm informs the controller when a meal is detected (flag) (Dassau et al., 2008; Ramkisson et al., 2018; Turksoy et al., 2016). (ii) Meal detection and meal size estimation: the algorithm estimates the consumed carbohydrates in meals (to compute a correction bolus according to the FIT rules or to inform to the controller) (Cameron et al., 2009; Lee et al., 2009; Samadi et al., 2017). (iii) State estimation robust to unannounced meals without/with meal-size estimation: the algorithm estimates the states used by the controller, employing an augmented model structure to account for plant-model mismatches related to unannounced meals (Gillis et al., 2007; Sala et al., 2018), and also, it estimates the consumed CHO (Lee et al., 2009).

The contributions of this chapter are enclosed in a combined scheme of impulsive estimation/control suitable for biomedical applications, focused on the T1DM dynamics since it is an actual relevant biomedical problem that includes many control challenges as the presence of unknown disturbances, non-Gaussian noise, and input/output constraints. The whole strategy is composed by the iZMPC with artificial variables (a stable zone formulation, with an enlarged domain of attraction) and a new observer structure (see Fig. 1 for an schematic plot of the control structure). Both strategies are based on a new T1DM model which was identified by using both UVA/Padova simulator and real patients' data.

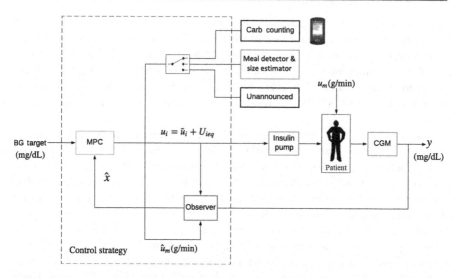

Fig. 1 Artificial pancreas. Combined scheme of iZMPC and observer.

The chapter is organized as follows: after Section 1, a description of impulsive systems is carried out in Section 2. There it is shown that the continuous system with short-duration inputs is seen as an impulsive system, and two discrete-time linear systems are extracted from that description. Based on these systems, the generalized equilibriums and their dynamical characterization are provided. In Section 3, the iZMPC strategy is formulated providing a full description of the cost function and the optimization problem to be solved at each time step. In Section 4, the case study is introduced. The selected application is the T1DM dynamics, and it consists in three parts: (i) a description of the model, including the identification results, (ii) the control and estimation formulations, and (iii) the results obtained when the combined strategy is used. A discussion of the results is given in Section 5, while the final conclusions and future works are provided in Section 6.

2 Dynamic systems with short-duration inputs

In many biomedical applications, the medication is administered to the patient in a discrete way, in the form of small pulses or impulses. For instance, in the AP problem, the insulin can be injected to the patient in both ways.

If a fixed time period is assumed, then it can be argued that an insulin dose is administered by a zero-order hold which maintains their rate all the period long, and changes its value only at the beginning of a new periods. This is in fact what is made in most of the approaches, although this modeling of the input is not realistic since insulin pumps are not able to do that. Actually, insulin doses are administered by pulses that covers only a short interval of these periods. If these pulses are extremely short in comparison with the period, they approach impulses (i.e., pulses of infinitesimal duration). These latter schemes work, for instance, to represent diseases that are controlled by taking pills one or two times per day. However, in the AP, it is not so clear if the (probably fast) insulin injection is short enough to have a good representation by the impulsive approaches. In any case, having a general pulse representation that account for both the zero-order hold input and impulsive input as particular cases is desired.

Consider the following affine continuous-time system:

$$\begin{aligned} \dot{x}(t) &= Ax(t) + B_u u(t) + B_r r(t) + E, \quad x(0) = x_0, \\ y(t) &= Cx(t), \end{aligned} \tag{1}$$

where x represents the state, u the control action, y the output of the system, and r the disturbances, with A, B_u, C, and B_r being the corresponding matrices, respectively, and E being a constant term. If the drug infusion u is injected to the system only at certain time instants given by kT (and quickly enough), where T is the fixed period, and $k \in \mathbb{N}$, then it is possible to work under the impulsive system framework recently developed by Rivadeneira et al. (2018) (see Fig. 2 for a schematic plot). Formally, assume that the input is given by

$$u(t) = u(kT)\delta(t - kT), \quad t \in [kT, (k+1)T), \quad k \in \mathbb{N},$$

where $\delta(t)$ is a generalized function (or distribution), Dirac delta,[a] that fulfills $\delta(0) = \infty$, $\delta(t) = 0$ for any $t \neq 0$, and

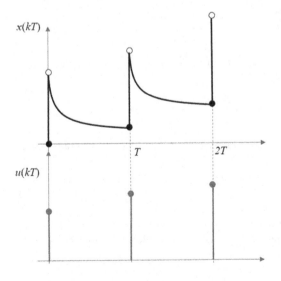

Fig. 2 Impulsive scheme.

---

[a] The Dirac delta is only an abstraction to formulate the impulsive problem. The idea behind this concept is that quick insulin injections can be properly approximated by impulses, if compared to T.

$$\int_{-\infty}^{\infty} g(\zeta)\delta(\zeta)d\zeta = g(0),$$

for all continuous, compactly supported functions g.

The solution of Eq. (1) at each periods T can be divided into two parts, the first one, describing the system in the period $[kT, kT + \Delta T]$, and the second one describing the system in the period $(kT + \Delta T, (k + 1)T)$, for a positive and arbitrary small Δ, $0 < \Delta < 1$:

$$\varphi(t; x(kT), u, r) = e^{A(t-kT)}x(kT) + \int_{kT}^{t} e^{A(t-\zeta)}B_u u(\zeta)\delta(\zeta - kT)d\zeta$$

$$+ \int_{kT}^{t} e^{A(t-\zeta)}B_r r(\zeta)d\zeta + \int_{kT}^{t} e^{A(t-\zeta)}d\zeta E,$$

for $t \in [kT, kT + \Delta]$, and

$$\varphi(t; x(kT + \Delta T), u, r) = e^{A(t-kT)}x(kT + \Delta T) + \int_{kT+\Delta T}^{t} e^{A(t-\zeta)}B_r r(\zeta)d\zeta$$

$$+ \int_{kT+\Delta T}^{t} e^{A(t-\zeta)}d\zeta E,$$

for $t \in (kT + \Delta, (k+1)T)$.

Now, if we consider the limits of this solution for $\Delta \to 0$, it follows that

$$x(kT^+) \triangleq \lim_{\Delta \to 0} \varphi(t; x(kT), u, r) = x(kT) + B_u u(kT), \tag{2}$$

where $x(kT^+) \triangleq \lim_{\Delta \to 0} x(kT + \Delta T)$, and

$$x(t) \triangleq \varphi(t; x(kT^+), u, r) = e^{A(t-kT)}x(kT^+) + \int_{kT}^{t} e^{A(t-\zeta)}B_r r(\zeta)d\zeta$$

$$+ \int_{kT}^{t} e^{A(t-\zeta)}d\zeta E,$$

for $t \in (kT, (k+1)T)$. This latter solution is the one corresponding to the continuous-time system

$$\dot{x}(t) = Ax(t) + B_r r(t) + E, \quad x(0) = x_0, \tag{3}$$

for the aforementioned period of time.

This way, merging Eqs. (2), (3) into a single model, the following impulsive system can be obtained:

$$\begin{cases} \dot{x}(t) = Ax(t) + B_r r(t) + E, & x(0) = x_0, \quad t \neq kT, \\ x(kT^+) = x(kT) + B_u u(kT), & k \in \mathbb{N}, \end{cases} \tag{4}$$

where $x(kT^+)$ denotes the limits of $x(t)$ when t approaches kT from the right. In Eq. (4), the latter equation describes the discontinuities that the impulsive input causes into the system, while the former describes the free response (only affected by the disturbance r) between the discontinuities.

Remark 1. Note that the affine impulsive system (4) has no formal equilibrium points. In fact, there is no triplet (x_s^i, u_s^i, r_s^i) fulfilling the condition:

$$Ax_s^i + B_r r_s^i + E = 0, \tag{5}$$

$$x_s^i = x_s^i + B_u u_s^i, \quad \text{(no discontinuity)}, \tag{6}$$

because the first equation has no solution, for $r_s^i = 0$ (vanishing disturbance).

An interpretation of the latter situation is that it is not possible to maintain the system in a given point by only applying impulsive inputs, even when this input is null. However, as it will be shown in the following section, it is possible to maintain the system (switching) inside a given region. The conditions for that are: (i) to find states before and after the discontinuity that, although different, remain constant and (ii) to ensure that the transient state trajectories between these states remain inside a given set.

2.1 Underlying discrete-time subsystem

A natural way to obtain a discretization (discrete-time system) of the continuous-time impulsive system (4) is by sampling it with a sampling time given by the period of the impulses, T. According to the results presented by Rivadeneira et al. (2015, 2018), the idea is to characterize two discrete-time subsystems describing the evolution of the states $x(kT)$ and $x(kT^+)$; that is, the evolution of the states just before and after the discontinuities, respectively. If the disturbance r is assumed to be piecewise constant, that is, $r(t) = r(kT)$, for $t \in [kT, (k+1)T)$, $k \in \mathbb{N}$, the subsystems are as follows:

$$x^\bullet(k+1) = A^\bullet x^\bullet(k) + B_u^\bullet u^\bullet(k) + B_r^\bullet r^\bullet(k) + E^\bullet, \quad x^\bullet(0) = x(0) = x_0, \tag{7}$$

$$x^\circ(k+1) = A^\circ x^\circ(k) + B_u^\circ u^\circ(k) + B_r^\circ r^\circ(k) + E^\circ, \quad x^\circ(0) = x(0^+) = x_0 + B_u u(0), \tag{8}$$

where $x^\bullet(k) \triangleq x(kT)$, $x^\circ(k) \triangleq x(kT^+)$, $A^\bullet = A^\circ = e^{AT}$, $B_u^\bullet = e^{AT} B_u$, and $B_u^\circ = B_u$. Furthermore, the input and disturbance of each subsystem are such that $u^\circ(k+1) = u^\bullet(k) = u(kT)$, $r^\circ(k) = r^\bullet(k) = r(kT)$, and $E^\bullet = E^\circ \triangleq \int_0^T e^{A\zeta} d\zeta E$, $B_r^\bullet = B_r^\circ \triangleq \int_0^T e^{A\zeta} d\zeta B_r$.

These two subsystems are not only useful to describe the evolution of the impulsive system (4) at the sampling times kT and kT^+—which is necessary to implement an

MPC—but also to characterize the equilibrium regions of Eq. (4) accounting for both the discontinuities and the free responses between them.

2.2   Extended equilibrium of the impulsive system

The idea now is to find an equilibrium set for Eq. (4), based on the underlying discrete-time subsystems (7), (8). To do that, we define first the equilibrium pairs of Eqs. (7), (8)—$(x_s^\bullet, u_s^\bullet)$ and (x_s°, u_s°), respectively—as the ones fulfilling the conditions[b]:

$$
\begin{aligned}
x_s^\bullet &= A^\bullet x_s^\bullet + B_u^\bullet u_s^\bullet + E^\bullet, \\
x_s^\circ &= A^\circ x_s^\circ + B_u^\circ u_s^\circ + E^\circ.
\end{aligned}
\tag{9}
$$

Given that $u^\circ(k+1) = u^\bullet(k)$ by definition (and $E^\circ = E^\bullet$), we have that at steady state it is $u_s^\circ = u_s^\bullet$. This means that one equilibrium input (defined as $u_s^\bullet(k)$ for simplicity) has to fulfill both equilibrium equations, that is, the last equation can be written as

$$
x_s^\circ = A^\circ x_s^\circ + B_u^\circ u_s^\bullet + E^\bullet.
\tag{10}
$$

This way, given that Eqs. (7), (8) are subsystems describing one single impulsive system, we need to find triplets $(x_s^\bullet, x_s^\circ, u_s^\bullet) \in \mathcal{X} \times \mathcal{X} \times \mathcal{U}$ fulfilling both Eqs. (9), (10). Furthermore, a final condition that the equilibrium triplet $(x_s^\bullet, x_s^\circ, u_s^\bullet)$ must fulfill to be a feasible extended equilibrium is that the free responses corresponding to them (orbits) must be feasible:

$$
o_s(x_s^\bullet, u_s^\bullet) \in \mathcal{X},
$$

where

$$
o_s(x_s^\bullet, u_s^\bullet) \triangleq \{ \varphi(t; x_s^\bullet, u_s^\bullet, 0), \; t \in (kT, (k+1)T], \; k \in \mathbb{N} \},
$$

and $\varphi(t; x_s^\bullet, u_s^\bullet, 0) = e^{At}(x_s + B_u u_s^\bullet) + \int_{kT}^{t} e^{A(t-\zeta)} d\zeta E$ (i.e., $\varphi(\cdot)$ it is the solution of the system (3), for an impulsive input of value $(u_s^\bullet)$ and considering no disturbances).

Summarizing, we can characterize the extended equilibrium set, $\mathcal{X}_s^\bullet$, as[c]:

$$
\mathcal{X}_s^\bullet \triangleq \{ x_s^\bullet \in \mathcal{X} : \exists u_s^\bullet \in \mathcal{U} \text{ such that } A^\bullet x_s^\bullet + B_u^\bullet u_s^\bullet + E^\bullet = x_s^\bullet, \; o_s(x_s^\bullet, u_s^\bullet) \in \mathcal{X} \}
\tag{11}
$$

Furthermore, the set of equilibrium inputs—that is, the set of inputs for which an equilibrium state exists—is denoted as $\mathcal{U}_s^\bullet$. Note that this latter impulsive equilibrium is different from the input equilibrium defined in Section 4.2. For details in the

---

[b] Given that it has no sense to consider a permanent disturbance (meal), it is assumed that $r^\bullet = r^\circ = 0$ in the steady state.

[c] Note that condition (10) is implicitly accounted for definition (11).

procedure to compute such an equilibrium sets $\mathcal{X}_s^\bullet$ and $\mathcal{U}_s^\bullet$, in a general context, see Rivadeneira et al. (2018).

2.3 Pulse input scheme

Here, a pulse input scheme (i.e., when the inputs are pulses of a given duration, instead of impulses) is introduced to be consistent with the real application of the control action that an actuator can supply in some biomedical cases, in which impulsive inputs are not allowed. See Fig. 3 for a schematic plot of the input pulses and state responses. To properly represents model (1) under the pulse scheme, some modifications need to be made on the underlying subsystems of Section 2.1. Specifically, matrix $B_u^\bullet$ must be redefined as

$$B_u^\bullet \triangleq e^{A(T-\Delta T)} \int_0^{\Delta T} e^{A(\Delta T - \zeta)} d\zeta B_u,$$

where ΔT represents now the input pulse duration. Furthermore, to characterize the extended equilibrium, matrix B_u° takes the form

$$B_u^\circ \triangleq \int_0^{\Delta T} e^{A(\Delta T - \zeta)} d\zeta B_u.$$

This way, the extended equilibrium sets $\mathcal{U}_s^\bullet$ and $\mathcal{X}_s^\bullet$ are different from the ones obtained for the impulsive case. Clearly, the latter representation is a generalization of the former, in such a way that the impulsive representation is a particular case of the pulsive one, when the pulse duration tends to zero.

Fig. 3 Pulse scheme.

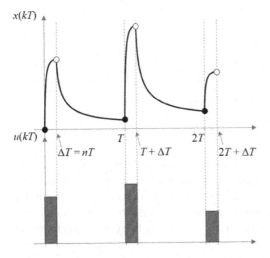

3 MPC formulation for impulsive systems

The idea now is to formulate an MPC based on the impulsive input scheme presented in Section 2. Let us define the impulsive state target set, $\mathcal{X}_s^{\bullet Tar} \subseteq \mathcal{X}_s^\bullet$, such that $\mathcal{Y}^{Tar} = C\mathcal{X}_s^{\bullet Tar}$, and the impulsive input target set as $\mathcal{U}_s^{\bullet Tar}$. The cost of the optimization problem that the proposed MPC solves online reads

$$V_N(x,r,\mathcal{X}_s^{\bullet Tar},\mathcal{U}_s^{\bullet Tar};\mathbf{u},u_a,x_a) \triangleq V_{dyn}(x,r;\mathbf{u},u_a,x_a) + V_f(\mathcal{X}_s^{\bullet Tar},\mathcal{U}_s^{Tar};u_a,x_a),$$

where

$$V_{dyn}(x,r;\mathbf{u},u_a,x_a) \triangleq \sum_{j=0}^{N-1} \|x(j)-x_a\|_Q^2 + \|u(j)-u_a\|_R^2,$$

with $Q > 0, R > 0$, and N the prediction horizon, is a term devoted to drive the system to the artificial equilibrium variable given by the artificial pair $(u_a,x_a) \in \mathcal{U}_s^\bullet \times \mathcal{X}_s^\bullet$, and

$$V_f(\mathcal{X}_s^{\bullet Tar},\mathcal{U}_s^{Tar};u_a,x_a) \triangleq p\left(dist_{C\mathcal{X}_s^{\bullet Tar}}(Cx_a) + dist_{\mathcal{U}_s^{\bullet Tar}}(u_a)\right)$$

with $p > 0$ and $dist_A(a)$ representing the distance from the point a to the set $\mathcal{A}$, is a terminal cost devoted to drive Cx_a to the set $C\mathcal{X}_s^{\bullet Tar}$ and u_a to $\mathcal{U}_s^{\bullet Tar}$, respectively.

Note that in the latter cost, the current state x, the current disturbance r, and the target sets $\mathcal{X}_s^{\bullet Tar}$ and $\mathcal{U}_s^{\bullet Tar}$ are optimization parameters, while $\mathbf{u} = \{u(0), u(1), \ldots, u(N-1)\}$, u_a, and x_a are the optimization variables. The optimization problem to be solved at time k by the MPC is given by

$$P_{MPC}(x,r,\mathcal{X}_s^{\bullet Tar},\mathcal{U}_s^{\bullet Tar})$$

$$\min_{\mathbf{u},u_a,x_a} \quad V_N(x,r,\mathcal{X}_s^{\bullet Tar},\mathcal{U}_s^{\bullet Tar};\mathbf{u},u_a,x_a),$$
$$s.t.$$

$$x(0) = x, \quad r(0) = r,$$
$$x(j+1) = A^\bullet x(j) + B_u^\bullet u(j) + B_r^\bullet r(j) + E^\bullet, \quad j \in \mathbb{I}_{0:N-1},$$
$$r(j+1) = 0, \quad j \in \mathbb{I}_{1:N-1}, \tag{12}$$
$$x(j) \in \mathcal{X}, \quad u(j) \in \mathcal{U}, \quad j \in \mathbb{I}_{0:N-1},$$
$$x(N) = x_a,$$
$$x_a = A^\bullet x_a + B_u^\bullet u_a + E^\bullet,$$

where x is the current state ($x(k)$), and the disturbance r is considered—without lose of generality—only for the first prediction time. Constraint $x(N) = x_a$ is the terminal constraint that forces the terminal state—at the end of control horizon N—to reach the artificial equilibrium state x_a. Furthermore, the last constraint forces the artificial variable pair (u_a, x_a) to be an equilibrium pair in $\mathcal{X}_s^\bullet \times \mathcal{U}_s^\bullet$. These two constraints force the state at the end of the horizon to be any feasible equilibrium state (and not a specific one in the target set), corresponding to the undisturbed scenario ($r \equiv 0$). These flexible constraints, which gives to the problem extra degrees of freedom, are necessary to ensure the closed-loop stability, as it is shown by Ferramosca et al. (2010).

Once the problem $P_{MPC}(x, r, \mathcal{X}_s^{\bullet Tar}, \mathcal{U}_s^{\bullet Tar})$ is solved, the (optimal) solution is denoted as $(\mathbf{u}^0, u_a^0, x_a^0)$, while the optimal cost function is given by $V_N^0(x, r, \mathcal{X}_s^{\bullet Tar}, \mathcal{U}_s^{\bullet Tar}) \triangleq V_N(x, r, \mathcal{X}_s^{\bullet Tar}, \mathcal{U}_s^{\bullet Tar}; \mathbf{u}^0, u_a^0, x_a^0)$. The control law, derived from the application of a receding horizon control policy, is given by $\kappa_{MPC}(x) = u^0(0; x)$, where $u^0(0; x)$ is the first element of the solution sequence $\mathbf{u}^0(x)$. The main benefits of the latter controller are: (i) the domain of attraction of the closed loop, $\mathcal{X}_N^{\bullet}$, is given by the controllable set in N steps to the entire equilibrium set $\mathcal{X}_s^{\bullet}$; (ii) the controller is recursively feasible and stable, and (iii) it ensures that upon convergence, the continuous-time system trajectories (i.e., jumps and free responses) will remain in the target set $\mathcal{X}_s^{\bullet Tar}$. The detailed proof of these results can be seen by Rivadeneira et al. (2018).

As it can be seen in problem (12), the MPC formulation needs to know the complete current state to generate the predictions and the control actions. However, in most of the cases just few of the system states can be measured. To tackle this problem, an estimator based on impulsive systems needs to be developed. In the case study section such an estimator is presented (in the context of the AP), and the combined scheme, that is, the estimator/control strategy is implemented and evaluated.

4 Case study: Type 1 diabetes mellitus

This section consists of the performance analysis of the control and observation methods proposed before. The chosen biomedical application is the dynamics of glucose, insulin, and carbohydrates in a type 1 diabetic patient. The first part is devoted to the description of the T1DM model which is a generalization of linear models found in the literature (Ruan et al., 2017). The model is identified by means of virtual patients extracted from the UVA/Padova simulator and with real signals recorded from patients in the Clãnica Integral de Diabetes in Medellãn, Colombia. Finally, the estimator is applied to recover insulin and digestion states even when the disturbance, that is, the meals are unannounced. The second part is dedicated to the combined scheme of observer/control with impulsive inputs including meal disturbances.

4.1 An enhanced model for type 1 diabetes patients

A long-term valid model is introduced here to be applied in sensor-augmented insulin pumps. The glucose-insulin-carbohydrates model comprises three subsystems describing insulin pharmacokinetics, meal absorption, and glucose dynamics. This leads to five differential equations with several parameters and two inputs (exogenous insulin and meal), which represent the control and disturbance input, respectively.

4.1.1 Glucose dynamics

To describe the glucose-insulin dynamics, three major factors are considered: glucose infusion, glucose production, and insulin-independent and insulin-dependent glucose utilization. The following differential equation is adapted from Ruan et al. (2017):

$$\frac{dG(t)}{dt} = -\theta_0 G(t) - \theta_1 Q_i(t) + \theta_2 Q_d(t) + \theta_3, \tag{13}$$

where G is the BG concentration (glycemia [mg dL^{-1}]), Q_i is the insulin delivery rate in plasma (U min^{-1}), and Q_d is the rate of glucose appearance delivery rate (g min^{-1}). Furthermore, θ_0 represents the glucose self-regulation fractional rate[d] (1 min^{-1}), θ_1 is the insulin action effectiveness (mg dL^{-1} U^{-1}), θ_2 is the carbohydrates (CHO) bio-availability (mg g^{-1} dL^{-1}), and θ_3 is the net balance between the liver endogenous glucose production and the insulin-independent glucose consumption (mg dL^{-1} min^{-1}).

This kind of linear model properly describes the physiological equilibria in the desired glucose and insulin range of interest and, furthermore, the FIT tools could be associated with the model parameters, as it was established in Hoyos et al. (2018).

4.1.2 Insulin and digestion dynamics

The insulin absorption/pharmacokinetic subsystems are the same for both insulin infusions, and consist of two compartments: the subcutaneous and blood compartments. The resulting differential equations are given by Magdelaine et al. (2015), Ruan et al. (2017), and Hoyos et al. (2018):

$$\begin{aligned}
\frac{dQ_i(t)}{dt} &= -\frac{1}{\theta_4} Q_i(t) + \frac{\theta_5}{\theta_4} Q_{i_{sub}}, \\
\frac{dQ_{i_{sub}}(t)}{dt} &= -\frac{1}{\theta_6} Q_{i_{sub}}(t) + \frac{\theta_7}{\theta_6} u_i(t),
\end{aligned} \tag{14}$$

where $Q_{i_{sub}}$ stands for the insulin delivery rate in the subcutaneous compartment (U min^{-1}), θ_4 and θ_6 are the insulin diffusion time constants in first and second insulin compartment, and the gains θ_5 and θ_7 are insulin diffusion effectiveness in each compartment. The input $u_i(t)$ is the insulin infusion rate (U min^{-1}). On the other hand, the digestion subsystem also consists of two compartments; the stomach and the duodenum. The resulting differential equations read:

$$\begin{aligned}
\frac{dQ_d(t)}{dt} &= -\frac{1}{\theta_8} Q_d(t) + \frac{\theta_9}{\theta_8} Q_{d_{sto}}, \\
\frac{dQ_{d_{sto}}(t)}{dt} &= -\frac{1}{\theta_{10}} Q_{d_{sto}}(t) + \frac{\theta_{11}}{\theta_{10}} u_m(t),
\end{aligned} \tag{15}$$

---

[d] This parameter could be explained by the superposition of two effects: the renal glucose clearance occurring at high glycemia levels and the counter-regulatory hormone effect at low glycemia level. Furthermore, θ_0 is generally given by a small value in comparison with the other parameters, which implies a tendency to the integrating behavior of the system (when $\theta_0 = 0$), as it is the case in Magdelaine et al. (2015).

where $Q_{d_{sto}}$ and Q_d stand for the glucose delivery rate (mg min^{-1}) from stomach and duodenum, respectively, θ_8 and θ_{10} are the CHO diffusion time constants in first and second digestion compartment; respectively. Similarly to Eq. (14), θ_9 and θ_{11} are meal absorption effectiveness in each compartment. Hence, $0 < (\theta_5, \theta_7, \theta_9, \theta_{11}) \leq 1$. The input $u_m(t)$ is the amount of carbohydrates CHO in meals (mg min^{-1}).

4.2 Affine state space model

Based on Eqs. (13)–(15), the following affine, continuous-time, state space model can be obtained:

$$
\begin{aligned}
\dot{x}(t) &= Ax(t) + B_u u_i(t) + B_m u_m(t) + E, \quad x(0) = x_0, \\
y(t) &= Cx(t),
\end{aligned}
\tag{16}
$$

where $x(t) = [x_1(t)\, x_2(t)\, x_3(t)\, x_4(t)\, x_5(t)]'$, with $x_1 = G$, $x_2 = Q_i$, $x_3 = Q_{i_{sub}}$, $x_4 = Q_d$, and $x_5 = Q_{d_{sto}}$. The output $y(t)$ represents the glycemia to be controlled. As before, $u_i(t)$ is the insulin infusion (U min^{-1}) and $u_m(t)$ is the amount of carbohydrates CHO in meals (mg min^{-1}). E is a constant term denoting the difference between the liver endogenous glucose production and the glucose absorption rate by the brain and erythrocytes. The model matrices are given by

$$
A = \begin{pmatrix}
-\theta_0 & -\theta_1 & 0 & \theta_2 & 0 \\
0 & -\dfrac{1}{\theta_4} & \dfrac{\theta_5}{\theta_4} & 0 & 0 \\
0 & 0 & -\dfrac{1}{\theta_6} & 0 & 0 \\
0 & 0 & 0 & -\dfrac{1}{\theta_8} & \dfrac{\theta_9}{\theta_8} \\
0 & 0 & 0 & 0 & -\dfrac{1}{\theta_{10}}
\end{pmatrix}, \quad B = (B_u \ B_m \ E), \tag{17}
$$

$$
B_u = \begin{pmatrix} 0 \\ 0 \\ \dfrac{\theta_7}{\theta_6} \\ 0 \\ 0 \end{pmatrix} \quad B_m = \begin{pmatrix} 0 \\ 0 \\ 0 \\ 0 \\ \dfrac{\theta_{11}}{\theta_{10}} \end{pmatrix}, \quad E = \begin{pmatrix} \theta_3 \\ 0 \\ 0 \\ 0 \\ 0 \end{pmatrix}, \tag{18}
$$

$$
C = (1\ 0\ 0\ 0\ 0). \tag{19}
$$

Constraints for both states and inputs are considered, in such a way that $u \in \mathcal{U}, x \in \mathcal{X}$, where $\mathcal{U}$ and $\mathcal{X}$ are assumed to be polyhedrons in the positive orthant (i.e., every quantity is positive). Constraints on x_1 are usually determined by regions of severe hyper and hypoglycemic episodes, which not only are dangerous for the subject, but also determine a region of validity of the linear model. Furthermore, it is also assumed that the amount of CHO is bonded by $0 \leq u_m \leq u_m^{max}$, for a positive physiological maximum ingest rate u_m^{max}.

4.2.1 Equilibrium and controllability characterization of the model

An equilibrium couple of model (16) corresponding to fasting (i.e., couples (u_{is}, x_s), such that $0 = Ax_s + B_u u_{is} + E$, considering $u_{ms} = 0$), with $\theta_0 > 0$ and a desired glycemia (or target) x_{1tg} is obtained as follows:

$$-\frac{1}{\theta_0}x_{1tg} - \theta_1 x_{2s} + \theta_3 = 0, \quad -\frac{1}{\theta_4}x_{2s} + \frac{\theta_5}{\theta_4}x_{3s} = 0,$$

$$-\frac{1}{\theta_6}x_{3s} + \frac{\theta_7}{\theta_6}u_s = 0, \quad \leftrightarrow u_{is} = \frac{\theta_3}{\theta_1\theta_5\theta_7} - \frac{x_{1tg}}{\theta_0\theta_1\theta_5\theta_7}. \tag{20}$$

Therefore, for each value of x_{1tg}, there exists a constant infusion rate u_s that stabilizes the glycemia; that is, the BIR depends on the glycemia target and vice versa. The state equilibrium results $x_{1,s} = x_{1tg}$, $x_{2,s} = \theta_5\theta_7 u_{is}$, $x_{3,s} = \theta_7 u_{is}$, $x_{4,s} = 0$, and $x_{5,s} = 0$.

Remark 2. Note that for the special case of $\theta_0 = 0$, the system is an integrator, and so, $x_{1,s} = x_{tg}$ (an arbitrary value), and $x_2 = \frac{\theta_3}{\theta_2}$, $x_{3,s} = \frac{\theta_3}{\theta_2\theta_5}$, $x_{4,s} = 0$, and $x_5 = 0$. Furthermore, the equilibrium input u_{is} is a fixed insulin value (denoted as BIR, u_b), given by $u_{is} = \frac{\theta_3}{\theta_2\theta_5\theta_7} = u_b$. Note that the value of $y_s = x_{1,s}$ (the equilibrium glycemia level) does not depend on u_{is}, but on the past values of the insulin infusion $u_i(t)$ and meal intake $u_m(t)$.

Regarding the controllability of model (16), it should be noted that only the first three states—accounting for the insulin-glycemia dynamic—are controllable, while the last two states from the digestion subsystem have their own dynamics not affected by the insulin, therefore, they are taken as disturbances.

4.3 Identification

The exogenous insulin infusion u_i is the way to reduce the BG level x_1 in order to maintain the normoglycemia of a patient. This drug administration occurs at different times only when it is required. An arbitrary administration of insulin would cause hyperglycemic and hypoglycemic events which are dangerous to the patient's health. As the health might be compromised, the possibility of manipulating the insulin infusion in order to obtain a persistent excitation is restricted. This excitation requirement is indispensable for some identification techniques (Ljung, 1999). In consequence, the data are registered during normal scenarios. The model presented here is identified using data from 33 in silico subjects (or virtual patients) of the UVA/Padova simulator and from data of 40 real patients.

The nonlinear UVA/Padova metabolic simulator is accepted by the FDA as a substitute for certain animal trials. The reliability of the models is tested in simulation using 33 virtual patients (11 children, 11 adults, and 11 adolescents) with the same meal scenario (Kovatchev et al., 2009). The chosen simulation scenario is applied to 33 virtual patients, the insulin infusion is configured with a generic pump and the measurement is performed with a CGM sensor. The simulation is started at midnight, using basal-bolus infusion regimes based on standard BIR and a bolus calculator (BC) provided by the simulator, where the bolus dose of insulin is $u_b = g_{CHO}/CIR$, with $CIR = TDI/500$, being CIR the carbohydrate-insulin ratio and TDI is the total daily insulin of the patient. The scenario of four meals per day is repeated every day during 7 days (total simulation time). In silico collected data consist of CGM measurements with a sampling time of 5 min ($T_s = 5$ min), insulin pump records: BIR, infused bolus times and sizes, and meal announcements (consumed CHO times and amounts).

Standard clinical data—CGM measurements, insulin delivery (basal-bolus), subject-recorded estimates of CHO intake times and amounts by using CC—from 40 type 1 diabetic patients from the "Clãnica Integral de Diabetes" of Medellãn (Colombia) were recorded in ambulatory conditions. All patients used the Paradigm Veo insulin pump augmented and CGM ($T_s = 5$ min), with a selected dataset spanned from 5 to 7 days. In vivo data were collected under a clinical protocol approved by the "Clãnica Integral de Diabetes." The inclusion requirements of the trial were patients without other illnesses which can influence the BG, with at least 6 months of treatment with an insulin pump.

In Fig. 4, the identification results for UVA/Padova virtual patients are illustrated. The goodness of calibration fit is 68% while in validation is 67.53%. In Fig. 5, a real patient is provided. The clinical used data were of 5 days lengths; the first 3 were taken for calibration of the model, while the last 2 were taken for validation. Although, the goodness of fit achieved during calibration and validation was lesser than the virtual patient case, it is good enough to represent T1DM patient. The complete results and discussion can be seen in Hoyos et al. (2018).

4.4 Impulsive scheme

The main objective of this work—in contrast to other existing strategies—is to develop an MPC control strategy able to control the glycemia of a diabetic patient (described by model (16)) by only using impulses (boluses) of insulin, without any continuous basal rate.[e]

---

[e] Given that from a technological point of view it could be desirable to have a minimal continuous flow to avoid catheter obstruction, a continuous small insulin flow can be considered.

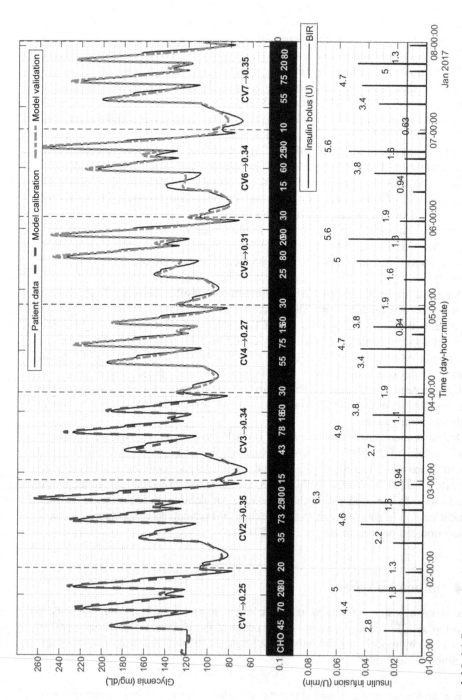

Fig. 4 Model fit a sample virtual patient child 11 during identification stage (first 3 days) and validation stage (next 4 days). *Top:* CGM with no additive sensor noise (*solid line*) and model simulation (*dashed line*). *Middle:* Amount of CHO intake. *Bottom:* Insulin delivery regime—BIR and insulin boluses.

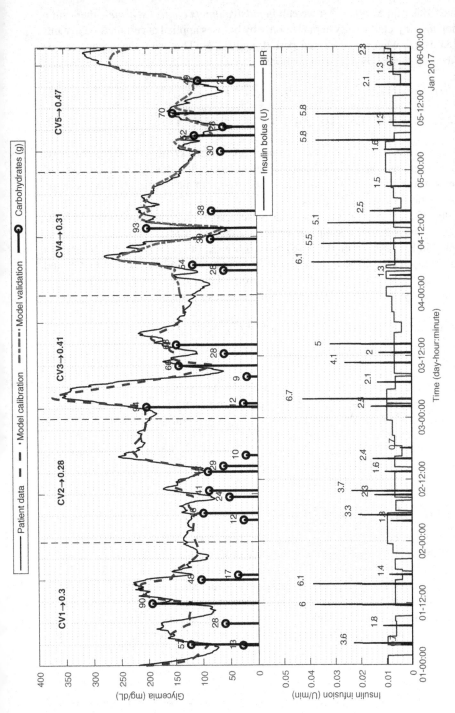

Fig. 5 *Top*: Comparison of glycemia provided by CGM sensor and predicted by identified model (with structure 1) for the real patient 22 during identification (3 days) and validation (2 days) stage. Meal announcements are also indicated. *Bottom*: Insulin delivery regime—BIR and insulin boluses.

Under this framework—that we call *impulsive inputs control scheme*—the insulin infusion u_i is injected into system (16) only by boluses applied at certain time instants denoted as kT, where T is a fixed period, and $k \in \mathbb{N}$. This way, the system can be described by the following two equations:

$$\dot{x}(t) = Ax(t) + B_m u_m(t) + E, \quad x(0) = x_0, \ t \neq kT, \tag{21}$$

$$x(kT^+) = x(kT) + B_u u_i(kT), \quad k \in \mathbb{N}, \tag{22}$$

where $x(kT^+)$ denotes the limits of $x(t)$ when t approaches kT from the right. This way, we have solutions containing discontinuities of first order (described by Eq. 22) and free responses affected only by the term associated with meals $(B_m u_m(t))$, between the discontinuities (described by Eq. 21).

System (21), (22) is a continuous-time system, so to implement the proposed MPC a discrete-time version (a sampled version, in fact) is needed. Based on the results presented in Rivadeneira et al. (2018) and Section 2, it is possible to have a sampled version of Eqs. (21), (22) that simultaneously accounts for both the discontinuities and the free responses. To proceed with the MPC formulation, the discrete-time subsystem representing system (21), (22) at the sampling times kT are given by

$$x(k+1) = A^\bullet x(k) + B_u^\bullet u_i(k) + B_m^\bullet u_m(k) + E^\bullet, \quad x(0) = x, \tag{23}$$

$$y(k) = Cx(k), \tag{24}$$

where index k represents the time kT, for $k \in \mathbb{N}$. The symbol $(\cdot)^\bullet$ denotes that the discrete-time system (23) describes the impulsive system (exactly) at the instants of the impulses.[f] Given that system (21) has no formal equilibria (see Remark 1, in Section 2), a nontrivial extended equilibrium region must be defined for system (23). The state and input equilibrium sets are denoted by $\mathcal{X}_s^\bullet$ and $\mathcal{U}_s^\bullet$, respectively, and the details of their definition are given in Section 2.2.

4.5 State observation schemes

As stated at the end of Section 3, a state observer is needed to estimate the internal states of the system. The function of the state observer (or estimator) is to reconcile the past and current measurements to estimate the current states of the system. A state observer requires knowledge of the deterministic and stochastic part of the model. Because the deterministic part (i.e., the discretized system matrices $A^\bullet$, $B^\bullet$, C, D) was already identified and computed according to the impulsive scheme introduced in Section 2, the optimality of the state observer is based on an accurate knowledge

---

[f] On the other hand, the symbol $(\cdot)^\circ$ is used to describe the impulsive system just an instant after the instant of the impulses.

of the actual noise statistics entering the plant (the real patient). In this regard, it is noticed that the output y has measurement noise associated with the sensor and the disturbance $u_m(t)$—the meals—has uncertainty due to the recording method. In general, there are three different scenarios for the available meal information (as shown in Fig. 1): (i) the meal can be perfectly known (by means of an accurate CC or an accurate "meal detection and size estimation" procedures), (ii) it can be partially known (when the latter procedures are—as it is usual—inaccurate[g]), and (iii) it can be completely unannounced/unknown.

Here, the stochastic part (or disturbance model) is obtained by proposing an initial conjecture and tuning its parameters. Consequently, three disturbance models (or three state observers) will be analyzed.

First notice that the measured state $y(k)$ can be expressed in terms of its rate of change $d(k)$ as follows: $y(k) = y(k-1) + d(k-1)$, and this rate of change can also be represented as a function of its own rate of change $e(k)$ as follows: $d(k) = d(k-1) + e(k-1)$ (Dassau et al., 2008). Then, the resulting general form of the extended model for the observer is given by

$$
\begin{bmatrix} x(k) \\ d(k) \\ e(k) \end{bmatrix} = \begin{bmatrix} A^{\bullet} & B^{*} & 0 \\ 0 & 1 & 1 \\ 0 & 0 & 1 \end{bmatrix} \begin{bmatrix} x(k-1) \\ d(k-1) \\ e(k-1) \end{bmatrix} + \begin{bmatrix} B^{\bullet} \\ 0 \\ 0 \end{bmatrix} \begin{bmatrix} u_i(k-1) \\ u_m(k-1) \\ 1 \end{bmatrix} + Gw(k-1),
$$

$$
y(k) = [C\,0\,0] \begin{bmatrix} x(k) \\ d(k) \\ e(k) \end{bmatrix} + v(k),
$$

(25)

where $A^{\bullet}, B^{\bullet} = [B^{\bullet}_u\, B^{\bullet}_m\, E^{\bullet}]$, and C are the impulsive matrices defined in Section 2, B^* is a matrix depending on the disturbance model, $G = [0_{1 \times n}\, 0\, 1]'$, $w(k-1)$ is a zero-mean- and Q-variance Gaussian process noise which enters in the acceleration term $e(k)$, and $v(k)$ is a Gaussian noise sequence with zero-mean and R-covariance. An estimate $x(k)$ of the state vector from model (25) is provided at each step k by a time-varying Kalman filter (or recursive Kalman observer) with gain projection (to consider the constraints $x \geq 0$) of the estimated physiological states (Simon, 2006). As the noise of CGM sensor can be simulated (Breton and Kovatchev, 2008), a reliable estimation of its standard deviation σ_{sensor} can be obtained. Then, considering $R = \sigma^2_{sensor}$ and $Q = (\sigma_{sensor}/200)^2$ as initial values, Q is retuned by a trial and error method.

In the first case (observer 1), the initial conjecture of the stochastic part of the model is that the state x_1 is corrupted by a double integrating state disturbance and the output $y(k)$ is corrupted by a white noise. Thus, the matrix $B^* = B_n = C'$.

---

[g] With the CC procedure, patients usually forget to inform the meal to the devise; the meal detection and size estimation procedure has always an inherent error.

The second case (observer 2) considers $d(k)$ as a double integrating input disturbance entering the state vector through the matrix $B^* = B^\bullet(:, 2)$ (i.e., the second column of $B^\bullet$) and $e(k)$ is an integrating input disturbance corrupted by the process noise $w(k)$. Finally, the last case (observer 3) combines the two first cases and then $B^* = B_n + B^\bullet(:, 2)$.

The performance of the three proposed observers is assessed next, using data from real patients. The ability to recover the states even when the meal input is not announced is evaluated and compared later.

Fig. 6 shows the time evolution of the state observer 1 corresponding to the model (25) for a sample real patient. Note that from 3450 min the meals intake information is omitted, but the states x_1, x_2, and x_3 seem not to be altered. In particular, state x_1 perfectly follows the CGM measurement. On the other hand, the states x_4 and x_5 go to zero, which is expected considering that no meal information is provided to the observer since this time. From these results, it is expected that this estimator will be useful for a control strategy which uses a prediction model considering only insulin and glycemia effects. Fig. 7 shows the time evolution of the observer 2 corresponding to the same real patient. Notice that the output (or state x_1) does not follow perfectly the CGM sensor, but it changes significantly in relation to the states x_2 and x_4. This observed behavior is the desired one. In addition, a forgetfulness of meal announcement (i.e., a zero intake) is simulated since time 3450 min. This is the worst-case scenario and still the five states are reliably estimated (see Fig. 7 since time 3450 min).

Fig. 8 illustrates the time state estimation for observer 3 corresponding to the same patient. The estimator follows almost perfectly the output, and maintains the better estimation of insulin and carbohydrates dynamics. Notice that the estimation of states x_4 and x_5 is almost identical during the interval with meal announcement and for the time period greater than 3450 min characterized by the meal missing information. Also, in Fig. 9, it can be seen that $d(k)$ becomes nonzero from 3800 to 4700 min (approximately). In conclusion, this third observer has the best performance and it will be used to close the loop with the MPC control.

To evaluate the robustness of the third-state observer to partially announced meals, it is compared with a conventional observer without augmented structure. In both cases, the observers are assessed considering different types of meal announcement errors, which are classified as: (i) announcement without uncertainty in meal size, (ii) announcement with meal size classification (SC) levels (small ≤ 35 gCHO, medium [35,66) gCHO, and large ≥ 65 gCHO), and (iii) announcement with 50% of error in meal size. In all cases, the area under the curve (AUC) of the rate of glucose appearance (Ra), which is estimated by $\theta_2 x_4$ in the proposed model is used to measure the goodness of estimation. In order to compare the Ra with the intake, u_m is multiplicated by the gain of the absorption subsystem $\theta_2 \theta_9 \theta_{11}$ (which is related to meal digestion and carbohydrate bioavailability). Fig. 10 shows the Ra estimated by the state observer 3 (upper panel) and by the conventional observer (bottom panel) for 1 day of simulation with five meals of 45, 70, 20, 80, and 10 gCHO consumed at 7.5, 12, 17, 20, and 22.30 hours, respectively, for a sample virtual patient. In both cases, it is observed that the Ra estimated with partially announced meals follows

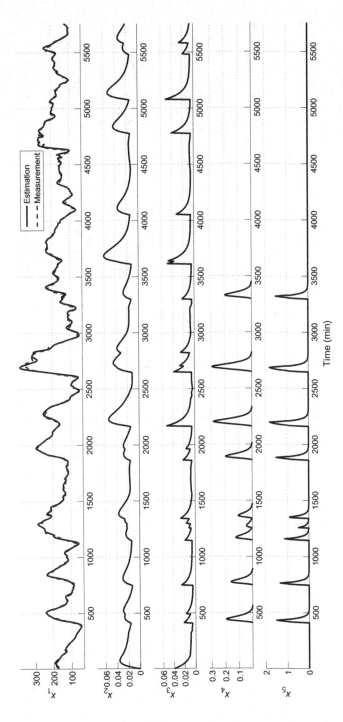

Fig. 6 State observer 1: Time evolution of the estimations for the real patient 1 with meal announcement omitted from time 3450 min.

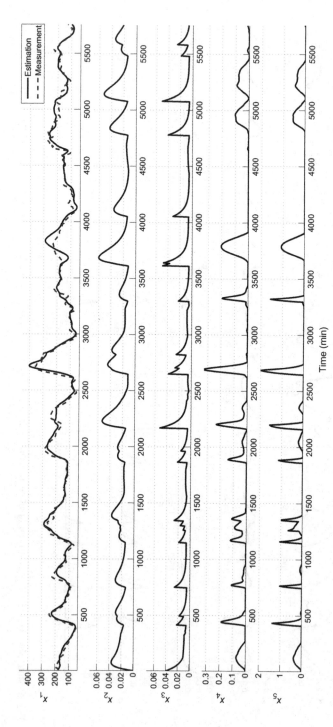

Fig. 7 State observer 2: Time evolution of the estimations for the real patient 1 with meal announcement omitted from time 3450 min.

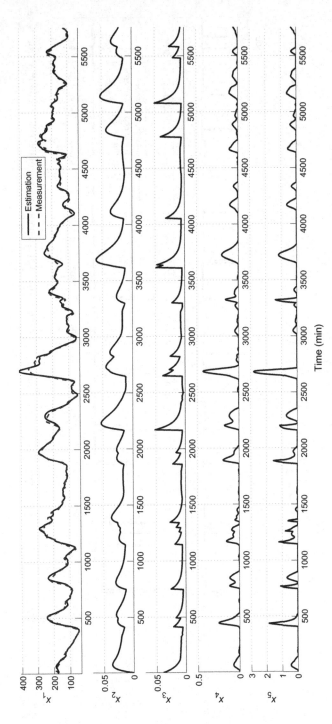

Fig. 8 State observer 3: Time evolution of the estimations for the real patient 1 with meal announcement omitted from time 3450 min.

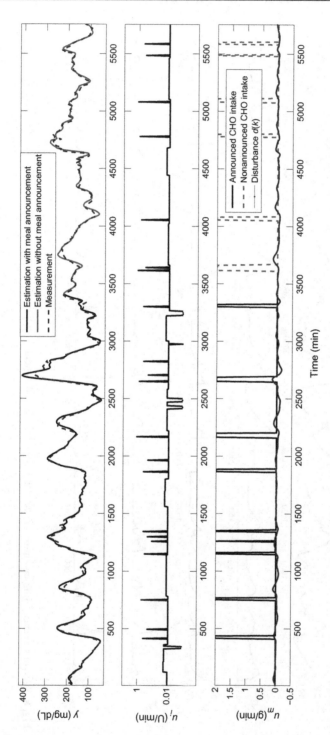

Fig. 9 Performance of the state observer 3 when the intake is not announced for the real patient 1. The announcement of the intake is omitted from time 3450 min.

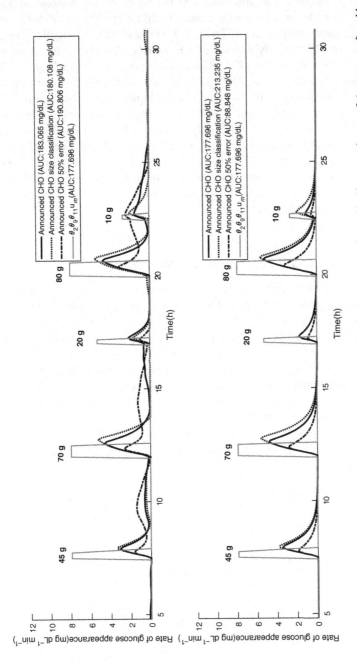

Fig. 10 In silico assessment of the state observer 3 robustness. Rate of glucose appearance estimated by the state observer 3 (*upper panel*) and by the conventional state observer (*bottom panel*) when meals intake is perfectly announced (*solid line*), announced with meal SC levels (*dotted line*), and announced with 50% error (*dashed line*). The plot corresponds to Adolescent 007.

the Ra estimated with perfectly announced meals. However, the AUCs for the state observer 3 are 183.065, 180.108, and 190.744 mg dL^{-1} for announced CHO, announced CHO with meal SC, and announced CHO with 50% error (50E); while for the conventional observer are 177.695, 213.235, and 124.387 mg dL^{-1}, respectively. Hence, for the partially announced CHO cases, the relative errors between AUCs and the overall intake (177.695 mg dL^{-1}) are: 1.36% (SC) and 7.34% (50E) for state observer 3, and 20% (SC) and 30% (50E) for conventional state observer, that is, the proposed one is more robust. Employing t-test on relative errors is determined a mean relative error of 4.9% ($P < .003$) for announced CHO, 6.3% ($P < .001$) for SC, and 8.4% ($P < .001$) for 50E, showing that the increment of sample relative error is nonsignificance when meal announcement is deteriorated. Also, in the extremely case of unannounced CHO the relative error becomes 10.4% ($P < .0008$). Therefore, the use of an augmented model with double integration disturbance to represent plant-model mismatch allowed an increase of the observer robustness to meal announcement uncertainty.

4.6 Impulsive model predictive control (iZMPC)

The general MPC control objective is to keep the system in an equilibrium that maintains the glycemia $y(t)$ in a target zone, by only manipulating the (positive and impulsive) insulin injection $u_i(t)$, while keeping the states $x_1(t)$, $x_2(t)$, and $x_3(t)$ fulfilling the constraints. The target zone is defined as $\mathcal{Y}^{Tar} \triangleq \{y \in \mathbb{R} : G_{\min} \leq y \leq G_{\max}\}$, where $G_{\min}$ and $G_{\max}$ are approximately given by 70 and 140 mg dL^{-1}, respectively. This zone is known as normoglycemia or euglycemia range. As part of the control objectives, it should be noted that hypoglycemic episodes (when G is below the euglycemia zone) are even more dangerous, in the short term, than hyperglycemic episodes (when G is above the euglycemia zone).

The optimization problem to be solved at time k by the MPC is a particular case of problem $P_{MPC}(x, r, \mathcal{X}_s^{\bullet Tar}, \mathcal{U}_s^{\bullet Tar})$ of Section 3, adapted for the AP problem.

The cost function is given by

$$V_N\left(x, \mathcal{X}_s^{\bullet Tar}, \mathcal{U}_s^{\bullet Tar}; \mathbf{u}, u_a, x_a\right) \triangleq V_{dyn}\left(x; \mathbf{u}, u_a, x_a\right) + V_f\left(\mathcal{X}_s^{\bullet Tar}, \mathcal{U}_s^{Tar}; u_a, x_a\right),$$

where

$$V_{dyn}(x; \mathbf{u}, u_a, x_a) \triangleq \sum_{j=0}^{N-1} \left\| \tilde{C}x(j) - \tilde{C}x_a \right\|_Q^2 + \left\| u(j) - u_a \right\|_R^2,$$

with $Q > 0, R > 0$, and N the prediction horizon, is a term devoted to drive the system to the artificial equilibrium variables, and

$$V_f\left(\mathcal{X}_s^{\bullet Tar}, \mathcal{U}_s^{Tar}; u_a, x_a\right) \triangleq p\left(dist_{C\mathcal{X}_s^{\bullet Tar}}(Cx_a) + dist_{\mathcal{U}_s^{\bullet Tar}}(u_a)\right)$$

with $p > 0$ and $dist_A(a)$ representing the distance from the point a to the set A, is a terminal cost devoted to drive Cx_a to the sets $C\mathcal{X}_s^{\bullet Tar} = \mathcal{Y}^{Tar}$ and u_a to $\mathcal{U}_s^{\bullet Tar}$, respectively. Matrix $\widetilde{C}$ is a matrix that selects the first three states of the system, which are the controllable states (the remaining two correspond to the meal effect, and they are not controllable by means of the input u).

The optimization problem then reads

$$\min_{\mathbf{u},u_a,x_a} V_N(x, \mathcal{X}_s^{\bullet Tar}, \mathcal{U}_s^{\bullet Tar}; \mathbf{u}, u_a, x_a),$$
$$s.t.$$

$$x(0) = \hat{x},$$
$$x(j+1) = A^{\bullet}x(j) + B_u^{\bullet}u_i(j) + E^{\bullet}, \quad j \in \mathbb{I}_{0:N-1},$$
$$\widetilde{C}x(j) \in \widetilde{C}\,\mathcal{X}, \, u_i(j) \in \mathcal{U}, \quad j \in \mathbb{I}_{0:N-1},$$
$$\widetilde{C}x(N) = \widetilde{C}x_a,$$
$$\widetilde{C}x_a = \widetilde{C}(A^{\bullet}x_a + B_u^{\bullet}u_a + E^{\bullet}),$$

where $\hat{x}$ is the estimated state at time k (coming from the observer). Note that no meal measurement or meal estimation is used for the predictions since the meal effect is entirely accounted for the observer and it is passed to the controller by means of $\hat{x}$. Furthermore, given that not all the states are controllable and constraints satisfaction for such states cannot be required, matrix $\widetilde{C} = [I_3 \; 0_{3 \times 2}]$ is defined to extract the controllable states x_1, x_2, x_3 from the complete state vector x.

In the case study, states are estimated by means of the observer 3, which has the best estimating states performance in the unannounced meal scenario. The scenario of the simulation is as follows:

1. The length of simulation is settled in 48 h.
2. An ideal pump and a CGM are used. The corresponding noise of the CGM is taken from Breton and Kovatchev (2008), with the following parameters: $PACF = 0.7$, $\gamma = -0.5444$, $\lambda = 7.9787$, $\delta = 1.6898$, and $\xi = -5.47$.
3. Meal disturbances are included. Five meals per day are simulated, at 7:00 hours with 50 g, 12:00 hours with 70 g, 17:00 hours with 20 g, 20:00 hours with 60 g, and 22:00 hours with 10 g. For the second day, the meal size changes to 55, 65, 30, 75, and 15, respectively. Recall that the iZMPC works with unannounced meals.

4.6.1 Results of the observer/control scheme

The application of the combined scheme is illustrated with the Adolescent 1 from the UVA/Padova simulator. The virtual patient model is given by Eqs. (16), (17), identified following the procedure described in Section 4.3 with data extracted from the simulator. For instance, the matrices obtained for the virtual patient labeled as Adolescent 1 are

$$A = \begin{bmatrix} -0.0077 & -73.2077 & 0 & 52.7822 & 0 \\ 0 & -0.0024 & 0.0010 & 0 & 0 \\ 0 & 0 & -0.0139 & 0 & 0 \\ 0 & 0 & 0 & -0.0699 & 0.0187 \\ 0 & 0 & 0 & 0 & -0.0631 \end{bmatrix},$$

$$B = \begin{bmatrix} 0 & 0 & 1.1431 \\ 0 & 0 & 0 \\ 0.0116 & 0 & 0 \\ 0 & 0 & 0 \\ 0 & 0.0165 & 0 \end{bmatrix}.$$

Once the model is identified and the scenario defined, it is necessary to tune the observer and the controller. In this work, the tuning is made by trial and error method. In Fig. 11, it is shown the time evolution of glycemia and the insulin corresponding to Adolescent 1. The control objective is partially fulfilled, since the patient does not fall to the hypoglycemia zone, even after the insulin injections that follows the intake of meals, but it reaches the hyperglycemia range during the meals, for about for 30–40 min. In Fig. 12, the overall behavior of the strategy is presented. In this case, the combined scheme shows a good behavior since (i) the estimator produces accurate estimations of the states without meal announcement and (ii) hypoglycemic episodes (which are the most dangerous in the short term) were avoided. Note that states x_4 and x_5 have nonnull time evolutions after the meal intakes, which indicates that the observer is able to account for (unannounced) meals by only measuring glycemia.

As it can be seen in Fig. 11, the aforementioned results were achieved by only injecting pulses of insulin, with periods of no insulin infusion. This strategy is in contrast with the well-known current infusion strategies—based on continuous infusion plus boluses—which permanently deliver insulin. See for instance, Fig. 5, where a (classical) FIT control strategy is shown.

5 Discussion

Biomedical processes have introduced several challenges to the control theory domain. One of them is the way the control actions are applied to the systems. In many cases, the control actions are manually implemented as pills or injections, as it is the case in HIV, malaria, influenza, hepatitis B, among others. On the other side, for type 1 diabetes, an infusion pump injects boluses of insulin. This kind of control action (input) has been mathematically modeled as impulses or pulses and, hence, a new class of systems has been established, being the impulsive systems the core of this chapter.

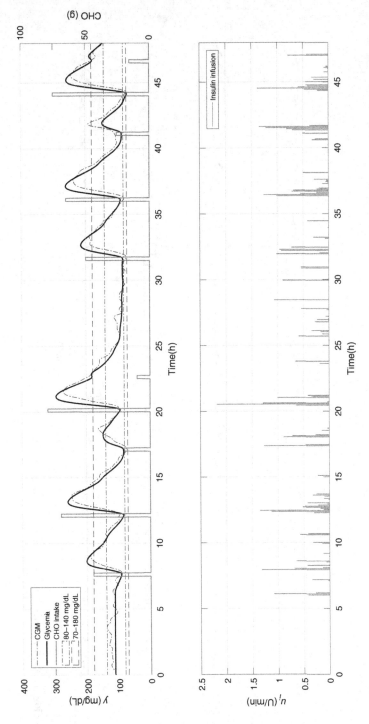

Fig. 11 Glycemia time evolution produced by the insulin administered for the combined scheme under the scenario of simulation applied to Adolescent 1.

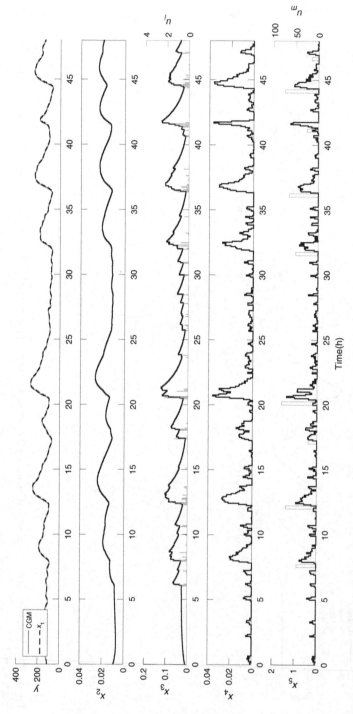

Fig. 12 Overall behavior of the strategy. State evolution for Adolescent 1.

Different control laws have been generated for impulsive systems, to show the potential to handle them when adjusted to reality. In this regard, MPC has played an important role because of the ability to incorporate constraints. In spite of the aforementioned advantages, the main drawback of iZMPC is the need for a prediction model and state estimation, which in many biomedical processes is difficult to achieve. The measurement of the output is not easy, it is expensive and sometimes requires a laboratory analysis, so more research in medical instrumentation is required. In type 1 diabetes, this problem has been overcome since by the use of CGM sensors. This device has facilitated the advances in controlling the illness generating the well-known AP development.

In this context, an estimation/MPC control strategy for impulsive systems has been substantiated and evaluated in the AP problem. This new formulation herein enjoys several properties that are to be mentioned:

(i) Due to the proper use of artificial variables, the MPC not only has an enlarged domain of attraction, but also ensures constraint satisfaction and stability (Ferramosca et al., 2012; Rivadeneira et al., 2015, 2018), for moderate-to-large disturbances and model-plant mismatches. Furthermore, the latter properties are achieved by only solving a sequence of quadratic programming problems (QP), which means that no computational complexity is added.

(ii) The use of a long-term model for predictions permits to better exploit the anticipative benefits of the strategy. The early prediction of possible hypoglycemic episodes that the insulin will not be able to correct by means of feasible actions (because of its positivity) is the key point to properly avoid them.

(iii) The use of impulses for both, basal and bolus doses, drastically increases the controllable set to the equilibrium (i.e., the set of states that can be effectively controlled by means of feasible control actions) in comparison to classical MPCs for diabetes, which only handle the boluses. In the context of T1DM, it also contributes to better avoid hypoglycemic episodes, since no continuous/permanent insulin infusion is used. The use of impulses/pulses also improves the performance of insulin pumps and, furthermore, classical injections can also be used to implement the control.

(iv) The MPC computes by itself the impulsive insulin basal rate (there is no need of an offline computation), which permits to easily cope with the frequent variations of its value. From the MPC point of view, even when there is model-plant mismatch, the basal rate is just the equilibrium of the closed loop.

(v) The double integrator system (which models the effect of disturbances in the estimator) permits to estimate adequately the state even when unknown meals are present. This feature critically enhances the quality of the predictions made by the MPC, and so, better control actions can be computed.

6 Conclusions

In this chapter, a combined scheme of estimation and control based on impulsive system for biomedical processes was discussed. The formulation permits to treat these problems in a realistic way and it is particularly suitable for T1DM treatments (in the context of the so-called AP). The control inputs seen as impulses or pulses has shown to be successful to reach satisfactory performances.

An important challenge not investigated is the variation of parameters during the day (according to the circadian rhythm). For this case, two options arises: to consider a robust MPC, or to consider a time-variant model. Both of them are, clearly, matter of future works.

Finally, the tuning of both formulations was made by the trial and error method, hence an automatic tuning methodology is highly desired in the development of AP. This matter will play an important role in the success of the applicability of these strategies.

References

Bellman, R., 1971. Topics in pharmacokinetics, III: repeated dosage and impulsive control. Math. Biosci. 12 (1–2), 1–5.

Bergenstal, R.M., et al., 2010. Effectiveness of sensor-augmented insulin-pump therapy in type 1 diabetes. N. Engl. J. Med. 363 (4), 311–320.

Bergenstal, R.M., et al., 2013. Threshold-based insulin-pump interruption for reduction of hypoglycemia. N. Engl. J. Med. 369 (3), 224–232.

Breton, M., Kovatchev, B., 2008. Analysis, modeling, and simulation of the accuracy of continuous glucose sensors. J. Diabetes Sci. Technol. 2 (5), 853–862.

Cameron, F., Niemeyer, G., Buckingham, B.A., 2009. Probabilistic evolving meal detection and estimation of meal total glucose appearance. J. Diabetes Sci. Technol. 3 (5), 1022–1030.

Dassau, E., Bequette, B.W., Bukingaham, B.A., Doyle III, F.J., 2008. Detection of a meal using continuous glucose monitoring. Emerg. Treat. Technol. 3, 295–300.

Dassau, E., et al., 2017. 12-Week 24/7 ambulatory artificial pancreas with weekly adaptation of insulin delivery settings: effect on hemoglobin A1c and hypoglycemia. Diabetes Care, 1–20. https://doi.org/10.2337/dc17-1188.

Ferramosca, A., Limon, D., González, A.H., Odloak, D., Camacho, E.F., 2010. MPC for tracking zone regions. J. Process Control 20 (4), 506–516.

Ferramosca, A., Limon, D., González, A.H., Alvarado, I., Camacho, E.F., 2012. Robust MPC for tracking zone regions based on nominal predictions. J. Process Control 22 (10), 1966–1974.

Food and Administration, 2016. A relevant glucose-insulin model: validation using clinical data. Recently approved devices: The 670G System—P160017.

Gillis, R., Palerm, C., Zisse, H.C., Jovanovic`, L., Seborg, D.E., Doyle III, F.J., 2007. Glucose estimation and prediction through meal responses usings ambulatory subject data for advisory mode model predictive control. J. Diabetes Sci. Technol. 1 (6), 825–833.

Hoyos, J.D., Bolanos, F., Vallejo, M., Rivadeneira, P.S., 2018. Population-based incremental learning algorithm for identification of blood glucose dynamics model for type-1 diabetic patients. In: International Conference on Artificial Intelligence. pp. 29–35.

González, A.H., Odloak, D., 2009. A stable MPC with zone control. J. Process Control 19 (1), 110–122.

González, A.H., Ferramosca, A., Bustos, G.A., Marchetti, J.L., Fiacchini, M., Odloak, D., 2014. Model predictive control suitable for closed-loop re-identification. Syst. Control Lett. 69, 23–33.

Godoy, J.L., Abuin, P., Rivadeneira, P.S., González, A., 2018. Modeling, identification and state estimation for artificial pancreas. Part I: description of the model and identification procedure. In: Proceedings of the Congreso Argentino de Control Automático AADECA 2018, pp. 1–6.

Jeandidier, N., et al., 2008. Treatment of diabetes mellitus using an external insulin pump in clinical practice. Diabetes Metab. 34 (4), 425–438.

Kirchsteiger, H., Johansson, R., Renard, E., Re, L., 2014. Continuous-time interval model identification of blood glucose dynamics for type 1 diabetes. Int. J. Control 87 (7), 1454–1466.

Kovatchev, B., Breton, M., Man, C.D., Cobelli, C., 2009. In silico preclinical trials: a proof of concept in closed-loop control of type 1 diabetes. J. Diabetes Sci. Technol. 3, 44–55.

Lee, H., Buckingham, B.A., Wilson, D.M., Bequette, B.W., 2009. A closed-loop artificial pancreas using model predictive control and a sliding meal size estimator. J. Diabetes Sci. Technol. 3, 1082–1090.

Ljung, L., 1999. System Identification: Theory for the User, second ed. Prentice Hall, Upper Saddle River, NJ.

Magdelaine, N., Chaillous, L., Guilhem, I., Poirier, J.Y., Krempf, M., Moog, C.H., Carpentier, E.L., 2015. A long-term model of the glucose-insulin dynamics of type 1 diabetes. IEEE Trans. Biomed. Eng. 62 (6), 1546–1552.

Ramkisson, C.M., Herrero, P., Bondia, J., Vehi, J., 2018. Unannounced meals in the artificial pancreas: detection using continuous glucose monitoring. Sensors 18, 1–18. https://doi.org/10.3390/s18030884.

Rawlings, J.B., Mayne, D.Q., 2009. Model Predictive Control: Theory and Design, first ed. Nob-Hill Publishing, Madison, WI.

Rivadeneira, P.S., Gonzalez, A.H., 2018. Non-zero set-point affine feedback control of impulsive systems with application to biomedical processes. Int. J. Syst. Sci. 49 (15), 3082–3093.

Rivadeneira, P.S., Ferramosca, A., González, A.H., 2015. MPC with state window target control in linear impulsive systems. IFAC-PapersOnline 48 (23), 507–512.

Rivadeneira, P.S., Ferramosca, A., González, A.H., 2018. Control strategies for non-zero set-point regulation of linear impulsive systems. IEEE Trans. Autom. Control 69 (9), 2994–3001.

Ruan, Y., Wilinska, M.E., Thabit, H., Hovorka, R., 2017. Modeling day-to-day variability of glucose-insulin regulation over 12-week home use of closed-loop insulin delivery. IEEE Trans. Biomed. Eng. 64 (6), 1412–1419.

Sala, I., Díez, J.L., Bondia, J., 2018. Generalized extended state observer design for the estimation or the rate or glucose appearance in artificial pancreas. In: 2018 European Control Conference (ECC), Limassol, Cyprus, pp. 2393–2398.

Samadi, S., Turksoy, K., Hajizadeh, I., Feng, J., Sevil, M., Cinar, A., 2017. Meal detection and carbohydrate estimation using continuous glucose sensor data. IEEE J. Biomed. Health Inform. 21 (3), 619–627.

Schmidt, S., Nørgaard, K., 2014. Bolus calculators. J. Diabetes Sci. Technol. 8 (5), 1035–1041.

Simon, D., 2006. Optimal State Estimation: Kalman, H Infinity, and Nonlinear Approaches. John Wiley & Sons, Hoboken, NJ.

Sindelka, G., et al., 1994. Effect of insulin concentration, subcutaneous fat thickness and skin temperature on subcutaneous insulin absorption in healthy subjects. Diabetologia 37 (4), 377–380.

Sopasakis, P., Patrinos, P., Sarimveis, H., Bemporad, A., 2015. Model predictive control for linear impulsive systems. IEEE Trans. Autom. Control 60 (8), 2277–2282.

Turksoy, K., Samadi, S., Feng, J., Littlejohn, E., Quinn, L., Cinar, A., 2016. Meal-detection in patients with type 1 diabetes: a new module for the multivariable adaptive artificial pancreas control system. IEEE J. Biomed. Health Inform. 20, 47–54. 10.1109/JBHI.2015.2446413.

Robust control applications in biomedical engineering: Control of depth of hypnosis

4

Mehdi Hosseinzadeh
Department of Control Engineering and System Analysis, The Free University of Brussels, Brussels, Belgium

1 Introduction

The goal of anesthetizing patients in operating rooms is to allow surgeons to operate in appropriate conditions while protecting the patients from the effects of the surgical procedure and maintaining homeostasis and hemodynamic stability as much as possible.

General anesthesia consists of three components (Bibian et al., 2005): (i) hypnosis, (ii) analgesia, and (iii) neuromuscular blockade. This means that anesthesiologists have to administer a number of drugs (one drug for each component) to the patients. Hence, general anesthesia is a multiinput multioutput problem (Bibian et al., 2003; Dumont, 2012). However, commercial availability of depth of hypnosis monitors and the lack of any reliable direct measure of analgesia have led to a focus on single-input single-output problem. Thus, this chapter studies only the hypnosis component of the general anesthesia induced by propofol, that is, propofol hypnosis.

Propofol hypnosis can be divided into three temporal phases (Soltesz et al., 2011): (i) induction, (ii) maintenance, and (iii) emergence. Changing a patient's consciousness from an alert to an anesthetized state involves a transition phase called induction. In other words, the aim of the induction phase is to bring the patient from total awareness to a desired depth of hypnosis. Once a stable depth of hypnosis is achieved, the maintenance phase begins. Surgery takes place during the maintenance phase, meaning that it is necessary to maintain an adequate depth of hypnosis during this phase. After completing the surgery, the emergence phase begins, when the administration of propofol is terminated. The emergence from anesthesia is simply achieved by turning off delivery of any administration devices.

Considering the aforementioned different phases, it can be concluded that the role of anesthesiologists is very similar to that of pilots, that is, after takeoff (induction phase in anesthesia), the pilots maintain an adequate flight trajectory (maintenance phase in anesthesia). Thus, the idea of designing an automated drug delivery system to regulate hypnotic drug administration in order to achieve desired depth of hypnosis and reduce the workload is very reasonable.

Control Applications for Biomedical Engineering Systems. https://doi.org/10.1016/B978-0-12-817461-6.00004-4

In addition to workload reduction, a closed-loop anesthesia has the following advantages (Bibian et al., 2005): (i) it can reduce drug consumption and lessen recovery time, which can improve the patient comfort while reducing drug-associated costs and bed occupancy, (ii) in closed-loop anesthesia the anesthesiologists have access to intravenous agents with very fast onset of action and fast metabolism, and (iii) the profound synergy and multivariable aspects which exist between intravenous anesthetics and opioids can be fully exploited by a closed-loop anesthesia.

A schematic of an automated hypnotic drug delivery system is shown in Fig. 1. As seen in this figure, first, a quantitative estimation of the depth of hypnosis is computed by the measurement unit. The control unit then automatically adjusts the infusion rate of propofol. Finally, the infusion pump takes the responsibility of administrating the needed amount of drug to achieve the set point. A monitoring system is also used to show the depth of hypnosis and infusion rate on a screen for anesthesiologists' consideration, and to store the data for later refers. Note that the anesthesiologists only set the operating set points and modify them during the maintenance phase based on the requirements of the surgical procedure and the patients' needs.

Designing an automated drug delivery system for controlling depth of hypnosis is practically challenging. In particular, significant interpatient and intrapatient variability causes large uncertainty in dose-response model. This uncertainty can cause overdosing or even instability when not accounted for. Therefore, not only the closed-loop control scheme should provide an acceptable performance in terms of set-point tracking, but also it should provide sufficient robustness against the mentioned uncertainty to prevent overdosing, and to guarantee a safe anesthesia.

This chapter is written with the idea of studying the most recent results in robust control of depth of hypnosis and of providing a perspective on some open research directions in this field. The remainder of this chapter is organized as follows. Section 2 studies techniques used to quantify the depth of hypnosis. Section 3 describes the models used in propofol delivery systems. Proposed robust control

Fig. 1 Schematic diagram for automated drug delivery system.

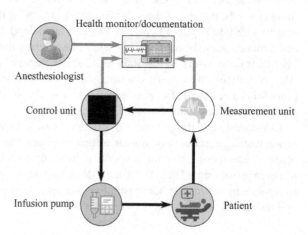

schemes are reported in Section 4. Finally, Section 5 concludes the chapter and highlights some opportunities and directions for future research.

2   Measurement of depth of hypnosis

Since the target organ of hypnotic drugs is the central nervous system (in particular the brain), it is reasonable to assume that electroencephalogram (EEG) signals can reflect the depth of drug-induced hypnosis (Rampil, 1998). Indeed, in the early 1940s, physiologists observed that the EEG of hypnotized patients contains slower waves with higher amplitudes. However, since these changes are difficult to detect, it is fairly impossible to interpret depth of hypnosis based on raw EEG signals. Due to this reason, a number of techniques have been proposed in the recent years to extract univariate features from the EEG to quantify the hypnotic component of anesthesia. In this section, bispectral and wavelet techniques, as two more promising techniques, will be discussed.

2.1   Bispectral analysis

Bispectral analysis is an advanced signal processing technique that quantifies quadratic nonlinearities and deviations from normality[a] by quantifying the interaction between the components that make up a signal.

The early results of using bispectral analysis in EEG to extract clinical information were reported by Ning and Bronzino (1989), where bispectral analysis was applied to characterize sleep pattern in rats. The results were validated later by Kearse et al. (1990), Sebel et al. (1990), and Vernon et al. (1992), indicating that an index based on bispectral analysis of EEGs can be used to monitor the depth of hypnosis. The most compelling result is reported by Bowles et al. (1994), where a dimensionless index, called Bispectral (BIS) index, is proposed. This index is scaled between 0 and 1, as shown in Fig. 2. A value of 0 represents a wakeful state and of 1 represents the maximum level of hypnosis.

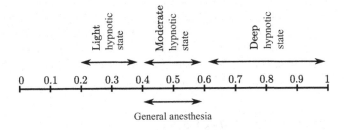

Fig. 2 BIS index scale and its meaning (Zikov et al., 2002).

---

[a] A signal is normal, or normally distributed, if its amplitude histogram is a normal [Gaussian] distribution.

In 1996, Aspect Medical Systems has developed a commercial monitor of consciousness, referred to as the BIS Monitor. In BIS Monitors, for the sake of simplicity and convenience in interpretation, the BIS index is rescaled between 0 and 100 (see Fig. 2), such that 100 means wakefulness and 0 means maximum level of hypnosis. The most publicized role of the BIS Monitor is the reduction of the incidence of intraoperative awareness (Bibian, 2006). Another well-publicized advantage of a BIS-guided titration is the reduction of drug usage and discharge time (Gan et al., 1997; Kreuer et al., 2003; Rosow and Manberg, 2001). This reduction has added the advantage of reducing anesthesia-associated costs, which can be seen as a major advantage in today's cost-sensitive society.

Apart from its advantages, the BIS Monitor has some serious drawbacks that can reduce significantly the performance of any closed-loop control scheme that relies on their measurements. One of the main drawbacks of these monitors is that they introduce a large inherent measurement delay (Hagihira et al., 2001), ranging from 15 to 55 s (Pilge et al., 2006). Furthermore, since the algorithm used in BIS Monitors to calculate the BIS index may need to switch between different analysis modes (Rampil, 1998), BIS Monitors introduce discontinuities that can be seen as non-linearity, which, if not coped with appropriately, can hamper the performance of control schemes.

2.2 Wavelet analysis

A discrete wavelet transform (DWT) is a transform that decomposes a given signal into a number of sets, where each set is a time series of coefficients describing the time evolution of the signal in the corresponding frequency band.

The first results of using wavelet analysis of EEG signal to derive a univariate descriptor of the depth of hypnosis have been presented by Bibian et al. (2001). The resulting index, called wavelet-based anesthetic value for central nervous system (WAV_{CNS}), has been validated clinically with success later by Bibian et al. (2004a, b), Lundqvist et al. (2004), and Zikov et al. (2006).

The WAV_{CNS} technology has been integrated into the NeuroSENSE Monitors. Similarly to BIS, the WAV_{CNS} uses the same 0–100 scale to represent the depth of hypnosis of the patient. Special care was taken in the WAV_{CNS} technology to avoid nonlinearities and the use of nonminimum phase elements. The WAV_{CNS} technology does not add additional computational delays that makes it to be virtually delay free. The dynamic behavior of the WAV_{CNS} is linear and can be fully captured by a second-order linear time invariant (LTI) transfer function. However, the WAV_{CNS} may provide inaccurate data in deeper hypnotic states (Zikov, 2002).

3 Dynamic model of hypnosis

In order to control the depth of hypnosis, the first step is to understand the relationship between dose and pharmacological effect of the intravenously administrated hypnotic drug. This relationship can be explained by two separate models: (1) the pharmacokinetic (PK) model, which relates the drug plasma concentration to the administered dose, and (2) the pharmacodynamic (PD) model, which relates the drug concentration to the observed effect. In the following sections, PK and PD models of propofol will be explained rigorously.

3.1 PK model of propofol

Although a large number of PK models for propofol have been presented in the literature (see Bibian, 2006 for a comprehensive survey), one of the more accurate models is the one presented by Schüttler and Ihmsen (2000), where the PK model of propofol is estimated with respect to the age, body weight, and gender, and for different modes of administration (bolus and infusion), and for different sampling sites (venous and arterial).

In Schüttler and Ihmsen (2000), in order to represent mathematically the processes and rates of drug movement within the body, the three-compartment model (Schüttler et al., 1988), as shown in Fig. 3, was used. In this model, the visceral group compartment consists of arterial blood and highly perfused tissues (e.g., brain, heart, kidney, liver), the lean group compartment contains muscles and skins, and the vessel-poor group compartment contains mainly fat and bones. In Fig. 3, V_1, V_2, and V_3 (in L) are the volumes of the visceral group, the lean group, and the vessel-poor group compartments, respectively. Note that PK volumes describe the extent of dilution of a drug in the corresponding compartments, and are defined as the ratio of the total amount of drug in each compartment divided by their initial concentration.

After an intravenous injection of propofol, drug concentration on the visceral group compartment, denoted by C_1 (in µg mL^{-1}), rises rapidly. Simultaneously, three following processes start: (i) the drug is distributed from the central to the two peripheral compartments at rates of k_{12} and k_{13} (in min^{-1}), respectively; (ii) when the drug concentrations C_1, C_2, and C_3 (in µg mL^{-1}) reach a state of equilibrium, the distribution process reverses and the drug returns back to the central compartment at rates of k_{21} and k_{31} (in min^{-1}); and (iii) the amount of drug injected into the central compartment is eliminated at a rate k_{10} (in min^{-1}).

Since arterial blood is grouped into the central compartment (i.e., the visceral group compartment), the drug plasma concentration is actually the concentration of the

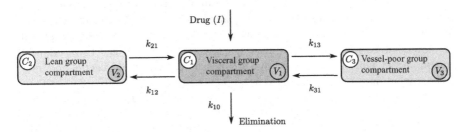

Fig. 3 Three-compartment PK model.

central compartment. As presented by Martinez (2005), the mass balance representation of the PK model can be written in the following state-space form:

$$
\begin{bmatrix} \dot{C}_1(t) \\ \dot{C}_2(t) \\ \dot{C}_3(t) \end{bmatrix} = \begin{bmatrix} -(k_{10}++k_{12}+k_{13}) & k_{21} & k_{31} \\ k_{12} & -k_{21} & 0 \\ k_{13} & 0 & -k_{31} \end{bmatrix} \begin{bmatrix} C_1(t) \\ C_2(t) \\ C_3(t) \end{bmatrix} + \begin{bmatrix} \frac{1}{V_1} \\ 0 \\ 0 \end{bmatrix} I(t),
\tag{1}
$$

where $I(t)$ is the propofol infusion rate (in mg s^{-1}).

In order to calculate the values of PK parameters, some relations are presented in the literature, for example, Schüttler and Ihmsen (2000) and Schnider et al. (1998, 1999). Here, we focus on the relations presented by Schüttler and Ihmsen (2000), that estimate the clearances Cl_1, Cl_2, and Cl_3 (in L min^{-1}) and volumes V_1, V_2, and V_3 depending on the patient's weight, age, administration type, and sampling site. These relations are

$$
Cl_1 = \begin{cases} 1.44\left(\dfrac{BW}{70}\right)^{0.75} & \text{if age} \le 60 \\ 1.44\left(\dfrac{BW}{70}\right)^{0.75} - \theta_{10}(\text{Age}-60) & \text{if age} > 60 \end{cases},
\tag{2}
$$

$$
Cl_2 = 2.25\left(\frac{BW}{70}\right)^{0.62}(1-0.4Ven)(1+2.02Bol),
\tag{3}
$$

$$
Cl_3 = 0.92\left(\frac{BW}{70}\right)^{0.55}(1-0.48Bol),
\tag{4}
$$

$$
V_1 = 9.3\left(\frac{BW}{70}\right)^{0.71}\left(\frac{\text{Age}}{30}\right)^{-0.39}(1+1.61Bol),
\tag{5}
$$

$$
V_2 = 4.2\left(\frac{BW}{70}\right)^{0.61}(1+0.73Bol),
\tag{6}
$$

$$
V_3 = 266,
\tag{7}
$$

where BW is body weight, $Ven = 1$ for venous sample and $Ven = 0$ for arterial sample, and $Bol = 1$ for bolus data and $Bol = 0$ for infusion data. Note that clearance Cl_1 represents the ability of body to eliminate the drug, and clearances Cl_2 and Cl_3 represent the ability of body to exchange the drug between the central and two peripheral compartments. Once the clearance and volumes are calculated, the central and inter-compartment rates can be calculated as

$$
k_{10} = \frac{Cl_1}{V_1},
\tag{8}
$$

$$k_{12} = \frac{Cl_2}{V_1}, \tag{9}$$

$$k_{21} = \frac{Cl_2}{V_2}, \tag{10}$$

$$k_{13} = \frac{Cl_3}{V_1}, \tag{11}$$

$$k_{31} = \frac{Cl_3}{V_3}. \tag{12}$$

3.2 PD model of propofol

The role of PD modeling of propofol is to mathematically express the observed effect of propofol as a function of plasma concentration at brain. The characteristic of propofol, like any other drug, can be represented by a dose-response curve, as shown in Fig. 4. The dose-response curve can be expressed with a sigmoid function (Bibian et al., 2005), as

$$\overline{E} = E_0 + (E_{max} - E_0)\left(\frac{\overline{C}_p^\gamma}{\overline{C}_p^\gamma + EC_{50}^\gamma}\right), \tag{13}$$

where $\overline{E}$ is the steady-state effect, $\overline{C}_p$ is the steady-state plasma concentration, EC_{50} is the concentration which yields 50% of the maximal effect, γ is the steepness factor, E_0 is the minimum effect, and E_{max} is the maximum effect.

The dose-response in Eq. (13) is obtained based on steady-state observation, and does not capture transients. The dynamic behavior between the drug plasma concentration and its effect can be expressed as a first-order plus time-delay function (Bibian, 2006), as follows:

Fig. 4 Dose-response model of propofol.

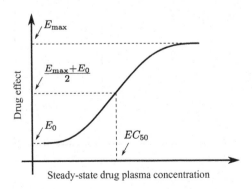

$$PD(s) = \frac{C_e(s)}{C_p(s)} = e^{-T_d s} \frac{k_{e0}}{s + k_{e0}}, \tag{14}$$

where T_d (in s) represents the arm-to-brain traveling time. Thus, in order to obtain the time course of drug effect, Eq. (13) can be updated as follows:

$$E(t) = E_0 + (E_{max} - E_0) \left(\frac{(C_e(t))^\gamma}{(C_e(t))^\gamma + EC_{50}^\gamma} \right). \tag{15}$$

Note that since PD models describe only one of the many different possible endpoints of a drug, they are not unique. This means that the PD model to predict the BIS index can be significantly different than that of WAV_{CNS} index. For more investigation, a summary of values of PD parameters is reported in Table 1,[b] where values of E_0 and E_{max} in all studies are 0 and 1, respectively.

3.3 PKPD model and its uncertainty

The relationship between infusion rate and pharmacological effect of propofol can be expressed by combining the PK model, as in Eq. (1), and PD model, as in Eqs. (14), (15). More precisely, infusion rate-response relation of propofol, called PKPD model, can be represented as a single-input-single-output (SISO) system followed by a nonlinear function, as depicted in Fig. 5.

Patient variability with respect to drug effect, which can be translated into PKPD model uncertainty, has always been a source of concern in pharmacology. In general,

Table 1 Propofol PD parameters (mean ± SD)

Study	No. of pop.	γ	EC_{50} (μg mL^{-1})	T_d (s)	k_{e0} 10^{-3} (s^{-1})
Billard et al. (1997) (BIS index)	12	5.30 ± 2.10	3.40 ± 0.00	0	3.30 ± 0.00
Kazama et al. (1999) (BIS index)	47	2.38 ± 0.39	7.06 ± 1.14	0	5.00 ± 0.20
Kuizenga et al. (2001) (BIS index)	8	4.20 ± 3.00	2.44 ± 1.37	0	3.40 ± 1.90
Bibian (2006) (WAV$_{CNS}$ index)	44	2.06 ± 0.66	3.49 ± 0.79	14.80 ± 12.12	44.57 ± 28.18

[b] SD stands for standard deviation.

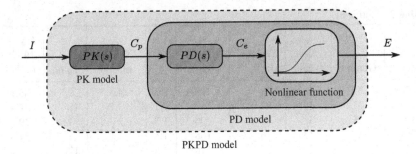

Fig. 5 Block diagram of infusion rate-response model of propofol.

two types of uncertainty exist in anesthesias: (i) the uncertainty caused by interpatient variability and (ii) the uncertainty caused by intrapatient variability. The former type is caused by differences between patient characteristics, such as age, weight, lean body mass, ethnicity, and genetic factors. These differences affect both PK and PD models.

In order to study the interpatient variability and resulting uncertainty in PKPD models, we use the real clinical dataset provided by Bibian (2006), as presented in Tables 2 and 3. This dataset is obtained during propofol induction while measuring the depth of anesthesia with WAV_{CNS} index. Considering the linear part of the PKPD model, the singular values of the frequency responses of the resulting

Table 2 Patient characteristics and their estimated PD parameters (Bibian, 2006)

Patient	Gender	Age (year)	Weight (kg)	Height (cm)	γ	EC_{50} (μg mL^{-1})	T_d (s)	k_{e0} 10^{-3} (s^{-1})
01	M	21	100	178	4.7	3.2	22	133.5
02	F	28	59	168	2.5	3.1	4	44.4
03	M	26	90	190	1.9	2.4	44	25.0
04	F	21	53	157	1.2	3.8	45	51.5
05	M	19	90	185	2.3	3.8	39	85.7
06	F	28	60	162	2.1	3.9	18	82.5
07	M	24	78	170	2.8	2.9	32	44.4
08	F	19	68	164	2.3	1.9	12	26.7
09	M	25	70	170	1.9	3.4	7	35.2
10	M	23	81	180	2.8	2.8	9	32.8
11	M	18	83	178	2.3	2.8	17	46.4
12	F	21	67	163	2.5	2.4	4	26.2
13	M	22	88	183	2.6	2.5	9	50.4
14	M	21	59	162	3.9	3.6	18	160.5
15	M	19	72	176	1.9	2.5	20	75.0
16	M	34	73	180	2.7	3.2	44	54.8

Continued

Table 2 Continued

Patient	Gender	Age (year)	Weight (kg)	Height (cm)	γ	EC_{50} (μg mL^{-1})	T_d (s)	k_{e0} 10^{-3} (s^{-1})
17	M	34	71	178	2.3	4.0	29	83.1
18	M	35	91	189	2.1	3.7	18	34.4
19	M	38	85	190	1.2	3.3	1	29.6
20	F	33	86	168	1.5	3.1	1	24.9
21	M	38	75	176	1.8	3.9	12	35.2
22	M	39	91	178	2.0	2.7	4	24.8

Table 3 Patient characteristics and their estimated PD parameters continued from Table 2

Patient	Gender	Age (year)	Weight (kg)	Height (cm)	γ	EC_{50} (μg mL^{-1})	T_d (s)	k_{e0} 10^{-3} (s^{-1})
23	M	34	78	178	2.2	2.8	12	28.7
24	F	36	65	163	2.1	2.8	5	27.0
25	F	34	58	157	2.0	3.6	4	67.2
26	M	33	77	180	1.8	3.1	12	29.3
27	M	39	105	185	1.8	3.7	13	29.1
28	F	46	66	168	1.3	6.1	11	36.6
29	M	45	99	182	1.3	4.7	2	32.6
30	M	48	77	173	1.4	4.5	12	35.0
31	M	47	98	189	2.0	3.9	10	28.7
32	M	46	79	172	1.8	3.9	12	34.8
33	M	40	96	193	1.9	3.2	9	36.6
34	M	40	77	176	2.3	3.4	8	35.6
35	M	48	97	190	1.9	3.0	13	30.0
36	M	53	80	177	1.8	3.3	35	38
37	F	60	63	165	2.3	3.5	3	31.5
38	M	52	91	185	2.2	4.4	29	42.0
39	M	52	100	182	1.4	4.7	2	21.8
40	M	52	97	176	1.1	3.7	16	28.8
41	M	50	95	183	1.9	4.0	10	26.4
42	F	52	56	173	1.5	5.0	6	58.0
43	M	54	100	180	1.5	4.2	6	32.2
44	M	60	95	176	1.8	3.1	12	24.3

44 models are presented in Fig. 6. As seen in this figure, interpatient variability causes an uncertainty of 10 and 20 dB at low and high frequencies, respectively. This uncertainty can affect the stability of any closed-loop control scheme, and needs to be coped with properly.

The latter type of uncertainty is caused by intrapatient variability, that is, the variability observed within one particular individual. This type of uncertainty has three

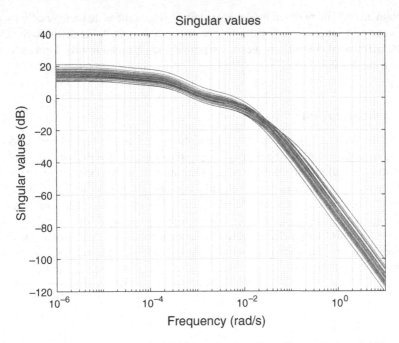

Fig. 6 Singular values of 44 PKPD models illustrating the effect of interpatient variability; each color for one patient.

main origins (Bibian et al., 2005): (i) changes in physiological processes, (ii) patient's state, and (iii) methods of administration. The former origin refers to parameter changes that can happen over time. Since the variability due to these changes is often limited as compared with interpatient variability, there is no need to take into account this variability in design procedure. Regarding the second origin, that is, patient's state, since the nonlinear function (15) can be seen as a variable gain depending on patient's state (depth of hypnosis), by analyzing the system separately for the induction and maintenance phases, one can cope with this variability. The latter origin refers to the fact that the PK parameters (2)–(6) depend on the method of administration. Thus, in order to avoid the variability caused by this fact, the method of administration should be kept identical during the surgery.

According to the aforementioned discussion, it is concluded that the major difficulty in closed-loop anesthesia is the interpatient variability, which results from the differences in demographics.

4   Control of depth of hypnosis

This section aims at providing detailed design procedure of robust proportional-integral-derivative (PID) and H_∞ control schemes in closed-loop anesthesia. To do so, first, the PKPD model is linearized around the operating point. Then, a method to obtain an optimal nominal model is explained. The indices that are broadly used to assess the performance of closed-loop anesthesia are studied. Finally, design details for the mentioned control schemes are reported and obtained results during a virtual surgery are discussed.

4.1 Linearization

Due to the nonlinear function (15), linear controller design techniques cannot be applied directly to control the depth of hypnosis. One possible way to cope with this problem is linearizing the nonlinear function around the operating point. Note that since operating points are different in induction and maintenance phases, the linearization should be done separately for the phases. However, since the operating point during the maintenance phase is the steady-state operating point, usually the gain obtained for maintenance phase is used to design a controller. In this case, the difference between the gains during induction and maintenance phases can be assumed as a parameter uncertainty.

Before continuing with the linearization, first, we rewrite the nonlinear function (15) as follows[c]:

$$E(t) = \frac{\hat{C}_e^{\gamma}}{\hat{C}_e^{\gamma} + 0.5^{\gamma}}, \tag{16}$$

where $\hat{C}_e = \frac{C_e}{2EC_{50}} \in [0, 1]$.

Optimum hypnosis level during maintenance phase is between 0.4 and 0.6; thus, 0.5 can be selected as the operating point. This level of hypnosis can be reached with a steady-state plasma concentration EC_{50}, which means that $\hat{C}_e = 0.5$ is the operating point. Thus, we have

$$E(t) \approx \frac{\partial}{\partial \hat{C}_e} \left(\frac{\hat{C}_e^{\gamma}}{\hat{C}_e^{\gamma} + 0.5^{\gamma}} \right) \Bigg|_{\hat{C}_e = 0.5} \cdot C_e(t) = \frac{\gamma}{2} \cdot C_e(t), \tag{17}$$

which means that the nonlinear function can be approximated with a gain depending only on the steepness factor γ.

4.2 Nominal model

Following the discussions presented in Section 4.1, the linearized PKPD model representing the relation between infusion rate and pharmacological effect of propofol can be expressed as follows:

---

[c]In Eq. (15), we have $E_0 = 0$ and $E_{max} = 1$.

$$G_p(s) = e^{-T_d s} K_{PKPD} \frac{(s+z_1)(s+z_2)}{(s+p_1)(s+p_2)(s+p_3)(s+p_4)}, \qquad (18)$$

where $K_{PKPD} = \gamma k_{e0}/2V_1$ is the gain of the system.

As explained in Section 3.3, due to the interpatient variability, values of parameters in Eq. (18) are uncertain. The actual patient's model can be expressed as a multiplicative uncertain structure, as follows:

$$G_p(s) = G_o(s)(1 + W_\Delta(s)\Delta(s)), \qquad (19)$$

whose block diagram structure is depicted in Fig. 7. In Eq. (19), $\Delta(s)$ is any stable transfer function that satisfies the condition $|\Delta(j\omega)| \leq 1$, $\forall\omega$, whereas $W_\Delta(s)$ is the weighting function that quantifies the magnitude of the unstructured uncertainty. It should be remarked that instead of multiplicative structure, one can also consider an additive structure.

As shown by Schnider et al. (1999) and Bibian et al. (2006), patients' age is a significant covariate in the PK and PD models. This means that by considering age brackets, uncertainty in patients' models can be reduced. Thus, we select a 10-year age bracket and divide the patient models of Tables 2 and 3 into four age groups (Group 1: 18–30 years, Group 2: 31–40 years, Group 3: 41–50 years, and Group 4: 51–60 years). Hence, a nominal model should be identified for each age group.

The most intuitive way is to construct the nominal model $G_o(s)$ is to use average values of parameters. Once the nominal model is identified, in order to guarantee that $|\Delta(j\omega)| \leq 1$, $\forall\omega$, the weighting function $W_\Delta(s)$ should be determined such that

$$|W_\Delta(j\omega)| \geq \max_n \left| \frac{G_{p,n}(j\omega) - G_o(j\omega)}{G_o(j\omega)} \right|, \quad n = 1, \ldots, N, \qquad (20)$$

where $G_{p,n}(s)$ is the model of nth patient, and N is the number of patients taken into account.

Although the averaging method quantifies the magnitude of the uncertainty, it is not optimal and injects conservatism to the system. Note that according to Eq. (20), improper nominal model can lead to high magnitude weighting function, and consequently hamper the robust performance of the closed-loop control scheme. Optimal nominal values can be identified through the following optimization problem:

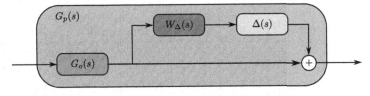

Fig. 7 Multiplicative uncertain structure of PKPD model.

$$\begin{cases} \min \sum_{n=1}^{N} \left| G_{p,n}(j\omega) - G_o(j\omega) \right|, \quad \dfrac{1}{T_{d,\max}} \leq \omega \leq \dfrac{1}{T_{d,\min}} \\ \text{s.t.} \quad T_{d,\min} \leq T_d \leq T_{d,\max} \\ \qquad K_{PKPD,\min} \leq K_{PKPD} \leq K_{PKPD,\max} \\ \qquad z_{i,\min} \leq z_i \leq z_{i,\max}, \quad i = 1,2 \\ \qquad p_{i,\min} \leq p_i \leq p_{i,\max}, \quad i = 1,\ldots,4, \end{cases} \tag{21}$$

where lower and upper limits of each parameter are minimum and maximum values of that parameter in each age group. The nominal parameters obtained via the optimization problem (21) are reported in Table 4.

Now, let $\tilde{G}_p(s)$ be defined as follows:

$$\tilde{G}_p(s) \triangleq (G_p(s) - G_o(s))/G_o(s). \tag{22}$$

The singular value of $\tilde{G}_p(s)$ calculated with optimal nominal models are presented in Figs. 8–11. From these figures, the best structure for the uncertainty weighting function $W_\Delta(s)$ to cover all $\tilde{G}_p(s)$s is

$$W_\Delta(s) = k_\Delta \frac{(s + z_{1,\Delta})(s + z_{2,\Delta})}{(s + p_{1,\Delta})(s + p_{2,\Delta})}, \tag{23}$$

where $p_{2,\Delta} > z_{2,\Delta} > z_{1,\Delta} > p_{1,\Delta}$, and can be selected as in Table 5.

4.3 Evaluation indices

The performance of an automated hypnotic drug delivery system during induction phase is usually evaluated with rise time (in min), settling time (in min), and overshoot (in %).

During the maintenance phase, the performance can be assessed by calculating the median performance error (MPE), median absolute performance error (MAPE), divergence, and wobble, as defined in the following:

Table 4 Nominal parameters obtained through optimization

Group	T_d (s)	K_{PKPD} 10^{-4} (mg s^{-1})	$z_1\ 10^{-3}$	$z_2\ 10^{-5}$	$p_1\ 10^{-2}$	$p_2\ 10^{-3}$	$p_3\ 10^{-4}$	$p_4\ 10^{-5}$
1	18.6	1.698	1.477	2.572	3.239	6.961	2.803	2.703
2	16.5	1.928	1.478	2.703	3.843	7.735	2.912	2.787
3	8.3	1.438	1.486	3.627	2.870	7.748	2.843	2.121
4	17.8	1.823	1.478	2.651	3.656	9.100	2.962	2.710

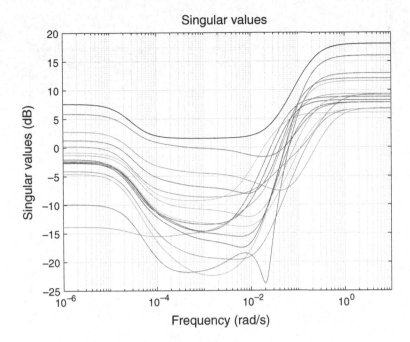

Fig. 8 Singular value of $\widetilde{G}_p(s)$ (in *color*; each color for one patient) and $W_\Delta(s)$ (in *black*) with optimal nominal model for Group 1.

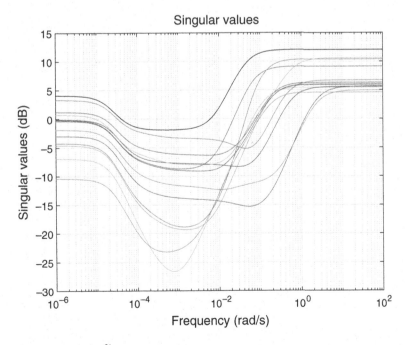

Fig. 9 Singular value of $\widetilde{G}_p(s)$ (in *color*; each color for one patient) and $W_\Delta(s)$ (in *black*) with optimal nominal model for Group 2.

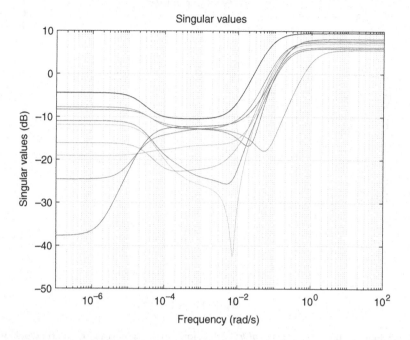

Fig. 10 Singular value of $\widetilde{G}_p(s)$ (in *color*; each color for one patient) and $W_\Delta(s)$ (in *black*) with optimal nominal model for Group 3.

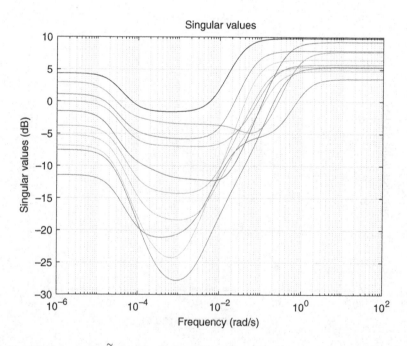

Fig. 11 Singular value of $\widetilde{G}_p(s)$ (in *color*; each color for one patient) and $W_\Delta(s)$ (in *black*) with optimal nominal model for Group 4.

Table 5 Parameters of weighting function $W_\Delta(s)$ for the case of optimal nominal models

Group	K_Δ	$z_{1, \Delta}$	$z_{2, \Delta}$	$p_{1, \Delta}$	$p_{2, \Delta}$
1	8	4×10^{-5}	3×10^{-2}	2×10^{-5}	2×10^{-1}
2	4	4×10^{-5}	8×10^{-3}	2×10^{-5}	4×10^{-2}
3	3	4×10^{-5}	8×10^{-3}	2×10^{-5}	8×10^{-2}
4	3.1	6×10^{-3}	8×10^{-3}	3×10^{-5}	3×10^{-2}

- MPE (in %): This measure, also referred as bias measure, describes whether the value of depth of hypnosis is systematically either above or below the reference value. The MPE for the ith patient can be calculated as

$$MPE_i = \text{median}\{PE_{ij}, j = 1, \ldots, J_i\}, \tag{24}$$

where J_j is the number of samples acquired for ith patient during the surgery, and the performance error (PE) is defined as

$$PE_{ij} = \frac{E(j) - r(j)}{r(j)} \times 100, \tag{25}$$

with $E(j)$ as the value of jth sample of the depth of hypnosis, and $r(j)$ as the value of jth sample of the reference signal.

- MAPE (in %): This measure, also referred as inaccuracy measure, indicates absolute value of the error between the depth of hypnosis and the reference value. For the ith patient, MAPE can be calculated as

$$MAPE_i = \text{median}\{|PE_{ij}|, \ j = 1, \ldots, J_i\}. \tag{26}$$

where PE_{ij} is as in Eq. (25).

- Divergence (in % min^{-1}): The divergence measure reflects the possible time-related trend of the depth of hypnosis in relation to the reference signal. For the ith patient, divergence can be calculated as

$$DIVERGENCE_i = \frac{\sum_{j=1}^{J_i} |PE_{ij}| t_{ij} - \dfrac{\sum_{j=1}^{J_i} |PE_{ij}| \sum_{j=1}^{J_i} t_{ij}}{J_i}}{\sum_{j=1}^{J_i} (t_{ij})^2 - \dfrac{\left(\sum_{j=1}^{J_i} t_{ij}\right)^2}{J_i}}, \tag{27}$$

where PE_{ij} is as in Eq. (25), and t_{ij} is the time (in min) at which the PE_{ij} is determined. A positive value of divergence shows that the depth of hypnosis diverges from the reference value, while a negative value shows that the depth of hypnosis converges to the reference value.

- Wobble (in %): This measure quantifies the oscillation of the depth of hypnosis around the reference value. The wobble for the ith patient can be calculated as

$$WOBBLE_i = \text{median}\{|PE_{ij} - MPE_i|, \ j = 1, \ldots, J_i\}, \tag{28}$$

where MPE_i and PE_{ij} are as in Eqs. (24), (25), respectively.

4.4 Robust PID control scheme

The first attempt to apply PID control scheme to anesthesia practice is reported by Absalom et al. (2002). After that many researchers attempted to modify the classic PID control scheme so to achieve a robust structure to control the depth of hypnosis (e.g., Heusden et al., 2018; Merigo et al., 2018; Padula et al., 2015, 2017; Pawlowski et al., 2018).

One of the promising robust PID control schemes is proposed by Dumont et al. (2008), which is clinically approved later by Heusden et al. (2014). The proposed robust PID control scheme is depicted in Fig. 12, where $v(r)$ is the applied reference, $E_o(t)$ is the depth of hypnosis observed by the monitor, $d_i(t)$ is the load disturbance, $d_o(t)$ is the output disturbance, $G_o(s)$ represents the nominal model developed in Section 4.2, $H(s)$ represents the model of sensor, $K_{fb}(s)$ represents the feedback controller, and $K_{ff}(s)$ represents the feed-forward controller. Note that $d_i(t)$ represents any external disturbances that affect the control signal (i.e., infusion rate) directly, such as obstruction of the infusion pump; whereas, $d_o(t)$ represents any external disturbances that affect the system output directly, such as incisions or body invasion.

Now, let $K_{fb}(s)$ and $K_{ff}(s)$ be as

$$K_{fb}(s) = k_p + \frac{k_i}{s} + k_d s, \tag{29}$$

$$K_{ff}(s) = k_p + \frac{k_i}{s}, \tag{30}$$

where k_p, k_i, and k_d are design parameters. In this case, the structure (Fig. 12) is equivalent to the classical PID structure when the applied reference is time-invariant, that is, $\dot{v}(t) = 0$. However, in the case of time-varying applied reference, the control scheme takes into account only the changes of the output to determine the control signal, which can cause greater controlling or dampening effect (Araki, 2009).

Suppose that the clinical hypnotic effect $E_o(t)$ is measured by a WAV$_{CNS}$ monitor. The dynamics of the WAV$_{CNS}$ monitor can be modeled as a second-order low-pass filter (Dumont et al., 2008), as

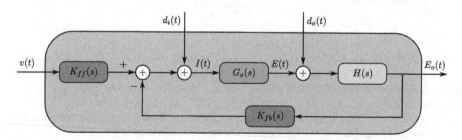

Fig. 12 Robust PID control scheme presented by Dumont et al. (2008).

$$H(s) = \frac{E_o(s)}{E(s)} = \frac{1}{(8s+1)^2},$$ (31)

where $E_o(s)$ is the Laplace transform of the WAV$_{CNS}$ index.

At current stage, consider the integrated error (IE) performance index, defined as

$$IE = \int_0^\infty (E_o(t) - r(t))dt.$$ (32)

As shown by Aström and Hägglund (1995), after applying a step disturbance $d_i(t)$, the value of IE is directly related to PID's integral gain; more precisely, IE$= 1/k_i$. This means that by maximizing the integral gain k_i, one can increase system's ability to attenuate disturbance $d_i(t)$. However, large values of k_i can cause instability. Therefore, an optimization problem should be formulated to maximize the integral gain k_i, while ensuring stability of the system. In the following, we will formulate the optimization problem based on the Nyquist method.

From Fig. 12, we have

$$E_o(s) = S(s)\big(G_o(s)K_{ff}(s)H(s)V(s) + G_o(s)H(s)D_i(s) + H(s)D_o(s)\big),$$ (33)

where $V(s), D_i(s)$, and $D_o(s)$ are Laplace transform of signals $v(t), d_i(t)$, and $d_o(t)$, respectively, and $S(s) \triangleq (1 + L(s))^{-1}$ is the sensitivity function with $L(s) \triangleq G_o(s)K_{fb}(s)H(s)$ as the loop transfer function.

From the Nyquist method, in order to ensure stability of the system, $L(s)$ should stay far away from the point -1. The shortest distance from the Nyquist curve of $L(s)$ to the critical point -1 is inversely proportional to the maximum of the sensitivity function.

Let M_s be the maximum value of the sensitivity function, defined as

$$M_s \triangleq \max_{\omega} |S(j\omega)|.$$ (34)

Thus, to ensure stability, the Nyquist curve of $L(s)$ should stay outside a circle of radius $1/M_s$ and centered at -1. In other words, the following inequality should hold true at all frequencies:

$$|1 + L(j\omega)|^2 = \left|1 + \left(k_p - j\frac{k_i}{\omega} + j\omega k_d\right)G_o(j\omega)H(j\omega)\right| \geq \frac{1}{M_s^2}, \quad \forall \omega.$$ (35)

Now, suppose the time domain specification of the system can be approximated with a second-order system. Furthermore, suppose consider that one of the design objectives is to have a disturbance response with an overshoot of less than 5%, which corresponds to a damping ratio of 0.69. Thus, as shown by Aström and Hägglund (1995), M_s can be calculated as

$$M_s = \sqrt{\left.\frac{1+8\xi^2+(1+4\xi^2)\sqrt{1+8\xi^2}}{1+8\xi^2+(-1+4\xi^2)\sqrt{1+8\xi^2}}\right|_{\xi=0.69}} = 1.2828. \tag{36}$$

Therefore, the design procedure can be formulated as an optimization problem to maximize the integral gain k_i, subject to constraint (35) with $M_s = 1.2828$, that is,

$$\begin{cases} \max_{k_p,k_i,k_d} k_i \\ \text{s.t.} \quad \left|1+\left(k_p-j\frac{k_i}{\omega}+j\omega k_d\right)G_o(j\omega)H(j\omega)\right| \geq 0.6077, \quad \forall \omega. \end{cases} \tag{37}$$

The obtained PID parameters by solving the optimization problem (37) are presented in Table 6. Note that although the optimization problem (37) is formulated by only considering input disturbance attenuation, since the output disturbance attenuation only depends on the maximum sensitivity value M_s, the optimization problem (37) guarantees a closed-loop system with good input and output disturbance rejection and robustness to modeling errors.

In order to study the effectiveness of the robust PID control scheme, all 44 virtual surgeries (i.e., one virtual surgery for each patient) are simulated in MATLAB. The total experiment time is 70 min, and includes an induction, maintenance, and emergence phase. The emergency phase starts 1 h after the beginning of the simulation. The reference signal $v(t)$ is assumed to be time invariant and equal to 0.5. Also, a disturbance profile for $d_o(t)$, as shown in Fig. 13, is assumed to emulate the patients arousal reflexes during a surgical procedure. Furthermore, to provide the determined drug administration regime for propofol, a Graseby 3400 Syringe Pump with an open communication protocol (that allows the pump to be controlled and monitored remotely) is used. The pump

Table 6 Parameters of the robust PID control scheme

Group	K_p	k_i	k_d
1	2.610	0.026	65.09
2	3.947	0.046	85.29
3	10.207	0.107	202.38
4	4.455	0.058	104.83

Fig. 13 Disturbance profile to simulate surgical stimulus.

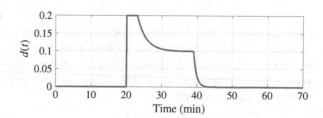

can deliver infusion rates up to 1200 mL h^{-1} (3.33 mg s^{-1} with propofol 10 mg mL^{-1}) with an accuracy flow rate of 2 %. It should be also noted that to protect the controller from integrator windup, the antiwindup method based on back calculation (Aström and Hägglund, 2006) is implemented to reset the integrator dynamically with a time constant T_t, where

$$T_t = \sqrt{\frac{k_d}{k_i}}. \tag{38}$$

Simulation results are shown in Figs. 14–17. Also, achieved performance indices are presented in Tables 7 and 8. In order to follow the same scale used in BIS and NeuroSENSE Monitors, the depth of hypnosis $DoH(t)$ is defined as

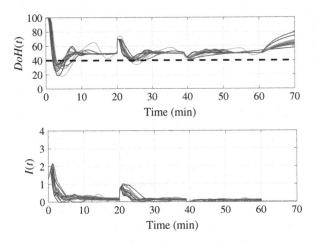

Fig. 14 Simulation results with the robust PID control scheme for Group 1; each color for one patient.

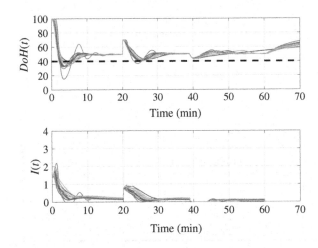

Fig. 15 Simulation results with the robust PID control scheme for Group 2; each color for one patient.

Fig. 16 Simulation results
with the robust PID control
scheme for Group 3; each
color for one patient.

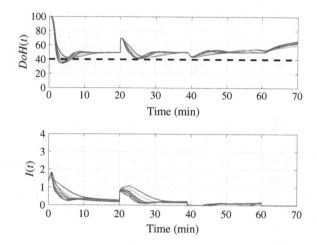

Fig. 17 Simulation results
with the robust PID control
scheme for Group 4; each
color for one patient.

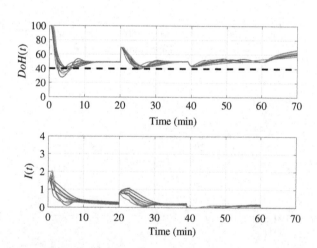

Table 7 Performance of the robust PID control scheme at induction phase (mean ± SD)

Group	Rise time (min)	Settling time (min)	Overshoot (%)
1	1.999 ± 0.284	8.550 ± 3.635	40.719 ± 11.480
2	2.143 ± 0.238	7.761 ± 1.479	33.440 ± 12.975
3	2.437 ± 0.443	8.945 ± 2.168	24.861 ± 6.245
4	2.399 ± 0.383	8.826 ± 1.732	27.165 ± 9.470

Table 8 Performance of the robust PID control scheme at maintenance phase (mean $\pm$ SD)

Group	MPE (%)	MAPE (%)	Divergence (% min^{-1})	Wobble (%)
1	0.053 ± 0.619	2.143 ± 2.622	-0.006 ± 0.001	2.205 ± 2.648
2	-0.016 ± 0.373	2.221 ± 1.568	-0.006 ± 0.001	2.164 ± 1.577
3	0.312 ± 0.977	3.158 ± 2.562	-0.006 ± 0.001	3.054 ± 2.569
4	0.279 ± 1.122	2.984 ± 1.964	-0.006 ± 0.001	2.916 ± 1.924

$$DoH(t) \triangleq 100(1 - E_o(t)), \tag{39}$$

where $DoH(t) = 100$ represents a wakeful state and $DoH(t) = 0$ represents the maximum level of hypnosis.

According to the results, the settling time is less than 10 min, which is acceptable in clinical anesthesia. However, the overshoot is pretty high, meaning that patients are in the danger of overdosing. Note that the overshoot should be less than 10% to meet typical anesthesiologists' response specifications. Regarding the maintenance phase, the results reveal that the robust PID control scheme ensures stability in the presence of parameter uncertainty, and attenuates the applied disturbance effectively.

As stated earlier, the main problem of the robust PID control scheme is that it does not prevent overdosing, such that as seen in Table 9, most of the patients are experiencing overdosing. Note that for the sake of safety, $E_o(t)$ should be less than 0.6 ($DoH(t)$ should be higher than 40%) at all times.

In order to cope with the overdosing problem, a set-point prefilter, as in Eq. (40), can be introduced:

$$F_{sp}(s) = \frac{V(s)}{R(s)} = \frac{1}{T_{sp}s + 1}, \tag{40}$$

where T_{sp} is the time constant of the filter, and $V(s)$ and $R(s)$ are Laplace transform of $v(t)$ and $r(t)$, respectively. The resulting structure is depicted in Fig. 18. In this structure, the set-point prefilter generates the applied reference $v(t)$ by smoothing (reducing the sharpness) the reference $r(t)$. This action decreases the overshoot,

Table 9 Number of overdosed patients with the robust PID control scheme

Group	Number of patients	Number of overdosed patients
1	15	15
2	12	11
3	8	5
4	9	5

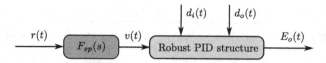

Fig. 18 Robust PID control scheme with set-point prefilter.

and consequently reduces the overdosing occurrences.[d] In the following, based on Bode method, a systematic way to compute the time constant T_{sp} will be presented.

Consider a disturbance-free situation, that is, $d_i(t) = d_o(t) = 0$. In this case, from Fig. 18, we have

$$E_o(t) = F_{sp}(s)G_{sp}(s)R(s), \tag{41}$$

where

$$G_{sp}(s) \triangleq S(s)G_o(s)H(s)K_{ff}(s) = \frac{G_o(s)H(s)K_{ff}(s)}{1 + G_o(s)H(s)K_{fb}(s)}. \tag{42}$$

Since in real surgeries, instead of $G_o(s)$, the PKPD models of patients are in the loop, in order to compensate the worst-case set-point response, the prefilter constant T_{sp} should be tuned based on the model that has the maximum resonance peak.

Let M_{sp} be the maximum magnitude of $G_{sp}(s)$, defined as

$$M_{sp} = \max_n \left| \frac{G_{p,n}(j\omega)H(j\omega)K_{ff}(j\omega)}{1 + G_{p,n}(j\omega)H(j\omega)K_{fb}(j\omega)} \right|, \quad n = 1, \dots, N, \tag{43}$$

where N is the number of patients. Also, let $\bar{n}$ be the patient that gives M_{sp}.

At this stage, suppose that the closed-loop system is desired to behave as a second-order system with damping ration of 0.69 and with natural frequency of 1.07×10^{-2} rad s^{-1}. One way to compute the constant F_{sp} is to use the Bode diagram of the ration between the desired second-order system and $G_{sp}(s)$ for the patient $\bar{n}$. In particular, the constant T_{sp} can be determined through the following equality:

$$\left| \frac{G_{ref}(j\frac{1}{T_{sp}})}{G_{sp,\bar{n}}(j\frac{1}{T_{sp}})} \right| = -3 \text{ dB}, \tag{44}$$

where $G_{ref}(s)$ is the reference system (i.e., a second-order system with damping ration of 0.69 and natural frequency of 1.07×10^{-2} rad s^{-1}), and $G_{sp,\bar{n}}$ is the transfer function $G_{sp}(s)$ for the patient $\bar{n}$.

---

[d]Recently, an active set-point prefilter is proposed to guarantee overdosing prevention. See Hosseinzadeh et al. (2019a, 2019b) and Hosseinzadeh and Garone (2019) for more details.

Table 10 Value of the set-point prefilter time constant for the robust PID control scheme

Parameter	Group 1	Group 2	Group 3	Group 4
T_{sp}	156.81	129.90	111.95	124.96

Calculated time constants for all groups are reported in Table 10. Furthermore, simulation results and achieved performance indices by using the robust PID control scheme with set-point prefilter, and following the aforementioned experiment to conduct virtual surgeries, are presented in Figs. 19–22 and Tables 11 and 12, respectively.

According to the results, although adding set-point prefilter to the robust PID control scheme increases the rising time around 2 min, it reduces the settling time 4 min

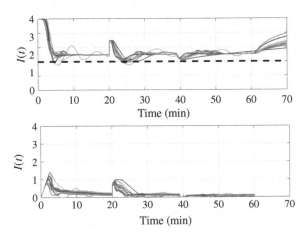

Fig. 19 Simulation results with the robust PID control scheme with set-point prefilter for Group 1; each color for one patient.

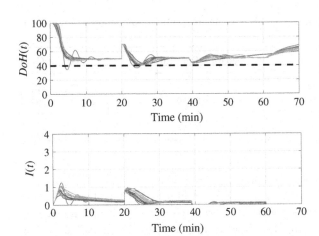

Fig. 20 Simulation results with the robust PID control scheme with set-point prefilter for Group 2; each color for one patient.

Fig. 21 Simulation results with the robust PID control scheme with set-point prefilter for Group 3; each color for one patient.

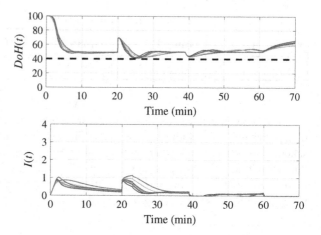

Fig. 22 Simulation results with the robust PID control scheme with set-point prefilter for Group 4; each color for one patient.

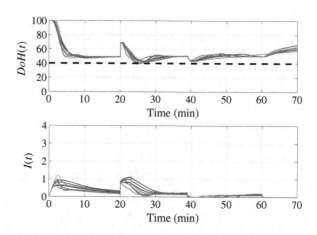

Table 11 Performance of the robust PID control scheme with set-point prefilter at induction phase (mean ± SD)

Group	Rise time (min)	Settling time (min)	Overshoot (%)
1	5.386 ± 1.865	6.236 ± 3.487	6.677 ± 8.886
2	6.500 ± 1.913	5.928 ± 1.851	4.909 ± 8.431
3	6.684 ± 0.743	6.914 ± 3.298	3.144 ± 1.552
4	6.967 ± 1.707	6.407 ± 2.375	3.594 ± 1.614

which is more interesting for anesthesiologists. Furthermore, adding set-point prefilter reduces the overshoot 27%, such that overshoot for all groups is less than 10%. This means that the set-point prefilter increases safety with respect to overdosing prevention. This fact can be seen in Table 13, where the number of overdosed patients by the

Table 12 Performance of the robust PID control scheme with set-point prefilter at maintenance phase (mean ± SD)

Group	MPE (%)	MAPE (%)	Divergence (% min^{-1})	Wobble (%)
1	0.079 ± 0.545	1.941 ± 2.117	−0.006 ± 0.0010	2.025 ± 2.227
2	−0.009 ± 0.343	2.121 ± 1.602	−0.006 ± 0.0009	2.083 ± 1.602
3	0.341 ± 0.988	3.073 ± 2.596	−0.006 ± 0.0006	3.025 ± 2.648
4	0.289 ± 1.113	2.909 ± 1.987	−0.006 ± 0.0007	2.849 ± 1.951

Table 13 Number of overdosed patients with the robust PID control scheme with set-point prefilter

Group	Number of patients	Number of overdosed patients
1	15	2
2	12	1
3	8	0
4	9	0

robust PID control scheme with set-point prefilter is reported. As seen in this table, the number of overdosed patients is significantly reduced, such that patients older than 40 years do not experience overdosing. Note that since the set-point prefilter is active only during induction phase, it has almost no effect on the performance of the system during maintenance phase.

4.5 Robust H$_\infty$ control scheme

The H_∞ norm of a stable transfer function $T(s)$ is its largest input/output root mean square gain, that is,

$$\| T(s) \|_\infty = \sup_{\substack{u \in L_2 \\ u \neq 0}} \frac{\| y(t) \|_{L_2}}{\| u(t) \|_{L_2}}, \tag{45}$$

where L_2 is the space of signals with finite energy and $y(t)$ is the output of the system for a given input $u(t)$. This norm also corresponds to the peak gain of the frequency response $G(j\omega)$, that is,

$$\| T(s) \|_\infty = \sup_\omega \sigma_1(G(j\omega)), \tag{46}$$

where σ_1 is the largest singular value. Since the H_∞ performance is convenient to enforce robustness to model uncertainty and external disturbances, it can be used

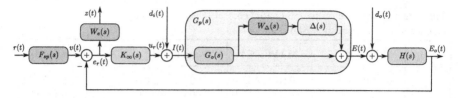

Fig. 23 Robust H_∞ control structure.

in closed-loop anesthesia to automatically control the depth of hypnosis, and consequently reduce the risk of overdosing in patients.

The robust H_∞ control structure to control the depth of hypnosis is shown in Fig. 23, where $G_o(s)$ is the nominal model developed in Section 4.2, $W_\Delta(s)$ is the weighting function as in Eq. (23), $K_\infty(s)$ denotes the controller to be designed, and $W_e(s)$ represents the performance weighting function.

The control objective is to minimize the error $e_r(t) = v(t) - E_o(t)$. The error to be penalized, denoted by $z(t)$, is a frequency-weighted version of the error, that is,

$$Z(s) = W_e(s)E_r(s), \tag{47}$$

where $Z(s)$ and $E_r(s)$ are Laplace transform of $z(t)$ and $e_r(t)$, respectively.

To keep the order of the resulting H_∞ controller as low as possible, a first-order weighting function is considered for $W_e(s)$, as

$$W_e(s) = k_w \frac{\omega_2^{-1}s + 1}{\omega_1^{-1}s + 1}, \tag{48}$$

where k_w specifies the upper bound of the error in the steady-state and low-frequency region (i.e., up to ω_1), while ω_2 is needed for the well posedness of the control design procedure. The values of the parameters of the weighting function $W_e(s)$ are reported in Table 14. Note that, here, only the H_∞ performance is taken into account in the design procedure. However, with a little modification, one can include other robust performances. For instant, by assuming measurement noise on $E_o(t)$, H_2 performance can be also considered in the design procedure to come up with a robust H_2/H_∞ control scheme (see Hosseinzadeh et al., 2014; Sadati et al., 2018 for more details).

Table 14 Parameters of the weighting function $W_e(s)$

Group	k_w	ω_1	ω_2
1	1×10^3	1×10^{-8}	1×10^4
2	1×10^1	1×10^{-5}	1×10^4
3	1×10^1	1×10^{-6}	1×10^3
4	1×10^3	1×10^{-8}	1×10^5

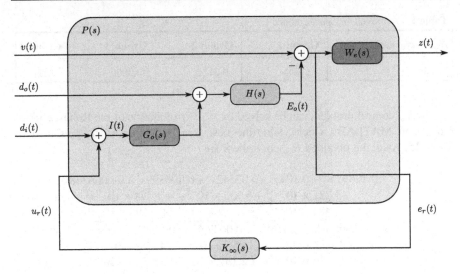

Fig. 24 LFT representation of the robust H_∞ control structure.

The robust H_∞ control structure, shown in Fig. 23, can be rearranged as a linear fractional transformation (LFT), as shown in Fig. 24. From this figure, state-space realization of $P(s)$ can be expressed as

$$
P(s): \begin{cases}
\dot{x}_P = \begin{bmatrix} A_o & 0 & 0 & 0 \\ B_H C_o & A_H & 0 & 0 \\ 0 & -B_e C_H & 0 & A_e \end{bmatrix} x_P \\
\qquad + \begin{bmatrix} 0 & 0 & B_o & B_o \\ 0 & B_H & B_H D_o & B_H D_o \\ B_e & 0 & 0 & 0 \end{bmatrix} \begin{bmatrix} r \\ d_o \\ d_i \\ u_r \end{bmatrix}, \\
\begin{bmatrix} z \\ e_r \end{bmatrix} = \begin{bmatrix} 0 & -D_e C_H & 0 & C_e \\ 0 & -C_H & 0 & 0 \end{bmatrix} x_P \\
\qquad + \begin{bmatrix} D_e & 0 & 0 & 0 \\ 1 & 0 & 0 & 0 \end{bmatrix} \begin{bmatrix} r \\ d_o \\ d_i \\ u_r \end{bmatrix}
\end{cases}
\tag{49}
$$

where (A_o, B_o, C_o, D_o), (A_H, B_H, C_H, D_H), and (A_e, B_e, C_e, D_e) are state-space realization matrices of $G_o(s)$, $H(s)$, and $W_e(s)$, respectively, and $x_P(t) = [x_o(t) \ x_H(t) \ x_e(t)]^T$ is the state variable of $P(s)$ with $x_o(t)$, $x_H(t)$, and $x_e(t)$ as the state variables of $G_o(s)$, $H(s)$, and $W_e(s)$, respectively. Note that to compute (A_o, B_o, C_o, D_o), the time-delay operator can be approximated by a second-order Padé approximant.

Table 15 Achieved H_∞ performance

H_∞ Performance	Group 1	Group 2	Group 3	Group 4
Value	0.1265	0.0122	0.0009	0.1254

The H_∞ control problem can be solved by `hinfsyn` function of the Robust control toolbox in MATLAB, which yields the best $H\infty$ performances as presented in Table 15. Also, the obtained H_∞ controllers are

$$K_{\infty,1} = \frac{\begin{array}{c}0.12897s^{10}+0.104s^9+0.03542s^8+0.006399s^7+0.0006348s^6\\+3.222\times10^{-5}s^5+6.735\times10^{-7}s^4+3.771\times10^{-9}s^3\\+2.481\times10^{-12}s^2+1.079\times10^{-16}s+1.195\times10^{-21}\end{array}}{\begin{array}{c}s^{11}+0.8606s^{10}+0.3168s^9+0.0637s^8+0.007397s^7+0.0004841s^6\\+1.653\times10^{-5}s^5+3.056\times10^{-7}s^4+4.293\times10^{-10}s^3\\+1.915\times10^{-14}s^2+2.137\times10^{-19}s+2.363\times10^{-28}\end{array}},$$

$$(50)$$

$$K_{\infty,2} = \frac{\begin{array}{c}0.9554s^{10}+0.67s^9+0.1974s^8+0.02989s^7+0.002386s^6\\9.541\times10^{-5}s^5+1.702\times10^{-6}s^4+9.399\times10^{-9}s^3\\+5.832\times10^{-12}s^2+2.593\times10^{-16}s+2.929^{-21}\end{array}}{\begin{array}{c}s^{11}+0.8241s^{10}+0.2973s^9+0.05948s^8+0.007081s^7+0.0005036s^6\\+2.235\times10^{-5}s^5+4.319\times10^{-7}s^4+6.197\times10^{-10}s^3\\+3.187\times10^{-14}s^2+4.862\times10^{-19}s+1.864\times10^{24}\end{array}},$$

$$(51)$$

$$K_{\infty,3} = \frac{\begin{array}{c}1.826s^{10}+1.993s^9+0.893s^8+0.1862s^7+0.01909s^6\\+0.0009327s^5+1.854\times10^{-5}s^4+1.076\times10^{-7}s^3\\+6.824\times10^{-11}s^2+3.473\times10^{-15}s+4.302\times10^{-20}\end{array}}{\begin{array}{c}s^{11}+1.199s^{10}+0.612s^9+0.1602s^8+0.02381s^7+0.002095s^6\\+0.0001157s^5+3.243\times10^{-6}s^4+4.749\times10^{-9}s^3\\+2.638\times10^{-13}s^2+3.566\times10^{-18}s+3.263\times10^{-24}\end{array}},$$

$$(52)$$

$$K_{\infty,4} = \frac{\begin{array}{c}0.1227s^{10}+0.08142s^9+0.02263s^8+0.00323s^7+0.0002416s^6\\+8.938\times10^{-6}s^5+1.487\times10^{-7}s^4+8.331\times10^{-10}s^3\\+4.777\times10^{-13}s^2+2.50\times10^{-17}s+3.443\times10^{-22}\end{array}}{\begin{array}{c}s^{11}+0.7091s^{10}+0.2157s^9+0.03496s^8+0.003162s^7+0.0001553s^6\\+4.345\times10^{-6}s^5+5.486\times10^{-8}s^4+7.471\times10^{-11}s^3\\+4.091\times10^{-15}s^2+5.696\times10^{-2}s+6.113\times10^{-29}\end{array}},$$

$$(53)$$

where $K_{\infty,i}(s)$, $i=1,\ldots,4$ is the H_∞ controller developed for the ith age group. One the H_∞ controller is designed, time constant of the set-point prefilter can be determined

Table 16 Value of the set-point prefilter time constant in the robust H_∞ control scheme

Parameter	Group 1	Group 2	Group 3	Group 4
T_{sp}	147.28	108.81	90.09	130.21

following the same procedure presented in Section 4.4. Calculated values are reported in Table 16.

Simulation results and performance indices by using the H_∞ controllers (50)–(53) are presented in Figs. 25–28 and Tables 17 and 18, respectively. According to the simulation results, the robust H_∞ control scheme provides safe induction with low overdosing risk. This can be seen in Table 19 which reports the number of overdosed

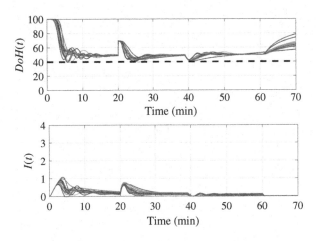

Fig. 25 Simulation result with the robust H_∞ controller for Group 1; each color for one patient.

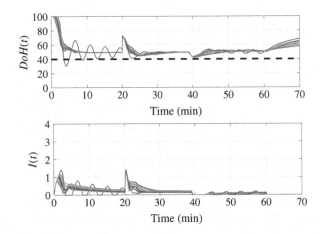

Fig. 26 Simulation result with the robust H_∞ controller for Group 2; each color for one patient.

Fig. 27 Simulation result
with the robust H_∞ controller
for Group 3; each color for
one patient.

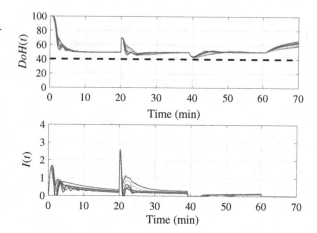

Fig. 28 Simulation result
with the robust H_∞ controller
for Group 4; each color for
one patient.

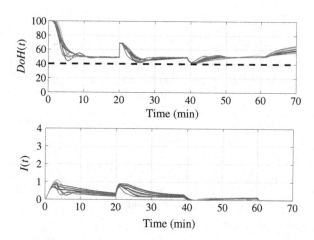

Table 17 Performance of the robust H_∞ control scheme at induction phase (mean $\pm$ SD)

Group	Rise time (min)	Settling time (min)	Overshoot (%)
1	7.082 $\pm$ 2.491	9.174 $\pm$ 3.056	7.266 $\pm$ 7.016
2	7.506 $\pm$ 1.501	6.778 $\pm$ 3.571	4.573 $\pm$ 10.774
3	7.323 $\pm$ 0.637	5.147 $\pm$ 0.597	1.289 $\pm$ 0.260
4	7.976 $\pm$ 2.535	7.299 $\pm$ 1.450	3.693 $\pm$ 3.562

patients. However, achieved rising and settling times are a bit high, especially for the
first group.

Regarding maintenance phase, the robust H_∞ control scheme keeps the depth of
hypnosis within safe interval during the virtual surgery, and even in the presence

Table 18 Performance of the robust H_∞ control scheme at maintenance phase (mean ± SD)

Group	MPE (%)	MAPE (%)	Divergence (% min^{-1})	Wobble (%)
1	1.453 ± 1.094	1.456 ± 1.456	−0.0064 ± 0.0008	1.935 ± 1.075
2	0.930 ± 0.414	1.761 ± 0.986	−0.0047 ± 0.0007	1.704 ± 1.039
3	0.916 ± 0.519	1.758 ± 1.056	−0.0039 ± 0.0006	1.810 ± 1.149
4	1.950 ± 1.233	2.799 ± 1.707	−0.0071 ± 0.0008	2.180 ± 1.075

Table 19 Number of overdosed patients with the robust H_∞ control scheme

Group	Number of patients	Number of overdosed patients
1	15	0
2	12	1
3	8	0
4	9	0

of disturbance. More precisely, according to the achieved divergence indices, the depth of hypnosis obtained by the robust H_∞ control scheme converges to the reference value well.

The aforementioned robust H_∞ control scheme has been studied widely in the literature, for example, Caiado et al. (2013a, b) and Lemos et al. (2014). It is noteworthy that using the philosophy of the H_∞ performance, it is possible to divide the controller into two components to come up with a two-degrees-of-freedom robust H_∞ control scheme: (i) feed-forward component and (ii) feedback component. It is discussed in Hahn et al. (2010, 2012) that two-degrees-of-freedom version may provide less chattering in infusion rate, better tracking, and better disturbance attenuation.

5 Conclusion

This chapter summarized the most recent and promising contributions dispersed regarding the robust control of depth of hypnosis. First, two EEG-based measurement indices (i.e., BIS and WAV$_{CNS}$) were studied, and their pros and cons with regard to needs in closed-loop anesthesia were mentioned. Then, the relationship between dose and pharmacological effect of propofol was studied. More precisely, it was shown that the mentioned relationship can be explained by two separate models: (i) PK model which relates the drug plasma concentration to the administrated does and (ii) PD model which relates drug concentration to the observed effect. Possible uncertainties in PK and PD models were also studied. Finally, two very promising robust schemes developed for control of depth of hypnosis, that is, robust PID and H_∞ schemes, were explained in detail. The mentioned schemes were simulated for 44 patients whose parameters have been identified in a real clinical study. The effectiveness of the schemes was investigated, and each one's pros and cons with respect to some performance indices were also reported.

While the subject of robust control in closed-loop anesthesia has been researched, some fundamental steps remain yet to be taken before widespread use of such control schemes, which can be considered as future research directions: (i) to ensure other safety factors (e.g., blood pressure drop), (ii) to guarantee robustness against possible effects of other drugs during surgery, and (iii) to ensure tolerance toward possible faults and failures. Once all control problems are addressed, large clinical trials on a very large cohort of patients will become possible in order to convince the clinicians of the benefits of closed-loop drug delivery.

References

Absalom, A.R., Sutcliffe, N., Kenny, G.N., 2002. Closed-loop control of anesthesia using bispectral index: performance assessment in patients undergoing major orthopedic surgery under combined general and regional anesthesia. Anesthesiology 96 (1), 67–73.

Araki, M., 2009. PID control. In: Unbehauen, H. (Ed.), Control Systems, Robotics and Automation, Volume II: System Analysis and Control: Classical Approaches II. EOLSS, UK, pp. 58–79.

Aström, K.J., Hägglund, T., 1995. PID Controllers: Theory, Design, and Tuning, second ed. International Society of Automation, United States.

Aström, K.J., Hägglund, T., 2006. Advanced PID Control. ISA—The Instrumentation, Systems, and Automation Society.

Bibian, S., 2006. Automation in Clinical Anesthesia (Ph.D. thesis). The University of British Columbia.

Bibian, S., Zikov, T., Dumont, G.A., Ries, C.R., Puil, E., Ahmadi, H., Huzmezan, M., MacLeod, B.A., 2001. Estimation of the anesthetic depth using wavelet analysis of electroencephalogram. In: Proceedings of the 23rd Annual International Conference of the IEEE Engineering in Medicine and Biology Society, Istanbul, Turkey, pp. 951–955.

Bibian, S., Ries, C.R., Huzmezan, M., Dumont, G.A., 2003. Clinical anesthesia and control engineering: terminology, concepts and issues. In: Proceedings of the 2003 European Control Conference, Cambridge, UK, pp. 2430–2440.

Bibian, S., Zikov, T., Voltz, D.M., Dumont, G.A., Huzmezan, M., 2004a. Pharmacodynamic intra- and inter-patient variability of processed electroencephalography variable during thiopental induction for ECT. In: Proceedings of the 85th Post Graduate Assembly of the New York State Society of Anesthesiologists, p. 9145.

Bibian, S., Zikov, T., Ries, C.R., Dumont, G.A., Huzmezan, M., 2004b. The wavelet-based anesthetic value (WAV): a novel alternative to the bispectral index (BIS). In: Proceedings of the American Society of Anesthesiologists, p. A-342.

Bibian, S., Ries, C.R., Huzmezan, M., Dumont, G.A., 2005. Introduction to automated drug delivery in clinical anesthesia. Eur. J. Control 11 (6), 535–557.

Bibian, S., Dumont, G.A., Huzmezan, M., Ries, C.R., 2006. Patient variability and uncertainty quantification in clinical anesthesia: part II—PKPD uncertainty. In: Proceedings of the 6th IFAC Symposium on Modelling and Control in Biomedical Systems, Reims, France, pp. 555–560.

Billard, V., Gambus, P.L., Chamoun, N., Stanski, D.R., Shafer, S.L., 1997. A comparison of spectral edge, delta power, and bispectral index as EEG measures of alfentanil, propofol, and midazolam drug effect. Clin. Pharmacol. Ther. 61 (1), 45–58.

Bowles, S.M., Sebel, P.S., Saini, V., Chamoun, N., 1994. Effects of anaesthesia on the EEG-bispectral analysis correlates with movement. In: Jones, S.J., Hetreed, M., Boyd, S., Smith, N.J. (Eds.), Handbook of Spinal Cord Monitoring. Springer, Dordrecht, pp. 247–252.

Caiado, D.V., Lemos, J.M., Costa, B.A., Silva, M.M., Mendonça, T.F., 2013a. Design of depth of anesthesia controllers in the presence of model uncertainty. In: Proceedings of the 21st Mediterranean Conference on Control and Automation, Chania, Greece, pp. 213–218.

Caiado, D.V., Lemos, J.M., Costa, B.A.D., 2013b. Robust control of depth of anesthesia based on H_∞ design. Arch. Control Sci. 23 (1), 41–59.

Dumont, G.A., 2012. Closed-loop control of anesthesia—a review. In: Proceedings of the 8th IFAC Symposium on Biological and Medical Systems, Budapest, Hungary, pp. 373–378.

Dumont, G.A., Martinez, A., Ansermino, J.M., 2008. Robust control of depth of anesthesia. Int. J. Adapt Control Signal Process. 23 (5), 435–454.

Gan, T., Glass, P., Windsor, A., 1997. Bispectral index monitoring allows faster emergence and improved recovery from propofol, alfentanil, and nitrous oxide anesthesia. BIS Utility Study Group. Anesthesiology 87 (4), 808–815.

Hagihira, S., Takashina, M., Mori, T., Mashimo, T., Yoshiya, I., 2001. Practical issues in bispectral analysis of electroencephalographic signals. Anesth. Analg. 93 (4), 966–970.

Hahn, J.O., Dumont, G.A., Ansermino, J.M., 2010. Robust closed-loop control of propofol administration using WAV_{CNS} index as the controlled variable. In: Proceedings of the 2010 Annual International Conference of the IEEE Engineering in Medicine and Biology, Buenos Aires, Argentina, pp. 6038–6041.

Hahn, J.O., Dumont, G.A., Ansermino, J.M., 2012. Robust closed-loop control of hypnosis with propofol using WAV_{CNS} index as the controlled variable. Biomed. Signal Process. Control 7, 517–524.

Heusden, K.V., Dumont, G.A., Soltesz, K., Petersen, C.L., Umedaly, A., West, N., Ansermino, J.M., 2014. Design and clinical evaluation of robust PID control of propofol anesthesia in children. IEEE Trans. Control Syst. Technol. 22 (3), 491–501.

Heusden, K.V., Ansermino, J.M., Dumont, G.A., 2018. Robust MISO control of propofol-remifentanil anesthesia guided by the NeuroSENSE monitor. IEEE Trans. Control Syst. Technol. 26 (5), 1758–1770.

Hosseinzadeh, M., Garone, E., 2019. Constrained control in closed-loop anesthesia. In: Proceedings of the 38th Benelux Meeting on Systems and Control, Lommel, Belgium.

Hosseinzadeh, M., Sayari, R., Mohammadi, S.A., 2014. H_2/H_∞ multi-model control scheme-architecture and performance evaluation. Intersciencia 39 (4), 238–241.

Hosseinzadeh, M., Dumont, G.A., Garone, E., 2019a. Constrained control of depth of hypnosis during induction phase. IEEE Trans. Control Syst. Technol. https://doi.org/10.1109/TCST.2019.2929489.

Hosseinzadeh, M., van Heusden, K., Dumont, G.A., Garone, E., 2019b. An explicit reference governor scheme for closed-loop anesthesia. In: Proceedings of the 17th European Control Conference, Napoli, Italy.

Kazama, T., Ikeda, K., Morita, K., Kikura, M., Doi, M., Ikeda, T., Kurita, T., Nakajima, Y., 1999. Comparison of the effect-site K_{e0}s of propofol for blood pressure and EEG bispectral index in elderly and younger patients. Anesthesiology 90, 1517–1527.

Kearse, L., Saini, V., deBros, F., Chamoun, N., 1990. Bispectral analysis of EEG may predict anesthetic depth using narcotic induction. Anesthesiology 3A, A175.

Kreuer, S., Biedler, A., Larsen, R., Altmann, S., Wilhelm, W., 2003. Narcotrend monitoring allows faster emergence and a reduction of drug consumption in propofol-remifentanil anesthesia. Anesthesiology 99 (1), 34–41.

Kuizenga, K., Proost, J.H., Wierda, J.M., Kalkman, C.J., 2001. Predictability of processed electroencephalography effects on the basis of pharmacokinetic-pharmacodynamic modeling during repeated propofol infusions in patients with extradural analgesia. Anesthesiology 95, 607–615.

Lemos, J.M., Caiado, D.V., Costa, B.A., Paz, L.A., Mendonca, T.F., Rabico, R., Esteves, S., Seabra, M., 2014. Robust control of maintenance-phase anesthesia [applications of control]. IEEE Control Syst. Mag. 34 (6), 24–38.

Lundqvist, K.L., Bibian, S., Ries, C.R., Yu, P.Y.H., 2004. Can the wavelet-based anesthetic value (WAV) predict airway motor response to LMA insertion? In: Proceedings of the American Society of Anesthesiologists, p. A-606.

Martinez, A., 2005. Robust Control: PID vs. Fractional Control Design, a Case Study (Master's thesis). University of British Columbia.

Merigo, L., Padula, F., Pawlowski, A., Dormido, S., Sánchez, J.L.G., Latronico, N., Paltenghi, M., Visioli, A., 2018. A model-based control scheme for depth of hypnosis in anesthesia. Biomed. Signal Process. Control 42, 216–229.

Ning, T., Bronzino, J.D., 1989. Bispectral analysis of the rat EEG during various vigilance states. IEEE Trans. Biomed. Eng. 36 (4), 497–499.

Padula, F., Ionescu, C., Latronico, N., Paltenghi, M., Visioli, A., Vivacqua, G., 2015. A gain-scheduled PID controller for propofol dosing in anesthesia. In: Proceedings of the 9th IFAC Symposium on Biological and Medical Systems, Berlin, Germany, pp. 545–550.

Padula, F., Ionescu, C., Latronico, N., Paltenghi, M., Visioli, A., Vivacqua, G., 2017. Optimized PID control of depth of hypnosis in anesthesia. Comput. Methods Programs Biomed. 144, 21–35.

Pawlowski, A., Merigo, L., Guzmán, J.L., Dormido, S., Visioli, A., 2018. Two-degree-of-freedom control scheme for depth of hypnosis in anesthesia. In: Proceedings of the 3rd IFAC Conference on Advances in Proportional-Integral-Derivative Control, Ghent, Belgium, pp. 72–77.

Pilge, S., Zanner, R., Schneider, G., Blum, J., Kreuzer, M., Kochs, E., 2006. Time delay of index calculation: analysis of cerebral state, bispectral, and Narcotrend indices. Anesthesiology 104 (3), 488–494.

Rampil, I.J., 1998. A primer for EEG signal processing in anesthesia. Anesthesiology 89 (4), 980–1002.

Rosow, C., Manberg, P.J., 2001. Bispectral index monitoring. Anesthesiol. Clin. North Am. 19 (4), 947–966.

Sadati, N., Hosseinzadeh, M., Dumont, G.A., 2018. Multi-model robust control of depth of hypnosis. Biomed. Signal Process. Control 40, 443–453.

Schnider, T.W., Minto, C.F., Gambus, P.L., Andresen, C., Goodale, D.B., Shafer, S.L., Youngs, E.J., 1998. The influence of method of administration and covariates on the pharmacokinetics of propofol in adult volunteers. Anesthesiology 88 (5), 1170–1182.

Schnider, T.W., Minto, C.F., Shafer, S.L., Gambus, P.L., Andresen, C., Goodale, D.B., Youngs, E.J., 1999. The influence of age on propofol pharmacodynamics. Anesthesiology 90 (6), 1502–1516.

Schüttler, J., Ihmsen, H., 2000. Population pharmacokinetics of propofol: a multicenter study. Anesthesiology 92 (3), 727–738.

Schüttler, J., Kloos, S., Schwilden, H., Stoeckel, H., 1988. Total intravenous anaesthesia with propofol and alfentanil by computer-assisted infusion. Anaesthesia 43, 341–345.

Sebel, P.S., Bowles, S., Saini, V., Chamoun, N., 1990. Accuracy of EEG in predicting movement at incision during isoflurane anesthesia. Anesthesiology 3A, A446.

Soltesz, K., Hahn, J.O., Dumont, G.A., Ansermino, J.M., 2011. Individualized PID control of depth of anesthesia based on patient model identification during the induction phase of anesthesia. In: Proceedings of the 50th IEEE Conference on Decision and Control and European Control Conference, Orlando, USA, pp. 855–860.

Vernon, J., Bowles, S., Sebel, P.S., Chamoun, N., 1992. EEG bispectrum predicts movement at incision during isoflurane or propofol anesthesia. Anesthesiology 77 (3A), A502.

Zikov, T., 2002. Monitoring the Anesthetic-Induced Unconsciousness (Hypnosis) Using Wavelet Analysis of the Electroencephalogram (Master's thesis). University of British Columbia.

Zikov, T., Bibian, S., Dumont, G.A., Huzmezan, M., Ries, C.R., 2002. A wavelet based de-noising technique for ocular artifact correction of the electroencephalogram. In: Proceedings of the Second Joint 24th Annual Conference and the Annual Fall Meeting of the Biomedical Engineering Society [Engineering in Medicine and Biology], Houston, TX, pp. 98–105.

Zikov, T., Bibian, S., Dumont, G.A., Huzmezan, M., Ries, C.R., 2006. Quantifying cortical activity during general anesthesia using wavelet analysis. IEEE Trans. Biomed. Eng. 53 (4), 617–632.

Robust control strategy for HBV treatment: Considering parametric and nonparametric uncertainties

Omid Aghajanzadeh[a], Mojtaba Sharifi[a], Ali Falsafi[b]
[a]Department of Mechanical Engineering, Sharif University of Technology, Tehran, Iran,
[b]Department of Mechanical Engineering, Swiss Federal Institute of Technology Lausanne, Lausanne, Switzerland

1 Introduction

Currently, hepatitis B virus (HBV) is one of the most widespread virions that causes severe health problems for the human being. Based on the statistics released in 2009, more than 2 billion people alive have been infected by hepatitis. Currently, 400 million individuals are virus carriers (based on WHO reports: 240 million people are chronically infected with hepatitis B and between 130 and 150 million people globally had chronic hepatitis C infection in 2016; World Health Organisation, 2017). Most of the reported hepatitis infections (about 75%) occurred in Asia. Unfortunately, hepatitis causes more than 1 million deaths annually as reported by Hattaf et al. (2009a).

In addition, chronic HBV can lead to the hepatocellular carcinoma, cirrhosis-induced liver, and liver cancer which are all extremely lethal. Therefore, necessary actions must be taken in order to overcome this deadly disease. It is notable that none of the available drugs on the market are capable of rooting out HBV completely. However, they can stop the virus replication and prevent or reduce liver damage. The treatment of chronic infection is required to reduce the risk of cirrhosis and liver cancer. For several patients infection takes a very aggressive course, hence an early antiviral treatment is obligatory according to Lai and Yuen (2007).

In order to study the HBV experimentally (using animal subjects), (Ganem et al., 2001) used a woodchuck HBV model whose sequence has 60% identity with the human HBV. In another work, conducted by Feitelson and Larkin (2001) ducks were infected with the HBV in order to study their symptoms. Nevertheless, some features of the human HBV such as specific parameters of infection, liver disease development, and human-specific immune responses make it extremely difficult to extend the employed animal model results to the human HBV as represented by Feitelson and Larkin (2001).

Mathematical modeling is an appropriate approach to address the HBV infection dynamics since they can provide a quantitative understanding of in vivo HBV replication behavior as explained by Nowak et al. (1996). Generally, these models seem to be a proper way of investigating disease behavior. Similar mathematical models have

Control Applications for Biomedical Engineering Systems. https://doi.org/10.1016/B978-0-12-817461-6.00005-6

also been employed for several other viral infections. For instance, the human immunodeficiency virus (HIV) (Gumel and Moghadas, 2004; Landi et al., 2008; Pinto and Carvalho, 2014) and hepatitis C virus (DebRoy et al., 2010) have been mathematically modeled. In these models, the viral and cell dynamics as well as the immune system behavior are reflected in a set of mathematical equations. The above-mentioned models have been improved in recent years to predict the realistic behavior of viral infection more precisely (Lv et al., 2014). The epidemiological models of infectious diseases have also been recently studied and controlled by vaccination and antiviral drug inputs (Kumar and Srivastava, 2017).

The employed mathematical model in this chapter is originally introduced by Nowak et al. (1996) after carrying out two clinical studies. First, 45 patients were treated and their states were observed for 28 days. Also, health states of 50 patients were studied for 24-week treatment interval in a subsequent study by Nowak et al. (1996). The obtained results were used to develop the mathematical model, which did not contain the contribution of drug usage as the control inputs. Hattaf et al. (2009b) have successfully enhanced this original model proposed by Nowak et al. (1996) by introducing the control inputs as the efficiency of drug therapy. This model has been widely used for the analysis of the HBV infection (Min et al., 2008; Ribeiro et al., 2002; Sheikhan and Ghoreishi, 2013a, b).

Recently, several different control strategies have been applied to the mathematical models of diseases. For example, De Souza et al. (2000) have designed an optimal controller for the HIV infection dynamics. Using this controller, they have minimized the accumulated side effects of drugs and viral loads. Ledzewicz and Schattler (2002) have proposed another optimal strategy for antiviral treatment of AIDS. In their proposed control strategy, the number of uninfected cells (CD4T cells) is maximized, while the dosage of the drug is minimized. Blayneh et al. (2009) have proposed an optimal strategy for malaria treatment in which the number of latent and infected cells were considered in the cost function. In another research, Lee et al. (2013) applied an optimal controller for the treatment of the flu, by minimizing the incidence and intervention costs. Okosun and Makinde (2014) also implemented an optimal strategy for incorporated time-dependent control of malaria and cholera diseases. In their work, Pontryagins maximum principle was utilized to derive the necessary condition of the optimal policy (Okosun and Makinde, 2014).

In particular, a wide range of control strategies have been applied to the HBV treatment. For instance, Ntaganda and Gahamanyi (2015) employed a fuzzy logic for optimal control of the HBV and a reasonable agreement has been observed with experimental data by Ntaganda and Gahamanyi (2015). Su and Sun (2015) compared the HBV treatment process using two different drugs. Furthermore, they employed an optimal method in their simulations and concluded that the optimal control strategy results in a more effective treatment compared to the constant drug dosage. Laarabi et al. (2013) applied another optimal control strategy in order to minimize the treatment costs and maximize the volume of healthy cells. However, realistic systems usually have modeling uncertainties that should be taken into account in the controller design. Therefore, some robust/adaptive control methods (Moradi et al., 2015;

Sharifi and Sayyaadi, 2015; Yazdi and Nagamune, 2010) have been adopted for different uncertain dynamic systems in the literature.

In this chapter, for the first time, a new nonlinear robust control strategy is developed for the treatment of the HBV infection. Unlike the previous optimal controllers (Laarabi et al., 2013; Su and Sun, 2015) suggested for the HBV, the employed dynamic model in this control strategy is subjected to both parametric and nonparametric uncertainties. Moreover, the proposed controller unlike previously suggested controllers (e.g., by Ntaganda and Gahamanyi, 2015) is designed based on the nonlinear HBV dynamics and fulfills its objectives without any linearization of the model. Therefore, for the first time, the proposed controller guarantees the system stability in the presence of parametric and nonparametric uncertainties of the nonlinear HBV model. It is shown that the proposed nonlinear robust control strategy has a remarkable performance in tracking the desired scenario for the virions' and infected cells' populations. As a result, the numbers of viruses and infected cells reduce and converge to desired decaying functions.

This chapter is organized as following. Section 2 addresses outlines of the employed nonlinear mathematical model. The proposed robust controller is designed in Section 3. Then, the dynamic systems stability is proved in Section 4 via the Lyapunov theorem. The obtained simulation results are represented in Section 5, which show the controllers performance. The obtained outcomes and the conclusion are expressed in Section 6.

2 HBV mathematical model

In the recent years, the methods of using antiviral drugs for the treatment of viral diseases like HBV has been studied via the mathematical equations of their dynamics (Sheikhan and Ghoreishi, 2013a). Accordingly, control methods can be employed in order to design the treatment strategies for these kinds of antiviral therapies.

According to the literature, a recently proposed and validated nonlinear mathematical model frequently utilized to study the HBV infection is (Hattaf et al., 2009a):

$$\frac{dx}{dt} = \lambda - dx - (1 - u_1)\beta xv \tag{1}$$

$$\frac{dy}{dt} = (1 - u_1)\beta xv - \delta y \tag{2}$$

$$\frac{dv}{dt} = (1 - u_2)py - cv \tag{3}$$

in which, x, y, and v represent the uninfected cells, infected cells, and virion numbers, respectively. u_1 and u_2 are control inputs expressing the rate of drug usage. In addition, d, δ, c, p, β, and λ are model parameters which are positive constants and described in details in Table 1. The third column of the table consists of the values of parameters extracted from the research conducted by Sheikhan and Ghoreishi (2013a) and

Table 1 Parameters definition and values

Parameter	Definition	Value
d	Death rate of target cells	$0.0038 \ \text{day}^{-1} \ \text{mL}^{-1}$
δ	Death rate of infected cells	$0.0125 \ \text{day}^{-1}$
c	Clearance rate of free virions	$0.067 \ \text{day}^{-1}$
p	Rate of production of virions per infected cell	$842.0948 \ \text{day}^{-1}$
β	Rate of infection of new target cells	$1.981 \times 10^{-13} \ \text{day}^{-1} \ \text{mL}^{-1}$
λ	Rate of production of new target cells	$2.5251 \times 10^{5} \ \text{day}^{-1} \ \text{mL}^{-1}$

Aghajanzadeh et al. (2017). It is worth mentioning that in Eqs. (1)–(3) uninfected cells are created at a rate λ, die at a rate dx, and become infected at a rate βvx; infected cells are created at a rate βvx and die at a rate δy; virion are created from infected cells at a rate py and are removed at a rate cv (Wang and Wang, 2007).

Eqs. (1)–(3) are differential equations originally proposed by Nowak et al. (1996) according to some experimental study through the examination of reasonable numbers of patients during their treatment intervals. In the first study, 45 patients were treated for 28 days; in the following experiment, 50 patients were treated for 24 weeks which were explained by Nowak et al. (1996). u_1 and u_2, as the control inputs, were added to this model by Hattaf et al. (2009a), which represent the efficiency of two different treatment mechanisms of the HBV. These mechanisms block new infection and inhibit viral production. Ciupe et al. (2007) have investigated a nonlinear mathematical model of HBV and also studied its drug therapy using two drug therapy controls. Yosyingyong and Viriyapong (2018) studied the efficiency of drug treatment in preventing new infections and efficiency of drug treatment in inhibiting viral production. In this study, numerical simulations are established to show the role of optimal therapy in controlling viral replication.

3   Robust controller design

As mentioned previously, the employed dynamic model of HBV has been obtained experimentally. However, a disease model obtained experimentally usually consists of different types of uncertainties (parametric and nonparametric) in different human bodies.

Parametric or structured uncertainties are rooted in measurement limitations and errors during the initial experiments. Furthermore, the probable differences (age, genetic diversity, and life style) among patients result in parametric uncertainty which is considered in this study for the utilized model.

Another type of uncertainty considered in this chapter is nonparametric or unstructured uncertainty. This type of uncertainty results from aspects of the system which is not taken into account in the developed model. The unstructured uncertainty may result from environmental conditions of the patient and external disturbances. Since we have an experimentally obtained model, it is highly probable that the model suffers from this type of uncertainty. Accordingly, two arbitrary disturbance functions D_1 and D_2 are taken into account as functions of nonparametric uncertainties in Eqs. (4)–(6) (Aghajanzadeh et al., 2018):

$$\frac{dx}{dt} = \lambda - dx - (1 - u_1)\beta xv - D_1\beta xv \tag{4}$$

$$\frac{dy}{dt} = (1 - u_1)\beta xv - \delta y + D_1\beta xv \tag{5}$$

$$\frac{dv}{dt} = (1 - u_2)py - cv + D_2py \tag{6}$$

The proposed controller for the drug usage (u_1 and u_2) is designed based on Eqs. (5), (6) of the HBV dynamics such that number of viruses (v) and infected cells (y) decrease. Using this strategy, the number of uninfected cells grows persistently according to Eq. (4) which is desired in the HBV treatment process.

In order to design the control laws, Eqs. (5), (6) are rearranged as

$$u_1(t) = -\frac{\dot{y}}{\beta xv} - \frac{\delta y}{\beta xv} + 1 + D_1 \tag{7}$$

$$u_2(t) = -\frac{\dot{v}}{py} - \frac{cv}{py} + 1 + D_2 \tag{8}$$

where $\dot{y}$ and $\dot{v}$ are time derivative of y and v, respectively. The control inputs are structured such that the desired values for the number of infected cells (y_{des}) and the number of viruses (v_{des}) are tracked while ensuring the stability of process. For these purposes, the proposed control inputs (u_1 and u_2) are defined as follows:

$$u_1(t) = J_1\hat{\theta}_1 + 1 + \frac{\gamma_1\text{sign}(\tilde{y})}{vx} \tag{9}$$

$$u_2(t) = J_2\hat{\theta}_2 + 1 + \frac{\gamma_2\text{sign}(\tilde{v})}{y} \tag{10}$$

where $\tilde{y} = y - y_{des}$ and $\tilde{v} = v - v_{des}$ are tracking errors, and $\gamma_1\text{sign}(\tilde{y})/vx$ and $\gamma_2\text{sign}(\tilde{v})/y$ represent the robust terms of the controller to overcome the modeling uncertainties. The sign $\wedge$ is used to clarify uncertain parameters. The matrices J_1 and J_2, and the vectors θ_1 and θ_2 in Eqs. (9), (10) are also defined as

$$J_1 = \left[-\frac{\phi_1}{vx}, -\frac{y}{vx} \right] \tag{11}$$

$$J_2 = \left[-\frac{\phi_2}{y}, -\frac{v}{y} \right] \tag{12}$$

$$\theta_1 = \left[\frac{1}{\beta}, \frac{\delta}{\beta} \right]^T \tag{13}$$

$$\theta_2 = \left[\frac{1}{p}, \frac{c}{p} \right]^T \tag{14}$$

where ϕ_1 and ϕ_2 are arbitrary variables that are set in the proposed controller as follows:

$$\phi_1 = \dot{y}_{des} - \alpha_1 (y - y_{des}) \tag{15}$$

$$\phi_2 = \dot{v}_{des} - \alpha_2 (v - v_{des}) \tag{16}$$

where α_1 and α_2 are positive constant parameters, and y_{des} and v_{des} are the desired values for y and v, respectively. The schematic structure of the nonlinear robust controller designed for the HBV drug usage is shown in Fig. 1. The control inputs are functions of the states (x, y, and v) as well as the tracking errors ($\tilde{y}$ and $\tilde{v}$).

Substitution of the control laws (9), (10) in the HBV model (7), (8) gives the following closed-loop dynamics:

$$-\frac{\dot{y}}{\beta xv} - \frac{\delta y}{\beta xv} + 1 + D_1 = J_1 \tilde{\theta}_1 + 1 + \frac{\gamma_1 \operatorname{sgn}(\tilde{y})}{xv} \tag{17}$$

$$-\frac{\dot{v}}{py} - \frac{cy}{py} + 1 + D_2 = J_2 \tilde{\theta}_2 + 1 + \frac{\gamma_2 \operatorname{sgn}(\tilde{v})}{y} \tag{18}$$

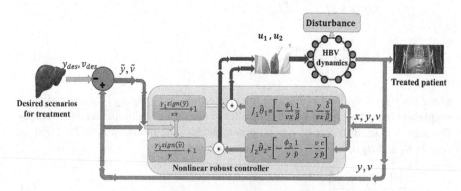

Fig. 1 The schematic structure of the proposed robust controller for the HBV treatment.

By rearranging Eqs. (17), (18), the tracking error dynamics are obtained as

$$\dot{\tilde{y}} = -\alpha_1 \tilde{y} - \beta xvJ_1\tilde{\theta}_1 - \beta\gamma_1 \operatorname{sgn}(\tilde{y}) + \beta xvD_1 \tag{19}$$

$$\dot{\tilde{v}} = -\alpha_2 \tilde{v} - pyJ_2\tilde{\theta}_2 - \gamma_2 p \operatorname{sgn}(\tilde{v}) + pyD_2 \tag{20}$$

The closed-loop systems stability will be proved in the following section using the Lyapunov theorem and according to the obtained error dynamics (19), (20).

4 Lyapunov stability

In order to prove the system stability and the tracking convergence using the robust controller presented in the previous section, the Lyapunov theorem is employed. For this purpose, a Lyapunov function candidate is considered as

$$V = \frac{1}{2}(\tilde{v}^2 + \tilde{y}^2) \geq 0 \tag{21}$$

which is positive definite including two quadratic terms. To verify the controlled systems' stability, the time derivative of the Lyapunov function (21) should be evaluated (Slotine and Li, 1991). Thus $\dot{V}(t)$, is obtained as

$$\dot{V}(t) = \tilde{v}\dot{\tilde{v}} + \tilde{y}\dot{\tilde{y}} \tag{22}$$

Substitution of $\dot{\tilde{v}}$ and $\dot{\tilde{y}}$ from Eqs. (19), (20) in $\dot{V}(t)$ (Eq. 22) and rewriting it gives

$$\dot{V} = -\alpha_2\tilde{v}^2 - \alpha_1\tilde{y}^2 - (\tilde{v}\,pyJ_2\tilde{\theta}_2 + \gamma_2 p|\tilde{v}| - \tilde{v}\,pyD_2) - (\beta\tilde{y}\,xvJ_1\tilde{\theta}_1 + \beta\gamma_1| \\ \tilde{y}| - \beta\tilde{y}\,xvD_1) \tag{23}$$

In order to obtain a negative definite $\dot{V}(t)$, the robust gains γ_1 and γ_2 are adjusted such that

$$\gamma_1 \geq |vxJ_1\tilde{\theta}_1| + |vxD_1| \tag{24}$$

$$\gamma_2 \geq |yJ_2\tilde{\theta}_2| + |yD_2| \tag{25}$$

which guarantee that the terms $\beta\tilde{y}\,vxJ_1\tilde{\theta}_1 + \beta\gamma_1|\tilde{y}| - \beta\tilde{y}\,vxD_1$ and $\tilde{v}\,pyJ_2\tilde{\theta}_2 + \gamma_2 p|\tilde{v}| - \tilde{v}\,pyD_2$ in Eq. (23) are positive definite. Therefore, the time derivative of the Lyapunov function is obtained as

$$\dot{V}(t) \leq -\alpha_2\tilde{v}^2 - \alpha_1\tilde{y}^2 \leq 0 \tag{26}$$

Proposition. Based on the Lyapunov theorem (Slotine and Li, 1991), the tracking convergence $(\tilde{y} \to 0, \tilde{v} \to 0$ as $t \to \infty)$ and the system stability are proved by implementing the designed robust control strategy. This means that application of the proposed controller for the uncertain HBV infection provides us with the appropriately adjusted drug dosage (u_1 and u_2) to achieve the desired scenarios $(\tilde{y} \to y_{des}$ and $\tilde{v} \to v_{des})$.

Proof. The Lyapunov function proposed in Eq. (21) is positive definite ($V(t) > 0$) in terms of $\tilde{v}$ and $\tilde{y}$. Its time derivative is negative semidefinite ($\dot{V} \leq 0$); therefore, $V(t)$ is bounded. Accordingly, one can conclude that $\tilde{y}$ and $\tilde{v}$ remain bounded. In addition, the desired numbers of virions v_{des} and infected cells y_{des} are defined bounded; thus, the boundedness of $v = v_{des} + \tilde{v}$ and $y = y_{des} + \tilde{y}$ is also concluded. Now, the function $p(t)$ is defined as

$$p(t) = \alpha_1 \tilde{v}^2 + \alpha_2 \tilde{y}^2 \geq 0 \tag{27}$$

such that $\dot{V}(t) \leq -p(t) \leq 0$, whose integration gives

$$V(0) - V(\infty) \geq \lim_{t \to \infty} \int_0^t p(\eta) d\eta \tag{28}$$

Considering that $\dot{V} = dV/dt$ is negative semidefinite, ($V(0) - V(\infty)$) is positive and finite. Therefore, $\lim_{t \to \infty} \int_0^t p(\eta) d\eta$ has a finite positive value. Moreover, according to the Barbalat's lemma explained by Slotine and Li (1991), if $p(t)$ is uniformly continuous and the limit of integral $\lim_{t \to \infty} \int_0^t p(\eta) d\eta$ exists, then

$$\lim_{t \to \infty} p(t) = 0 \tag{29}$$

Based on Eq. (27) and considering that $p(t)$ tends to zero as time tends to infinity, it is concluded that

$$\lim_{t \to \infty} (\alpha_2 \tilde{v}^2 + \alpha_1 \tilde{y}^2) = 0 \tag{30}$$

It is known that $\alpha_1 > 0$ and $\alpha_2 > 0$ are nonzero positive constants, $\tilde{v}^2 \geq 0$ and $\tilde{y}^2 \geq 0$ are positive. As a result, Eq. (30) implies that $\tilde{v} \to 0$ and $\tilde{y} \to 0$ as $t \to \infty$. Consequently, the numbers of virions and infected cells converge to their corresponding desired values ($v \to v_{des}$ and $y \to y_{des}$). If the desired scenario for the reduction of the virions and infected cells are defined such that $v_{des} \to 0$ and $y_{des} \to 0$ as $t \to \infty$; then, $v \to 0$ and $y \to 0$ as $t \to \infty$ due to the convergence of $v \to v_{des}$ and $y \to y_{des}$.

Now, for the evaluation of the uninfected cells behavior, Eq. (4) can be rearranged as

$$\frac{dx}{dt} = \lambda - x(d - (1 - u_1)\beta v - D_1 \beta v) \tag{31}$$

Considering the convergence of $v \to 0$, the term $(d - (1 - u_1)\beta v - D_1 \beta v)$ in Eq. (31) converges to d as $t \to \infty$. Therefore, the dynamics of uninfected (healthy) cells approaches to

$$\dot{x} \to \lambda - dx \tag{32}$$

Since $\lambda \gg dx$ (according to the values of λ and d in Table 1 and the maximum of value of x), Eq. (32) implies that the population of uninfected cells increases during the treatment time which is favorable from the clinical point of view.

5 Numerical results

In this section, in order to verify the capability of the proposed controller, some simulations are conducted considering different desired scenarios and various levels of uncertainties. A wide range of uncertainties is taken into account as the real dynamics may differ from the nominal model identified for a specific person. Accordingly, three different levels of uncertainty are considered for all model parameters (introduced in Table 1). Moreover, some disturbances with three different frequencies are assumed to exist in the dynamics of all state variables as the unstructured (nonparametric) uncertainty. In the following simulations, two treatment scenarios (which differ in their treatment rate) are studied as the desired physician's objectives for the reduction of hepatitis B viruses and infected cells. The time interval between these treatment scenarios is chosen to be 80 days. The MATLAB-Simulink environment is employed in order to perform simulations by implementing the designed controller on the uncertain process.

5.1 Untreated HBV infection

Fig. 2A–C illustrates the behavior of the HBV state variables in the absence of treatment (without drug usage). In other words, the controller inputs u_1 and u_2 are considered to be zero in Eqs. (1)–(3), similar to the original model proposed by Nowak et al. (1996). The nominal dynamic parameters (presented in Table 1) are employed without any uncertainty and disturbances (D_1 and D_2). The initial conditions for the states of the HBV infection (x and y at $t = 0$) are considered to be the same as ones presented by Sheikhan and Ghoreishi (2013a) and represented in Table 2. Note that the time unit for all reported results is a day.

From Fig. 2A–C for untreated HBV infection, the trend of healthy cells population is decreasing. In contrary, the population of infective cells and virions increase, which means that the HBV disease develops in the course of time. The HBV disease

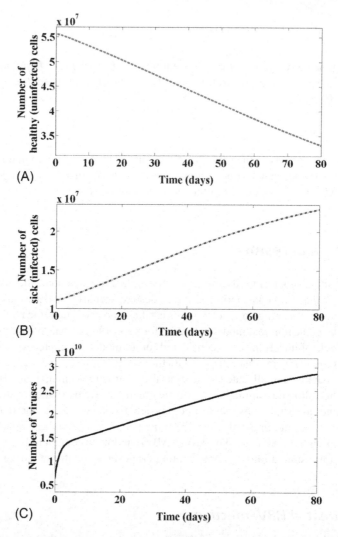

Fig. 2 (A) Number of healthy cells, (B) number of sick cells, and (C) number of viruses, without treatment during 80 days.

Table 2 Initial conditions of the HBV dynamic states (Sheikhan and Ghoreishi, 2013a).

Parameter	Value
$x(0)$	5.5556×10^7 cells
$y(0)$	1.1111×10^7 cells
$v(0)$	6.3096×10^5 copies mL^{-1}

development may lead to the patient's death in the absence of treatment. Therefore, it is necessary to provide an effective treatment in order to prevent the disease development.

5.2 Treated HBV infection using the proposed robust strategy

5.2.1 Desired scenarios

In order to treat the HBV infection, two descending desired scenarios are defined for the number of infected cells and virions which are tracked using the designed controller. Initial values of the desired scenarios are set according to the initial states of the patient (initial populations of infected cells and virions). Employing these treatment scenarios, the proposed controller performance is evaluated and corresponding results are presented in the following sections.

The functionality of the desired scenario denoted by v^i_{des} and y^i_{des} for viruses and infected cells are defined with respect to time as

$$v^i_{des} = (v_0 - v_f)\exp(-g_{1i} \times t) + v_f \tag{33}$$

$$y^i_{des} = (y_0 - y_f)\exp(-g_{2i} \times t) + y_f \tag{34}$$

In Eqs. (33), (34), v_t and y_t are the final desired values of exponentially decaying v^i_{des} and y^i_{des}, respectively, which are assumed to be zero. v_0 and y_0 are the initial conditions of the dynamic states (v and y). g_{1i} and g_{2i} are appropriate rates for population reduction of the virions and infected cells, respectively. Two cases of these rates for the desired treatment scenarios are considered for Eqs. (33), (34) as

$$v^1_{des}, y^1_{des} : g_{11} = g_{21} = 6.5 \tag{35}$$

$$v^2_{des}, y^2_{des} : g_{12} = g_{22} = 5.5 \tag{36}$$

Due to Eqs. (33)–(36), the first desired scenario ($i = 1$) has a steeper reduction in comparison with the second one ($i = 2$). The physician can choose appropriate reduction rates (g_1 and g_2) depending on the patient's conditions and the HBV disease progression.

5.2.2 Results of the first desired scenario

In this section, simulation results of the designed controller using the first desired scenario (Eqs. 33–35) for the HBV treatment are presented. The initial conditions of the states are listed in Table 2. Furthermore, different levels of parametric uncertainties are taken into account (-40%, -10%, and 30% of nominal parameter magnitudes listed in Table 1) in order to evaluate the robustness of the proposed controller. Moreover, in these simulations, the unstructured (nonparametric) uncertainties or disturbances (and in Eqs. 4–6) are considered as

$$D_1 = k_{11}\sin(2\pi \times t) + k_{12}\sin\left(2\pi \times \frac{t}{7}\right) + k_{13}\sin\left(2\pi \times \frac{t}{30}\right) \tag{37}$$

$$D_2 = k_{21} \sin\left(2\pi \times t\right) + k_{22} \sin\left(2\pi \times \frac{t}{7}\right) + k_{23} \sin\left(2\pi \times \frac{t}{30}\right) \tag{38}$$

where the time unit is day. Frequencies of the functions in Eqs. (37), (38) are chosen such that they represent daily, weekly, and monthly fluctuations for unmodeled dynamics (disturbance). The amplitude of the functions are k_{ij} with the same order as the control inputs. The employed coefficients for above-mentioned disturbances are considered $k_{11} = k_{12} = k_{13} = 0.1$ and $k_{21} = k_{22} = k_{23} = 0.3$, in this section. The controller gains are also chosen as $\alpha_1 = 8$, $\alpha_2 = 10$, $\gamma_1 = 0.050$, and $\gamma_2 = 0.045$ using a trial and error method in order to achieve the appropriate tracking and robustness performance for the treatment process.

Fig. 3A–C depicts the states (uninfected and infected cells, and virions) of the controlled HBV dynamics with respect to the time, employing the first desired scenario and considering three different percentages of parametric uncertainties.

It is observed that the number of healthy cells increases as time elapses, regardless of how much parametric uncertainty is considered in the model dynamics. As shown in Fig. 3B, the populations of infected cells and virions are controlled to track their desired scenarios (v_{des}^1 and y_{des}^1 in Eqs. 33–35), in the presence of different parametric uncertainty levels. As seen, the number of infected cells decreases to almost 10% of its initial value after 80 days of treatment. Even though different parametric uncertainty ranges change the behavior of closed-loop HBV dynamics slightly and despite the existence of external disturbances, the overall performance of the controller is favorable.

In order to investigate the tracking performance of the controller, normalized tracking errors of the infected cells and virions populations (($y - y_{des}^1)/y$ and ($v - v_{des}^1)/v_0$) are plotted in Fig. 4A and B. As observed, the tracking errors decrease after overshoots and the maximum errors are less than 4% for infected cells and less than 5% for virions. Thus, the first desired scenario could be tracked well using the proposed robust control strategy.

The control inputs (drug usage u_1 and u_2) of the HBV model are shown with respect to the time in Fig. 5A and B for three levels of parametric uncertainty. It is observed that the control inputs are in the allowable range (between 0 and 1). These inputs (u_1 and u_2) are adjusted based on the control laws (9), (10) in terms of the tracking errors of infected cells ($\tilde{y}$) and virions ($\tilde{v}$), respectively. Although the drug usage in initial days has some differences for various uncertainties (Fig. 5A and B), the control objective, which tracks the desired treatment scenario, is satisfied (Fig. 3A–C) due to the robustness of the controller. Therefore, the proposed control strategy prescribes different time-dependent drug usages according to differences in patient's characteristics (considered as parametric uncertainty in the dynamic model).

5.2.3 Effect of reduction rate on desired treatment scenario

In this section, the proposed strategy is evaluated for the second desired scenario (Eqs. 33, 34, and 36) and the corresponding results are represented in comparison with the first scenario. Initial conditions of the dynamic states are the same as ones

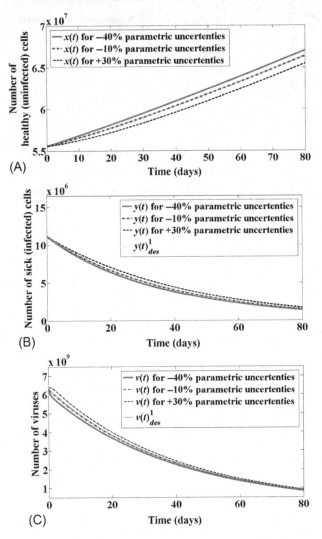

Fig. 3 (A) Number of healthy cells, (B) number of infected cells in comparison with its desired value, and (C) number of virions in comparison with its desired value, within 80 days of treatment period, employing the first desired treatment scenario and considering − 40%, − 10%, and + 30% of parametric uncertainties.

used in the previous section (presented in Table 2). Also, the disturbance coefficients are considered similar to those mentioned in Section 5.2.2 as $k_{11} = k_{12} = k_{13} = 0.1$ and $k_{21} = k_{22} = k_{23} = 0.3$ which are used in Eqs. (37), (38) as the nonparametric (unstructured) uncertainties. The robust gains of the control laws (9), (10) are adjusted as $\gamma_1 = 0.050$ and $\gamma_2 = 0.045$ using a trial and error method such that the tracking

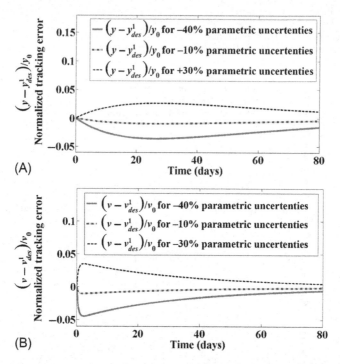

Fig. 4 Normalized tracking errors for populations of (A) infected cells and (B) virions using the first desired treatment scenario and different levels of parametric uncertainty and nonparametric uncertainty.

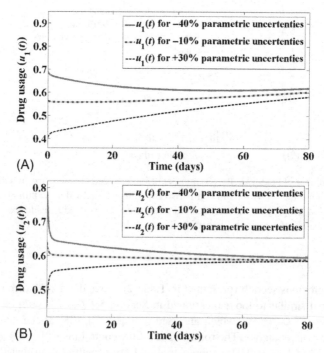

Fig. 5 Drug usage (A) u_1 and (B) u_2 as the control inputs for tracking the first desired scenario in the presence of different levels of parametric uncertainty.

convergence is achieved by satisfying inequalities (24), (25). In addition, -25% parametric uncertainty is taken into account in these simulations. The other controller gains are considered the same as ones ($\alpha_1 = 8$ and $\alpha_2 = 10$) used in the previous section for the first scenario.

Fig. 6A–C depicts the number of healthy cells, infected cells, and virions during the HBV treatment using both the first and second desired scenarios. As observed, the

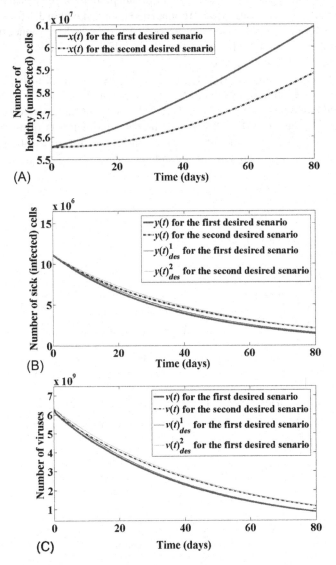

Fig. 6 (A) Number of healthy cells, (B) number of infected cells, and (C) number of virions for the first and second desired scenarios, within 80 days of treatment period.

number of infected cells and virions track their corresponding desired values and have steeper reductions in the first treatment scenario in comparison with the second one. Moreover, the number of healthy cells increases much more in the first desired scenario compared to the second desired scenario. As shown in Fig. 6B and C, regardless of the uncertainties and treatment rates, the desired tracking convergence ($y \rightarrow y_{des}$ and $v \rightarrow v_{des}$) is achieved as proved in Section 4.

The normalized tracking errors for the number of infected cells (y) and virions (v) with respect to their desired values (y_{des} and v_{des}) are plotted for 80 days in Fig. 7A and B. As observed these tracking errors are small and bounded within 3% for the infected cells and 4% for the virions population. Accordingly, despite the existence of different treatment rates, the control strategy successfully fulfilled the HBV treatment objectives by tracking desired scenarios.

The control inputs u_1 and u_2 as the drug usage are illustrated in Fig. 8A and B, which are bounded in the allowable range (between 0 and 1). These drug regulations are obtained based on the control laws (Eqs. 9, 11) in terms of the dynamic states and their errors. As shown in Fig. 8A and B, higher control inputs (drug usages) are required for the first scenario that has larger rates of reduction for the virions and infected cells in comparison with the second desired scenario (see Eqs. 33–36). This

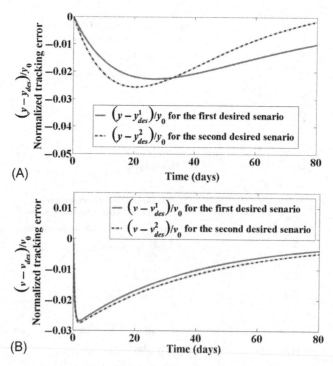

Fig. 7 Normalized tracking errors for (A) the number of infected cells and (B) the number of virions considering both the first and second desired scenarios.

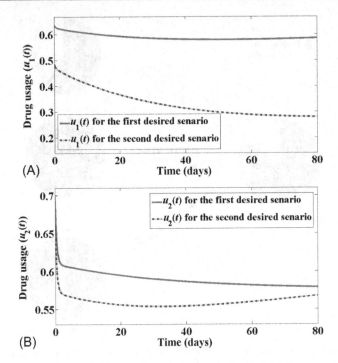

Fig. 8 Drug usage (A) u_1 and (B) u_2 as the control inputs for tracking the second desired scenario in the presence of different disturbance levels.

is a natural behavior of the HBV disease dynamics that requires even more drugs for faster treatments.

The above-mentioned results confirm that the proposed strategy can robustly control the HBV disease in the presence of different parametric and nonparametric (unstructured) uncertainties and treatment rates. In order to briefly illustrate the treatment outcome using the proposed robust control strategy at a glance, the initial and final percentages of infected (sick) and uninfected (healthy) cells are shown in Fig. 9. Initial contributions of the infected and uninfected cells obtained from Table 2 are demonstrated in Fig. 9A. The final percentages of different cells are represented in Fig. 9B without any treatment, after an 80-day interval. Fig. 9C and D illustrates the cell percentages for 80 days of treatment using the first and second desired scenarios, respectively. These simulation results are described in detail in Sections 5.2.2 and 5.2.3.

5.2.4   Discussion and interpretation of the results (comparison of untreated and treated HBV using two scenarios)

The above-mentioned results confirm that the proposed strategy can robustly control the HBV disease in the presence of different parametric and nonparametric (unstructured) uncertainties and treatment rates. In order to briefly illustrate the treatment outcome using the proposed robust control strategy in a glance, the initial and final

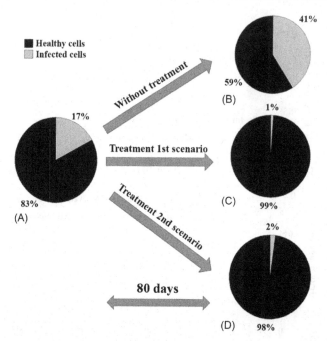

Fig. 9 Proportions of healthy cells and infected cells for (A) initial state and final state (B) without treatment, (C) with the first treatment scenario and (D) with the second treatment scenario, after 80 days.

percentages of infected (sick) and uninfected (healthy) cells are shown in Fig. 9. Initial contributions of the infected and uninfected cells obtained from Table 2 are demonstrated in Fig. 9A. The final percentages of different cells are represented in Fig. 9B without any treatment, after an 80-day interval. Fig. 9C and D illustrates the cell percentages for 80 days of treatment using the first and second desired scenarios, respectively. These simulation results are described in detail in Sections 5.2.2 and 5.2.3.

As shown in Fig. 9 the HBV treatment on employing the proposed control strategy leads to a drastic decrease in infected cell percentage to less than 1% and 2% using the first and second scenarios, respectively. This achievement can be interpreted as the successful cure of the HBV infection. However, the infected cell percentage grows to 41% in the absence of treatment during the same time period (80 days). This implies the necessity of controlling the HBV spread using an appropriate drug regulation.

5.3 Limitation of the study

Although this study has been proposed according to the reliable facts, it is still hard to implement this method on a human being without precautions. To employ the proposed method on the patients, it is recommended to use it on the laboratory samples at the beginning.

6 Conclusion

In this study, an efficient robust control strategy was presented for the treatment of the HBV infection with nonlinear dynamics. This strategy was developed such that the HBV treatment is achieved considering parametric uncertainties and disturbances (nonparametric uncertainties). The proposed controller reduces the virions and infected cell populations by tracking their prespecified desired scenarios while the population of uninfected cells grows persistently. This objective was achieved by prescribing an appropriate time-varying drug usage by employing the proposed feedback control laws. The tracking convergence to the desired scenario and the stability of closed-loop dynamics were proved using the Lyapunov theorem and Barbalat's Lemma.

In the conducted simulations, two rates were considered for exponential reduction in the virions and infected cells in desired treatment scenarios. It was observed that both desired scenarios could be tracked in a reasonable fashion using the proposed strategy. However, even more drug usages were required for the first scenario that has larger rates of virus reduction (steeper treatment) in comparison with the second desired scenario. The performance of the controller was also evaluated in the presence of different levels of parametric uncertainties and disturbances. These uncertainties may have resulted from a variety of genetics, ages, life styles, and environmental factors for different patients in realistic treatments. It was shown that the controller can robustly adapt to different levels of uncertainties and disturbances by adjusting the control inputs (the drug usage) to achieve the treatment objective.

6.1 Future directions of research

In future works, developed robust controller can be extended to a discretized form in order to be employed in realistic clinical treatments. In this condition, physicians can prescribe daily and weekly drug schedules for their patients dealing with the HBV infection. Moreover, the proposed nonlinear robust control strategy can be structured and applied for the treatment of other diseases with either linear or nonlinear mathematical model.

References

Aghajanzadeh, O., Sharifi, M., Tashakori, S., Zohoor, H., 2017. Nonlinear adaptive control method for treatment of uncertain hepatitis B virus infection. Biomed. Signal Process. Control 38, 174–181.

Aghajanzadeh, O., Sharifi, M., Tashakori, S., Zohoor, H., 2018. Robust adaptive Lyapunov-based control of hepatitis B infection. IET Syst. Biol. 12 (2), 62–67.

Blayneh, K., Cao, Y., Kwon, H.-D., 2009. Optimal control of vector-borne diseases: treatment and prevention. Discrete Contin. Dyn. Syst. B 11 (3), 587–611.

Ciupe, S.M., Ribeiro, R.M., Nelson, P.W., Perelson, A.S., 2007. Modeling the mechanisms of acute hepatitis B virus infection. J. Theor. Biol. 247 (1), 23–35. https://doi.org/10.1016/j.jtbi.2007.02.017.

De Souza, J.A.M.F., Caetano, M.A.L., Yoneyama, T., 2000. Optimal control theory applied to the anti-viral treatment of AIDS. In: Proceedings of the 39th IEEE Conference on Decision and Control, vol. 5, pp. 4839–4844.

DebRoy, S., Kribs-Zaleta, C., Mubayi, A., Cardona-Meléndez, G.M., Medina-Rios, L., Kang, M., Diaz, E., 2010. Evaluating treatment of hepatitis C for hemolytic anemia management. Math. Biosci. 225 (2), 141–155.

Feitelson, M.A., Larkin, J.D., 2001. New animal models of hepatitis B and C. ILAR J. 42 (2), 127–138.

Ganem, D., Knipe, D.M., Howley, P.M., Griffin, D.E., Lamb, R.A., Martin, M.A., Roizman, B., et al., 2001. Fields Virology, vol. 1. EUA, Philadelphia, PA.

Gumel, A.B., Moghadas, S.M., 2004. HIV control in vivo: dynamical analysis. Commun. Nonlinear Sci. Numer. Simul. 9 (5), 561–568.

Hattaf, K., Rachik, M., Saadi, S., Tabit, Y., Yousfi, N., 2009a. Optimal control of tuberculosis with exogenous reinfection. Appl. Math. Sci. 3 (5), 231–240.

Hattaf, K., Rachik, M., Saadi, S., Yousfi, N., 2009b. Optimal control of treatment in a basic virus infection model. Appl. Math. Sci. 3 (17–20), 949–958.

Kumar, A., Srivastava, P.K., 2017. Vaccination and treatment as control interventions in an infectious disease model with their cost optimization. Commun. Nonlinear Sci. Numer. Simul. 44, 334–343.

Laarabi, H., Abta, A., Rachik, M., Bouyaghroumni, J., 2013. Optimal antiviral treatment strategies of HBV infection model with logistic hepatocyte growth. ISRN Biomath. https://doi.org/10.1155/2013/912835.

Lai, C.-L., Yuen, M.-F., 2007. The natural history and treatment of chronic hepatitis B: a critical evaluation of standard treatment criteria and end points. Ann. Intern. Med. 147 (1), 58–61.

Landi, A., Mazzoldi, A., Andreoni, C., Bianchi, M., Cavallini, A., Laurino, M., Ricotti, L., Iuliano, R., Matteoli, B., Ceccherini-Nelli, L., 2008. Modelling and control of HIV dynamics. Comput. Methods Prog. Biomed. 89 (2), 162–168.

Ledzewicz, U., Schattler, H., 2002. On optimal controls for a general mathematical model for chemotherapy of HIV. In: Proceedings of the 2002 American Control Conference, vol. 5, pp. 3454–3459.

Lee, J., Kim, J., Kwon, H.-D., 2013. Optimal control of an influenza model with seasonal forcing and age-dependent transmission rates. J. Theor. Biol. 317, 310–320.

Lv, C., Huang, L., Yuan, Z., 2014. Global stability for an HIV-1 infection model with Beddington-DeAngelis incidence rate and CTL immune response. Commun. Nonlinear Sci. Numer. Simul. 19 (1), 121–127.

Min, L., Su, Y., Kuang, Y., 2008. Mathematical analysis of a basic virus infection model with application to HBV infection. Rocky Mt. J. Math. 38, 1573–1585.

Moradi, H., Sharifi, M., Vossoughi, G., 2015. Adaptive robust control of cancer chemotherapy in the presence of parametric uncertainties: a comparison between three hypotheses. Comput. Biol. Med. 56, 145–157.

Nowak, M.A., Bonhoeffer, S., Hill, A.M., Boehme, R., Thomas, H.C., McDade, H., 1996. Viral dynamics in hepatitis B virus infection. Proc. Natl. Acad. Sci. USA 93 (9), 4398–4402.

Ntaganda, J.M., Gahamanyi, M., 2015. Fuzzy logic approach for solving an optimal control problem of an uninfected hepatitis B virus dynamics. Appl. Math. 6 (9), 1524.

Okosun, K.O., Makinde, O.D., 2014. A co-infection model of malaria and cholera diseases with optimal control. Math. Biosci. 258, 19–32.

Pinto, C.M.A., Carvalho, A., 2014. Mathematical model for HIV dynamics in HIV-specific helper cells. Commun. Nonlinear Sci. Numer. Simul. 19 (3), 693–701.

Ribeiro, R.M., Lo, A., Perelson, A.S., 2002. Dynamics of hepatitis B virus infection. Microbes Infect. 4 (8), 829–835.

Sharifi, M., Sayyaadi, H., 2015. Nonlinear robust adaptive Cartesian impedance control of UAVs equipped with a robot manipulator. Adv. Robot. 29 (3), 171–186.

Sheikhan, M., Ghoreishi, S.A., 2013a. Antiviral therapy using a fuzzy controller optimized by modified evolutionary algorithms: a comparative study. Neural Comput. Appl. 23 (6), 1801–1813.

Sheikhan, M., Ghoreishi, S.A., 2013b. Application of covariance matrix adaptation-evolution strategy to optimal control of hepatitis B infection. Neural Comput. Appl. 23 (3–4), 881–894.

Slotine, J.J.E., Li, W., 1991. Applied Nonlinear Control. Prentice Hall. ISBN 9780130408907 Available from: https://books.google.com/books?id=cwpRAAAAMAAJ.

Su, Y., Sun, D., 2015. Optimal control of anti-HBV treatment based on combination of traditional Chinese medicine and western medicine. Biomed. Signal Process. Control 15, 41–48.

Wang, K., Wang, W., 2007. Propagation of HBV with spatial dependence. Math. Biosci. 210 (1), 78–95.

World Health Organisation, 2017. HBV. http://www.who.int/mediacentre/factsheets/fs282/fr/ (Accessed 11 November 2017).

Yazdi, E.A., Nagamune, R., 2010. Multiple robust H_∞ controller design using the nonsmooth optimization method. Int. J. Robust Nonlinear Control 20 (11), 1197–1212.

Yosyingyong, P., Viriyapong, R., 2018. Global stability and optimal control for a hepatitis B virus infection model with immune response and drug therapy. J. Appl. Math. Comput. https://doi.org/10.1007/s12190-018-01226-x.

A closed loop robust control system for electrosurgical generators

NasimUllah[a], Muhammad Mohsin Rafiq[b], M. Ishfaq[c], Mumtaz Ali[d], Asier Ibeas[e], Jorge Herrera[f]
[a]Department of Electrical Engineering, College of Engineering, Taif University, Taif, Kingdom of Saudi Arabia, [b]Aprus Technologies Pvt Ltd, Peshawar, Pakistan, [c]South Al Qunfudhah General Hospital, Al Qunfudhah, Saudi Arabia, [d]Neurosurgery Department LRH, Peshawar, Pakistan, [e]Escola d'Enginyeria, Autonomous University of Barcelona, Barcelona, Spain, [f]Departamento de Ingeniería, Universidad de Bogotá Jorge Tadeo Lozano, Bogotá, Colombia

1 Introduction

Electrosurgery is a term referred for utilizing high-power and high-frequency alternating current (AC) signal to bring the tissues' temperature up to a level so that due to the process of vaporization drying up or coagulation is achieved. As a result of the phenomena cutting of tissues, hemostasis, blocking of lumen-containing structures, and removal of huge volumes of tissue is achieved. Radiofrequency electrosurgery technology differs from the commonly used cautery devices. The cautery devices utilize the direct current (DC), while the ESU utilizes the AC power for heating of the tissues. The ESU units are utilized for cutting or to stop bleeding by causing coagulation (hemostasis). The target area of the tissues receives thermal energy from the tip of the electrode due to which drying up, vaporization, and scorching of the tissues is achieved.

The concept of electrosurgery has emerged since the start of the 19th century. The first electrosurgical generator was introduced by Bovieas in 1928. With this invention, the utilization and proof of concept of the high-frequency and high-power AC current in surgical procedures was proved (Bovie and Cushing, 1928). The basic working idea was to heat the tissues in an area of focus using the radio-frequency high-power alternating current signals applied through active electrodes, which is driven by an inverter unit. As a result, several functionalities, including cutting, removal of tissues, and altering of blood flow across the damaged tissues, are achieved (Webste, 1992). In order to have safe electrosurgery, proper handling of the ESU unit is vital and the operators must be well trained.

The neural cells show excitation and stimulated behavior when it is exposed to electric fields or current passes through it. However, this excitation is observed at low signal frequencies. The excitation phenomena triggered in muscles and neural cells at low frequency currents can cause adverse effects in patients such as pain, heart

Control Applications for Biomedical Engineering Systems. https://doi.org/10.1016/B978-0-12-817461-6.00006-8

failure, etc. The cell membranes are associated with voltage-gated ion channels. Due to the presence of the voltage-gated ion channels, the neural cells show sensitivity to the electric field or current. Typically, least normal value for neural cells frequency is in the range of 0.1–10 ms. Thus, the operating frequency of the electrosurgical generators is chosen in the range of radiofrequency (RF) range of 100 kHz to 5 MHz (Polk, 1992).

The interaction of the radiofrequency high-power currents with the human tissues changes its impedance; thus, the impedance offered at the output terminal of ESU unit is variable and uncertain. The specified range of tissue impedance is between 0 Ω to 4k Ω. Due to this variation, the ESU unit must be designed in such a way as to incorporate the uncertainty of the tissue impedance and be able to deliver power in specified limit. Besides the power signal, maximum limit for current and voltage signals also needs to be specified (Dodde et al., 2008, 2012). The current and voltage limits are applied so that the thermal damage of the tissues is prevented (Gerhard, 1984). As a result of electrosurgery, the after effects of warmth development inside the tissue result in its thermal damage. Thus, the device must be operated in controlled manner. There are some issues such as scorch at the return electrode side and surgical flames and these have been reported in the literature (Gardner, 1994; Schneider et al., 2010). All the operators must be given training on health safety in order to avoid serious injuries as a result of these issues.

In order to operate the ESU unit in an efficient way, it is necessary to operate the inverter and converters units in close loop manner. To achieve the desired output impedance characteristics, the average power of ESU machines is measured in several cycles and the converters duty cycle is controlled using low-bandwidth controller (Thompson, 2005; Becker and Klicek, 1998; Pearce, 1986). A control method using adaption of peak current mode is reported by (Erickson and Maksimovic, 2004; Tan and Middlebrook, 1995). To regulate the fixed frequency power, the switching waveforms are generated by comparing the inductor current to its reference value. However, these controllers lead to instability when the duty cycle increases more than 50% (Shambroom, 2006). In addition, Palanker et al. (2008) reported the similarity of human tissue to the chicken or porcine; therefore, in the R&D stage and for calibrations, the ESU machine can be tested on chicken tissues before testing it on the actual patient.

ESU devices are commonly built using the resonant inverter configuration (Jensen et al., 2011). The design proposed in Jensen et al. (2011) has a simple architecture and the proposed design has the capability to naturally limit the output current and the voltage of the ESU device. However, the major disadvantage of the resonant inverter-based topology includes the lack of flexibility for shaping the frequency of the output power over a wide range of operations (Erickson and Maksimovic, 2001). In order to address the problem, a new design was proposed by Jensen et al. (2015) using the GaN fast switches. The new proposed design has the flexibility to adjust the output frequency of the power signal over a wide range of operations. Friedrichs et al. (2012) proposed a buck-boost converter-based topology for the ESU device, which is able to minimize the steady-state error between the instantaneous power and average power with square wave output.

Apart from the proposed topologies, closed loop control plays crucial role to accurately adjust the output power of the ESU unit. In the discussed literature, only classical control system has been reported for ESU devices (Friedrichs et al., 2012; Jensen et al., 2011, 2015). As the ESU device is used in surgical procedures, its control performance must not degrade with variable tissue impendence and other applied disturbances. In order to ensure minimal thermal damage to the tissues of the patient, the control loop must be robust to regulate the output power, voltage, and current signals to the respective desired values. Fractional order control offers additional benefits over the conventional integer-order control system in terms of high degree of freedom, robustness against noise, and enhanced stability margins (Ullah et al., 2017a, b, c). Some fractional-order sliding mode controllers have been proposed in the literature such as for the DFIG-based wind energy system (Ullah et al., 2017a, b, c), static series synchronous compensator (Ullah et al., 2017a, b, c), uncertain system (Ullah et al., 2015a, b), and aerospace actuation system (Ullah et al., 2015a, b). Apart from the additional advantages of the fractional order control, the implementation of the fractional systems over the hardware platform is still an open research problem. The implementation of the fractional order control system on FPGA platform is discussed in detail in Ullah et al. (2017a, b, c). Based on the foregoing literature survey, this chapter is focused on the derivation of the mathematical model for the ESU unit. Based on the derived model, nonlinear fractional-order control system based on sliding mode concepts is derived for the voltage and current loops of the ESU device, respectively. Finally, the ESU device is simulated for constant power mode with both fractional-order and integer-order controllers.

The rest of the chapter is organized as following. Section 2 explains basic working principle and design specifications of the ESU unit, Section 3 discusses mathematical modeling, Section 4 shows controller formulation, and Section 5 discusses the results. Finally, the chapter is concluded in Section 6.

2   Working and design specifications of electrosurgical unit

Basic working diagram of the electrosurgical unit is shown in Fig. 1. The output of the high-frequency and high-power inverter forms an arc between the output probe of ESG unit and the patient; thus, it introduces joule heating effect. As shown in Fig. 1, the current flows from the output of the inverter through the probe. For the return current, a plate is placed below the back of the patient. The patient tissues offer impedance as a result of flow of electrons at high frequency produces heating phenomena across it. The surgical procedures are done by passing the charges through the tissues in controlled manner.

Fig. 2 shows the electrical power stages of the ESU unit. Generally, the ESU unit consists of three stages. The first stage is the rectification stage, which is used to convert the AC power to DC power. This stage is omitted in Fig. 2. In the second stage, a buck converter is used to step down the magnitude of the DC-rectified voltage signal. The third stage consists of the push pull inverter, which is used to convert the DC

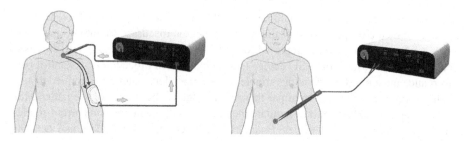

Fig. 1 Working diagram of electrosurgical unit.

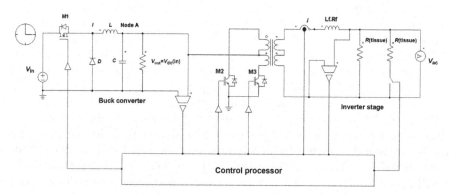

Fig. 2 Power stages of electrosurgical unit.

signal to high-power and high-frequency AC power signal. The output of the inverter is then connected with the probe through a high frequency transformer, which interacts with the human tissues for surgical procedures. Apart from the converters topologies, different sensors such as output voltage and current sensors are utilized for measuring the output quantities of the inverter unit. The power of the device is indirectly calculated from the output voltage and current measurement. The sensors require outputting RMS values of the quantities measured in order to calculate the RMS power of the ESU unit. Depending on the mode of operation, either of three measured signals is used as feedback to the controller board for operation of the unit in closed loop manner. The main controller board is used to read feedback signals from both the buck converter and the inverter circuit and based on it controls output power of the inverter.

The input voltage source of the ESU unit is AC. The AC source is step down to 24 V and then rectified to 24-V DC. The 24-V DC is input to the buck converter, while the output voltage and current of the buck converter are regulated at levels of 11.11 V and 11 A, respectively.

Fig. 3 shows the ideal output characteristics of the electrosurgical unit. The ideal output characteristics graph is a function of the output voltage and current of the inverter unit. The device is operated in three modes, including constant current, constant voltage, and constant power modes. Constant voltage mode imposes restriction on the

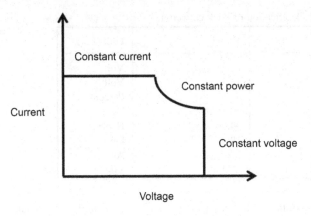

Fig. 3 Output characteristics of electrosurgical unit.

maximum output voltage of the inverter unit, which in turns limits the arc length, which is usually formed between the ESU unit and the patient due to the interaction of high impedance tissues at high powers. Similarly, the constant current mode limits the maximum output current of the inverter unit, while in the constant power mode; the current and voltage of the unit are changing such that the product of the two quantities is always constant within the upper limits. Generally speaking, in order to ensure the reliable and safe surgical procedures, the ESU unit must be intelligently programmed so that the device can switch automatically between all the three regions based on the patient tissue measurement. As explained earlier, the constant voltage mode is used when the device senses high impedance tissues of the patient. Constant current mode is used when the tissue impedance is smaller. Forgoing in view, the constant current mode region represents the low impedance range, constant voltage mode represents the high impedance range and for constant power range, the tissue impedance is medium.

Apart from the above explanations, it is necessary to understand the design specifications of the different stages of the ESU unit. As shown in Fig. 2, the first stage of the ESU unit represents the buck converter and the design specifications of its power stage elements, i.e., (inductor and capacitor) are crucial to understand.

The inductor is an inertial element, which is used to store and pump energy to other components connected with buck converter. The voltage drop across the inductor is given as

$$V_L = L \frac{dI}{dt} \tag{1}$$

In Eq. (1), L represents the self-inductance, V_L is the voltage across inductor, and I represents the current flowing through it. For the buck converter, the inductance is specified as following.

$$L = \frac{V_{out} \times (V_{in} - V_{out})}{\Delta I \times f_s \times V_{in}} \tag{2}$$

Table 1 Parameter specification of buck converter

Parameter	Value	Unit
Input voltage	24	Volt
Output voltage	11.11	Volt
D (duty ratio)	46	Percent
P_{wm} frequency	50×10^3	Hz
L	54.24×10^{-6}	Henry
C	10×10^{-6}	Farad
Current ripple (I_{ripple})	2.2	Ampere
Voltage ripple (V_{ripple})	0.5556	Volt

In Eq. (2), V_{out} represents the output voltage, V_{in} is the input voltage, and f_s represents the switching frequency. Based on Eq. (2) and the current ripples given in Table 1, the calculated value of inductor at switching frequency of 50 kHz is $L = 52.24 uH$.

For the buck converter, the capacitance is calculated as following.

$$C = \frac{\Delta I}{8 * f_s * \Delta V_{out}} \tag{3}$$

Based on Eq. (3), switching frequency of 50 kHz and the voltage ripples tabulated in Table 1, the required amount of the capacitance is calculated as $C = 10 uF$. The maximum duty cycle of the buck converter is calculated as $d = \frac{11.1}{24} = 46.3\%$. The design specifications of the buck converter are tabulated in Table 1.

From Fig. 2, the second stage is the inverter unit. The design specifications of the inverter are explained as following. The inverter unit utilizes the DC power available at the output of the buck converter as input source. The inverter unit is operated at a switching frequency of 500 kHz. The maximum output voltage of the inverter is specified as 100 V. A high-frequency step-up transformer is connected at the output of the inverter to boost up the voltage output of the inverter. The maximum limit of the output current of the inverter is specified as 1.1 A. The design specifications of the push pull inverter unit are shown in Table 2.

3 Mathematical modeling of electro surgical unit

In this section, a detailed mathematical model of the ESU unit is derived. The derived mathematical model is later on utilized for the derivation of the robust control system. As shown in Fig. 2, the basic ESU device consists of two units. The first unit represents a buck converter and the second unit consists of a DC-AC converter.

Referring to Fig. 2, the switch M1 of the buck converter unit can be operated either in closed or open stage. When the switch M1 is closed, the diode D is reverse biased, and the inductor L and capacitor C will store energy in their respective fields, and it will form a series loop. In order to derive the dynamics of the converter with the switch M1 is closed, apply Kirchhoff's voltage law (KVL) and current law (KCL) to the

Table 2 parameters specification of the inverter unit

Parameter	Value
Input voltage	11.11 V
Output voltage	100 V
Efficiency	90%
n	10
Switching frequency of inverter	500 kHz
Output power	110 W
Input DC current	11 A
Input power	122 W
R_{min}	40 Ω
R_{max}	200 Ω
Reference command	50 W
At 40 Ω	$V_{in} = 5.05$ V, $I_{in} = 11$ A
At 200 Ω	$V_{in} = 11.11$ V, $I_{in} = 5$ A

series loop and node A, respectively. The resultant expressions are derived as following.

$$V_{in} = L\dot{I} + V_{out} \tag{4}$$

$$C\dot{V}_{out} = I - \frac{V_{out}}{R_L} \tag{5}$$

Referring to Fig. 2, Eqs. (4), (5), V_{in} and V_{out} represent the input DC voltage of the buck converter, L and C represent the inductance and capacitance of the power stage elements, I represents the inductor current, and R_L is the load resistance.

When the switch M1 is opened, the primary source of voltage V_{in} is disconnected from the circuit and the current flowing due to stored energy in the capacitor is circulated in the loop to make diode D as forward biased. The system's dynamics with switch M1 open state are expressed as

$$L\dot{I} = -V_{out} \tag{6}$$

$$C\dot{V}_{out} = I - \frac{V_{out}}{R_L} \tag{7}$$

All the parameters of Eqs. (6), (7) have already been defined in Eqs. (4), (5). By combining Eqs. (4)–(7), the Average model of the buck converter is derived as following.

$$L\dot{I} = uV_{in} - V_{out} \tag{8}$$

$$C\dot{V}_{out} = I - \frac{V_{out}}{R_L} \tag{9}$$

In Eq. (8), u represents the switching function and it can have two states, i.e., one or zero.

Output voltage of a single-phase inverter is expressed as following

$$V_{ac} = V_{DC(in)} \times m \times \sin(wt) \tag{10}$$

In Eq. (10), V_{ac} represents the root mean square (RMS) output voltage of the inverter, $V_{DC(in)}$ is the output DC voltage of the buck converter m represents the modulation index, and w is the switching frequency of the inverter in rad/sec. Before deriving the dynamics of the inverter, as a first step, the output of the inverter connected to the secondary of high-frequency transformer is explained in detail. A series inductor filter having inductance L_f is connected between the secondary of the transformer and the ESU probe. This filter is used to remove the undesired ripples in the output power signal, which is applied to the tissues of the patient. Referring to Fig. 2, alternating current i flows from the secondary side of the transformer through the filter having small resistance R_f to the tissues. By applying KVL to the secondary loop of the inverter unit, one obtains the following expression.

$$L_f \frac{di}{dt} = V_{sec} - i(R_f + R_{tissues}) \tag{11}$$

In Eq. (11), L_f represents the total inductance of the filter and transformer, i is the ac current, R_f represents the resistance of the filter unit, and V_{sec} represents the secondary voltage of the high-frequency transformer. With turn ratio n and control switching function u_{inv}, Eq. (11) is expressed in modified form as following.

$$L_f \frac{di}{dt} = n u_{inv} V_{DC(in)} - i(R_f + R_{tissues}) \tag{12}$$

In Eq. (12), $V_{DC(in)}$ represents the input DC voltage at the input of the inverter unit and u_{inv} represents the switching function and it can have the value of $+1$ and -1 for half time period of the switching signal. In the next section, the controllers are derived using Eqs. (9), (12). It is worth to mention that for automatic operation of the ESU unit the tissue resistance in Eq. (12) is either estimated or calculated online based on the voltage and current measurements. Similarly, the operator can manually input the tissue resistance to the main controller board for the mode selection and updating Eq. (12).

4 Controller formulation for electro surgical unit

This section derives the robust fractional order and integer order controllers for both the buck converter and the inverter units of the ESU device. Based on the reference power command, the reference voltage and reference current commands are calculated for buck converter and inverter units, respectively. The detailed power loop control block diagram is shown in Fig. 4.

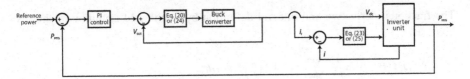

Fig. 4 Constant power control loop.

From the given diagram, the external loop represents the power loop with a proportional, integral (PI) control loop. Output of the power loop calculates the reference voltage command for the buck converter's controller loop governed by Eq. (20). Moreover, the output current of the buck converter is treated as input current command for the inverter unit. The inverter current control loop is derived in Eq. (23). Before presenting the detailed derivation, some basic definitions for fractional systems are included and are given here.

The fractional-order operator is symbolized by $a D_t^\alpha$ as generalization of integer order operator. It can be defined as following (Ullah et al., 2017a, b, c):

$$
aD_t^\alpha \cong D^\alpha = \begin{cases} \dfrac{d^\alpha}{dt^\alpha}, & R(\alpha) > 0 \\[2mm] 1, & R(\alpha) = 0 \\[2mm] \displaystyle\int_a^t (d\tau)^{-\alpha}, & R(\alpha) < 0 \end{cases}
\tag{13}
$$

Here a and t are the limits of operation, α is the order of fractional operator, and R is the real part of alpha. Eq. (13) represents the fractional-order derivative operator, when alpha is positive, and it represents the fractional-order integral operator when alpha is negative. The operator is equal to one if alpha is zero.

The fractional differential integral operator defined by Riemann Liouville definition is given as following (Ullah et al., 2017a, b, c):

$$
{}_aD_t^\alpha f(t) = \frac{d^\alpha}{dt^\alpha} f(t) = \frac{1}{\Gamma(m-\alpha)dt^m} \int_a^t \frac{f(\tau)}{(t-\tau)^{\alpha-m+1}} d\tau
\tag{14}
$$

$$
aD_t^{-\alpha} f(t) = I^\alpha f(t) = \frac{1}{\Gamma(\alpha)} \int_a^t \frac{f(\tau)}{(t-\tau)^{1-\alpha}} d\tau
\tag{15}
$$

In Eq. (14), $\Gamma(.)$ represents Euler's gamma function and can be defined as:

$$
\Gamma(x) = \int_0^\infty e^{-t} t^{(x-1)} dt, \quad x > 0
\tag{16}
$$

The fractional derivative and integral definition by Grunewald Letnikov method is given as:

$$a^{GL}D_t^\alpha f(t) = \lim_{h \to 0} \frac{1}{h^\alpha} \sum_{j=0}^{[(t-\alpha)/h]} (-1)^j \binom{\alpha}{j} f(t - jh) \tag{17}$$

To derive the voltage loop controller loop for buck converter, the following derivations are presented. Let a fractional-order voltage error surface S_v be defined as following.

$$S_v = c_1 e_v + c_2 D^{-\alpha} e_v \tag{18}$$

Referring to Fig. 4 and from Eq. (18), $e_v = V_r - V_{out}$ represents the voltage tracking error, c_1, c_2 are the constants, and V_r is the reference command generated from the power control loop. Differentiating Eq. (18) and combining it with Eqs. (8), (9), one obtains the following expression.

$$\dot{S}_v = c_1 \dot{V}_r - \frac{c_1 I}{C} + \frac{u c_1 V_{in}}{CR_L} - \frac{c_1 L \dot{i}}{CR_L} + c_2 D^{1-\alpha} e_v \tag{19}$$

Using Eq. (19), the voltage control law is derived as following

$$u = \frac{CR_L}{c_1 V_{in}} \left(-c_1 \dot{V}_r + \frac{c_1 I}{C} + \frac{c_1 L \dot{i}}{CR_L} - c_2 D^{1-\alpha} e_v - K_1 \operatorname{sgn}(S_v) \right) \tag{20}$$

The stability of the proposed controller is ensured by choosing the Lyapunov function $V_v = \frac{1}{2} S_v^2$. By combing the first derivative of the Lyapunov function with Eq. (20) and by letting $K_1 > 0$, the expression $\dot{V}_v \leq 0$.

Now to derive the current mode controller for single-phase inverter, the sliding surface is chosen as following.

$$S_i = c_3 e_i + c_4 D^{-\alpha} e_i \tag{21}$$

In Eq. (21) $e_i = i_r - i$ represents the current tracking error of the inverter unit, c_3, c_4 are the constants, and i_r is the reference command generated (output current of the buck converter). By differentiating Eq. (21) and combining it with Eq. (12) one obtains the following expression.

$$\dot{S}_i = c_3 \frac{di_r}{dt} - c_3 \frac{n}{L_f} u_{inv} V_{DC(in)} + c_3 \frac{(R_f + R_{tissues})}{L_f} i + c_4 D^{1-\alpha} e_i \tag{22}$$

Using Eq. (22), the control law is derived as following

$$u_{inv} = \left(c_3 \frac{n}{L_f} V_{DC(in)} \right)^{-1} \left(c_3 \frac{di_r}{dt} + c_3 \frac{(R_f + R_{tissues})}{L_f} i + c_4 D^{1-\alpha} e_i + K_2 \operatorname{sgn}(S_i) \right) \tag{23}$$

To prove the stability of the proposed current loop controller, the Lyapunov candidate function is chosen as $V_i = \frac{1}{2}S_i^2$. By combining the first derivative of the Lyapunov function with Eq. (23) and by letting $K_2 > 0$, the first derivative of the Lyapunov function is less than zero, i.e., $\dot{V}_i \leq 0$.

The derived fractional-order voltage and current controllers for buck converter and inverter unit are expressed in Eqs. (20), (23). In order to compare the performance of the proposed controllers with the integer-order robust control system, the fractional-order α used in the derivation of these controllers is set to unity. The simplified expressions for Eqs. (18), (20), (21), (23) represent the integer-order version of the proposed control system. The resultant integer-order voltage controller is given as following.

$$u = \frac{CR_L}{c_1 V_{in}}\left(-c_1\dot{V}_r + \frac{c_1 I}{C} + \frac{c_1 L\dot{I}}{CR_L} - c_2 e_v - K_1\,\mathrm{sgn}(S_v)\right) \tag{24}$$

Similarly, the integer-order current control system of the inverter unit is expressed as following.

$$u_{inv} = \left(c_3\frac{n}{L_f}V_{DC(in)}\right)^{-1}\left(c_3\frac{di_r}{dt} + c_3\frac{(R_f + R_{tissues})}{L_f}i + c_4 e_i + K_2\,\mathrm{sgn}(S_i)\right) \tag{25}$$

From the analysis, it was proved that the first derivative of the voltage and current Lyapunov functions is always equal to or less than zero, i.e., $\dot{V}_v \leq 0$ and $\dot{V}_i \leq 0$, which proves the reaching condition of the sliding surfaces are satisfied is satisfied and the sliding surfaces of Eqs. (18), (21) are equated to zero, i.e., $S_i = 0$, $S_v = 0$. In order to prove that the states of the system reach the equilibrium points in finite time, consider the lemma given in Eqs. (26), (27).

Lemma 1. *Fractional integral of the fractional derivative $_aD_t^\alpha$ of a function F(t) is expressed as following (Ullah et al., 2015a, b)*

$$_aD_t^{-\alpha}\left(_aD_t^\alpha F(t)\right) = F(t) - \sum_{J=1}^{K}\left[_aD_t^{\alpha-J}F(t)\right]_{a=t}\frac{(t-a)^{\alpha-J}}{\Gamma(\alpha-J+1)} \tag{26}$$

Here $K - 1 \leq \alpha < K$ and Γ represents the standard gamma function.

Lemma 2. *The fractional integral $_aD_t^{-\alpha}$ of a function F(t) is bounded such that the following expression is true (Ullah et al., 2015a, b)*

$$\left\|_aD_t^{-\alpha}F\right\|_P \leq K\|F\|_P; \quad 1 \leq P \leq \infty; \quad 1 \leq K \leq \infty \tag{27}$$

Since the sliding surfaces given in Eqs. (18), (21) are of the same type, we choose Eq. (18) to prove the finite time convergence of the states. The second case can be proved in the same way. With $S_v = 0$, Eq. (18) is expressed as following.

$$e_v = -\frac{c_2}{c_1} D^{-\alpha} e_v \tag{28}$$

Using the fractional-order property of $D^{-\alpha} D^{\alpha} = 1$, Eq. (28) is rewritten as following.

$$D^{-\alpha}(D^{\alpha} e_v) = -\frac{c_2}{c_1} D^{-\alpha} e_v \tag{29}$$

By applying Lemma 1 to the left-hand side of Eq. (29), one obtains the following expression.

$$e_v - \left[{}_{t_r} D_t^{(\alpha-1)} e_v \right]_{t=t_r} \frac{(t-t_r)^{\alpha-1}}{\Gamma(\alpha)} = -\frac{c_2}{c_1} D^{-\alpha} e_v \tag{30}$$

By choosing time, $t = t_r$, the second term in the left side of Eq. (30) is equated to zero, so the remaining expression of Eq. (30) is simplified as following.

$$e_v = -\frac{c_2}{c_1} D^{-\alpha} e_v \tag{31}$$

Eq. (31) is rewritten as.

$$D^{-2}(D^2 e_v) = -\frac{c_2}{c_1} D^{-\alpha} e_v \tag{32}$$

By applying Lemmas 1 and 2 to the left-hand and right-hand sides of Eq. (32), respectively, one obtains the following resultant expression.

$$\left\| e_v(t) - \left[{}_{t_r} D_t^{(1)} e_v(t) \right]_{t=t_r} \frac{(t-t_r)}{2} \right\| - \| e_v(t_r) \| \leq -\frac{c_2}{c_1} K \| e_v(t) \| \tag{33}$$

If $S_v = 0$ and $e_v(t) = 0$, at time $t = t_s$, then the necessary condition of convergence of the states is calculated as following.

$$t_r \leq t_s - \frac{2\| e_v(t_r) \|}{\| [\dot{e}_v(t)]_{t=t_r} \|} \tag{34}$$

Eq. (34) proves the finite time convergence property of the proposed fractional-order control system.

5 Results and discussion

The fractional-order controllers derived in Eqs. (20), (23) are tested in simulation environment with both constant and variable tissue impedance. The controllers are tested in constant power loop mode. A proportional integral (PI) controller is used to regulate the external power loop. The details of different control loops are shown in Fig. 4. The fractional operator is implemented using the Oustaloup approximation method. Initially, the tissue impedance is chosen as 200 Ω. Output of the power-mode controller is fed as input to the voltage loop controller (Eq. 20 or 24) of the buck converter to adjust the output DC voltage of the buck converter. The output current of the buck converter serves as reference current command for the current loop of the inverter unit. The inverter's current loop controllers are derived in Eqs. (23), (25). For voltage and current loop controllers, the discontinuous signum function is approximated using tangent hyperbolic function. The parameters of PI, integer, and fractional-order controllers are tabulated in Table 3.

In the first simulation test, the tissues impedance is kept constant at 200 Ω. The reference command of the power loop is selected is 50 W. Fig. 5 shows the comparative performance of the power-tracking control loop of the ESU device. From the numerical simulations, it is clear that under the fractional-order control the power output of the ESU device is less corrupted by high-frequency oscillations. Moreover under the fractional order control, the output power signal of the ESU device quickly settles down to the reference value as compared to the integer-order control. The maximum overshoot with integer order control is recorded as 80 W and with

Table 3 Parameters specification of the control loops

Parameter	Value
K_p	50
K_i	30
C_1, C_2, C_3, C_4	1.5, 0.5, 1.5, 0.5
K_1	10
K_2	5.5
α	0.9

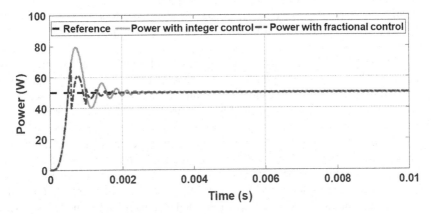

Fig. 5 Power tracking performance under fixed tissue impedance.

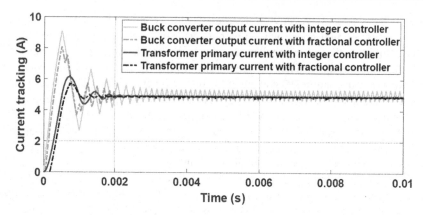

Fig. 6 Current (RMS) tracking performance of inverter unit under fixed tissue impedance.

fractional-order controller it reaches a maximum value of 70 W. Similarly with fractional-order controller, the output power of the ESU unit settles to its steady value at time $t = 0.0018$ s, while in case of integer-order control it settles at time $t = 0.0022$ s.

Fig. 6 shows the current tracking performance of the inverter unit under the fractional- and integer-order controllers. Since the output current of the buck converter is set as a reference command for the inverters current loop, the performance of the inverters control loop is dependent on the smoothness of the input command. The results presented in Fig. 6 show that the generated reference command with fractional-order control contains less oscillations as compared to the integer-order control. Maximum overshoot of 9 A is recorded in the output current of the buck converter with integer-order control, while with fractional-order control it reaches a peak value of 8 A. Similarly, the current tracking performance of the inverter unit under the fractional-order control is superior as compared to the integer-order control in terms of the overshoots and high-frequency oscillations. With fractional-order control, output current of the inverter unit settles to steady state at time $t = 0.0018$ s, while with integer-order control it approaches the steady state at time $t = 0.0022$ s output current.

Fig. 7 shows the RMS current measured at the secondary side of the high-frequency transformer. From the simulation results, it is clear that the secondary current of the ESU unit with fractional-order controller's exhibits small overshoots and it settles to the steady state in smaller time interval as compared to the integer-order controller. Figs. 8 and 9 show the output voltage of the buck converter and the RMS secondary voltage measured across the high-frequency transformer, respectively. From Fig. 2, it is already known that the RMS voltage measured at the primary side of the high-frequency transformer is equal to the DC voltage output of the buck converter. From Fig. 8, a maximum overshoot of 13 V is observed in the buck converter's output voltage with integer order control, while it is restricted to a maximum value of 12 V with fractional-order control. From Fig. 9, similar behavior is noted in the secondary RMS

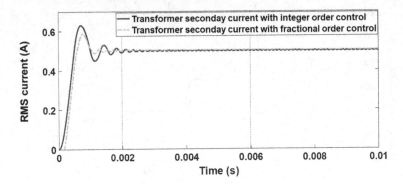

Fig. 7 Transformer secondary RMS current.

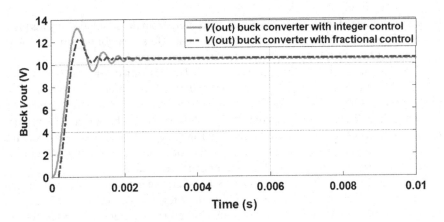

Fig. 8 Buck voltage output/Inverter input voltage.

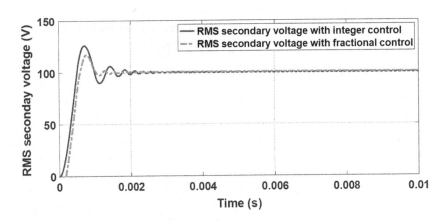

Fig. 9 Transformer secondary RMS voltage.

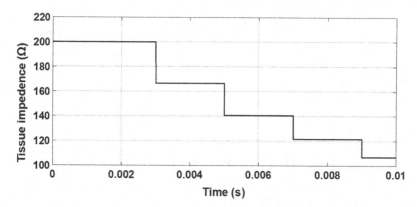

Fig. 10 Tissue impedance variation.

measured voltage of the high-frequency transformer. From the simulation results presented in Figs. 5–9, it is concluded that the ESU unit offers superior performance with fractional-order control in comparison to the integer-order control.

In the second simulation test, the tissues impedance is varied between 200 and 100 Ω in the time interval between $t = 0.003$ and 0.009 s. The reference command of the power loop is selected is 50 W. Fig. 10 shows the tissue impedance variation with respect to time. Fig. 11 shows the comparative performance of the power tracking control loop of the ESU device with both fractional and integer order controllers. From the numerical simulations, it is clear that the system under fractional-order control offers more robust behavior with variable tissue impedance condition.

From Fig. 11, a maximum overshoot of 8 V and an undershoot of 4 V is recorded in the power-tracking signal of ESU unit with integer order control, while with fractional-order control the system exhibits negligible overshoot and undershoot in the power tracking signal.

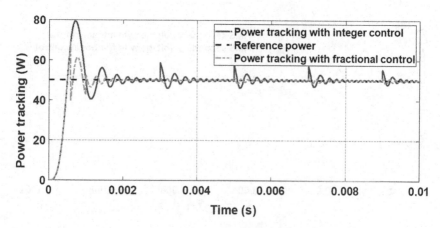

Fig. 11 Power tracking performance under variable tissue impedance.

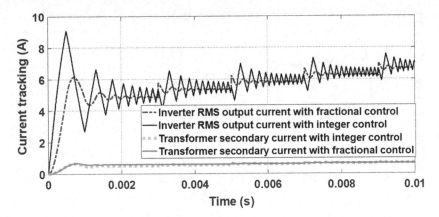

Fig. 12 Current tracking performance of inverter unit under variable tissue impedance.

Fig. 12 shows the current tracking performance of the inverter unit under the fractional- and integer-order controllers with variable tissue impedance condition. The results presented in Fig. 12 show that the generated variable reference command with fractional-order control contains less oscillations as compared to the integer order control. Maximum overshoot of 9 A is recorded in the output current of the buck converter with integer-order control, while with fractional-order control it reaches a peak value of 6 A. Referring to Fig. 10, as the tissue impedance is decreased at equal time interval, the magnitude of the inverter's current increases in the same ratio and in the same time interval. From Fig. 12, it is evident that the inverter current under fractional order control is smoother, and it contains less high-frequency oscillations. Under the action of the integer-order control, the inverter's current contains high-frequency oscillations; thus, it may limit the practical application of the device. Figs. 13 and 14 show the RMS primary and secondary voltages. From the results presented, it is

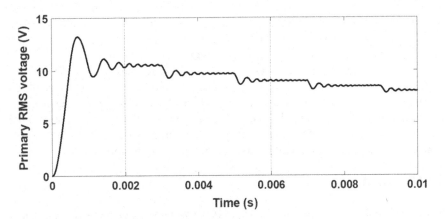

Fig. 13 Inverter output voltage under variable tissue impedance.

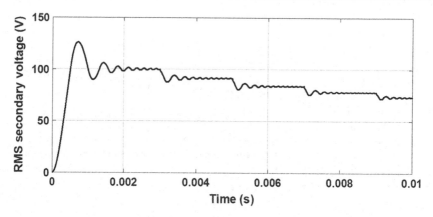

Fig. 14 Inverter output RMS voltage under variable tissue impedance.

concluded that both the primary and secondary voltages decrease in a proportion, which reflects the increase in the tissue impedance. Moreover, the measured voltage signals contain few oscillations under the fractional-order control, so the fractional-order control is feasible from the practical implementation point of view.

6 Conclusion

This chapter briefly introduced the basic concepts, mathematical modeling, and closed-loop control system for an electrosurgical generator unit. As a test case, the design of a 50-W prototype is discussed, and the device is tested in constant power region with both fixed and variable tissue impedance. Fractional-order robust control scheme has been derived for voltage and current loop control of the buck converter and inverter unit, respectively. The controllers have been tested in the simulation environment and from the numerical simulation results, it is concluded that in comparison to the integer-order controllers, the ESU unit offers superior performance with fractional-order controllers. In power-tracking simulations, a maximum overshoot of 30 volts and undershoot of 10 V is recorded in the transient time. Moreover, the measured RMS output power of the inverter unit converges to steady state at time $t = 0.0018$ s with fractional-order control, while it reaches steady state at time $t = 0.0022$ s with integer-order control. Similarly with the application of variable tissue impedance, an overshoot of 6 V and undershoot of 4 V is observed in the measured output power signal when the ESU unit is operated under the action of integer order control, while with fractional order control, it shows negligible variation in the power signal under the same condition of the variable tissue impedance.

References

Becker, D.J., Klicek, M.S., 1998. Electrosurgical Generator Power Control Circuit and Method. U.S. Patent 5 772 659.

Bovie, W.T., Cushing, H., 1928. Electro-surgery as an aid to the removal of the intracranial tumors with a preliminary note on a new surgical-current generator. Surg. Gynecol. Obstet. 47, 751–784.

Dodde, R.E., Gee, J.S., Geiger, J.D., Shih, A.J., 2008. Thermal-electric finite element analysis and experimental validation of bipolar electrosurgical cautery. J. Manuf. Sci. Eng. 130 (2), 1–8.

Dodde, R.E., Gee, J.S., Geiger, J.D., Shih, A.J., 2012. Mono-polar electro-surgical thermal management for minimizing tissue damage. IEEE Trans. Biomed. Eng. 59 (1), 167–173.

Erickson, R.W., Maksimovic, D., 2001. Fundamentals of Power Electronics, second ed. Springer, LLC, New York.

Erickson, R.W., Maksimovic, D., 2004. Fundamentals of Power Electronics, second ed. Springer, New York, pp. 439–482.

Friedrichs, D.A., Erickson, R.W., Gilbert, J., 2012. A new dual current mode controller improves power regulation in electrosurgical generators. IEEE Trans. Biomed. Circ. Syst. 6 (1), 39–44.

Gardner, J., 1994. Practical problems of electro-surgery. In: IEEE Colloquium on Electromagnetic Interference in Hospitals, London. pp. 6/1–6/3.

Gerhard, G.C., 1984. Surgical electro technology. IEEE Trans. Biomed. Eng. BME-31 (12), 787–792.

Jensen, S., Corradini, L., Rodriguez, M., Maksimovic, D., 2011. Modeling and digital control of LCLC resonant inverter with varying load. In: IEEE Energy Conversion Congress and Exposition, USA, pp. 3823–3829.

Jensen, S., Maksimovic, D., Friedrichs, D., Gilbert, J., 2015. Fast tracking electrosurgical generator using GaN switches. In: IEEE Applied Power Electronics Conference and Exposition (APEC). https://doi.org/10.1109/apec.2015.7104531.

Palanker, D.V., Vankov, A., Huie, P., 2008. Electrosurgery with cellular precision. IEEE Trans. Biomed. Eng. 55 (2), 838–841.

Pearce, J.A., 1986. Electro Surgery. Wiley, New York, pp. 44–59.

Polk, C., 1992. Biological effects of low frequency electric and magnetic fields: an overview. In: Proceedings of the Eighteenth IEEE Annual Northeast Bioengineering Conference, Kingston, RI, USA, pp. 55–57.

Schneider, B., Dias, E., Abatti, P.J., 2010. How can electrosurgical sparks generate undesirable effects. In: 1st IEEE Latin American Symposium on Circuits and Systems (LASCAS), Fozdo Iguacu, pp. 93–96.

Shambroom, J.R., 2006. A true to life simulation of electrosurgery. In: Proc. 28th IEEE EMBS Annu. Int. Conf, pp. 5105–5108.

Tan, F.D., Middlebrook, R.D., 1995. A unified model for current-programmed converters. IEEE Trans. Power Electron. 10 (4), 397–408.

Thompson, R., 2005. Electrosurgical Generator and Method With Voltage and Frequency Regulated High-Voltage Current Mode Power Supply. U.S. Patent 6 939 347.

Ullah, N., Ali, M.A., Ahmad, R., Khattak, A., 2017b. Fractional order control of static series synchronous compensator with parametric uncertainty. IET Gener. Transm. Distrib. 11 (1), 289–302.

Ullah, N., Ali, M.A., Ibeas, A., Herrera, J., 2017a. Adaptive fractional order terminal sliding mode control of a doubly fed induction generator-based wind energy system. IEEE Access 5, 21368–21381.

Ullah, N., Han, S., Khattak, M.I., 2015a. Adaptive fuzzy fractional-order sliding mode controller for a class of dynamical systems with uncertainty. Trans. Inst. Meas. Control. 38 (4), 402–413.

Ullah, N., Ullah, A., Ibeas, A., Herrera, J., 2017c. Improving the hardware complexity by exploiting the reduced dynamics-based fractional order systems. IEEE Access 5, 7714–7723.

Ullah, N., Wang, S.P., Khattak, M.I., Shafi, M., 2015b. Fractional order adaptive fuzzy sliding mode controller for a position servo system subjected to aerodynamic loading and nonlinearities. Aerosp. Sci. Technol. 43, 381–387.

Webste, J.G., 1992. Important devices in biomedical engineering. In: Proceedings of the 1992 International Biomedical Engineering Days, Istanbul, pp. 1–9.

Application of a T-S unknown input observer for studying sitting control for people living with spinal cord injury

7

Mathias Blandeau, Thierry-Marie Guerra, Philippe Pudlo
Université Polytechnique Hauts-de-France CNRS, Valenciennes, France

1 Introduction

People living with a complete spinal cord injury (SCI) lose all sensibility and mobility below their injury level. A complete injury located in the lumbar region would result in the loss of the lower limbs and any higher injury will affect the abdominal belt, the back, and intervertebral muscles, which are required to finely adjust the inherently unstable human spine (Crisco et al., 1992; Silfies et al., 2003) or stabilize it in the presence of perturbation. After an SCI, sitting control in everyday life becomes one of the most crucial goals of rehabilitation (Janssen-Potten et al., 1999) and people living with SCI can use their upper limbs and head, instead of the lumbar muscle, as a compensatory strategy to maintain equilibrium (Grangeon et al., 2012). Despite its importance in the life of people living with SCI, sitting control is still little known (Vette et al., 2010). The human anatomical complexity impels us to consider various mechanical actions to sitting control [active joint force, passive joint resistance, etc. (Panjabi, 1992)]. The use of a dynamic mathematical model allows the representation of all these contributions and the experimental observation of their impact on sitting stability. Biomechanical modeling aims at defining a theoretical framework in order to test behavior hypothesis through simulation or experimental data (Reeves and Cholewicki, 2003). One advantage of modeling in biomechanics is to be able to estimate internal variables where it is hard, or impossible, to put sensors as it is the case for human internal joint torques.

The majority of existing models for the study of sitting stability are linked-segment model allowing simple mechanical representation of the body. Nevertheless, the-representation is often inappropriate for the application to the SCI sitting stability because the trunk and upper limbs are considered rigid (Cholewicki et al., 1999a, b; Vette et al., 2010), because the equation are linearized around and equilibrium point (Reeves et al., 2009) or because the muscular activity is concentrated at the lumbar joint (Tanaka et al., 2010). The H2AT model (for "Head, 2 Arms, and Trunk") was created to provide a better representation to the sitting stability analysis, a variation of the inverted

Control Applications for Biomedical Engineering Systems. https://doi.org/10.1016/B978-0-12-817461-6.00007-X

pendulum, taking into account the action of the upper limbs (Blandeau et al., 2016). One advantage of this representation when analyzing sitting stability for people living with complete SCI is the opportunity to differentiate the voluntary forces generated at the uppers limbs to stabilize the position from the passive torques produced at the lumbar level by the stretching of soft tissues (Bergmark, 1989) and internal organs compression.

Modeling the dynamics of mechanical systems leads naturally to nonlinear differential equations in descriptor form with an invertible inertia matrix (Lewis et al., 2003). Several observation techniques can come at hand such as sliding modes, high gain observers, and so on. Nevertheless, an important argument, in our case, is to keep a representation that stays close to the initial model, in order to keep the physical interpretation of the variables (forces, torques). Therefore, an interesting representation can be obtained via quasi-LPV (linear parameter-varying) models (Boyd et al., 1994) or so-called Takagi-Sugeno (T-S) model via the sector nonlinearity approach (Ohtake et al., 2001; Taniguchi et al., 1999) and, because it represents the natural differential equation, is better suited for reducing the conservatism of the system (Sala et al., 2005). This approach has already been applied to biomechanical systems (Guelton et al., 2008) for the study of standing postural stability. The main idea is to design an observer that has the capability, according to the measurements, to reconstruct the unknown (unmeasurable) variables and to prove that the estimation error on both measured and unmeasured variables converges asymptotically toward the equilibrium point. This family of observers is called unknown input observer (UIO) and the convergence proof uses the direct Lyapunov method. On a technical point of view, the results need to be written as linear matrix inequality (LMI) constraints problem (Boyd et al., 1994) to be efficiently solved via convex optimization techniques. The observer design for the descriptor T-S models with proof of asymptotic convergence follows the results presented in Guerra et al. (2015).

A last, when dealing with such descriptor mechanical systems within the quasi-LPV, T-S framework, well-known important lock appears with the nonmeasurable nonlinearities included in the descriptor and/or the state matrices. This problem, in its general form, is still an open problem for the design of T-S control and observer (Ichalal et al., 2016b). Generally, due to these nonmeasurable nonlinearities, the result has to be downgraded, from asymptotic convergence to restricted convergence in a domain (a ball containing the origin) via assumptions such as Lipchitz conditions. In our particular case, this problem has been solved efficiently using a robust-like control approach for the convergence of the state error estimation maintaining an asymptotic convergence toward the equilibrium point.

This chapter is organized as follows: Section 2 presents the modeling of H2AT via Lagrangian techniques; Section 3 deals with the design and implementation of the control law; Section 4 presents the T-S modeling and the way to derive a UIO for the states and input observation; Section 5 provides both simulation and experimental results on human subjects and discusses the obtained results; and Section 6 concludes the paper and gives future works.

2 Modeling

The goal of this section is to derive a model that represents a good tradeoff between a too high complexity (for example, including model for the muscles) and inappropriate assumptions (any linearization and/or neglecting nonlinearities). The solution adopted is called *Head 2 Arms and Trunk* (H2AT) system and it is a variation of the 2D double-inverted pendulum represented in the sagittal plane. It is very important to understand that if we want to estimate the unmeasurable variables such as the forces or the passive lumbar torque represented, respectively, as $F(t)$ and $M(t)$ in Fig. 1, any assumption that would lead to reduction of the physics of the H2AT model will directly impact the results of estimation, bringing important deviations. Therefore, the model has been built following an incremental complexity approach in order to test the different existing methods from the quasi-LPV field. In Blandeau et al. (2016), the model and associated UIO were introduced in the continuous-time framework. More recent works were dedicated to tackle the nonmeasurable premise variables (see Section 2.3.3) in the discrete time through a slow variation hypothesis (Blandeau et al., 2017) and in the continuous time (Blandeau et al., 2018) using a robust technique.

2.1 Model description

The *Head 2 Arms and Trunk* (H2AT) takes into account the action of the head and upper limbs in the stabilization of people living with SCI in seated position unlike classical model using only and inverted pendulum (Reeves et al., 2009; Tanaka et al., 2010; Vette et al., 2012). It is composed of two rods, the first one represents the trunk as a classical inverted pendulum rotating at the lumbar joint (point T) while the second rod, representing the head and upper limbs, slides at the top of the first one with a constant angle a and crosses the first rod at point E. The output of the system is

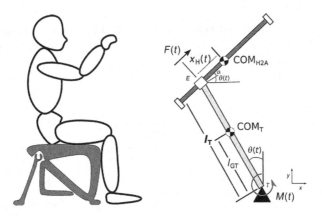

Fig. 1 H2AT model.

the position $x_H(t)$ of the COM_{H2A}, i.e., the head and upper limbs barycenter of mass, and the angular position of the trunk $\theta(t)$. Inputs of the system are the controlling force $F(t)$, which makes the upper rod slide over the lower one and the passive lumbar torque $M(t)$, which represents the passive resistance in the lumbar region, Fig. 1 shows the H2AT system scheme.

This model is generic and just requires a minimum of biomechanical parameters. For the simulations, we consider an 80-kg male subject. Recent work of Fang et al. (2017) is used to obtain the segments mass and regression rules provide the other inertial parameters (Dumas et al., 2007). The notation is $m_H = 16.1$ kg standing for the mass of the upper segment, corresponding to the head, neck, and upper limbs; $m_T = 26.64$ kg is the mass of the trunk; $l_T = 477$ mm is the length of the trunk; and $l_{G_T} = 276.66$ mm is the length of the center of mass of the trunk. A full neck flexion with both arms stretched gives a value of $x_H = 105.27$ mm, whereas an extension of the neck and upper limbs gives $x_H = -75.18$ mm (Kapandji, 2005). Experimental data of people living with SCI when exposed to perturbation (Bjerkefors et al., 2007) allow us to define the following compact set, which is used for the T-S modeling,

is $\Omega_X = \left\{ \begin{array}{l} -0.075 \leq x_H \leq 0.105 \\ -0.087 \leq \theta \leq 0.122 \\ -0.209 \leq \dot{\theta} \leq 0.209 \end{array} \right\}$ in meters, radian, and radian/s.

2.2 Euler Lagrange formulation

To derive the equations of motion, the Lagrangian of the model $L = K - U$ is calculated. The kinetic energy of the system K equals the sum of K_i of each segment according to the following formula (Khalil and Dombre, 2004):

$$K_i = \frac{1}{2} \left(\vec{\omega}_i^T I_{G_i} \vec{\omega}_i + m_i \vec{V}_{G_i}^T \vec{V}_{G_i} \right) \tag{1}$$

where $\vec{V}_{G_i}$ is the velocity of the center of mass (COM_i), I_{G_i} represents the moment of inertia, and $\vec{\omega}_i$ is the angular velocity of segment $i \in \{H, T\}$. The same applies for U, the potential energy of the system:

$$U_i = m_i \vec{g}^T \vec{TG}_i \tag{2}$$

where $\vec{g}$ is the gravity vector. Calculating and adding for each segment, the Lagrangian of the H2AT system equals:

$$L = \frac{m_H l_T^2 + m_T l_{G_T}^2}{2} \dot{\theta}^2 + \frac{m_H}{2} \dot{x}_H^2 + \frac{m_H}{2} x_H^2 \dot{\theta}^2 + m_H l_T x_H \dot{\theta}^2 \sin(a) - m_H l_T \dot{x}_H \dot{\theta} \cos(a)$$
$$- \left(m_H g l_T + m_T g l_{G_T} \right) \cos(\theta) - m_H g x_H \sin(\theta + a) \tag{3}$$

With $J(t) = \left(I_{G_T} + m_T l_{G_T}^2 + I_{G_H} + m_H l_T^2 + m_H x_H^2 + 2 m_H x_H l_T \sin(a) \right)$, the dynamic of the H2AT model is given by:

$$
\begin{cases}
\dfrac{d}{dt}\dfrac{\partial L}{\partial \dot{x}_H} - \dfrac{\partial L}{\partial x_H} = F(t) \\[2ex]
\dfrac{d}{dt}\dfrac{\partial L}{\partial \dot{\theta}} - \dfrac{\partial L}{\partial \theta} = M(t)
\end{cases}
\tag{4}
$$

which yields the following system:

$$
\begin{cases}
F(t) = m_H \ddot{x}_H - m_H \ddot{\theta} l_T \cos(a) - \dfrac{1}{2}\dot{\theta}^2 (m_H 2 x_H + 2 m_H l_T \sin(a)) + m_H g \sin(\theta + a) \\[2ex]
M_T(t) = \ddot{\theta}\left(I_{G_T} + m_T l_{G_T}^2 + I_{G_H} + m_H l_T^2 + m_H x_H^2 + 2 m_H x_H l_T \sin(a) \right) - m_H l_T \ddot{x}_H \cos(a) \\[1ex]
\quad + \dot{\theta}(m_H 2 x_H \dot{x}_H + 2 m_H \dot{x}_H l_T \sin(a)) - (m_H l_T + m_T l_{G_T})g \sin(\theta) + m_H g x_H \cos(\theta + a)
\end{cases}
\tag{5}
$$

Using the following state vector $x(t) = [x_H(t)\ \theta(t)\ \dot{x}_H(t)\ \dot{\theta}(t)]^T \in \mathbb{R}^4$, system (5) writes:

$$
\begin{cases}
E(x(t))\dot{x}(t) = A(x(t))x(t) + Bu(t) \\
y(t) = Cx(t)
\end{cases}
\tag{6}
$$

With the input vector $u(t) = \begin{bmatrix} F(t) - m_H g \sin(a)\cos(\theta(t)) \\ M(t) \end{bmatrix}$, the output vector $y(t) \in \mathbb{R}^2$,

and the following matrices, $E(x(t)) = \begin{bmatrix} 1 & 0 & 0 & 0 \\ 0 & 1 & 0 & 0 \\ 0 & 0 & m_H & -m_H l_T \cos(a) \\ 0 & 0 & -m_H l_T \cos(a) & J(x(t)) \end{bmatrix}$,

$B = \begin{bmatrix} 0 & 0 \\ 0 & 0 \\ 1 & 0 \\ 0 & 1 \end{bmatrix}$, $C = \begin{bmatrix} 1 & 0 & 0 & 0 \\ 0 & 1 & 0 & 0 \end{bmatrix}$ and

$$
A(x(t)) = \begin{bmatrix}
0 & 0 & 1 & 0 \\[1ex]
0 & 0 & 0 & 1 \\[1ex]
0 & \begin{pmatrix} -m_H g \cos(a) \\ \times \dfrac{\sin(\theta(t))}{\theta(t)} \end{pmatrix} & 0 & \begin{pmatrix} m_H x_H(t)\dot{\theta}(t) \\ + m_H l_T \dot{\theta}(t)\sin(a) \end{pmatrix} \\[3ex]
-m_H g \cos(\theta(t) + a) & \begin{pmatrix} (m_H l_T + m_T l_{G_T})g \\ \times \dfrac{\sin(\theta(t))}{\theta(t)} \end{pmatrix} & \begin{pmatrix} -2 m_H x_H(t)\dot{\theta}(t) \\ -2 m_H \dot{\theta}(t) l_T \sin(a) \end{pmatrix} & 0
\end{bmatrix}
$$

The term $m_H g \sin(a) \cos(\theta(t))$ in $u(t)$ is included in the input signal as it only depends on measured variables, i.e., the estimation of $F(t)$ will be directed given via $u(t)$.

2.3 Remarks

2.3.1 Unstable model

The H2AT model, as a variation of an inverted pendulum, is naturally unstable in open loop. Therefore, obviously, for simulation it is necessary to design a control for the model; it is the topic of the next section.

2.3.2 Descriptor form justification

The dynamic equation (6) is written in *descriptor form* (Luenberger, 1977). It is often obtained when creating a mechanical model where, in this case, $E(x(t))$ represents the inertia matrix of the system and has the characteristic of being both positive definite and invertible $\forall x_H \in \mathbb{R}$ (Lewis et al., 2003). One could rightfully argue that given the fact that $E(x(t))$ being well defined, it is possible to turn back to a classical state-space representation $\dot{x} = A(\cdot)x + B(\cdot)u$. Nevertheless, it will increase the complexity by multiplying matrix B by the inverse of matrix $E(x(t))$ and thus drastically reduce the set of solutions, this problem has been issued in Guelton et al. (2008) and Seddiki et al. (2010) and new solutions for LMI formulation have been recently proposed in Estrada-Manzo et al. (2016) and Guerra et al. (2015).

2.3.3 Unmeasured premise variables

In Eq. (6), the matrix $A(x(t))$ is defined with the angular speed, which is not measured in the output vector y. This situation is the well-known problem of nonmeasurable premise variables and this problem, in its general form is still open, in the sense finding conditions of convergence for the estimation (Ichalal et al., 2016a). Previous works (Yoneyama, 2009) intended to bound the estimation error based on the mean value theorem, on Lipschitz assumption or on model uncertainties and robustness (Ichalal et al., 2012). This issue is addressed in the observation part.

2.3.4 Notations

Convex sums of matrix expressions, in discrete time:

$$A_h = \sum_{i=1}^{r} h_i(z(k)) A_i, \quad A_{hv} = \sum_{i=1}^{r} \sum_{j=1}^{r} h_i(z(k)) v_j(z(k)) A_{ij}, \quad A_{h^-} = \sum_{i=1}^{r} h_i(z(k-1)) A_i$$

An asterisk (*) will be used in matrix expressions to denote the transpose of the symmetric element; for inline expressions, it will denote the transpose of the terms on its left side, for example:

$$\begin{bmatrix} A & B^T \\ B & C \end{bmatrix} = \begin{bmatrix} A & (*) \\ B & C \end{bmatrix}, \quad A + B + A^T + B^T + C = A + B + (*) + C$$

When a variable is underlined, i.e., $\underline{x}$, it represents the minimal value, when there is an upper line, i.e., $\bar{x}$, it represents the maximum value.

3 Stabilization

3.1 Discrete-time and Takagi-Sugeno framework

The main goal of the H2AT model is to allow the creation of an observer to estimate human internal torques during experimental acquisition. The discrete-time framework is chosen because it is better fitted with experimental (discrete) data and essentially because it allows coping with more elaborated Lyapunov functions than the classical quadratic one. Especially past samples information can come at hand (Guerra et al., 2012) together with the Lyapunov function and the control law or the observation gains. In order to express (6) in discrete time, the classical Euler's approximation is used by introducing the following discrete state vector $x_k = \begin{bmatrix} x_{H_k} & \theta_k & \dot{x}_{H_k} & \dot{\theta}_k \end{bmatrix}^T \in \mathbb{R}^4$ so that $\dot{x}(t) \approx \frac{x_{k+1} - x_k}{s}$, and the discrete system is obtained:

$$\begin{cases} E(x_k)x_{k+1} = A_d(x_k)x_k + sBu(k) \\ y_k = Cx_k \end{cases} \tag{7}$$

with $s = 0.01s$ is the sampling time and $A_d(x_k) = sA(x_k) + E(x_k)$ in the form:

$$A_d(x_k) = \begin{bmatrix} 1 & 0 & s & 0 \\ 0 & 1 & 0 & s \\ 0 & s\left(\frac{-m_H g \cos(a)}{\times \frac{\sin(\theta_k)}{\theta_k}}\right) & m_H & s\left(\begin{matrix} m_H{}^x H_k \dot{\theta}_k \\ +m_H l_T \dot{\theta}_k \sin(a) \\ -m_H l_T \cos(a) \end{matrix}\right) \\ -s m_H g \cos(\theta_k + a) & s\left(\frac{(m_H l_T + m_T l_{G_T})g}{\times \frac{\sin(\theta_k)}{\theta_k}}\right) & s\left(\begin{matrix} -2m_H{}^x H_k \dot{\theta}_k \\ -2m_H \dot{\theta}_k l_T \sin(a) \\ -m_H l_T \cos(a) \end{matrix}\right) & J(x_k) \end{bmatrix}$$

The sector nonlinearity methodology is applied on the discrete-time nonlinear descriptor model (7), four nonlinear terms are present in Eq. (7), $z_1 = \sin(\theta_k)/\theta_k$, $z_2 = \cos(\theta_k + a)$, $z_3 = x_{H_k}\dot{\theta}_k + l_T\dot{\theta}_k \sin(a)$, and $z_4 = x_{H_k}{}^2 + 2x_{H_k}l_T \sin(a)$. Each nonlinear term is bounded, due to the anatomical constraints defined previously and grouped in the premise variable vector $z \in \mathbb{R}^4$:

$z_1 = \sin(\theta_k)/\theta_k \in [0.9975, 1]$, $\qquad z_2 = \cos(\theta_k + a) \in [0.9782, 1]$

$z_3 = x_{H_k}\dot{\theta}_k + \dot{\theta}_k l_T \sin(a) \in [-0.0568, 0.0568]$, $\quad z_4 = x_{H_k}{}^2 + 2x_{H_k}l_T \sin(\alpha) \in [0, 0.0134]$

$$\tag{8}$$

As a result, an equivalent T-S descriptor model form (Taniguchi et al., 1999) is obtained:

$$
\begin{cases}
E_h x_{k+1} = A_{d_h} x_k + sBu(k) \\
y_k = Cx_k
\end{cases}
\tag{9}
$$

with $E_h = \sum_{i=1}^{r} h_i(z(k))E_i$ and $A_{d_h} = \sum_{i=1}^{r} h_i(z(k))A_{di}$ are convex sums of matrices coming from the sector nonlinearity methodology. Now, the membership functions (MFs) can be constructed with the following weighting functions:

$$
w_0^j(\cdot) = \frac{z_j(\cdot) - \underline{z}_j}{\overline{z}_j - \underline{z}_j}, \quad w_1^j(\cdot) = 1 - w_0^j(\cdot), \quad j \in \{1, 2, 3, 4\}
\tag{10}
$$

where $\underline{z}_j$ and $\overline{z}_j$ are the minimum and the maximum of their corresponding premise variable defined in Eq. (8). The MFs are calculated according to the following notation:

$$
h_i = h_{d_1^i \times 2^{p-1} + \cdots + d_3^i \times 2 + d_4^i + 1} = \prod_{j=1}^{4} w_{d_j^i}^j (z_j)
\tag{11}
$$

where $i = d_1^i \times 2^{p-1} + \cdots + d_3^i \times 2 + d_4^i + 1 \in \{1, \ldots, 2^4\}$ is the number of rules (16 therein), $k \in \{1, \ldots, 4\}$ and $d_k^i \in \{0, 1\}$ is the kth digit of the p-digit representation of $i - 1$. For example $h_{10} = w_1^1 \times w_0^2 \times w_0^3 \times w_1^4$ and the corresponding matrices are $(E_{10}, A_{d10}) = (E(\underline{z}_4), A_d(\underline{z}_1, \overline{z}_2, \overline{z}_3, \underline{z}_4))$.

Note that the MFs hold the convex sum property in Ω_x, i.e., $\sum_{i=1}^{16} h_i(z(k)) = 1$ and $0 \leq h_i(z(k)) \leq 1$. The resulting T-S descriptor model (9) is an *exact* representation of the nonlinear one (7).

3.2 Design of the control law

Obviously, the H2AT model being open-loop unstable, a control law has to be designed. Nevertheless, it has an important particularity that state space feedback is "enough." What is stressed here is that a human with complete SCI has the knowledge of his/her internal variables. Thus, the goal here is to design a state feedback that "behaves" like a human, or at least that its behavior is compatible with observed data. Notice also, that people living with a complete thoracic SCI cannot recruit their abdominal muscle and lower back muscles; thus, position will depend only on the displacement of their head and upper limbs. Therefore, whereas for a noninjured person the controls would be $u(k) = \begin{bmatrix} u_F(k) \\ u_M(k) \end{bmatrix}$, it reduces for SCI to $u(k) = \begin{bmatrix} u_F(k) \\ 0 \end{bmatrix}$, i.e., under actuated control. This is resumed for control design by replacing the input matrix $B = \begin{bmatrix} 0 & 0 & 1 & 0 \\ 0 & 0 & 0 & 1 \end{bmatrix}^T$ in the case of SCI by: $B = [0\ 0\ 1\ 0]^T$. The design of the control law is based on a systematic method to synthesize controllers for nonlinear discrete-time regular descriptor systems (Estrada-Manzo et al., 2015a, b) with the following non PDC control law:

$$u(k) = K_{hh^-} G_h^{-1} x_k. \tag{12}$$

where K_{hh^-} and G_h^{-1} are matrix of appropriate dimensions to be determined. Obviously, to respect causality, these matrices only contain past samples. The closed-loop of the T-S model (9) under control law (12) is:

$$\begin{cases} E_h x_{k+1} = \left(A_{d_h} + sBK_{hh^-} G_h^{-1} \right) x_k \\ y_k = C x_k \end{cases} \tag{13}$$

The analysis of the closed-loop model (13) is achieved through the direct Lyapunov method considering a nonquadratic Lyapunov function and a convex sum of Lyapunov matrix P_h (Guerra and Vermeiren, 2004):

$$V_h = x_k^T P_h^{-1} x_k \tag{14}$$

Theorem 1. *The closed-loop T-S descriptor model (13) is asymptotically stable if there exist $P_h \in \mathbb{R}^{4 \times 4}$ with $P_h = P_h^T > 0$, $K_{hh^-} \in \mathbb{R}^{1 \times 4}$, $G_h \in \mathbb{R}^{4 \times 4}$ and $F_h \in \mathbb{R}^{4 \times 4}$ such that:*

$$\begin{bmatrix} -G_h - G_h^T + P_{h^-} & (*) & 0 \\ A_h G_h + BK_{hh^-} & -E_h F_h - F_h^T E_h^T & (*) \\ 0 & F_h & -P_h \end{bmatrix} < 0 \tag{15}$$

Remember that every subscript implies a 16-element sum. Therefore, it can be important according to the solvers capabilities to reduce the number of sums, which also reduces the number of variables, for a fruitful discussion about this issue, the reader can refer to Lendek et al. (2015). On the other side, the structure itself of the matrices can be used to "optimize" the degrees of freedom. Effectively, $A_d(x_k)$ and $E(x_k)$ do not include nonlinearities at each entries. Thus, considering these two previous remarks, both the G_h and F_h matrices of (15) are designed in order to avoid double products while increasing the number of free variables: $G_h = \begin{bmatrix} G_{1,hh^-} \\ G_{2,h^-} \end{bmatrix}$ and $F_h = \begin{bmatrix} F_{1,hh^-} \\ F_{2,h^-} \end{bmatrix}$, with $G_{1,\,hh^-} \in \mathbb{R}^{2 \times 4}$, $G_{2,\,h^-} \in \mathbb{R}^{2 \times 4}$, $F_{1,\,hh^-} \in \mathbb{R}^{3 \times 4}$ and $F_{2,\,h^-} \in \mathbb{R}^{1 \times 4}$.

Proof of theorem 1. The variation of the nonquadratic Lyapunov function (14) writes the following inequality constraint:

$$\Delta V_h = \begin{bmatrix} x_k & x_{k+1} \end{bmatrix} \begin{bmatrix} P_h^{-1} & 0 \\ 0 & P_h^{-1} \end{bmatrix} \begin{bmatrix} x_k \\ x_{k+1} \end{bmatrix} < 0 \tag{16}$$

System (13) is written as an equality constraint:

$$\left[A_{d_h} + sBK_{hh^-}G_h^{-1} \quad -E_h\right]\begin{bmatrix} x_k \\ x_{k+1} \end{bmatrix} = 0 \tag{17}$$

From here, Finsler's Lemma (de Oliveira and Skelton, 2001) is used to combine Eqs. (16), (17):

$$\begin{bmatrix} P_{h^-}^{-1} & 0 \\ 0 & P_h^{-1} \end{bmatrix} + M_{(\cdot)}\left[A_{d_h} + sBK_{hh^-}G_h^{-1} \quad -E_h\right]\begin{bmatrix} x_k \\ x_{k+1} \end{bmatrix} + (*) < 0 \tag{18}$$

The congruence property is used with matrix $\begin{bmatrix} G_h^T & 0 \\ 0 & F_h^T \end{bmatrix}$ and using $M_{(\cdot)} = \begin{bmatrix} 0 \\ F_h^{-T} \end{bmatrix}$ it yields:

$$\begin{bmatrix} -G_h^T P_{h^-}^{-1} G_h & (*) \\ A_h G_h + BK_{hh^-} & -E_h F_h - F_h^T E_h^T + F_h^T P_h^{-1} F_h \end{bmatrix} < 0 \tag{19}$$

Because $P_h = P_h^T > 0$ then $G_h^T P_{h^-}^{-1} G_h \geq G_h + G_h^T - P_{h^-}$, which gives:

$$\begin{bmatrix} -G_h - G_h^T + P_{h^-} & (*) \\ A_h G_h + BK_{hh^-} & -E_h F_h - F_h^T E_h^T + F_h^T P_h^{-1} F_h \end{bmatrix} < 0 \tag{20}$$

Finally, a Schur's complement (Boyd et al., 1994) is used on Eq. (20), which results directly (15); hence, the control law (12) asymptotically stabilizes the closed loop of the system (13). ∎

Remark. Different performances can be directly added to the LMI problem (15). For instance, a decay rate β, with $0 < \beta \leq 1$ to control the speed convergence, can be directly obtained via:

$$\begin{bmatrix} -G_h - G_h^T + \beta P_{h^-} & (*) & 0 \\ A_h G_h + BK_{hh^-} & -E_h F_h - F_h^T E_h^T & (*) \\ 0 & F_h & -P_{h^-} \end{bmatrix} < 0 \tag{21}$$

Preliminary simulation results are presented in Fig. 2 to show the state feedback control that represents the stabilized person control. Initial conditions are taken as $x(0) = [0\ 0\ 0\ 0.01]^T$ and a perturbation, simulating an $0.02\,\text{rad/s}^2$ impulsion on the acceleration during 1 s, is used at time 3 s (Fig. 2, dashed blue signal). Fig. 2, part left, shows the variable $\dot{\theta}_k$ and, part right, shows u_k issued from non-PDC control law (12). The controller as implemented is perfectly able to stabilize the nonlinear model for these conditions.

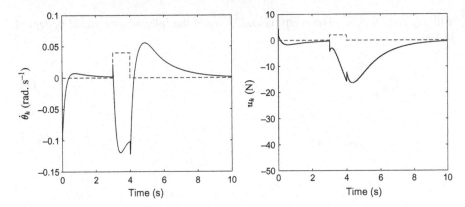

Fig. 2 Stabilization of (6) with control law (12) with nonzero initial condition and perturbation (dashed blue—dark gray in print version) occurring at 2 s (left) angular speed, (right) control law.

The internal control law is implemented and can stabilize the system; the goal is now to derive an observer that allows the estimation of the state and furthermore the internal input $u(t)$ considered unknown.

4 Observation

4.1 Takagi-Sugeno formalism

The UIO is based on the T-S model (7), but, in the context of observation, the control law $u(k)$ is replaced by an unknown input $d_k = \begin{bmatrix} d_{F_k} & d_{M_k} \end{bmatrix}^T$ that has to be estimated:

$$\begin{cases} E_h x_{k+1} = A_{d_h} x_k + sB d_k \\ y_k = C x_k \end{cases} \tag{22}$$

Now, to capture the dynamic of the discrete-time unknown input d_k, a second-order discrete integrator is used, meaning its dynamic writes $(z-1)^{p+1} d_k \approx 0$ with $p = 1$. This choice has been made after some real data trials and is a good compromise between accuracy of the results and complexity of the observer, especially in terms of gains and order. Thus, its dynamics writes:

$$D_{F^+} = \Gamma \times D_F \Leftrightarrow \begin{bmatrix} d_{F_{k+1}} \\ d_{F_{k+2}} \end{bmatrix} = \begin{bmatrix} 0 & 1 \\ -1 & 2 \end{bmatrix} \begin{bmatrix} d_{F_k} \\ d_{F_{k+1}} \end{bmatrix},$$

$$D_{M^+} = \Gamma \times D_M \Leftrightarrow \begin{bmatrix} d_{M_{k+1}} \\ d_{M_{k+2}} \end{bmatrix} = \begin{bmatrix} 0 & 1 \\ -1 & 2 \end{bmatrix} \begin{bmatrix} d_{M_k} \\ d_{M_{k+1}} \end{bmatrix} \tag{23}$$

with D_F and $D_M \in \mathbb{R}^2$. Thus, Eqs. (22), (23) yield the following extended system of order 8:

$$\begin{cases} E_h^e x_{k+1}^e = A_{d_h}^e x_k^e \\ y_k = C^e x_k^e \end{cases} \tag{24}$$

with the extended vector $x_k^e = \begin{bmatrix} x_k \\ D_F \\ D_M \end{bmatrix} \in \mathbb{R}^8, \quad E_h^e = \begin{bmatrix} E_h & 0_{4\times4} \\ 0_{4\times4} & I_{4\times4} \end{bmatrix},$

$$A_{d_h}^e = \begin{bmatrix} A_{d_h} & [sB_F \ 0_{4\times1}] & [sB_M \ 0_{4\times1}] \\ 0_{2\times4} & \Gamma & 0_{2\times4} \\ 0_{2\times4} & 0_{2\times4} & \Gamma \end{bmatrix}, \quad B_F = \begin{bmatrix} 0 \\ 0 \\ 1 \\ 0 \end{bmatrix}, \quad B_M = \begin{bmatrix} 0 \\ 0 \\ 0 \\ 1 \end{bmatrix}, \quad \text{and}$$

$C^e = [C \ 0_{2\times4}]$. Obviously E_h^e is still regular for all $x \in \Omega_x$.

4.2 Design of the unknown input observer with uncertainties

We remind that the angular speed $\dot{\theta}_k$ is not measured and must be estimated to be used in the MFs computation. In order to avoid making a slow variation assumption as it has been done in Blandeau et al. (2017), we choose to adapt the robust-like technique from Blandeau et al. (2018) and adapt it to the discrete-time framework. The discrete-time T-S descriptor system (24) is employed for the UIO design.

Therefore, using the previous works of Estrada-Manzo et al. (2015a, b), the following observer is defined:

$$\begin{cases} E_\alpha^e \hat{x}_{k+1}^e = A_{\alpha\beta}^e \hat{x}_k^e + G_{\hat{\beta}}^{-1} K_{\alpha\hat{\beta}}(y_k - \hat{y}_k) \\ \hat{y}_k = C^e \hat{x}_k^e \end{cases} \tag{25}$$

with $\hat{x}_k^e = \begin{bmatrix} \hat{x}_k^T & \hat{D}_F & \hat{D}_M^T \end{bmatrix}^T \in \mathbb{R}^8$ as the extended estimation vector. In order to be clear $\alpha \in \{\theta_k, x_{H_k}\}, \beta \in \{x_{H_k} \times \dot{\theta}_k\}$:

$$A_{\alpha\beta}^e x_d^e = \sum_{i=1}^{2} \sum_{j=1}^{2} \sum_{k=1}^{2} \sum_{l=1}^{2} w_{d^i}^i(\theta_k) w_{d^j}^j(\theta_k) w_{d^k}^k(x_{H_k}) w_{d^l}^l(x_{H_k} \times \dot{\theta}_k) A_{ijkl} x_d^e$$

$$A_{\alpha\hat{\beta}}^e x_d^e = \sum_{i=1}^{2} \sum_{j=1}^{2} \sum_{k=1}^{2} \sum_{l=1}^{2} w_{d^i}^i(\theta_k) w_{d^j}^j(\theta_k) w_{d^k}^k(x_{H_k}) w_{d^l}^l(x_{H_k} \times \hat{\dot{\theta}}_k) A_{ijkl} x_d^e \tag{26}$$

Matrices $G_{\hat{\beta}} \in \mathbb{R}^{8\times8}$ et $K_{\alpha\hat{\beta}} \in \mathbb{R}^{8\times2}$ are nonlinear gains to be designed. To this end, let us define the estimation error $e_k = x_k^e - \hat{x}_k^e$, and its dynamics via the descriptor form as:

$E_\alpha^e e_{k+1} = E_\alpha^e x_{k+1}^e - E_\alpha^e \hat{x}_{k+1}^e$, from which: $E_\alpha^e e_{k+1} = A_{\alpha\beta}^e x_k^e - A_{\hat{\alpha}\hat{\beta}}^e \hat{x}_k^e - G_{\hat{\beta}}^{-1} K_{\alpha\hat{\beta}}(y_k - \hat{y}_k)$,

or equivalently:

$$E_\alpha^e e_{k+1} = \left(A_{\alpha\beta}^e - G_{\hat{\beta}}^{-1} K_{\alpha\hat{\beta}} C^e\right) e_k + \left(A_{\alpha\beta}^e - A_{\hat{\alpha}\hat{\beta}}^e\right)\hat{x}_k^e \tag{27}$$

Notice that $\left(A_{\alpha\beta}^e - A_{\hat{\alpha}\hat{\beta}}^e\right)\hat{x}_k^e = \begin{bmatrix} A_{\alpha\beta} - A_{\hat{\alpha}\hat{\beta}} & 0_4 \\ 0_4 & 0_4 \end{bmatrix}\begin{bmatrix} \hat{x}_k \\ \hat{D}_F \\ \hat{D}_M \end{bmatrix}$ and:

$$\left(A_{\alpha\beta} - A_{\hat{\alpha}\hat{\beta}}\right)\hat{x}_k = sm_H \begin{bmatrix} 0 & 0 & 0 & 0 \\ 0 & 0 & 0 & 0 \\ 0 & 0 & 0 & \left(x_{H_k}\dot{\theta}_k + l_T\dot{\theta}_k \sin(a)\right) \\ & & & \left(-x_{H_k}\dot{\theta}_k - l_T\dot{\theta}_k \sin(a)\right) \\ 0 & \left(\begin{matrix}-2x_{H_k}\dot{\theta}_k - 2l_T\dot{\theta}_k \sin(a) \\ +2x_{H_k}\hat{\dot{\theta}}_k + 2l_T\hat{\dot{\theta}}_k \sin(a)\end{matrix}\right) & 0 & 0 \end{bmatrix}\begin{bmatrix} \hat{x}_{H_k} \\ \hat{\theta}_k \\ \hat{\dot{x}}_{H_k} \\ \hat{\dot{\theta}}_k \end{bmatrix}$$

$$= sm_H\left(\dot{\theta}_k - \hat{\dot{\theta}}_k\right)\begin{bmatrix} 0 & 0 & 0 & 0 \\ 0 & 0 & 0 & 0 \\ 0 & 0 & 0 & x_{H_k} + l_T \sin(a) \\ 0 & -2x_{H_k} - 2l_T \sin(a) & 0 & 0 \end{bmatrix}\begin{bmatrix} \hat{x}_{H_k} \\ \hat{\theta}_k \\ \hat{\dot{x}}_{H_k} \\ \hat{\dot{\theta}}_k \end{bmatrix}$$

$$= sm_H\left(\dot{\theta}_k - \hat{\dot{\theta}}_k\right)\begin{bmatrix} 0 \\ 0 \\ x_{H_k}\hat{\dot{\theta}}_k + l_T \sin(a)\hat{\dot{\theta}}_k \\ -2x_{H_k}\hat{\dot{x}}_{H_k} - 2l_T \sin(a)\hat{\dot{x}}_{H_k} \end{bmatrix}$$

Hence Eq. (27) can be written as:

$$E_\alpha^e e_{k+1} = \left(A_{\alpha\beta}^e - G_{\hat{\beta}}^{-1} K_{\alpha\hat{\beta}} C^e\right) e_k + sm_H\left(\dot{\theta}_k - \hat{\dot{\theta}}_k\right)\begin{bmatrix} 0 \\ 0 \\ x_{H_k}\hat{\dot{\theta}}_k + l_T \sin(a)\hat{\dot{\theta}}_k \\ -2x_{H_k}\hat{\dot{x}}_{H_k} - 2l_T \sin(a)\hat{\dot{x}}_{H_k} \\ 0_{4\times 1} \end{bmatrix} \tag{28}$$

and as $\dot{\theta}_k - \hat{\dot{\theta}}_k$ is the fourth term of e_k, Eq. (28) writes:

$$E_\alpha^e e_{k+1} = \left(A_{\alpha\beta}^e - G_{\hat{\beta}}^{-1} K_{\alpha\hat{\beta}} C^e + sm_H \Delta_k\right) e_k \tag{29}$$

With $\Delta_k \in \mathbb{R}^{8\times 8}$ a null matrix excepted for the terms: $\Delta_{k(3,4)} = x_{H_k}\hat{\dot{\theta}}_k + l_T \sin(a)\hat{\dot{\theta}}_k$ and $\Delta_{k(4,4)} = -2x_{H_k}\hat{\dot{x}}_{H_k} - 2l_T \sin(a)\hat{\dot{x}}_{H_k}$. Consider that for any bounded uncertainty $\mu(t) \in \left[\underline{\mu}, \overline{\mu}\right]$ we have: $\mu(t) = c + r \times \delta(t)$ with $\delta(t) \in [-1,1]$, the center: $c = 0.5\left(\overline{\mu} + \underline{\mu}\right)$ and the radius $r = 0.5\left(\bar{\mu} - \underline{\mu}\right)$. Therefore, defining accordingly $c_{(3,4)}$, $r_{(3,4)}$, $c_{(4,4)}$, and $r_{(4,4)}$ Eq. (29) becomes:

$$E_\alpha^e e_{k+1} = \left(A_{\alpha\beta}^e - G_{\hat{\beta}}^{-1} K_{\alpha\hat{\beta}} C^e + sm_H A_c + H\delta(t)F\right)e_k \tag{30}$$

with: $H = \begin{bmatrix} 0 & 0 \\ 0 & 0 \\ 1 & 0 \\ 0 & 1 \\ 0 & 0 \\ 0 & 0 \\ 0 & 0 \\ 0 & 0 \end{bmatrix}$, $F = \begin{bmatrix} 0 & 0 & 0 & sm_H \times r_{(3,4)} & 0 & 0 & 0 & 0 \\ 0 & 0 & 0 & -2sm_H \times r_{(4,4)} & 0 & 0 & 0 & 0 \end{bmatrix}$ and A_c being a null

matrix excepted: $A_c(3,4) = c_{(3,4)}$, $A_c(4,4) = c_{(4,4)}$. Let us express $A_{\alpha\beta}' = A_{\alpha\beta} + sm_H A_c$ in a more convenient way.

$$A_{\alpha\beta}' = A_{\alpha\beta} + \begin{bmatrix} 0 & 0 & 0 & 0 \\ 0 & 0 & 0 & 0 \\ 0 & 0 & 0 & sm_H c_{(3,4)} \\ 0 & 0 & 0 & sm_H c_{(4,4)} \end{bmatrix} \tag{31}$$

Thus, $A_{\alpha\beta}'^e = \begin{bmatrix} A_{\alpha\beta}' & [sB_F \ 0_{4\times 1}] & [sB_M \ 0_{4\times 1}] \\ 0_{2\times 4} & \Gamma & 0_{2\times 4} \\ 0_{2\times 4} & 0_{2\times 4} & \Gamma \end{bmatrix}$. Therefore, we are faced with the convergence of:

$$E_\alpha^e e_{k+1} = \left(A_{\alpha\beta}'^e - G_{\hat{\beta}}^{-1} K_{\alpha\hat{\beta}} C^e + H\delta(t)F\right)e_k \tag{32}$$

At last, the remarkable property is that the nonmeasurable premise problem has been transformed in the convergence problem of Eq. (32) that can be solved via robustness classical tools.

4.3 Proof of convergence

Let us consider the following nonquadratic Lyapunov function candidate:

$$V(e_k) = e_k^T P_{\alpha^-} e_k, \quad P_{\alpha^-} = P_{\alpha^-}^T > 0 \tag{33}$$

Remark: many other choices can come at hand, for example, $P_{\alpha\beta}$, $P_{\alpha\beta\hat{\beta}}$, $P_{\alpha^-\beta^-}$, $P_{\alpha^-\beta^-\hat{\beta}^-}$. Nevertheless, the reader has to keep in mind that due to the variation of Eq. (33), the next samples will be introduced also in the matrices, thus increasing the number of LMI constraints (see again the discussion in (Lendek et al., 2015). In the case of

Eq. (33), next step will resume in $V(e_{k+1}) = e_{k+1}^T P_\alpha e_{k+1}$ and represents a good compromise between complexity and quality of the solutions. Notice also that even modifying the Lyapunov function, the extension of the solution and the proof of theorem 2 would be straightforward.

The following result establishes the LMI design for the UIO (25).

Theorem 2. *The origin of the error dynamics (27) is asymptotically stable if there exists matrix* $P = P^T > 0$, $G_m \in \mathbb{R}^{8 \times 8}$, *and* $K_{ijkm} \in \mathbb{R}^{8 \times 2}$, *i, j, k, i', j', k', l, m* $\in \{1,2\}$ *and scalars* $\eta_{ijklm} > 0$ *and* $0 < \lambda < 1$ *such that the following inequality holds:*

$$
\begin{bmatrix}
-\lambda P_{i'j'k'} + \eta_{ijklm} F^T F & (*) & 0 \\
G_m A_{ijkl}^{le} - K_{ijkm} C^e & P_{ijk} - G_m E_k^e - (E_k^e)^T G_m^T & (*) \\
0 & H^T G_m^T & -\eta_{ijklm} I_{2\times 2}
\end{bmatrix} < 0
\tag{34}
$$

Proof. Eq. (32) can also be written as:

$$
\left[A_{\alpha\beta}^{le} - G_{\hat{\beta}}^{-1} K_{\alpha\hat{\beta}} C^e + H\delta(t)F - E_\alpha^e \right]
\begin{bmatrix} e_k \\ e_{k+1} \end{bmatrix} = 0
\tag{35}
$$

Similar for the difference of the Lyapunov function (33) $\Delta V(e_k) = V(e_{k+1}) - \lambda V(e_k)$ with $0 < \lambda < 1$ a decay rate to take into account the convergence speed of the observer:

$$
\Delta V(e_k) = \begin{bmatrix} e_k \\ e_{k+1} \end{bmatrix}^T
\begin{bmatrix} -\lambda P_{\alpha^-} & 0 \\ 0 & P_\alpha \end{bmatrix}
\begin{bmatrix} e_k \\ e_{k+1} \end{bmatrix} < 0
\tag{36}
$$

Then, by Finsler's Lemma (de Oliveira and Skelton, 2001), Eq. (36) holds under equality constraint (35) if:

$$
\begin{bmatrix} 0 \\ G_{\hat{\beta}} \end{bmatrix}
\left[A_{\alpha\beta}^{le} - G_{\hat{\beta}}^{-1} K_{\alpha\hat{\beta}} C^e + H\delta(t)F - E_\alpha^e \right] + (*) +
\begin{bmatrix} -\lambda P_{\alpha^-} & 0 \\ 0 & P_\alpha \end{bmatrix} < 0
\tag{37}
$$

which can be written as:

$$
\begin{bmatrix}
-\lambda P_{\alpha^-} & (*) \\
G_{\hat{\beta}} A_{\alpha\beta}^{le} - K_{\alpha\hat{\beta}} C^e + G_{\hat{\beta}} H\delta(t)F & P_\alpha - G_{\hat{\beta}} E_\alpha^e - (*)
\end{bmatrix} < 0
\tag{38}
$$

The uncertainty is rewritten as $\begin{bmatrix} 0 & (*) \\ G_{\hat{\beta}} H\delta(t)F & 0 \end{bmatrix} = \begin{bmatrix} 0 \\ G_{\hat{\beta}} H \end{bmatrix} \delta(t)[F \ 0] + (*)$ and considering that $\delta(t)\delta^T(t) \leq I_2$, we can define $\eta_{\alpha\hat{\beta}\hat{\beta}} > 0$ such as:

$$
\begin{bmatrix} 0 & (*) \\ G_{\hat{\beta}} H\delta(t)F & 0 \end{bmatrix} \leq \eta_{\alpha\hat{\beta}\hat{\beta}}^{-1}
\begin{bmatrix} 0 \\ G_{\hat{\beta}} H \end{bmatrix}
\begin{bmatrix} 0 & H^T G_{\hat{\beta}}^T \end{bmatrix} + \eta_{\alpha\hat{\beta}\hat{\beta}}
\begin{bmatrix} F^T \\ 0 \end{bmatrix} [F \ 0]
\tag{39}
$$

which means that Eq. (38) holds if:

$$
\begin{bmatrix}
-\lambda P_{\alpha^-} + \eta_{\alpha\beta\hat{\beta}} F^T F & (*) \\
G_{\hat{\beta}} A_{\alpha\beta}^{\prime e} - K_{\alpha\hat{\beta}} C^e & P_\alpha - G_{\hat{\beta}} E_\alpha^e - \left(E_\alpha^e\right)^T G_{\hat{\beta}}^T + \eta_{\alpha\beta\hat{\beta}}^{-1} G_{\hat{\beta}} H H^T G_{\hat{\beta}}^T
\end{bmatrix} < 0 \qquad (40)
$$

And via Schur's complement:

$$
\begin{bmatrix}
-\lambda P_{\alpha^-} + \eta_{\alpha\beta\hat{\beta}} F^T F & (*) & 0 \\
G_{\hat{\beta}} A_{\alpha\beta}^{\prime e} - K_{\alpha\hat{\beta}} C^e & P_\alpha - G_{\hat{\beta}} E_\alpha^e - \left(E_\alpha^e\right)^T G_{\hat{\beta}}^T & (*) \\
0 & H^T G_{\hat{\beta}}^T & -\eta_{\alpha\beta\hat{\beta}} I_{2\times2}
\end{bmatrix} < 0 \qquad (41)
$$

As α corresponds to the measured premise variables, see Eq. (26), three indexes are associated to them i, j, k, which means also three more indexes for α^- due to the non-quadratic nature of the Lyapunov function, i.e., i', j', k'; and for the nonmeasured premises l corresponds to β and m to $\hat{\beta}$. Thus, Eq. (41) corresponds to an 8-sums of the following generic terms to be negative:

$$
\begin{bmatrix}
-\lambda P_{i'j'k'} + \eta_{ijklm} F^T F & (*) & 0 \\
G_m A_{ijkl}^{\prime e} - K_{ijkm} C^e & P_{ijk} - G_m E_k^e - \left(E_k^e\right)^T G_m^T & (*) \\
0 & H^T G_m^T & -\eta_{ijklm} I_{2\times2}
\end{bmatrix} < 0 \qquad (42)
$$

Then the desired result (34) is obtained. ∎

5 Validation results

5.1 Numerical simulations

5.1.1 Simulation protocol

Fig. 3 presents the simulation protocol. A generalized stabilized model includes the discrete H2AT model (6), its stabilization via control law (12) considering an arbitrary passive internal torque to be defined. This generalized model represents a human with an SCI and, seen from the observer, it is a "black box" from which the only measurements $\begin{bmatrix} x_{H_k} \\ \theta_k \end{bmatrix}$ are extracted and added to a random white noise filtered with a 5-Hz low pass-band second-order filter to replicate experimental noise. The obtained signals are the ones from which the UIO (25) estimates both the state vector and the unknown input d_{F_k} and d_{M_k} drawn with a "?" on. Of course, the simulation process can validate the entire procedure as we can exhibit the unknown input error for the force and torque estimation. The initial condition of all simulation are set to $x_0 = [0 \ \ 0 \ \ 0 \ \ 0.01]^T$, the UIO decay rate performance is $\lambda = 0.7$, and the sampling period is $s = 0.1$.

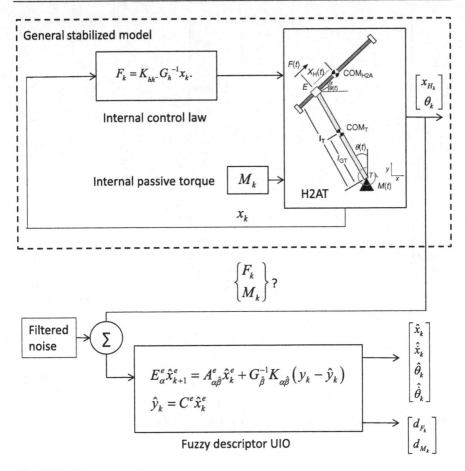

Fig. 3 Simulation protocol.

5.1.2 Simulation results

The first two simulations are performed to assess the impact of the passive torque on the stabilization of the H2AT model, first of all with a null torque then with a constant value of $M = 0.2\,\mathrm{N\,m}$. In Fig. 4, the measured values of x_{H_k} and θ_k are displayed for both simulations (above: without torque, below: with constant torque). The constant torque induces a shift in the measurement values when compared to the value without lumbar torque obtained after stabilization, i.e., the equilibrium position of the model is changed.

Fig. 5 presents results from the UIO, i.e., estimation of the internal efforts (controlling force and passive torque) during the first second of simulation with a constant torque. The spikes at the first samples of the simulation are due to initial conditions problems that will not occur for real-time experiments. The difference between simulated and estimated results converges to zero after 5 s of simulation.

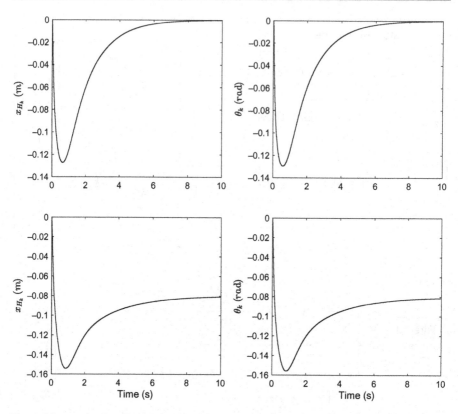

Fig. 4 Measured values in simulation without (above) and with passive torque (below).

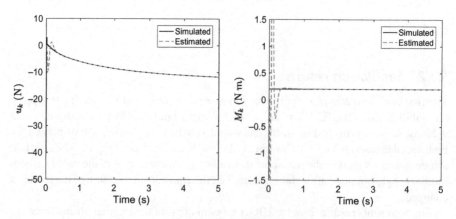

Fig. 5 Simulated (black) and estimation (dashed gray) of both unknown inputs, i.e., the force computed with the control law (12) (left) and the constant passive torque (right).

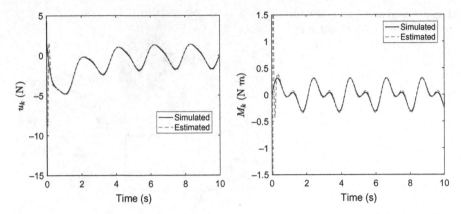

Fig. 6 Simulated (black) and estimation (dashed gray) of both unknown inputs, i.e., the force computed with the control law (12) (left) and the time-varying torque (right).

The last simulation was done with a time-varying internal torque defined by the sum of two sinusoidal signal with 0.2 N m amplitude and of frequency 3 and 6Hz. Estimation of the force and internal torque for this simulation is provided in Fig. 6; once again, convergence occurs rapidly and follow both unknown inputs during the simulation.

This part shown that the UIO is able to reconstruct both state and unknown inputs; therefore, next part discusses experimental results.

5.2   Experimental protocol

5.2.1   Protocol

Four subjects who sustained a complete SCI (ASIA-A, see Table 1) maintained a sitting positon on a height-adjustable table without back support with hip and knees flexed to 90°, feet resting on the floor and upper limbs flexed to 90° at the elbow level. When sitting stability is reached, a light destabilizing force is randomly applied at the level of the third thoracic vertebra between the scapulas. The destabilizing force is generated via an impact with a foam-coated wooden pole to which a pressure sensor is added on the tip to define the contact instant (see Figs. 7 and 8). Each subject completed a minimum of nine acquisitions.

Table 1 Subjects anthropometric parameters

Subjects	Age (y)	Sex	Masse (kg)	Height (cm)	Injury level	Injury age (y)
S1	32	F	54.8	162.6	T6	2.3
S2	37	M	73.1	187	T6	10.4
S3	32	F	57.5	163	T8	2.8
S4	53	M	100	180	T11	3.3

Fig. 7 Experimental setup with the pressure sensor (PS), placed at the tip of the destabilizing device.

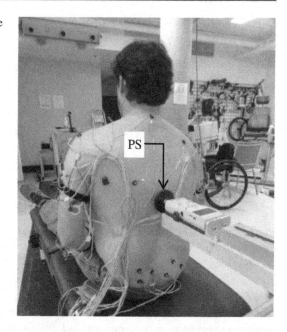

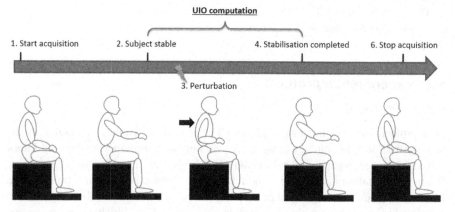

Fig. 8 Experimental procedure.

To record the kinematics, 39 skin-fixed infrared light-emitting diodes are attached on the participant's pelvis, trunk, head, and upper limbs (Wu et al., 2005) and 16 anatomical bony landmarks are digitized using a 6 LEDs probe. Kinematics data are recorded at 60 Hz using a four-camera OptoTrak system (NDI, Waterloo, Canada). The location of joints centers is estimated using the methodology and regressions tables proposed by Dumas et al. (2007) and Fang et al. (2017). The positions of points COM_{H2A} and T are calculated as the barycenter of the center of mass of upper limbs and head, respectively, and intersection of all lines passing through lumbar joint centers and point E during the acquisition. To reduce measurement noise, all signals are filtered with a low-pass four-order Butterworth filter with a cutoff frequency of 10 Hz (Gagnon et al., 2012). After verifying that the displacement of the COM$_{H2A}$ is linear related to the trunk segment, sagittal angle θ_k and COM$_{H2A}$ displacement x_{H_k} are calculated and injected in the UIO (25) with appropriate mass and segment length. The observers gains were calculated using a decay rate of $\lambda = 0.85$. The estimation error is then calculated for both measured data.

Ethical approval has been obtained from the Research Ethics Committee of the Center for Interdisciplinary Research in Rehabilitation of Greater Montreal (CRIR-1083-0515R). The participants read and signed the informed consent form prior to initiating the measurements. We wish to thank Pr. Dany H. Gagnon from the Pathokinesiology Laboratory of Montreal, Canada, for giving us the opportunity to use the technical equipment and apply our theoretical tools to subjects living with an SCI.

5.2.2   Results

When using UIO, the first step is to calculate the estimation error between measured and estimated data to verify the UIO's ability to follow the input signal and thus make a good estimation of the unknown inputs. Fig. 9 presents an example of angle θ_k and displacement x_{H_k} and relative estimation error during one acquisition, while Fig. 10 shows the root mean square error (RSME) for both measured variables during all acquisitions.

The small RMSE values show that both measured data are compatible with the UIO defined on the H2AT model, we can now analyze the estimated internal forces.

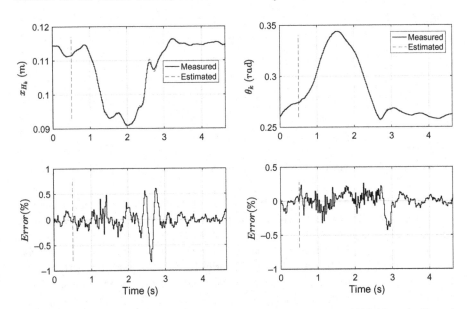

Fig. 9 (Up) Experimental input measurements (black) and estimation (dashed gray). (Down) Estimation error. The vertical dashed line represents the perturbation instant.

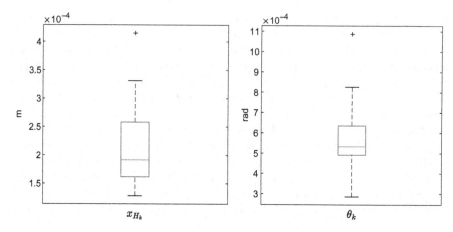

Fig. 10 Experimental input estimation error root mean square for all acquisitions.

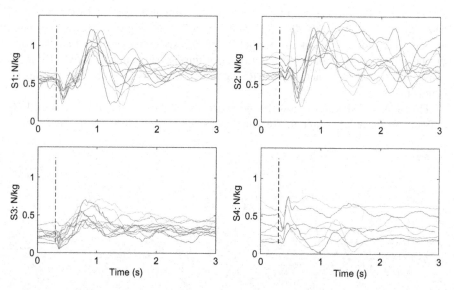

Fig. 11 Experimental input estimation error root mean square for all acquisitions.

Figs. 11 and 12 present the estimated internal force (resp. torque) for all acquisitions of the four subjects. In order to compare internal efforts between subjects, the data are normalized by the mass of each subjects (Moisio et al., 2003). In Fig. 11, the normalized force amplitude of subjects S1 and S2 is greater than for subjects S3 and S4. Moreover, the estimated force of S1 and S2 varies for a longer time after perturbation. These results mean that the first two subjects rely more on their uppers limbs, making both wider and longer stabilization movements. In Fig. 12, the normalized torques of subjects S3 and S4 decrease faster than S1 and S2 in the first instants after the perturbation and reach a steady state 1 s after perturbation, whereas subjects S1 and S2's normalized torques are still varying until 3 s after perturbations. In both figures, force

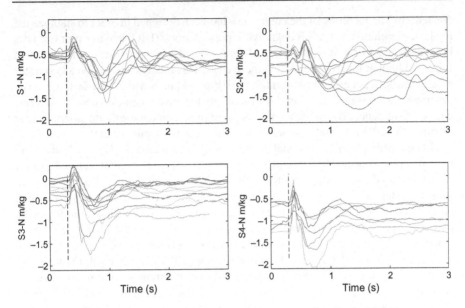

Fig. 12 Experimental input estimation error root mean square for all acquisitions.

and torque do not always return to their original position before perturbation; this result can be induced by two physiological reasons: (1) the subject chose a new position of its upper limbs after the perturbation, thus changing the equilibrium position and (2) a lack of precision in the human sensors resulting in a steady-state like error. Steady-state errors after trunk perturbations have been studied in healthy subjects (Thrasher et al., 2010) and were estimated to be less than 5% (i.e., the final position is $\pm 5\%$ of the initial position). Such a small value compared to the difference exposed in Figs. 11 and 12 suggests rejecting this latter hypothesis in favor of a new equilibrium position. Moreover, changing the upper body position during voluntary movements has been observed for people living with SCI before (Gabrielli et al., 2016; Milosevic et al., 2017). Because a change of static equilibrium position is impossible to obtain with a single inverted pendulum, this result highlights the relevance of the H2AT model.

These results show an important difference in the stabilization strategies used by the four subjects. They also suggest that subjects S1 and S2 are less stable than subjects S3 and S4; this can be explained by the fact that S1 and S2 have the highest injury level, which is associated with a poorer trunk sensorimotor control. Moreover, subjects S3 and S4 can recruit to a certain extent some of their lower back muscles, which explains the torque estimation differences in Fig. 12.

6   Conclusions and future works

The goal of this work was the modeling and characterization of sitting stability for people living with a complete SCI. To this end, a systematic procedure has been stated: (a) The H2AT model, a variation of a double inverted pendulum has been created; (b) The feedback of the control law has been designed in order to stabilize the model in simulation; (c) A T-S unknown-inputs-observer (UIO) in descriptor form has been designed to estimate the internal forces (head and upper limbs displacement force and lumbar torque) to stabilize the system. This method applied on a descriptor system has been put under LMI conditions (Estrada-Manzo et al., 2016, 2015a, b). The observation results in simulation show the ability to converge quickly despite the presence of disturbance. By the mean of a robust approach to cope with the nonmeasured premise, the UIO estimates the states and the unknown input correctly.

When creating the H2AT model, some assumptions are made. First of all, the main movement (rotation and displacement) is assumed to be in the sagittal plan and a linear displacement of the COM_{H2A} related to the trunk segment is defined. For obvious reasons of model correctness, both these assumptions are verified with the experimental data before going any further in the estimation process. A second assumption is the pure joint (i.e., without any friction or damping) defined at the lumbar level without any passive reaction against flexion and extension. This kind of assumption has often been used in biomechanics for the ankle (Peterka, 2000) or the trunk (Vette et al., 2010), but it is known that flexion and extension of the trunk lead to resistive forces created by passive physiological phenomenon (Cholewicki et al., 1999a, b; Panjabi, 1992). Taking into account these parameters (i.e., a viscoelastic definition of the torque at the lumbar level) would allow us to derive and UIO with a better coherence with human behavior. The resulting UIO built on this more detailed model could not only estimate torque from active joints but also the contribution of passive elements on the trunk region and posterior postural chain. Previous works attempted to estimate the contribution of active and passive torques in joints during sitting control (Reeves et al., 2009); however, this work used a linear mechanical representation, which only yields constant parameters estimation. Time-invariant parameters are not consistent in this case with the knowledge that passive stiffness contributions are time varying (Beach et al., 2005; De Carvalho and Callaghan, 2011). This highlights the absolute necessity to work with nonlinear models.

On the point of view of control, a next step is to adapt the control design to individual's ability by adding constraint like decay rate, D-stability, rising time, or the control rate maximum speed. The closed-loop behavior can also be estimated and some parameters of the control law identified from experiments. This point will allow us obtaining realistic patterns according to the injury and the individual.

Logically, a future direction is to extend the H2AT model to three dimensions and this will bring new problems to tackle, one of them being to deal with high-sized T-S models. The use of the T-S formalism gives the opportunity to deal with an exact nonlinear model to the prize of an increasing complexity according to the number of nonlinearities. Thus, we can be faced with limitation of the actual solvers. Being in the context of observation, where the point is to keep a model as close as possible to the real model, a way would be to consider cascaded T-S systems (Lendek et al., 2011) with low levels of nonlinear coupling. This method reduces the global complexity because the LMIs are solved for each subsystem but there is a need to find a common Lyapunov function for the complete system.

We wish to emphasize the method of the nonlinear UIO as another way to estimates unmeasurable values in biomechanics research. Future research are directed to experimental validation of the observer, geometrical adaptation of the model, and increasing the complexity of the model in order to take into account physiological phenomena.

References

Beach, T.A.C., Parkinson, R.J., Stothart, J.P., Callaghan, J.P., 2005. Effects of prolonged sitting on the passive flexion stiffness of the in vivo lumbar spine. Spine J. 5, 145–154.

Bergmark, A., 1989. Stability of the lumbar spine. A study in mechanical engineering. Acta Orthop. Scand. Suppl. 230, 1–54.

Bjerkefors, A., Carpenter, M.G., Thorstensson, A., 2007. Dynamic trunk stability is improved in paraplegics following kayak ergometer training. Scand. J. Med. Sci. Sports 17, 672–679.

Blandeau, M., Estrada-Manzo, V., Guerra, T.M., Pudlo, P., Gabrielli, F., 2018. Fuzzy unknown input observer for understanding sitting control of persons living with spinal cord injury. Eng. Appl. Artif. Intell. 67, 381–389.

Blandeau, M., Estrada-Manzo, V., Guerra, T.M., Pudlo, P., Gabrielli, F., 2016. Unknown input observer for understanding sitting control of persons with spine cord injury. In: IFAC-Papers OnLine, 4th IFAC Conference on Intelligent Control and Automation Sciences ICONS 2016, Reims, France, 1–3 June 2016. vol. 49, pp. 175–181.

Blandeau, M., Guerra, T.M., Estrada-Manzo, V., Pudlo, P., Gabrielli, F., 2017. Unknown input observer based on discrete-time nonlinear descriptor model for understanding sitting control. IFAC-Pap. OnLine 50, 820–825.

Boyd, S., Ghaoui, L.E., Feron, E., Balakrishnan, V., 1994. Linear Matrix Inequalities in System and Control Theory. Society for Industrial and Applied Mathematics.

Cholewicki, J., Juluru, K., McGill, S.M., 1999a. Intra-abdominal pressure mechanism for stabilizing the lumbar spine. J. Biomech. 32, 13–17.

Cholewicki, J., Juluru, K., Radebold, A., Panjabi, M.M., McGill, S.M., 1999b. Lumbar spine stability can be augmented with an abdominal belt and/or increased intra-abdominal pressure. Eur. Spine J. 8, 388–395.

Crisco, J.J., Panjabi, M.M., Yamamoto, I., Oxland, T.R., 1992. Euler stability of the human ligamentous lumbar spine. Part I. Theory. Clin. Biomech. (Bristol, Avon) 7, 19–26.

De Carvalho, D.E., Callaghan, J.P., 2011. Passive stiffness changes in the lumbar spine and effect of gender during prolonged simulated driving. Int. J. Ind. Ergon. 41, 617–624.

Dumas, R., Chèze, L., Verriest, J.-P., 2007. Adjustments to McConville et al. and Young et al. body segment inertial parameters. J. Biomech. 40, 543–553.

Estrada-Manzo, V., Lendek, Z., Guerra, T.M., 2016. Generalized LMI observer design for discrete-time nonlinear descriptor models. Neurocomputing 182, 210–220.

Estrada-Manzo, V., Lendek, Z., Guerra, T.-M., 2015a. Unknown input estimation of nonlinear descriptor systems via LMIs and Takagi-Sugeno models. In: Presented at the 54th IEEE Conference on Decision and Control, Osaka, Japan.

Estrada-Manzo, V., Lendek, Z., Guerra, T.M., Pudlo, P., 2015b. Controller design for discrete-time descriptor models: a systematic LMI approach. IEEE Trans. Fuzzy Syst. 23, 1608–1621.

Fang, Y., Morse, L.R., Nguyen, N., Tsantes, N.G., Troy, K.L., 2017. Anthropometric and biomechanical characteristics of body segments in persons with spinal cord injury. J. Biomech. 55, 11–17.

Gabrielli, F., Molenaar, C., Blandeau, M., Pudlo, P., 2016. Impact of Spatial Hindrance on Sit-to-Stand and Exit Strategies of Low Mobility Passengers. vol. 77 AMSE IFRATH Publication—2016—Modelling C, pp. 120–129.

Gagnon, D., Duclos, C., Desjardins, P., Nadeau, S., Danakas, M., 2012. Measuring dynamic stability requirements during sitting pivot transfers using stabilizing and destabilizing forces in individuals with complete motor paraplegia. J. Biomech. 45, 1554–1558.

Grangeon, M., Gagnon, D., Gauthier, C., Jacquemin, G., Masani, K., Popovic, M.R., 2012. Effects of upper limb positions and weight support roles on quasi-static seated postural stability in individuals with spinal cord injury. Gait Posture 36, 572–579.

Guelton, K., Delprat, S., Guerra, T.-M., 2008. An alternative to inverse dynamics joint torques estimation in human stance based on a Takagi-Sugeno unknown-inputs observer in the descriptor form. Control. Eng. Pract. 16, 1414–1426.

Guerra, T.M., Estrada-Manzo, V., Lendek, Z., 2015. Observer design for Takagi-Sugeno descriptor models: an LMI approach. Automatica 52, 154–159.

Guerra, T.M., Kerkeni, H., Lauber, J., Vermeiren, L., 2012. An efficient Lyapunov function for discrete T-S models: observer design. IEEE Trans. Fuzzy Syst. 20, 187–192.

Guerra, T.M., Vermeiren, L., 2004. LMI-based relaxed nonquadratic stabilization conditions for nonlinear systems in the Takagi-Sugeno's form. Automatica 40, 823–829.

Ichalal, D., Mammar, S., Ragot, J., 2016a. Auxiliary dynamics for observer design of nonlinear TS systems with unmeasurable premise variables. In: IFAC-Papers OnLine, 4th IFAC Conference on Intelligent Control and Automation Sciences ICONS 2016 Reims, France, 1–3 June 2016.vol. 49, pp. 1–6.

Ichalal, D., Marx, B., Maquin, D., Ragot, J., 2016b. A method to avoid the unmeasurable premise variables in observer design for discrete time TS systems. In: International Conference on Fuzzy Systems, FUZZ-IEEE. Vancouver, Canada.

Ichalal, D., Marx, B., Maquin, D., Ragot, J., 2012. Observer design and fault tolerant control of Takagi-Sugeno nonlinear systems with unmeasurable premise variables. In: Fault Diagnosis in Robotic and Industrial Systems, pp. 1–21.

Janssen-Potten, Y.J.M., Seelen, H.A.M., Drukker, J., Reulen, J.P.H., Drost, M.R., 1999. Postural muscle responses in the spinal cord injured persons during forward reaching. Ergonomics 42, 1200–1215.

Kapandji, A.I., 2005. Anatomie fonctionnelle I: Membres supérieurs. Physiologie de l'appareil locomoteur, sixth ed. Maloine.

Khalil, W., Dombre, E., 2004. Modeling, Identification and Control of Robots. Butterworth-Heinemann.

Lendek, Z., Guerra, T., Lauber, J., 2015. Controller design for TS models using delayed nonquadratic Lyapunov functions. IEEE Trans. Cybernet. 45, 439–450.

Lendek, Z., Guerra, T.M., Babuska, R., de Schutter, B., 2011. Stability Analysis and Nonlinear Observer Design Using Takagi-Sugeno Fuzzy Models, Studies in Fuzziness and Soft Computing. Springer.

Lewis, F.L., Dawson, D.M., Abdallah, C.T., 2003. Robot Manipulator Control: Theory and Practice. CRC Press.

Luenberger, D., 1977. Dynamic equations in descriptor form. IEEE Trans. Autom. Control 22, 312–321.

Milosevic, M., Gagnon, D.H., Gourdou, P., Nakazawa, K., 2017. Postural regulatory strategies during quiet sitting are affected in individuals with thoracic spinal cord injury. Gait Posture 58, 446–452.

Moisio, K.C., Sumner, D.R., Shott, S., Hurwitz, D.E., 2003. Normalization of joint moments during gait: a comparison of two techniques. J. Biomech. 36, 599–603.

Ohtake, H., Tanaka, K., Wang, H.O., 2001. Fuzzy modeling via sector nonlinearity concept. In: Presented at the IFSA World Congress and 20th NAFIPS International Conference. Joint 9th, pp. 127–132.

de Oliveira, M.C., Skelton, R.E., 2001. Stability tests for constrained linear systems. In: Perspectives in Robust Control, Lecture Notes in Control and Information Sciences. Springer, London, pp. 241–257.

Panjabi, M.M., 1992. The stabilizing system of the spine. Part I. Function, dysfunction, adaptation, and enhancement. J. Spinal Disord. Tech. 5, 383–389.

Peterka, R.J., 2000. Postural control model interpretation of stabilogram diffusion analysis. Biol. Cybern. 82, 335–343.

Reeves, N.P., Cholewicki, J., 2003. Modeling the human lumbar spine for assessing spinal loads, stability, and risk of injury. Crit. Rev. Biomed. Eng. 31.

Reeves, N.P., Cholewicki, J., Narendra, K.S., 2009. Effects of reflex delays on postural control during unstable seated balance. J. Biomech. 42, 164–170.

Sala, A., Guerra, T.M., Babuška, R., 2005. Perspectives of fuzzy systems and control. In: Fuzzy Sets and Systems, 40th Anniversary of Fuzzy Sets 156, pp. 432–444.

Seddiki, L., Guelton, K., Zaytoon, J., 2010. Concept and Takagi-Sugeno descriptor tracking controller design of a closed muscular chain lower-limb rehabilitation device. IET Control Theory Appl. 4, 1407–1420.

Silfies, S.P., Cholewicki, J., Radebold, A., 2003. The effects of visual input on postural control of the lumbar spine in unstable sitting. Hum. Mov. Sci. 22, 237–252.

Tanaka, M.L., Ross, S.D., Nussbaum, M.A., 2010. Mathematical modeling and simulation of seated stability. J. Biomech. 43, 906–912.

Taniguchi, T., Tanaka, K., Yamafuji, K., Wang, H.O., 1999. Fuzzy descriptor systems: stability analysis and design via LMIs. In: American Control Conference. IEEE, pp. 1827–1831.

Thrasher, T.A., Sin, V.W., Masani, K., Vette, A.H., Craven, B.C., Popovic, M.R., 2010. Responses of the trunk to multidirectional perturbations during unsupported sitting in normal adults. J. Appl. Biomech. 26, 332–340.

Vette, A.H., Masani, K., Sin, V., Popovic, M.R., 2010. Posturographic measures in healthy young adults during quiet sitting in comparison with quiet standing. Med. Eng. Phys. 32, 32–38.

Vette, A.H., Yoshida, T., Thrasher, T.A., Masani, K., Popovic, M.R., 2012. A comprehensive three-dimensional dynamic model of the human head and trunk for estimating lumbar and cervical joint torques and forces from upper body kinematics. Med. Eng. Phys. 34, 640–649.

Wu, G., Van Der Helm, F.C., Veeger, H.E.J., Makhsous, M., Van Roy, P., Anglin, C., Nagels, J., Karduna, A.R., McQuade, K., Wang, X., 2005. ISB recommendation on definitions of joint coordinate systems of various joints for the reporting of human joint motion. Part II. Shoulder, elbow, wrist and hand. J. Biomech. 38, 981–992.

Yoneyama, J., 2009. H∞ filtering for fuzzy systems with immeasurable premise variables: an uncertain system approach. Fuzzy Sets Syst. Theme: Control Eng. 160, 1738–1748.

Epidemic modeling and control of HIV/AIDS dynamics in populations under external interactions: A worldwide challenge

Paolo Di Giamberardino, Daniela Iacoviello
Department of Computer, Control and Management Engineering, Sapienza University of Rome, Rome, Italy

1 Introduction

In this chapter the interactions between populations affected by different diseases are discussed. In the last years, in the globalized world it has become increasingly important to propose prompt suitable actions whenever an infectious disease starts to spread. In fact it has been estimated that, whereas in the 19th century an infectious disease took about 18 months to spread all over the world, in the recent years it takes less than 36 h, less than the incubation period of most of the diseases. Moreover, the world population has increased in the same period by about five times and almost 400 million people travel to another country or region every year, thus becoming possible transmission vehicles. A classical classification for epidemics refers to epidemic disease, when it is limited both from geographic and temporal point of view, pandemic disease, limited from a temporal point of view but potentially spreading in an entire continent, sporadic case, an irregular (in time and space) presence of an epidemic-like disease, and endemic disease, bounded from a space point of view but not referring to a specific time. Various attempts have been made in the last years to study the spread of epidemic diseases; the main problem is to understand and describe the modalities of the geographic spread among nations with different social characteristics and health status, as well as the effects of migration phenomena.

The problem being so complex, in this work, is efficiently studied in sequential steps: first, a single population with a disease is analyzed, along with the effects of the different control actions when acting separately; then, by using such results, the effects of changes introduced on that population when interacting with external world are deeply discussed. These interactions may be suitably modeled as additional changes in the parameter values and by introducing additional inputs as sum of the effects of migration.

The case study analyzed in this chapter is the human immunodeficiency virus (HIV), which is responsible for the acquired immune deficiency syndrome (AIDS); it infects cells of the immune system, destroying or impairing their function: the

Control Applications for Biomedical Engineering Systems. https://doi.org/10.1016/B978-0-12-817461-6.00008-1

immune system becomes weaker, and the person is more susceptible to infections. AIDS is the most advanced stage of the HIV infection and could be reached in 10–15 years from the infection. It can be transmitted only by some body fluids: blood, semen, preseminal fluid, rectal fluid, vaginal fluid, and breast milk. The data from the World Health Organization (WHO) have confirmed the dangerousness of the HIV, an estimated 1.2 million of people have died of AIDS-related illnesses worldwide in 2014 (last update); in the same period the number of people living with HIV-AIDS was about 37 million. A significant aspect is that only 54% of people with HIV is aware of the infection. Currently no vaccine exists and the treatment consists in standard antiretroviral therapy to maximally suppress the HIV virus and stop its progression; use of condom and regular analysis of the subjects belonging to risk categories could help in contrasting the spread of this pandemia. The HIV/AIDS virus suitably represents the situation at hand: a virus spreading all over the world but with different levels of dangerousness depending on the population interested; the social, economic, and cultural habits may strongly vary the dynamics among the different categories of subjects. The HIV/AIDS is generally studied by considering compartmental modeling, that is, dividing the population into groups homogeneous with respect to the level of disease, as discussed in the following sections. In general susceptible individuals and infected ones are considered, splitting each class depending on the specific characteristics to be discussed; due to the long nonsymptom period, among the infected individuals there are those who are not aware of their situation and the ones in the pre-AIDS and AIDS conditions. Due to its dangerousness, an effective control is required to interrupt the spread, by putting together two characteristics: wise behavior and the fast detection of the infection so that the virus does not spread. Three possible actions are suggested by the WHO; they represent different levels of prevention:

- *Primary prevention*: It is designed for healthy people to reduce the possibility of new infections.
- *Secondary prevention*: It is devoted to a fast identification of new infections and risk conditions to improve the percentage of subjects who become aware of their illness by regular blood tests.
- *Tertiary prevention*: It is the medication to the infected subjects who are aware of infection.

The importance of such actions is discussed in this work by analyzing the effects of each control on the number of subjects in the different categories introduced and enhancing their effectiveness when the interaction among different populations is present.

This chapter is organized as follows: in Section 2, a brief review of the relevant literature is proposed, referring to the general HIV/AIDS modeling aspects and spreading processing. In Section 3, mathematical model referring to a single society is described, whereas in Sections 4 and 5, the model analysis is developed in the absence of control and with constant inputs, respectively. The interactions between populations are deeply described in Section 6, whereas the effects of migration parameters on the individual's evolutions are developed in Section 7. A final discussion of the proposed results is in Section 8; and some conclusions along with future work outlines are in Section 9.

2   Related works

When dealing with epidemic diseases, an important role is played by the development of mathematical models at different levels; for a useful description of the social implications of the infectious spread, various models, referred to different epidemic diseases, such as SIR (Susceptible, Infected and Removed), SARS (Severe Acute Respiratory Syndrome), SIRC (Susceptible, Infected, Removed and Cross-immune), SEIR (Susceptible, Exposed, Infected and Recovered), HIV (Human Immunodeficiency Virus), have been introduced and analyzed (Casagrandi et al., 2006; Chalub and Souza, 2011; Cinati et al., 2004; Kuniya and Nakata, 2012; Yan and Zou, 2008), and, on the basis of such analysis, possible control strategies, such as vaccination, drug medication, and quarantine, are studied (Behncke, 2000; Di Giamberardino and Iacoviello, 2017a, b, 2018; Di Giamberardino et al., 2018; Iacoviello and Liuzzi, 2008a, b; Iacoviello and Stasio, 2013; Joshi, 2002; Tanaka and Urabe, 2014; Wodarz, 2001). Moreover, epidemic models can be applied in different scenarios, such as biological and social networks (Dadlani et al., 2014; Kryftis et al., 2017).

The HIV/AIDS virus spread is considered in this chapter as the disease that could affect a population. Available models of HIV-AIDS infection may be divided into two main groups: one focuses on the dynamics at cell levels (Chang and Astolfi, 2009; Mascio et al., 2004; Wodarz, 2001; Wodarz and Nowak, 1999) and the other deals with the dynamic of subject interactions (Naresh et al., 2009; Pinto and Rocha, 2012). In the first approach, it is considered that in an HIV positive subject the virus infects the CD4 T-cells in the blood; when the number of these cells is below 200 per mm^3 the HIV patient is diagnosed with AIDS. The model by Wodarz (2001) includes the uninfected and the infected CD4 T-cells, the concentration of helper-independent and of the helper-dependent cells, and the concentration of the precursors. The attention is focused on the analysis of the two equilibrium points, one corresponds to the AIDS status and the other to the long-term nonprogressor (LTNP). The same model has been simplified by Chang and Astolfi (2009), by introducing the effects of cytotoxic T lymphocyte, to drive the HIV patient state into the LTNP region of attraction. A double control action aimed at delaying the virus progression and boosting the immune system was proposed by Joshi (2002), whereas by Zhou et al. (2014) have established an idea to reduce the number of virus particles, besides increasing the number of uninfected CD4 T-cells.

The second approach is the one followed in this chapter; generally four categories are introduced: the susceptible one (S), that is, the subjects who are not yet infected but may contract the virus, the infected one (I) containing subjects with HIV but are not aware of the infection, the HIV subjects who are the pre-AIDS patients (P), and the AIDS patients (A). The four-class model by Naresh et al. (2009) considers a constant inflow of HIV-infected subjects assuming birth balancing death and migration. Also natural death is introduced; from the S class a subject could go to the I or P class, whereas in the I class the subjects could discover to be in the pre-AIDS class (P) or in the AIDS one (A). A dynamic compartmental simulation model for Botswana and India by Nagelkerke et al. (2002) considers sex behavioral compartment, high risk and low risk, to identify the best strategies for preventing the spread of HIV/AIDS. As discussed in Section 3, in this work the model adopted is the one introduced by Di Giamberardino et al. (2018); two classes of noninfected subjects are introduced by splitting the S class considering the subjects who are not aware of irresponsible acts and, therefore, could contract the virus, and the subjects representing the wise population who, suitably informed, *try to avoid* dangerous behaviors. Therefore, in the considered approach, the first level of prevention corresponds to the information effort and the use of wise attitudes to assist the noninfected subjects in avoiding the acquisition of the HIV infection. The interaction between populations with different diseases is an interesting topic in a globalized world and requires prompt suitable action (Dadlani et al., 2014; Naresh et al., 2009; Tanaka and Urabe, 2014). A survey on the possible approaches to face the problem of spreading processes was presented by Nowzari et al. (2016); in particular, the concepts of network and metapopulation models are discussed, as well as deterministic and stochastic ones. It is also emphasized that the same kind of modeling could be efficiently applied to spreading processing regarding information propagation through social network, viral marketing, and malware spreading.

A specific work on the role of population interactions and HIV/AIDS spread was performed by Crush et al. (2005). It is evident that a deep understanding of the social, behavioral, and economical elements is important in the analysis of the spread of this virus in order to yield the most effective actions.

3 The single society mathematical model

The mathematical model adopted here has been introduced by Di Giamberardino et al. (2018). It considers two classes of susceptible individuals, divided according to the difference in the probability of being contagious due to different social attitudes and behavior: the first class, S_1, represents people who are not aware of unprotected sex acts and then can easily contract the virus; the second one, S_2, denotes the part of healthy population which, suitably informed, gives a great attention to the partners and to the protections. With such a classification, the first group is mainly responsible for the virus propagation. In addition to the susceptible individuals, the model considers the infected subjects, I, the individuals who are infected but do not know their illness status. They represent the most dangerous class, since they can have sexual relationships with the unwise susceptible individuals S_1 and so spread the infection. The diagnosed patients are divided into two classes: the P class and the A class, the first one represents the individuals who are diagnosed with HIV (pre-AIDS) condition, the latter contain the ones with a diagnosis of AIDS.

Using for the state variables the same names as for the classes, for a more intuitive description, the five-dimensional dynamical model describing the evolution of the population in each of the classes is designed.

On the basis of the following parameters:

- β regulates the interaction responsible of the infectious propagation;
- γ takes into account the fact that a wise individual in $S_2(t)$ can, accidentally, assume a incautious behavior as the $S_1(t)$ persons;
- δ weights the natural rate of $I(t)$ subjects becoming aware of their status;
- α characterizes the natural rate of transition from $P(t)$ to $A(t)$ due to the evolution of the infectious disease;

- ψ determines the effect of the test campaign on the unaware individuals $I(t)$;
- ϕ is the fraction of individuals in $I(t)$ which become, after test, classified as $P(t)$ (ϕ) or $A(t)$ $(1-\phi)$;
- ε is the fraction of individuals $I(t)$ who are discovered to be in the pre-AIDS condition or in the AIDS one;
- d is responsible of the natural death rate, assumed the same for all the classes, while μ is the additional death factor for individuals $A(t)$

the model can be written as

$$\dot{S}_1(t) = Q - dS_1(t) - \frac{c\beta S_1(t)I(t)}{N_c(t)} + \gamma S_2(t) - S_1(t)u_1(t) \tag{1a}$$

$$\dot{S}_2(t) = -(\gamma+d)S_2(t) + S_1(t)u_1(t) \tag{1b}$$

$$\dot{I}(t) = \frac{c\beta S_1(t)I(t)}{N_c(t)} - (d+\delta)I(t) - \psi\frac{I(t)}{N_c(t)}u_2(t) \tag{1c}$$

$$\dot{P}(t) = \varepsilon\delta I(t) - (\alpha_1+d)P(t) + \phi\psi\frac{I(t)}{N_c(t)}u_2(t) + P(t)u_3(t) \tag{1d}$$

$$\dot{A}(t) = (1-\varepsilon)\delta I(t) + \alpha_1 P(t) - (\alpha+d)A(t) + (1-\phi)\psi\frac{I(t)}{N_c(t)}u_2(t) - P(t)u_3(t) \tag{1e}$$

whose compact form is

$$\dot{\xi}(t) = f(\xi(t)) + g_1(\xi(t))u_1(t) + g_2(\xi(t))u_2(t) + g_3(\xi(t))u_3(t) \tag{2}$$

once $\xi(t) = (S_1(t)S_2(t)I(t)P(t)A(t))^T$ denotes the five-dimensional state vector of Eq. (1), and the vector fields have the expressions

$$f(\cdot) = \begin{pmatrix} Q - dS_1(t) - \frac{c\beta S_1(t)I(t)}{N_c(t)} + \gamma S_2(t) \\ -(\gamma+d)S_2(t) \\ \frac{c\beta S_1(t)I(t)}{N_c(t)} - (d+\delta)I(t) \\ \varepsilon\delta I(t) - (\alpha_1+d)P(t) \\ (1-\varepsilon)\delta I(t) + \alpha_1 P(t) - (\alpha+d)A(t) \end{pmatrix}, \quad g_1(\cdot) = \begin{pmatrix} -S_1(t) \\ S_1(t) \\ 0 \\ 0 \\ 0 \end{pmatrix}$$

$$g_2(\cdot) = \begin{pmatrix} 0 \\ 0 \\ -\psi\frac{I(t)}{N_c(t)} \\ \phi\psi\frac{I(t)}{N_c(t)} \\ (1-\phi)\psi\frac{I(t)}{N_c(t)} \end{pmatrix}, \quad g_3(\cdot) = \begin{pmatrix} 0 \\ 0 \\ 0 \\ P(t) \\ -P(t) \end{pmatrix} \tag{3}$$

More in details, the interactions between the classes in model (1) are given by the following terms:

(i) $\gamma S_2(t)$, which represents the fraction of the wise uninfected individuals who accidentally or occasionally behaves like the unwise ones, so move from S_2 to S_1;

(ii) $S_1(t)u_1(t)$, which models the fact that, under a suitable informative campaign whose strength is given by $u_1(t)$, a proportional part of S_1 becomes wise and then moves to S_2 class;

(iii) $\frac{c\beta S_1(t)I(t)}{N_c(t)}$ models the interaction between the unwise people S_1 and the infected individuals who are not conscious of their condition. The term is normalized with respect to $N_c(t) = S_1(t) + S_2(t) + I(t)$, the total population that can have sexual relationships. People who become infected go from S_1 to I, until the infection is diagnosed;

(iv) δI, the fraction of unaware infected people who were discovered to be ill, some of them, $\varepsilon \delta I$, with HIV (pre-AIDS) infection, and the remaining, $(1 - \varepsilon)\delta I$, positive for the AIDS;

(v) $\psi \frac{I(t)}{N_c(t)} u_2(t)$, proportional to the blood analysis campaign $u_2(t)$ operated on the potentially infected $N_c(t)$, for which a positive response is given for the actual $I(t)$ subjects; as for the previous term, the results can prove a HIV infection, $\phi \psi \frac{I(t)}{N_c(t)} u_2(t)$, or an AIDS one, $(1 - \phi)\psi \frac{I(t)}{N_c(t)} u_2(t)$;

(vi) $\alpha_1 P(t)$ and $P(t)u_3(t)$ describe the natural degeneration of the illness from HIV to AIDS (first term) and a therapy action to prevent such transition (second term);

(vii) a natural death rate d is added to all the classes; an additional rate α is introduced for the AIDS-infected individuals to consider the increased probability to die in such a critical conditions.

The external actions, which for the isolated group correspond to control inputs for the dynamical model, are defined by u_1, the informative actions to prevent dangerous relationships, u_2, the blood analysis to discover infected individuals, and u_3, the therapy action for the diagnosed patients.

A deeper explanation of the meaning and the role of each coefficient and each term, along with the motivation of their introduction was provided by Di Giamberardino et al. (2018).

4   Stability analysis

Before starting with the analysis of the behavior of the dynamics (1) under additional external immigration and emigration, some considerations on the equilibrium conditions and the stability for the isolated systems are briefly recalled to better understand the system behavior.

The internal stability analysis is performed on the uncontrolled system (1), with $u_1(t) = u_2(t) = u_3(t) = 0$. First, the equilibrium points are computed and then the local stability is studied.

4.1   Equilibrium points

The equilibrium points of dynamics (1) are computed solving the nonlinear system

$$f(S_1^e, S_2^e, I^e, P^e, A^e) = 0 \tag{4}$$

with $f(\cdot)$ given in Eq. (3) (Di Giamberardino et al., 2018). Computations give two solutions

$$
\xi_1^e = \begin{pmatrix} \frac{1}{d} \\ 0 \\ 0 \\ 0 \\ 0 \end{pmatrix} Q, \quad \xi_2^e = \begin{pmatrix} \frac{1}{c\beta - \delta} \\ 0 \\ \frac{c\beta - (d+\delta)}{(d+\delta)(c\beta - \delta)} \\ \frac{\varepsilon\delta[c\beta - (d+\delta)]}{(\alpha_1 + d)(d+\delta)(c\beta - \delta)} \\ \frac{\delta[c\beta - (d+\delta)][(1-\varepsilon)d + \alpha_1]}{(\alpha_1 + d)(\alpha + d)(d+\delta)(c\beta - \delta)} \end{pmatrix} Q
\tag{5}
$$

ξ_1^e is always a solution, while the existence of ξ_2^e depends on the fulfillment of the condition

$$
c\beta - (d+\delta) > 0
\tag{6}
$$

which is necessary and sufficient for the nonnegativeness of the equilibrium values in ξ_2^e (Di Giamberardino et al., 2018).

4.2 Stability analysis

Local stability of the equilibrium points is studied making use of the linear approximations in a neighborhood of the equilibrium point ξ_1^e and, if it exists, ξ_2^e in Eq. (5). The Jacobian matrix $J_f(\xi) = \frac{\partial f}{\partial \xi}$ must be computed and then evaluated at each equilibrium point. For the equilibrium point ξ_1^e, one gets

$$
A_1 = \frac{\partial f}{\partial \xi}\bigg|_{\xi = \xi_1^e} = \left(\begin{array}{ccc|cc} -d & \gamma & -c\beta & 0 & 0 \\ 0 & -(\gamma + d) & 0 & 0 & 0 \\ 0 & 0 & c\beta - (\delta + d) & 0 & 0 \\ \hline 0 & 0 & \varepsilon\delta & -(\alpha_1 + d) & 0 \\ 0 & 0 & (1-\varepsilon)\delta & \alpha_1 & -(\alpha + d) \end{array} \right)
\tag{7}
$$

The block triangular structure allows to find easily its eigenvalues

$$
\lambda_1 = -d, \lambda_2 = -(\gamma + d), \lambda_3 = c\beta - (d+\delta), \lambda_4 = -(\alpha_1 + d), \lambda_5 = -(\alpha + d)
\tag{8}
$$

and the equilibrium point ξ_1^e is locally asymptotically stable if and only if

$$
c\beta - (d+\delta) < 0
\tag{9}
$$

As far as the second equilibrium point ξ_2^e in Eq. (5) is concerned, the block triangular structure of the linear approximating dynamical matrix

$$A_2 = \left.\frac{\partial f}{\partial \xi}\right|_{\xi=\xi_2^e} = \begin{pmatrix} A_2^{11} & 0 \\ A_2^{21} & A_2^{22} \end{pmatrix}$$

(10)

simplifies the analysis, since the eigenvalues of matrix A_2, denoted by $\sigma(A_2)$, are given by $\sigma(A_2) = \sigma(A_2^{11}) \cup \sigma(A_2^{22})$. Then, once the two matrices

$$A_2^{11} = \left. \begin{pmatrix} -d - \frac{c\beta I^e}{N_c^e} + \frac{c\beta S_1^e I^e}{(N_c^e)^2} & \gamma + \frac{c\beta S_1^e I^e}{(N_c^e)^2} & -\frac{c\beta S_1^e}{N_c^e} + \frac{c\beta S_1^e I^e}{(N_c^e)^2} \\ 0 & -(\gamma+d) & 0 \\ \frac{c\beta I^e}{N_c^e} - \frac{c\beta S_1^e I^e}{(N_c^e)^2} & -\frac{c\beta S_1^e I^e}{(N_c^e)^2} & -(d+\delta) + \frac{c\beta S_1^e}{N_c^e} - \frac{c\beta S_1^e I^e}{(N_c^e)^2} \end{pmatrix} \right|_{\xi=\xi_2^e}$$

(11)

and

$$A_2^{22} = \begin{pmatrix} -(\alpha_1+d) & 0 \\ \alpha_1 & -(\alpha+d) \end{pmatrix}$$

(12)

are computed, the stability of the equilibrium point results from the stability of dynamical matrix (Eq. 11), being the eigenvalues of Eq. (12), $\lambda_4 = -(\alpha_1 + d)$ and $\lambda_5 = -(a + d)$ are always negative.

The eigenvalues of Eq. (11) can be computed solving the equation

$$\lambda^2 + (c\beta - \delta)\lambda + \frac{(d+\delta)(c\beta-(d+\delta))(c\beta-\delta)}{c\beta} = 0$$

(13)

since $\lambda_3 = -(\gamma + d)$ is straightforwardly obtained.

For the Descartes' rule of signs, the equilibrium point is locally asymptotically stable if

$$c\beta - (d+\delta) > 0$$

(14)

This means that the equilibrium point ξ_2^e is locally asymptotically stable if and only if Eq. (14) holds. Therefore, it is possible to conclude that if ξ_1^e is the only equilibrium point (Eq. 9 satisfied), it is also locally asymptotically stable. If the second point ξ_2^e exists, then ξ_1^e becomes unstable while ξ_2^e is locally asymptotically stable.

Such a behavior depends on the values of parameters $c\beta$, d, and δ. On the basis of their meanings, the existence and the stability of the two different equilibrium points depend on the relative values of probability of the infection transmission ($c\beta$) and the velocity of reduction of the infected $I(t)$ by natural death (d) or by infection diagnosis (δ).

5 Equilibria and stability analysis under constant inputs

In this section, the effects of the presence of constant inputs on the existence of equilibrium conditions for dynamics (1) and the corresponding stability properties are studied. The aim is to compute where the new equilibria are located and if the stability properties are changed or not, in order to investigate the effects of the inputs in view of the addition of possible external input arising from the interaction between groups.

This analysis is motivated by observing that interactions with other groups can be also represented by suitable inputs (migrations, travels, daily movements across the borders, etc.). Then, preliminary, the inner controls u_1 and u_2 are introduced and their effects are studied. For sake of clarity, in order to put in evidence the contribution and the effects of each control input introduced, the study is performed analyzing the dynamics under the action of one input at a time.

5.1 Analysis of the case $u_1(t) = u_1 = const$, $u_2(t) = u_3(t) = 0$

The study of the changes in the equilibrium points and their stability conditions under the hypothesis that the control u_1 assumes a constant value different from zero is performed here. First, the equilibria are computed and the relationships with the uncontrolled equilibria are investigated. Then, their stability conditions are checked.

5.1.1 Computation of the equilibrium points for $u_1 \neq 0$

In this case, the system to be solved is

$$f(S_1^e, S_2^e, I^e, P^e, A^e) + g_1(S_1^e, S_2^e, I^e, P^e, A^e)u_1 = 0 \tag{15}$$

which is the same as Eq. (4), with the addition of $g_1(\xi(t))u_1$, with g_1 as in Eq. (3).

The solutions can be computed analytically. After some manipulations, one gets the two equilibrium points

$$\xi_1^e(u_1) = \left(\frac{1}{d\left(1 + \frac{u_1}{\gamma + d}\right)} \quad \frac{\frac{u_1}{\gamma + d}}{d\left(1 + \frac{u_1}{\gamma + d}\right)} \quad 0 \quad 0 \quad 0 \right)^T Q \tag{16}$$

and

$$
\xi_2^e(u_1) = \begin{pmatrix} \dfrac{1}{c\beta - \delta\left(1 + \frac{u_1}{\gamma + d}\right)} \\[2em] \dfrac{\frac{u_1}{\gamma + d}}{c\beta - \delta\left(1 + \frac{u_1}{\gamma + d}\right)} \\[2em] \dfrac{1}{d + \delta}\left(1 - \dfrac{d\left(1 + \frac{u_1}{\gamma + d}\right)}{c\beta - \delta\left(1 + \frac{u_1}{\gamma + d}\right)}\right) \\[2em] \dfrac{\varepsilon\delta}{(\alpha_1 + d)(d + \delta)}\left(1 - \dfrac{d\left(1 + \frac{u_1}{\gamma + d}\right)}{c\beta - \delta\left(1 + \frac{u_1}{\gamma + d}\right)}\right) \\[2em] \dfrac{\delta\left(1 - \frac{\varepsilon d}{\alpha_1 + d}\right)}{(\alpha + d)(d + \delta)}\left(1 - \dfrac{d\left(1 + \frac{u_1}{\gamma + d}\right)}{c\beta - \delta\left(1 + \frac{u_1}{\gamma + d}\right)}\right) \end{pmatrix} Q \tag{17}
$$

consistent with the uncontrolled case (5).

Moreover, as in the uncontrolled case, the solutions are admissible if they have nonnegative components. The first point, $\xi_1^e(u_1)$, clearly satisfies such condition for every $u_1 \geq 0$. For the second point, condition

$$
c\beta - (d + \delta)\left(1 + \frac{u_1}{\gamma + d}\right) \geq 0 \tag{18}
$$

must be verified; it is the extension of Eq. (6) under u_1.

5.1.2 Stability analysis for $u_1 \neq 0$

The study of local stability for the equilibrium points (16), (17) can be performed, as in the unforced case in Section 4.2, replacing f with $f + u_1 g_1$ (as in Eq. 3). Then, the Jacobian matrix to be computed can be denoted by

$$
J_{f + u_1 g_1}(\xi) = \frac{\partial(f + u_1 g_1)}{\partial \xi} \tag{19}
$$

The evaluation of Eq. (19) in the equilibrium point $\xi_1^e(u_1)$ gives

$$
A_1(u_1) = \left(\begin{array}{cc|ccc} -d - u_1 & \gamma & -c\beta\frac{1}{1 + \frac{u_1}{\gamma + d}} & 0 & 0 \\ u_1 & -(\gamma + d) & 0 & 0 & 0 \\ \hline 0 & 0 & c\beta\frac{1}{1 + \frac{u_1}{\gamma + d}} - (\delta + d) & 0 & 0 \\ 0 & 0 & \varepsilon\delta & -(\alpha_1 + d) & 0 \\ 0 & 0 & (1 - \varepsilon)\delta & \alpha_1 & -(\alpha + d) \end{array}\right) \tag{20}
$$

The block structure helps once again to study the signs of the real parts of the eigenvalues. For the upper left block, the eigenvalues are obtained solving the characteristic equation

$$\lambda^2 + (\gamma + 2d + u_1)\lambda + d(\gamma + d + u_1) = 0 \tag{21}$$

The solutions $\lambda_1(u_1)$ and $\lambda_2(u_1)$ of Eq. (21) have always negative real part because the three coefficients are all positive for any admissible u_1. So

$$Re(\lambda_1(u_1)) < 0 \quad \text{and} \quad Re(\lambda_2(u_1)) < 0 \quad \forall u_1 \in [0, +\infty)$$

For the bottom right triangular block, the eigenvalues are

$$\lambda_3(u_1) = c\beta \frac{1}{1 + \frac{u_1}{\gamma + d}} - (\delta + d), \quad \lambda_4 = -(\alpha_1 + d), \quad \lambda_5 - (\alpha + d) \tag{22}$$

While λ_4 and λ_5 are always real negative, the condition $c\beta \frac{1}{1 + \frac{u_1}{\gamma + d}} - (\delta + d) < 0$ must be verified for $\lambda_3(u_1)$ as u_1 varies. It can be rewritten as

$$c\beta - (d + \delta)\left(1 + \frac{u_1}{\gamma + d}\right) < 0 \tag{23}$$

This condition and the one in Eq. (18) are mutually exclusive. Then, as in the unforced case, if $\xi_2^e(u_1)$ does not exist (Eq. 18 not verified), then $\xi_1^e(u_1)$ is locally asymptotically stable; otherwise, if $\xi_2^e(u_1)$ exist, $\xi_1^e(u_1)$ is an unstable equilibrium point. In this forced case, existence and stability of equilibrium points depend on u_1. Recalling the stability condition on the system parameters for the unforced dynamics in Eq. (9), it can be stated that if for the unforced system ξ_1^e is the only, stable, equilibrium point, then the same holds for $\xi_1^e(u_1)$, $\forall u_1$; on the other hand, if $c\beta - (d + \delta) > 0$, that is, ξ_2^e exists and ξ_1^e is unstable, there exists the value

$$u_{1,0} = (\gamma + d)\left(\frac{c\beta}{d + \delta} - 1\right) \tag{24}$$

such that

$$c\beta - (\delta + d)\left(1 + \frac{u_1}{\gamma + d}\right) < 0 \quad \forall u_1 > u_{1,0} \tag{25}$$

Changing u_1, the transition from the conditions in which Eq. (23) holds, with one asymptotically stable equilibrium point $\xi_1^3(u_1)$, to the conditions with Eq. (23) not true, with two equilibria $\xi_1^3(u_1)$ and $\xi_2^3(u_1)$, presents a bifurcation (Di Giamberardino et al., 2018) at the point

$$\xi_{12}^e = \lim_{u_1 \to u_{1,0}} \xi_1(u_1) = \lim_{u_1 \to u_{1,0}} \xi_2(u_1) = \left(\frac{d + \delta}{c\beta d} \quad \frac{c\beta - (d + \delta)}{c\beta d} \quad 0 \quad 0 \quad 0\right)^T Q \tag{26}$$

In order to put in evidence such behaviors, a numerical case is introduced to depict the time histories of the state variables as u_1 changes.

The numerical values chosen for the model parameters are the same as used in Di Giamberardino et al. (2018).

The initial conditions $\xi_0 = (92,000 \ \ 0 \ \ 8000 \ \ 0 \ \ 0)^T$ are used in all the numerical simulations, corresponding to the situation in which an infection is present but nobody knows it yet.

To help the interpretation of the results in the next figures, the numerical values of the equilibrium points, for the set of parameters taken, are given:

$$\xi_1^e(u_1) = \left(\frac{1}{0.02\left(1+\frac{u_1}{0.22}\right)} \quad \frac{\frac{u_1}{0.22}}{0.02\left(1+\frac{u_1}{0.22}\right)} \quad 0 \quad 0 \quad 0 \right)^T Q \tag{27}$$

$$\xi_2^e(u_1) = \begin{pmatrix} \frac{1}{1.5-0.4\left(1+\frac{u_1}{0.22}\right)} \\[2mm] \frac{\frac{u_1}{0.22}}{1.5-0.4\left(1+\frac{u_1}{0.22}\right)} \\[2mm] 2.3810\left(1 - \frac{0.02\left(1+\frac{u_1}{0.22}\right)}{1.5-0.4\left(1+\frac{u_1}{0.22}\right)}\right) \\[2mm] 1.0989\left(1 - \frac{0.02\left(1+\frac{u_1}{0.22}\right)}{1.5-0.4\left(1+\frac{u_1}{0.22}\right)}\right) \\[2mm] 0.9122\left(1 - \frac{0.02\left(1+\frac{u_1}{0.22}\right)}{1.5-0.4\left(1+\frac{u_1}{0.22}\right)}\right) \end{pmatrix} Q \tag{28}$$

with $c\beta - (d+\delta) = 1.08 > 0$, so that the uncontrolled dynamics has two equilibrium points, the first unstable and the second, the endemic condition, asymptotically stable. With $u_1 \neq 0$, the threshold value $u_{1,0}$ is $u_{1,0} = 0.5657$ and the bifurcation point ξ_{12}^e is

$$\xi_{12}^e = \left(1.4 \times 10^5 \quad 3.6 \times 10^5 \quad 0 \quad 0 \quad 0\right)^T \tag{29}$$

Two sets of figures show, separately, the different behaviors for values of u_1 smaller and greater than $u_{1,0}$: in Figs. 1–4 the case in which $0 \leq u_1 \leq u_{1,0}$ is reported, while in Figs. 5–8 the case in which $u_1 \geq u_{1,0}$ is referred.

For the first set, the values $u_1 \in [0, 0.1, 0.20.30.4, 0.5, u_{1,0}]$ are considered. In Figs. 1–4, the dashed line denotes the evolutions for $u_1 = 0$, while the dotted line represents the behaviors when $u_1 = u_{1,0}$. The five solid lines are associated with the intermediate values: they show a behavior that uniformly goes from the uncontrolled one, dashed line, to the bifurcation value for the control, dotted line. Such a variation increases for $S_1(t)$ and $S_2(t)$, while it decreases for $I(t)$. This means that as u_1 increases, the parts of the population in $S_1(t)$ and $S_2(t)$, after an initial decrement, increases as well, showing, at each time, greater numbers for greater control amplitude and reaching, at steady state, greater numbers. Correspondingly, the infected population in $I(t)$ evolves following smaller values at each time, the steady-state value decreases as u_1 increases, till assuming zero as asymptotic value when u_1 is equal to the bifurcation value. The overall effect is the increment in the fraction of healthy population, depicted in Fig. 4, reaching the full health condition as $u_1 \to u_{1,0}$.

In the second set, the values $u_1 \in [u_{1,0}, 0.6, 0.7, 0.8, 0.9, 1]$ are considered. In this case also, in Figs. 5–8, the dotted line represents the behavior when $u_1 = u_{1,0}$. As previously proved, for $u_1 > u_{1,0}$ the system has only one equilibrium point, the first one, and it is locally asymptotically stable. Then, according to the expression (27), the behavior of the state variable $S_1(t)$, for its steady-state value, decreases as u_1 increases, and for the higher value simulated, $u_1 = 1$, it reaches the value of 9.0164×10^4. At the same time, S_2 increases, as shown in Fig. 6; for $u_1 = 1$, its steady-state value is 4.0984×10^5, as expression (27) gives. The time history of $I(t)$, depicted in Fig. 7, shows a faster convergence to zero as u_1 increases, always having zero as steady-state value, according to Eq. (27). Then, all the population tends to be healthy, as shown in Fig. 8.

For the study of local stability of $\xi_2^e(u_1)$ an analogous analysis must be performed, starting with the evaluation of Eq. (19) in such an equilibrium point. The result is the block triangular matrix

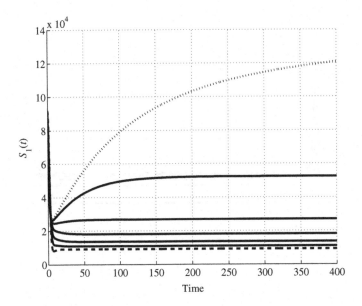

Fig. 1 Time history of $S_1(t)$ for values $0 \leq u_1 \leq u_{1,0}$; the *dashed line* corresponds to $u_1 = 0$, the *dotted line* denotes the case $u_1 = u_{1,0}$.

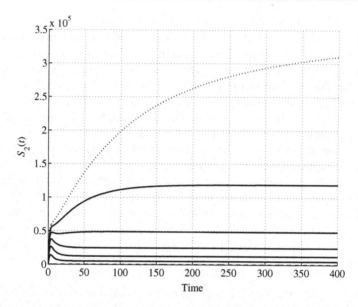

Fig. 2 Time history of $S_2(t)$ for values $0 \leq u_1 \leq u_{1,0}$; the *dashed line* corresponds to $u_1 = 0$, the *dotted line* denotes the case $u_1 = u_{1,0}$.

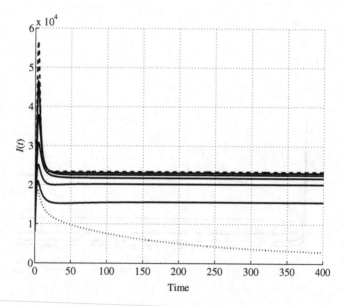

Fig. 3 Time history of $I(t)$ for values $0 \leq u_1 \leq u_{1,0}$; the *dashed line* corresponds to $u_1 = 0$, the *dotted line* denotes the case $u_1 = u_{1,0}$.

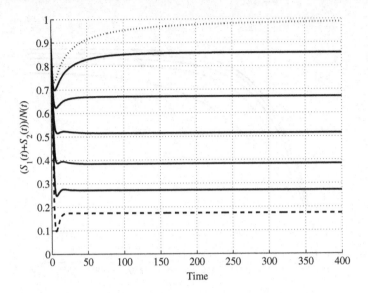

Fig. 4 Time evolution of the relative number of healthy individuals $S_1(t) + S_2(t)$ w.r.t. the total population for values $0 \leq u_1 \leq u_{1,0}$; the *dashed line* corresponds to $u_1 = 0$, the *dotted line* denotes the case $u_1 = u_{1,0}$.

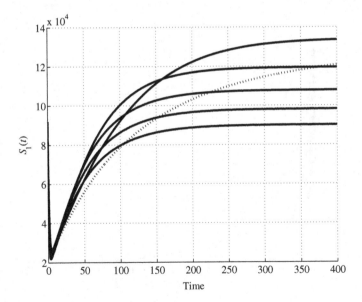

Fig. 5 Time history of $S_1(t)$ for values $u_1 \geq u_{1,0}$.

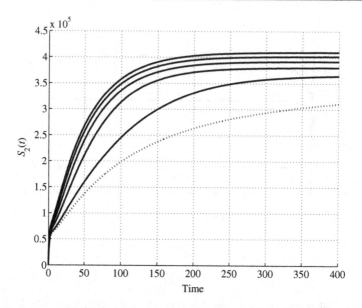

Fig. 6 Time history of $S_2(t)$ for values $u_1 \geq u_{1,0}$.

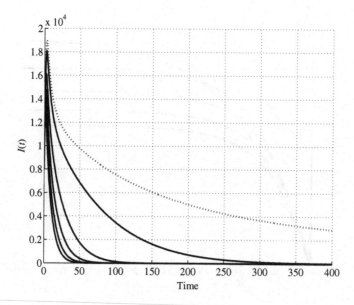

Fig. 7 Time history of $I(t)$ for values $u_1 \geq u_{1,0}$.

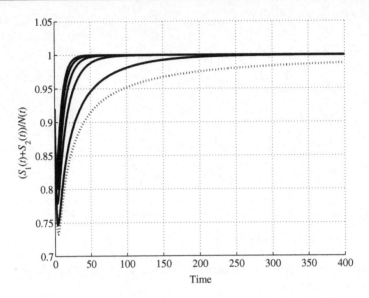

Fig. 8 Time evolution of the relative number of healthy individuals $S_1(t) + S_2(t)$ w.r.t. the total population for values $u_1 \geq u_{1,0}$.

$$A_2(u_1) = \left(\begin{array}{c|c} A_2^{11} & 0 \\ \hline A_2^{21} & A_2^{22} \end{array} \right) \tag{30}$$

with

$$A_2^{11}(u_1) = \begin{pmatrix} -(d+u_1) - c\beta\frac{I^e}{N_c^e}\left(1-\frac{S_1^e}{N_c^e}\right) & \gamma + c\beta\frac{S_1^e}{N_c^e}\frac{I^e}{N_c^e} & -c\beta\frac{S_1^e}{N_c^e}\left(1-\frac{I^e}{N_c^e}\right) \\ u_1 & -(\gamma+d) & 0 \\ c\beta\frac{I^e}{N_c^e}\left(1-\frac{S_1^e}{N_c^e}\right) & -c\beta\frac{S_1^e}{N_c^e}\frac{I^e}{N_c^e} & -(d+\delta) + c\beta\frac{S_1^e}{N_c^e}\left(1-\frac{I^e}{N_c^e}\right) \end{pmatrix} \tag{31}$$

$$A_2^{21} = \begin{pmatrix} 0 & 0 & \varepsilon\delta \\ 0 & 0 & (1-\varepsilon)\delta \end{pmatrix} \quad A_2^{22} = \begin{pmatrix} -(\alpha_1+d) & 0 \\ \alpha_1 & -(\alpha+d) \end{pmatrix} \tag{32}$$

After some computations, the characteristic polynomial of matrix (31) can be written as

$$p(\lambda) = \lambda^3 + m_2\lambda^2 + m_1\lambda + m_0$$

with

$$m_2 = \left(c\beta - \delta + (\gamma + d) + \frac{\gamma - \delta}{\gamma + d}u_1\right)$$

$$m_1 = (c\beta - \delta)\left((\gamma + d) + \frac{(c\beta - (d+\delta))(d+\delta)}{c\beta}\right)$$

$$+ \delta u_1 - \frac{(d+\delta)^2(c\beta - \delta)}{c\beta(\gamma + d)}u_1 \tag{33}$$

$$m_0 = -\frac{(d+\delta)(c\beta - (d+\delta))(\gamma + d)(c\beta - \delta)}{c\beta}$$

$$+ \frac{d+\delta}{c\beta}(dc\beta + 2\delta(c\beta - (d+\delta)))u_1 - \delta\frac{(d+\delta)^2}{c\beta(\gamma + d)}u_1^2$$

The conditions for the local stability of the equilibrium point (17) are

$$m_i > 0, \quad i = 0,1,2 \text{ and } m_1 m_2 > m_0 \tag{34}$$

For the numerical case referred here, the equilibrium point is Eq. (28) and the Jacobian matrix (31) becomes

$$A_2^{11}(u_1) = \begin{pmatrix} -0.7976 + 0.3745u_1 & 0.5024 - 0.5345u_1 & -0.1176 - 0.5345u_1 \\ u_1 & -0.2200 & 0 \\ 0.7776 - 1.3745u_1 & -0.3024 + 0.5345u_1 & -0.3024 + 0.5345u_1 \end{pmatrix} \tag{35}$$

In Fig. 9, the real part of the three eigenvalues of matrix (35) for values of $u_1 \in [0, 1]$ are reported. It can be observed that, for $u_1 < u_{1,0} = 0.5657$, highlighted in the figure with the vertical dotted line, the linearized dynamics is asymptotically stable, then the equilibrium point $\xi_2^e(u_1)$ (Eq. 17) exists as a second equilibrium point and is asymptotically stable (the endemic condition). Above that value, one eigenvalue becomes positive. The results of this analysis are that the equilibrium and the stability conditions of the dynamics (Eq. 1) can be affected by external inputs, u_1 in this case, even leading to the loss of some stability properties.

In the following section, the same study is carried on for the input u_2, aiming to show whether the influence of the input on the stability is present for the second input too.

5.2 Analysis of the case $u_1(t) = u_3(t) = 0$ and $u_2(t) = u_2$

The case of a constant input $u_2 \neq 0$ is now considered in Section 5.2. As in Section 5.1, first the equilibrium points are computed and then their stability properties are studied.

5.2.1 Equilibria computation for $u_2 \neq 0$

In this case, to compute the new equilibria, the nonlinear system

$$f(S_1^e, S_2^e, I^e, P^e, A^e) + g_2(S_1^e, S_2^e, I^e, P^e, A^e)u_2 = 0 \tag{36}$$

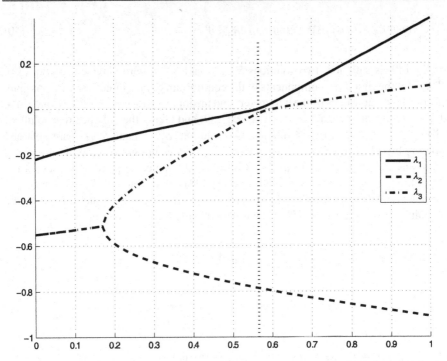

Fig. 9 Real part of the eigenvalues of Eq. (35) for $u_1 \in [0, 1]$ are reported. The *vertical dotted line* denotes $u_{1,0} = 0.5657$, the bifurcation value.

has to be solved. It is the same as Eq. (4), with the addition of $g_2(\xi(t))u_2$, with g_2 as in Eq. (3). Performing the computations, the solution

$$\xi_1^e(u_2) = \left(\tfrac{1}{d} \ 0 \ 0 \ 0 \ 0 \right)^T Q \tag{37}$$

is obtained; it does not depend on u_2 and it is the same as the unforced case ξ_1^e in Eq. (5). Moreover, once $S_2^e = 0$ is obtained, computations give also the expressions

$$I^e = \frac{c\beta - (d+\delta)}{d+\delta} S_1^e - \frac{\psi u_2}{d+\delta} \tag{38}$$

$$c\beta(\delta - c\beta)(S_1^e)^2 + (c\beta Q + \psi u_2(d + c\beta))S_1^e - \psi Q u_2 = 0 \tag{39}$$

Note that if $u_2 = 0$, from Eq. (38) one has the same condition on the uncontrolled case and, in fact, Eq. (39) reduces to a first-order one, yielding the same solution as in Eq. (5). If $(\delta - c\beta) > 0$, the two roots of Eq. (39) are real, one positive and one negative; then, one value for the equilibrium point can be obtained, the positive root, say $S_1^{e0}(u_2)$. On the other hand, if $(\delta - c\beta) < 0$, both the solutions have positive real part. They are real, then they are equilibrium points, if and only if

$$(Qc\beta + \psi u_2(d + c\beta))^2 > 4c\beta Q\psi u_2|\delta - c\beta| \tag{40}$$

If Eq. (40) is satisfied, two solutions $S_1^{e_1}(u_2)$ and $S_1^{e_2}(u_2)$ are obtained for Eq. (39). Then, for the first three components of the equilibrium points, three cases are possible, according to the values of the parameters and the input u_2: (i) for $\delta - c\beta > 0$, one value for $S_1^e(u_2)$ is obtained, $S_1^{e_0}(u_2)$, solving Eq. (39) and taking the only positive solution; (ii) for $\delta - c\beta < 0$, if Eq. (40) holds, two solutions $S_1^{e_1}(u_2)$ and $S_1^{e_2}(u_2)$ are obtained solving Eq. (39); (iii) for $\delta - c\beta < 0$, if Eq. (40) does not hold, no admissible (real) solutions are obtained from Eq. (39). Consequently, the corresponding values for the equilibrium $I^e(u_2)$ can be computed from Eq. (38) and denoted by $I^{e_i}(u_2)$, $i = 0, 1, 2$ correspondingly to the value $S_1^{e_i}(u_2)$, while $S_2^e(u_2) = 0$ holds in any case. For the remaining two components $P^e(u_2)$ and $A^e(u_2)$ using Eq. (38), one has

$$P^{e_i}(u_2) = \left(\frac{\varepsilon \delta}{\alpha_1 + d} + \frac{\phi \psi}{(\alpha_1 + d)\left(S_1^{e_i}(u_2) + I^{e_i}\right)} u_2 \right) I^{e_i} \tag{41}$$

$$A^{e_i}(u_2) = (1 - \varepsilon)\delta I^{e_i}(u_2) + \alpha_1 P^{e_i}(u_2) + \frac{(1 - \phi)\psi I^{e_i}(u_2)}{(\alpha + d)\left(S_1^{e_i}(u_2) + I^{e_i}(u_2)\right)} u_2 \tag{42}$$

The equilibrium points $\xi_2^{e_i}(u_2)$ can then be written as

$$\xi_2^{e_i}(u_2) = \left(S_1^{e_i}(u_2) \quad 0 \quad I^{e_i}(u_2) \quad P^{e_i}(u_2) \quad A^{e_i}(u_2) \right)^T \tag{43}$$

5.2.2 Stability analysis for the equilibrium $\xi_1^e(u_2)$

Once again, the study of local stability for the equilibrium points (37), (43) can be performed referring to their linear approximations. The study for $\xi_1^e(u_2)$ in Eq. (37) provides with the matrix

$$A_1(u_2) = \left(\begin{array}{ccc|cc} -d & \gamma & -c\beta & 0 & 0 \\ 0 & -(\gamma + d) & 0 & 0 & 0 \\ 0 & 0 & c\beta - (\delta + d) - \psi u_2 & 0 & 0 \\ \hline 0 & 0 & \varepsilon\delta + \phi\psi u_2 & -(\alpha_1 + d) & 0 \\ 0 & 0 & (1 - \varepsilon)\delta + (1 - \phi)\psi u_2 & \alpha_1 & -(\alpha + d) \end{array} \right) \tag{44}$$

Then, in addition to the four constant negative eigenvalues $\lambda_1 = -d$, $\lambda_2 = -(\gamma + d)$, $\lambda_3 = -(\alpha_1 + d)$, and $\lambda_4 = -(\alpha + d)$, there is one eigenvalue which is the function of the input u_2, $\lambda_5 = c\beta - (d + \delta) - \psi u_2$. If $c\beta - (d + \delta) < 0$, the same as Eq. (9), then $\lambda_5 < 0 \ \forall u_2 \geq 0$. On the other hand, if condition (9) is not verified for the uncontrolled system, there exists a value $u_{2,0} = \frac{c\beta - (d + \delta)}{\psi}$ such that $\lambda_5 < 0 \ \forall u_2 > u_{2,0}$. This means that a system, for which the parameters are such that the equilibrium ξ_1^e is unstable, under the action of an input u_2 can become stable. Once again, the effect of a control input, in this case, u_2, affects the stability properties of a system (1).

5.2.3   Stability analysis for the equilibrium $\xi_2^{ei}(u_2)$

On the basis of the discussion in Section 5.2.1, the existence, and hence the stability, of the equilibrium points $\xi_2^{ei}(u_2)$ (Eq. 43), is highly dependent on the values of the system parameters. Thus, an analytical study of their stability involves long and complicated expressions, not useful for qualitative discussions. To show the effects of an input u_2 on the existence and the stability of the points (43), a numerical analysis is performed, making use of the values introduced in Table 1.

Numerical simulations are performed for $u_1 = u_3 = 0$ and for values of u_2 in the set $u_2 \in \{0, 0.2, 0.4, 0.6, 0.8, 1\}$. Some of the results are reported in Figs. 10–17. In all the figures, the case $u_2 = 0$, the uncontrolled case, is depicted by the dashed line, for comparative purpose with the other increasing values. Figs. 10 and 11 show the time histories of the two state components $S_1(t)$ and $I(t)$, for different input values. Although monotonical increment in the individuals $S_1(t)$ as u_2 increases is expected, as well as the corresponding decrement in the infected subjects $I(t)$, there are two interesting results which the figures show. The first is the appearance of the oscillatory behavior which is related to the amplitude of the input u_2. This kind of time evolution is well evidenced referring to Fig. 14, where the trajectories in the plane $I(t) + P(t) + A(t)$ versus $S_1(t) + S_2(t)$ is plotted: the spiral trajectory while converging to the equilibrium point is clearly visible. This means that, under input $u_2 \neq 0$, there are time instants in which the number of the infected individuals $I(t)$ can seem greater than expected, but it is related to the transient behavior only. The second result puts in evidence from the numerical simulations is that the number of diagnosed infected individuals $P(t)$ and $A(t)$ do not decrease as $I(t)$ decreases. This is due to the fact that the action of the control u_2 aims at increasing the number of diagnoses, so that while $I(t)$ decreases, people moves to $P(t)$ (Fig. 12) and $A(t)$ (Fig. 13). The overall effects on the total population is shown in Fig. 15, where its increment is clear. While the total population increases, the fraction of healthy people $S_1(t) + S_2(t)$ with respect to the total population strongly increases, as reported in Fig. 16 and, at the same time, the fraction of diagnosed patients $P(t) + A(t)$ significantly decreases (Fig. 17).

Table 1 Numerical values of the model parameters

Parameter	Value		Parameter	Value
d	0.02		c	10
β	0.15		γ	0.2
δ	0.4		ψ	100,000
ε	0.6		α_1	0.5
ϕ	0.95		α	1
Q	10^4			

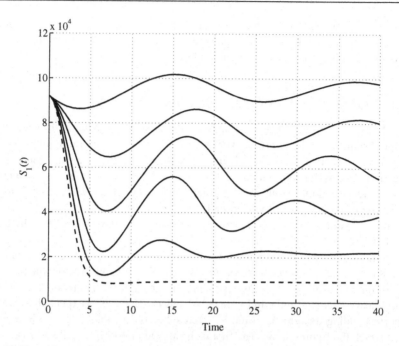

Fig. 10 Time history of $S_1(t)$ for $u_2 \in \{0, 0.2, 0.4, 0.6, 0.8, 1\}$. The *dashed line* depicts the case $u_2 = 0$, the behaviors change monotonically with u_2.

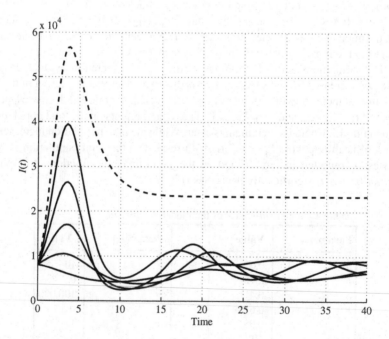

Fig. 11 Time history of $I(t)$ for $u_2 \in \{0, 0.2, 0.4, 0.6, 0.8, 1\}$. The *dashed line* depicts the case $u_2 = 0$, the behaviors change monotonically with u_2.

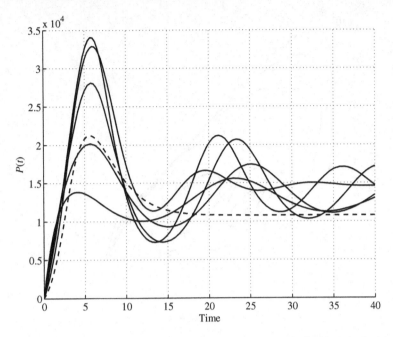

Fig. 12 Time history of $P(t)$ for $u_2 \in \{0, 0.2, 0.4, 0.6, 0.8, 1\}$. The *dashed line* depicts the case $u_2 = 0$, the behaviors change monotonically with u_2.

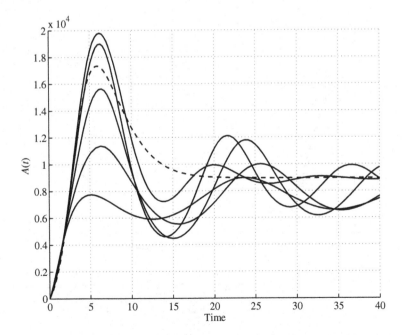

Fig. 13 Time history of $A(t)$ for $u_2 \in \{0, 0.2, 0.4, 0.6, 0.8, 1\}$. The *dashed line* depicts the case $u_2 = 0$, the behaviors change monotonically with u_2.

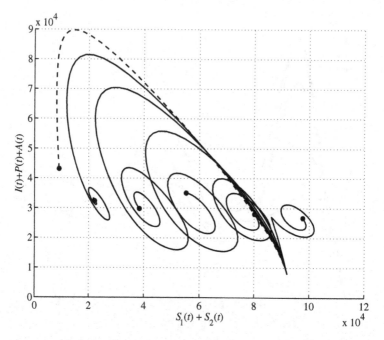

Fig. 14 Time history of $A(t)$ for $u_2 \in \{0, 0.2, 0.4, 0.6, 0.8, 1\}$. The *dashed line* depicts the case $u_2 = 0$, the behaviors change monotonically with u_2.

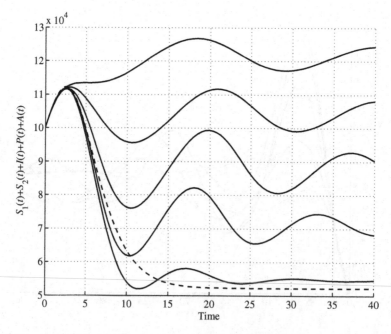

Fig. 15 Time evolution of the total population for $u_2 \in \{0, 0.2, 0.4, 0.6, 0.8, 1\}$. The *dashed line* depicts the case $u_2 = 0$, the behaviors change monotonically with u_2.

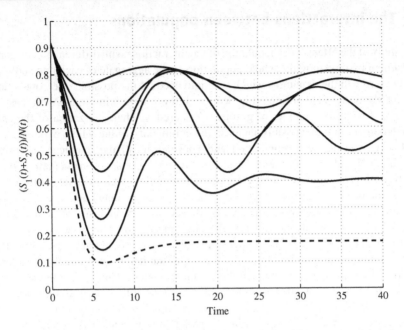

Fig. 16 Time evolution of the healthy individuals $S_1(t) + S_2(t)$ with respect to the total population for $u_2 \in \{0, 0.2, 0.4, 0.6, 0.8, 1\}$. The *dashed line* depicts the case $u_2 = 0$, the behaviors change monotonically with u_2.

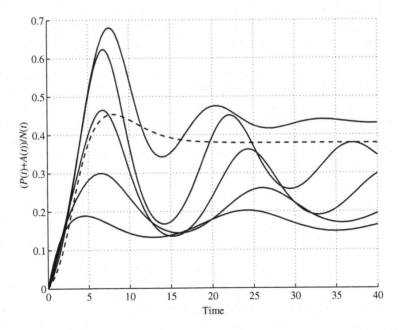

Fig. 17 Time evolution of the diagnosed patients $P(t) + A(t)$ with respect to the total population for $u_2 \in \{0, 0.2, 0.4, 0.6, 0.8, 1\}$. The *dashed line* depicts the case $u_2 = 0$, the behaviors change monotonically with u_2.

6 The interactions between populations

The study of the effects on a population of the interactions with different groups can be performed by modeling the whole population as the aggregate of each one and introducing terms in the dynamics which take into account the possible interconnections. For the dynamics (1) considered in Section 3, this would bring to a $5N$-dimensional system, N being the number of groups considered. A characterization of the entire population could then possibly increase the complexity of the whole expression.

Taking one of the N population and considering all the interactions with the remaining $N - 1$ ones by means of global averaged terms, the mathematical model can be strongly simplified introducing changes in Eq. (1) which can take into account the combined effects of the external interactions. Clearly, many details are lost since some contributions are averaged, but a useful preliminary analysis can be performed. On the basis of the results obtained for the single group model in previous sections, the evidence and the effects of the changes introduced in a population by the interactions with external world can be obtained once such interactions are suitably modeled as additional inputs and/or changes in the parameter values. Concerning the additional inputs, these can be assumed as the sum of the effects of migration, positive and negative, between the group under investigation with respect to all the other ones. Such effects can be modeled as constant fluxes, which introduce additive constant terms in the dynamics, or can be supposed to be driven by some characteristic of the group behavior, such as the public health level.

Migrations can also affect the value of some parameters, representing changes in the behavior of some individuals consequent to some cultural modifications. For examples, c and β for the number of the contagious interactions depend also on the cultural and educational facts, γ for the attitude of ignoring the dangerous effects of some kind of relationships, and so on.

In the present study, the assumption that the health status of a population can be an indicator of its attractiveness, being related to the richness, the higher social level, and so on, the changes introduced in model (1) to take into account the interactions with other groups are represented by the introduction of additional terms of the form $f_i(S_1, S_2, I, P, A)$, $i = 1$ for S_1, $i = 2$ for S_2, and $i = 3$ for I, assuming that known infected individuals P and A are not allowed to migrate. Then, Eq. (1) becomes

$$\dot{S}_1(t) = Q - dS_1(t) - \frac{c\beta S_1(t)I(t)}{N_c(t)} + \gamma S_2(t) - S_1(t)u_1(t) + f_1(S_1, S_2, I, P, A) \qquad (45a)$$

$$\dot{S}_2(t) = -(\gamma + d)S_2(t) + S_1(t)u_1(t) + f_2(S_1, S_2, I, P, A) \qquad (45b)$$

$$\dot{I}(t) = \frac{c\beta S_1(t)I(t)}{N_c(t)} - (d + \delta)I(t) - \psi \frac{I(t)}{N_c(t)}u_2(t) + f_3(S_1, S_2, I, P, A) \qquad (45c)$$

$$\dot{P}(t) = \varepsilon \delta I(t) - (\alpha_1 + d)P(t) + \phi \psi \frac{I(t)}{N_c(t)}u_2(t) + P(t)u_3(t) \qquad (45d)$$

$$\dot{A}(t) = (1-\varepsilon)\delta I(t) + \alpha_1 P(t) - (\alpha+d)A(t) + (1-\phi)\psi \frac{I(t)}{N_c(t)} u_2(t) - P(t)u_3(t)$$

$$(45e)$$

Some choices for $f_i(S_1, S_2, I, P, A)$, under the assumption previously mentioned, can be, for example, proportional to $\frac{S_1+S_2+I}{S_1+S_2+I+P+A}$, meaning that the more the group is healthy, the more it is attractive, or proportional to $-\frac{P+A}{S_1+S_2+I+P+A}$, meaning that the higher is the number of the infectious patients, the more it is repulsive. Simplified versions can be considered neglecting the normalizing denominators. In following section, the choice

$$f_i(S_1, S_2, I, P, A) = m_i \frac{S_1+S_2+I}{N} - n_i \frac{P+A}{N} \qquad (46)$$

is adopted, $i = 1, 2, 3$, where $N = S_1 + S_2 + I + P + A$.

This corresponds to the introduction, in the most general formulation, of both an immigration term

$$m_i \frac{S_1+S_2+I}{S_1+S_2+I+P+A} \qquad (47)$$

and an emigration one

$$n_i \frac{P+A}{S_1+S_2+I+P+A} \qquad (48)$$

to each of the involved dynamics, that is, for S_1, S_2, and I. In this case, the effects of the migration fluxes vary according to the values of the six parameters m_i and n_i, for $i = 1, 2, 3$.

6.1   Equilibria under external interactions

In this section, the effects on the equilibrium conditions of the external interactions just introduced are studied. A full analysis should be performed to study the effect of the changes of the new six parameters on the dynamics behavior. In order to better identify the relationships between each contribution and the corresponding effect, the analysis is performed isolating some particular fluxes. In particular, two cases are separately addressed. The first supposes that the contribution of the immigration/emigration flux acts on the susceptible group S_1 only, aiming at analyzing what may happen when the number of uninfected people changes due to migrations. The second one is, in some sense, the dual case: the effects of the migration are evaluated under the hypothesis that all the flux is concentrated on the infected but not diagnosed individuals I. With respect to the model introduced here for the interactions (45), in the terms (46), $m_2 = m_3 = n_2 = n_3 = 0$ are chosen for the first case investigated, while $m_1 = m_3 = n_1 = n_3 = 0$ are assumed for the second case.

Equilibrium points and their stability are first analyzed, in order to characterize the dynamical properties and to be able to compare the results with the isolated group case mentioned in Section 3.

6.1.1 Equilibrium points and stability properties for the case of healthy population migration

Under the hypothesis previously described, in this case the equilibrium points are computed solving the nonlinear system

$$Q - dS_1^e - \frac{c\beta S_1^e I^e}{N_c^e} + \gamma S_2^e + m_1 \frac{S_1^e + S_2^e + I^e}{N^e} - n_1 \frac{P^e + A^e}{N^e} = 0 \tag{49}$$

$$-(\gamma + d)S_2^e = 0 \tag{50}$$

$$\frac{c\beta S_1^e I^e}{N_c^e} - (d + \delta)I^e = 0 \tag{51}$$

$$\varepsilon \delta I^e - (\alpha_1 + d)P^e = 0 \tag{52}$$

$$(1 - \varepsilon)\delta I^e + \alpha_1 P^e - (\alpha + d)A^e = 0 \tag{53}$$

One solution is given by

$$\xi_1^e(m_1, n_1) = \left(\frac{Q + m_1}{d} \quad 0 \quad 0 \quad 0 \quad 0 \right)^T \tag{54}$$

Then, setting

$$P^e + A^e = \frac{\delta}{\alpha_1 + d}\left(\frac{\varepsilon \alpha + \alpha_1 + d}{\alpha + d}\right)I^e = \pi I^e \tag{55}$$

and performing some computations, the second solution

$$\xi_2^e(m_1, n_1) = \begin{pmatrix} \frac{Q}{c\beta - \delta} + \frac{m_1 c\beta - n_1 \pi(c\beta - (d+\delta))}{(c\beta(1+\pi) - \pi(d+\delta))(c\beta - \delta)} \\[2mm] 0 \\[2mm] \frac{c\beta - (d+\delta)}{d+\delta}\left(\frac{Q}{c\beta - \delta} + \frac{m_1 c\beta - n_1 \pi(c\beta - (d+\delta))}{(c\beta(1+\pi) - \pi(d+\delta))(c\beta - \delta)}\right) \\[2mm] \frac{\varepsilon \delta}{\alpha_1 + d} \frac{c\beta - (d+\delta)}{d+\delta}\left(\frac{Q}{c\beta - \delta} + \frac{m_1 c\beta - n_1 \pi(c\beta - (d+\delta))}{(c\beta(1+\pi) - \pi(d+\delta))(c\beta - \delta)}\right) \\[2mm] \frac{\delta(\alpha_1 + d(1-\varepsilon))}{(\alpha + d)(\alpha_1 + d)} \frac{c\beta - (d+\delta)}{d+\delta}\left(\frac{Q}{c\beta - \delta} + \frac{m_1 c\beta - n_1 \pi(c\beta - (d+\delta))}{(c\beta(1+\pi) - \pi(d+\delta))(c\beta - \delta)}\right) \end{pmatrix} \tag{56}$$

is obtained. The existence of this solution depends on the coefficients m_1 and n_1, since it must be verified that $S_1^e \geq 0$. However, once the condition on S_1^e is satisfied, from $I^e \geq 0$ the condition $c\beta - (d + \delta) \geq 0$ must hold. This is the same as Eq. (6) for ξ_2^{ge} in Section 4.1. Expression (56) can be simplified writing

$$\xi_2^e(m_1, n_1) = \begin{pmatrix} C_0 + C_m m_1 - C_n n_1 \\ 0 \\ K_I(C_0 + C_m m_1 - C_n n_1) \\ K_P(C_0 + C_m m_1 - C_n n_1) \\ K_A(C_0 + C_m m_1 - C_n n_1) \end{pmatrix} \tag{57}$$

once the following coefficients are defined:

$$C_0 = \frac{Q}{c\beta - \delta} \tag{58}$$

$$C_m = \frac{c\beta}{(c\beta(1 + \pi) - \pi(d + \delta))(c\beta - \delta)} \tag{59}$$

$$C_n = \frac{\pi(c\beta - (d + \delta))}{(c\beta(1 + \pi) - \pi(d + \delta))(c\beta - \delta)} \tag{60}$$

$$K_I = \frac{c\beta - (d + \delta)}{d + \delta} \tag{61}$$

$$K_P = \frac{e\delta}{\alpha_1 + d} K_I \tag{62}$$

$$K_A = \frac{\delta(\alpha_1 + d(1 - \varepsilon))}{(\alpha + d)(\alpha_1 + d)} K_I \tag{63}$$

The structure of expression (57) shows that the equilibrium point linearly change w.r.t. both m_1 and n_1.

The study of the stability characteristics for the two equilibrium points is performed, once again, analyzing the stability of the linearized dynamics in the neighborhood of the equilibria. For the first point $\xi_1^e(m_1, n_1)$, the Jacobian matrix computed in the equilibrium point is the same as for the isolated unforced case ξ_1^e in Eq. (5). Its expression is given in Eq. (7) and the same stability conditions can be obtained: the equilibrium point $\xi_1^e(m_1, n_1)$ is locally asymptotically stable under condition (9). For the second equilibrium point $\xi_2^e(m_1, n_1)$ in Eq. (57), the dynamical matrix of the linear approximation has the structure

$$A_2(m_1, n_1) = \left(\begin{array}{c|c} A_2^{11}(m_1, n_1) & 0 \\ \hline A_2^{21}(m_1, n_1) & A_2^{22}(m_1, n_1) \end{array} \right) \tag{64}$$

with $A_2^{22}(m_1,n_1)$ equal to the one in Eq. (12) for the isolated unforced case, for which the two eigenvalues, $\lambda_4 = -(\alpha_1 + d)$ and $\lambda_5 = -(\alpha + d)$, are real negative. Then, for the stability conditions, matrix

$$A_2^{11}(m_1,n_1) = \begin{pmatrix} -d - c\beta\frac{K_I^2}{(1+K_I)^2} & \gamma + c\beta\frac{K_I}{(1+K_I)^2} & -c\beta\frac{1}{(1+K_I)^2} \\ 0 & -(\gamma+d) & 0 \\ c\beta\frac{K_I^2}{(1+K_I)^2} & -c\beta\frac{K_I}{(1+K_I)^2} & c\beta\frac{1}{(1+K_I)^2} - (d+\delta) \end{pmatrix} \tag{65}$$

must be analyzed. It is interesting to note that it does not depend on the coefficients m_1 or n_1; then, the same holds for the stability of the equilibrium points obtained by varying such coefficients. One of the eigenvalues of Eq. (65), $\lambda_3 = -(\gamma + d) < 0$, is evident from the structure. The other two can be computed as the roots of the polynomial equation

$$\lambda^2 + \left(d + (d+\delta) - c\beta\frac{1-K_I^2}{(1+K_I)^2}\right)\lambda + \left(d(d+\delta) + c\beta\frac{(d+\delta)K_I^2 - d}{(1+K_I)^2}\right) \tag{66}$$

or, by replacing the full expression for K_I and performing all the simplifications,

$$\lambda^2 + (c\beta - \delta)\lambda + (d+\delta)(c\beta - \delta)\frac{c\beta-(d+\delta)}{c\beta} = 0 \tag{67}$$

which is the same as Eq. (13). On the basis of the discussion in Section 4.2, it is possible to conclude that the condition for the local asymptotic stability for $\xi_2^e(m_1,n_1)$ does not depend on m_1 and n_1 and it is the same as given in Eq. (14). Moreover, the same relationships between existence and stability of the two equilibrium points, as in Section 4.2 for ξ_1^e and ξ_2^e, hold.

6.1.2 Equilibrium points and stability properties for the case of infected population migration

Following the same procedure as in the previous cases, the equilibrium points are computed solving the nonlinear system

$$Q - dS_1^e - \frac{c\beta S_1^e I^e}{N_c^e} + \gamma S_2^e = 0 \tag{68}$$

$$-(\gamma + d)S_2^e = 0 \tag{69}$$

$$\frac{c\beta S_1^e I^e}{N_c^e} - (d+\delta)I^e + m_3\frac{S_1^e + S_2^e + I^e}{N^e} - n_3\frac{P^e + A^e}{N^e} = 0 \tag{70}$$

$$\varepsilon\delta I^e - (\alpha_1 + d)P^e = 0 \tag{71}$$

$$(1-\varepsilon)\delta I^e + \alpha_1 P^e - (\alpha + d)A^e = 0 \tag{72}$$

From Eq. (69) $S_2^e = 0$ is immediately obtained. Once $N_c^e = S_1^e + I^e$ and $N^e = S_1^e + (1+\pi)I^e = N_c^e + \pi I^e$ are introduced, the system to be solved can be reduced to Eqs. (68), (70), and rewritten as

$$Q - dS_1^e - \frac{c\beta S_1^e I^e}{S_1^e + I^e} = 0 \tag{73}$$

$$\frac{c\beta S_1^e I^e}{S_1^e + I^e} - (d+\delta)I^e + m_3 \frac{S_1^e + I^e}{S_1^e + (1+\pi)I^e} - n_3 \frac{\pi I^e}{S_1^e + (1+\pi)I^e} = 0 \tag{74}$$

From Eq. (73), one can write

$$I^e = \frac{(dS_1^e - Q)}{Q - (c\beta + d)S_1^e} S_1^e \tag{75}$$

while, for S_1^e, after computations one gets the polynomial equation

$$a_3 \left(S_1^e\right)^3 + a_2 \left(S_1^e\right)^2 + a_1 S_1^e + a_0 = 0 \tag{76}$$

whose coefficients are as follows:

$$
\begin{aligned}
a_3 &= d(c\beta - \delta)(c\beta - \pi d)\\
a_2 &= 2\pi Q d(c\beta - \delta) - (c\beta + d)(c\beta(Q + m_3) + \pi d n_3) + Q c\beta\delta + \pi Q d^2\\
a_1 &= c\beta Q(Q(1 - \pi) + m_3 + \pi n_3) + \pi Q^2 \delta - 2\pi Q d(Q - n_3)\\
a_0 &= \pi Q^2(Q - n_3)
\end{aligned} \tag{77}
$$

The feasible solutions correspond to the real positive roots of polynomial (76). Conditions should be given for the coefficients to characterize the roots in order to have admissible solutions but they are not easy to be written and their analysis requires a numerical evaluation of such conditions. Hence, from now on, the numerical values reported in Table 1 are used to show a possible behavior and also for comparative purpose with all the results and the discussions given in Section 5.

With the parameters presented in Table 1, coefficients (77) assume the following expression depending on the two parameters m_3 and n_3:

$$a_3 = 0.0326 \tag{78}$$

$$a_2 = -2.2800 m_3 - 0.0257 n_3 - 1.6425 \times 10^4 \tag{79}$$

$$a_1 = 15{,}000 m_3 + 1.3008 \times 10^4 n_3 + 5.3710 \times 10^7 \tag{80}$$

$$a_0 = -8.4465 \times 10^7 n_3 + 8.4465 \times 10^{11} \tag{81}$$

The three solutions of Eq. (76) are computed for different values of $m_3 \in [0, 5 \times 10^4]$ and $n_3 \in [0, 5 \times 10^4]$. In Figs. 18–20, the variations of such solutions with respect to

m_3 are reported, keeping $n_3 = 0$, while Figs. 21–23 depict how the solutions vary according to different values of n_3 with $m_3 = 0$. In each figure, the solid line represents the real part of the solution and the dashed line denotes the imaginary part, so that it is easy to reject the solutions which are not consistent with the real condition of the state variables. In all the six figures, it can be observed that the values of the equilibrium components S_1^e and I^e are all real. Moreover, in order to be admissible values, the components of the state must be nonnegative. In this case, for the range of values for the parameters m_3 and n_3, it can be seen that keeping $n_3 = 0$, the three roots S_1^e of the polynomial (76) are real, two positive, Figs. 18 and 19, and one negative, Fig. 20, for all the values given to $m_3 \in [0, 5 \times 10^4]$. The corresponding values for the equilibrium I^e are negative for the first solution in Fig. 18 and positive for the remaining two cases, Figs. 19 and 20. While for $m_3 = n_3 = 0$ the first and second solutions are admissible, as m_3 increases, only the second is a feasible solution; in this case, only one equilibrium point exists and its components S_1^e and I^e are given in Fig. 19 for each value of m_3 in the range considered.

On the other hand, varying n_3 while $m_3 = 0$, the first solution, depicted in Fig. 21, shows that n_3 does not influence one of the equilibrium point of the noninteracting case, defined in Section 4.1 as ξ_1^e. Also the second solution, plotted in Fig. 22, is admissible, since both S_1^e and I^e are positive for all the values considered for n_3. On the contrary, the third solution cannot be assumed as an equilibrium point, because when S_1^e is positive, I^e is negative. Moreover, the value $S_1^e = 0$ is never acceptable since it does not satisfy Eq. (73).

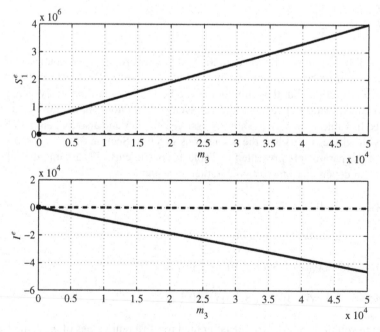

Fig. 18 Values of the first solution S_1^e in Eq. (76) and I^e in Eq. (75) for $m_3 \in [0, 5 \times 10^4]$.

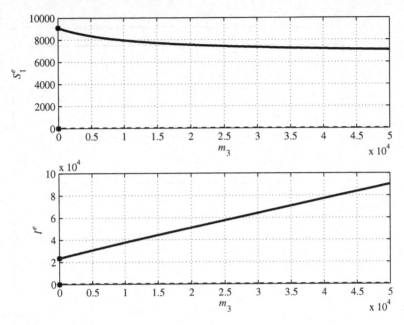

Fig. 19 Values of the second solution S_1^e in Eq. (76) and I^e in Eq. (75) for $m_3 \in [0, 5 \times 10^4]$.

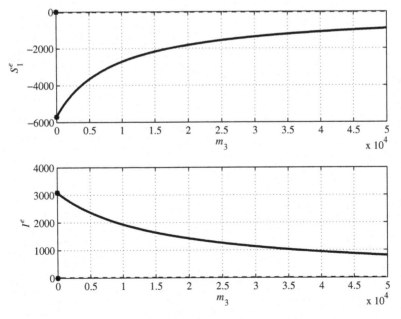

Fig. 20 Values of the third solution S_1^e in Eq. (76) and I^e in Eq. (75) for $m_3 \in [0, 5 \times 10^4]$.

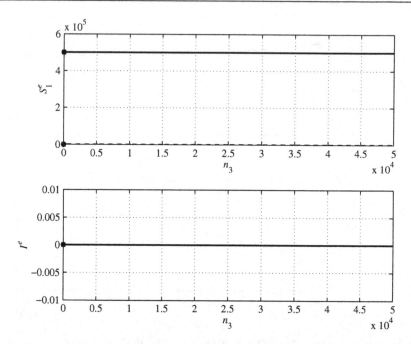

Fig. 21 Values of the first solution S_1^e in Eq. (76) and I^e in Eq. (75) for $n_3 \in [0, 5 \times 10^4]$.

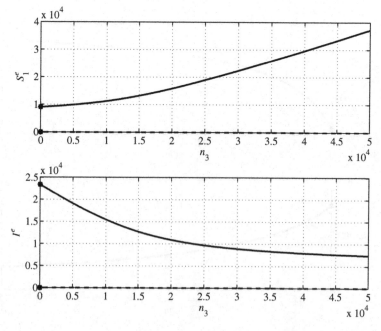

Fig. 22 Values of the second solution S_1^e in Eq. (76) and I^e in Eq. (75) for $n_3 \in [0, 5 \times 10^4]$.

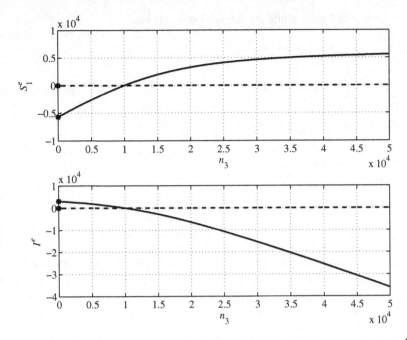

Fig. 23 Values of the third solution S_1^e in Eq. (76) and I^e in Eq. (75) for $n_3 \in [0, 5 \times 10^4]$.

A stability analysis for the equilibrium points just computed is performed necessarily on the numerical case addressed till now. The Jacobian matrix (19) $J_f(\xi) = \frac{\partial f}{\partial \xi}$ must be computed and then evaluated at each equilibrium point. Since the structure of the system is the same as in Eq. (1) with the additional term

$$m_3 \frac{S_1 + S_2 + I}{N} - n_3 \frac{P + A}{N} \tag{82}$$

in the dynamics of the infected subjects I, expression (19) for the present case is the same as for Eq. (1) except for the third row depending on f_3, in which the elements in Eq. (82) must be considered. What is then obtained is

$$\frac{\partial f_1}{\partial S_1} = -d - \frac{c\beta I}{N_c} + \frac{c\beta S_1 I}{N_c^2}, \quad \frac{\partial f_1}{\partial S_2} = \gamma + \frac{c\beta S_1 I}{N_c^2} \tag{83}$$

$$\frac{\partial f_1}{\partial I} = -\frac{c\beta S_1}{N_c} + \frac{c\beta S_1 I}{N_c^2}, \quad \frac{\partial f_2}{\partial S_2} = -(\gamma + d) \tag{84}$$

$$\frac{\partial f_4}{\partial I} = \varepsilon\delta, \quad \frac{\partial f_4}{\partial P} = -(\alpha_1 + d) \tag{85}$$

$$\frac{\partial f_5}{\partial I} = (1 - \varepsilon)\delta, \quad \frac{\partial f_5}{\partial P} = \alpha_1, \quad \frac{\partial f_5}{\partial A} = -(\alpha + d) \tag{86}$$

as for the isolated case, while for the third row one has

$$\frac{\partial f_3}{\partial S_1} = \frac{c\beta I}{N_c} - \frac{c\beta S_1 I}{N_c^2} + \frac{(m_3 + n_3)(P + A)}{N^2} \tag{87}$$

$$\frac{\partial f_3}{\partial S_2} = -\frac{c\beta S_1 I}{N_c^2} + \frac{(m_3 + n_3)(P + A)}{N^2} \tag{88}$$

$$\frac{\partial f_3}{\partial I} = -(d + \delta) + \frac{c\beta S_1}{N_c} - \frac{c\beta S_1 I}{N_c^2} + \frac{(m_3 + n_3)(P + A)}{N^2} \tag{89}$$

$$\frac{\partial f_3}{\partial P} = -\frac{(m_3 + n_3)(S_1 + S_2 + I)}{N^2} \tag{90}$$

$$\frac{\partial f_3}{\partial A} = -\frac{(m_3 + n_3)(S_1 + S_2 + I)}{N^2} \tag{91}$$

The corresponding Jacobian matrix is computed for each of the admissible equilibrium points obtained in Section 6.1. Then, for the case $n_3 = 0$, only one solution is considered, whose value changes with m_3 (Fig. 19). For each value of m_3, one equilibrium point is obtained as well as one corresponding Jacobian matrix. For each of them five eigenvalues are computed and are plotted in Fig. 24. It can be noted that all the five eigenvalues, three real and one conjugate complex pair, have their real part negative until $m_3 = 2.285 \times 10^4$, marked with the dotted

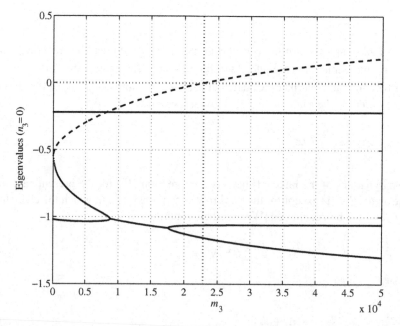

Fig. 24 Real part of the five eigenvalues of the Jacobian matrix computed in the first solution for $n_3 = 0$ and $m_3 \in [0, 5 \times 10^4]$. The *dotted line* denotes the real part of a couple of conjugate complex roots.

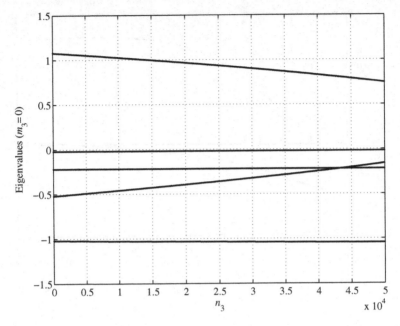

Fig. 25 Real part of the five eigenvalues of the Jacobian matrix computed in the first solution for $m_3 = 0$ and $n_3 \in [0, 5 \times 10^4]$.

vertical line. Above that value, one eigenvalue becomes positive and the local stability is lost.

The same computations are performed for the two admissible equilibrium points reported in Figs. 21 and 22. The eigenvalues of the Jacobian matrix which locally approximates the nonlinear dynamics in a neighborhood of the n_3-dependent first equilibrium point are reported in Fig. 25, where it is possible to see that for the range considered for n_3, one root is always positive and so the equilibrium point is unstable. On the contrary, for the second equilibrium point, the one in Fig. 22, the situation is equivalent to the case in Fig. 24: three real eigenvalues and one conjugate complex pair, whose real parts are depicted in Fig. 26, solid line for the real roots, dashed line for the complex ones. They are all negative for $n_3 < 8.4 \times 10^5$, so yielding to local asymptotic stability of the equilibrium points; for values $n_3 > 8.4 \times 10^5$ stability condition is no more satisfied.

7 The effects of migration parameters on the individuals evolutions

This section is devoted to report some behavior of the system for different values of the coefficients m_1, n_1, m_3, and n_3, according to the stability results discussed in Section 6. The numerical values of the system parameters are, obviously, the ones

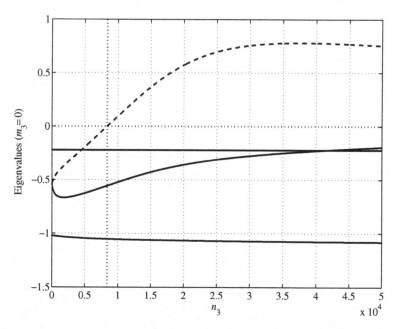

Fig. 26 Real part of the five eigenvalues of the Jacobian matrix computed in the second solution for $n_3 = 0$ and $m_3 \in [0, 5 \times 10^4]$. The *dotted line* denotes the real part of a couple of conjugate complex roots.

illustrated in Table 1; the initial conditions for all the simulations have been chosen as $S_1(0) = 92,000$, $I(0) = 8000$, $S_2(0) = P(0) = A(0) = 0$, aiming at modeling a population in which, initially, the epidemic is not known but some infected individuals are already present.

The first case reported refers to the time evolution of the system where m_1 varies while $n_1 = m_3 = n_3 = 0$. The results are reported in Figs. 27–31. In particular, Fig. 27 depicts the evolution of the number of unwise healthy population $S_1(t)$ when m_1 changes and Fig. 28 reports the same cases for m_1, starting from zero (corresponding to the dashed lines) and increasing its value. The evolution of $S_2(t)$ is not reported because for the initial condition chosen it is constant and equal to zero. Also the time histories of $P(t)$ and $A(t)$ are omitted since they have the same shape of $I(t)$. The obvious result that can be evidenced from Fig. 27 is that the number of healthy people increases as m_1 grows; unfortunately, this increment makes the infected individuals $I(t)$ grow too. However, a very interesting behavior of the system can be observed once the evolution of the fraction of diagnosed infected patients $P(t) + A(t)$ w.r.t. the total population is determined, varying m_1, as reported in Fig. 29. From these plots it can be seen that, under the hypothesis of an immigration of healthy people driven by the health of the total population, the different values of m_1 affects the transient behavior only, reaching, for all m_1, the same steady-state value. The same result can be observed in Fig. 30, where the fraction of the healthy population $S_1(t) + S_2(t)$ w.r.t. the total one is reported: also in this case, the steady-state value is the same. Thus,

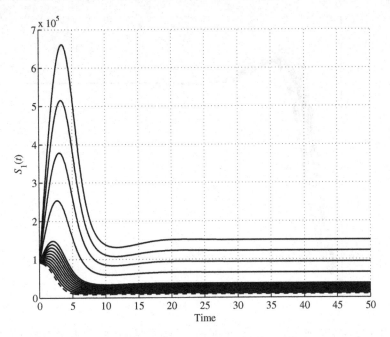

Fig. 27 Time history of $S_1(t)$ for different values of m_1. The *dotted line* corresponds to $m_1 = 0$; values increases according to m_1.

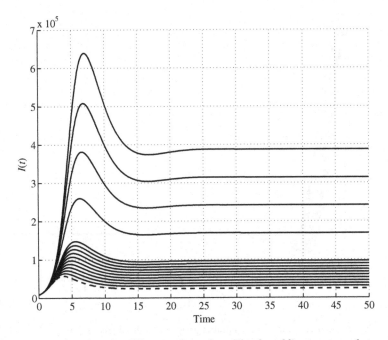

Fig. 28 Time history of $I(t)$ for different values of m_1. The *dotted line* corresponds to $m_1 = 0$; values increases according to m_1.

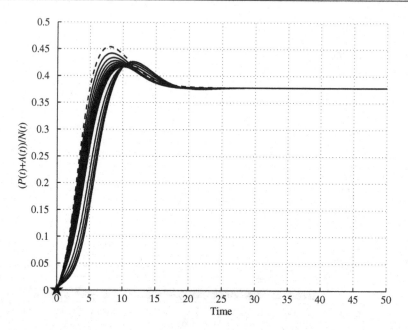

Fig. 29 Time evolution of the relative number of diagnosed patients $P(t) + A(t)$ w.r.t. the total population for different values of m_1. The *dotted line* corresponds to $m_1 = 0$.

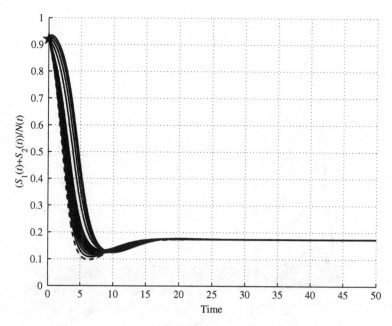

Fig. 30 Time evolution of the relative number of healthy individuals $S_1(t) + S_2(t)$ w.r.t. the total population for different values of m_1. The *dotted line* corresponds to $m_1 = 0$.

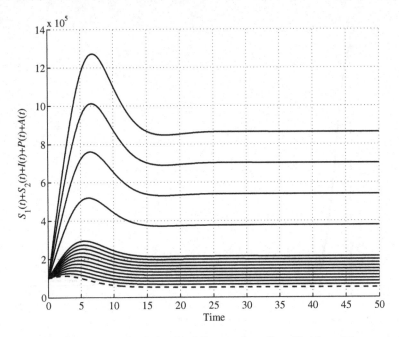

Fig. 31 Time evolution of the population for different values of m_1. The *dotted line* corresponds to $m_1 = 0$.

it is possible to conclude that, for the model considered, an immigration of healthy individuals makes the population grow, as evidenced in Fig. 31; each of the classes grows (Figs. 27 and 28) but the fraction of healthy population as well as the fraction of infected one tends to remain unchanged even if the rate of immigration (m_1) increases.

In Figs. 32–36, the results of simulations for different values of n_1 while $m_1 = m_3 = n_3 = 0$ are reported. This represents an emigration of healthy people based on the presence of infection in the population. As expected, both $S_1(t)$ and $I(t)$ decrease, as reported in Figs. 32 and 33, respectively, $I(t)$ more sensibly than $S_1(t)$. However, in this case also the percentage of infected subjects, plotted in Fig. 34, and the one of healthy individuals, depicted in Fig. 35, tend to remain unchanged as n_1 changes, having the same steady-state values. Differently from the case of variation of m_1, in this case the transients present a higher amplitudes as n_1 increases. This result can be used to better understand the actual dangerousness of the phenomenon, since high values of relative infected individuals can be limited to finite time intervals, during the transient, and may not represent a real social alarm situation. Fig. 36 shows that in this case the total population decreases as the emigration rate increases, but with oscillations during the transient with increasing amplitude.

A second set of numerical simulations addresses the case of migration which affects the infected population $I(t)$; in this case, the couple of coefficients m_1 and

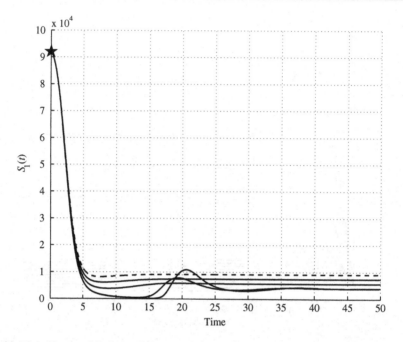

Fig. 32 Time history of $S_1(t)$ for different values of n_1. The *dotted line* corresponds to $n_1 = 0$.

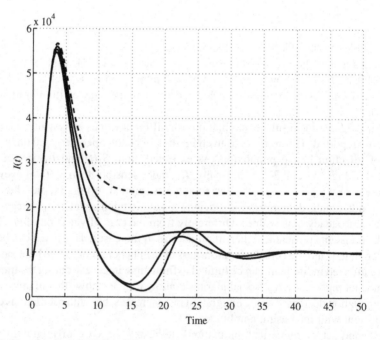

Fig. 33 Time history of $I(t)$ for different values of n_1. The *dotted line* corresponds to $n_1 = 0$.

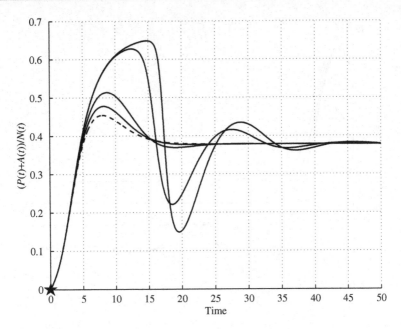

Fig. 34 Time evolution of the relative number of diagnosed patients $P(t) + A(t)$ w.r.t. the total population for different values of n_1. The *dotted line* corresponds to $n_1 = 0$.

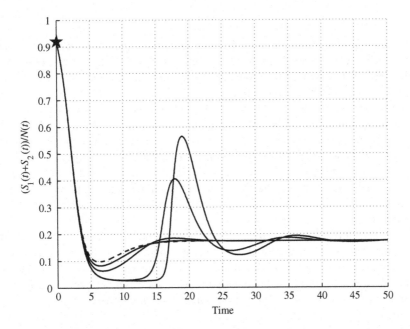

Fig. 35 Time evolution of the relative number of healthy individuals $S_1(t) + S_2(t)$ w.r.t. the total population for different values of n_1. The *dotted line* corresponds to $n_1 = 0$.

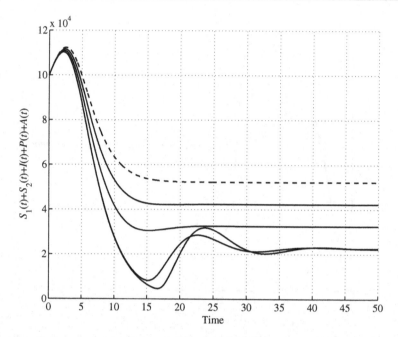

Fig. 36 Time evolution of all the population for different values of n_1. The *dotted line* corresponds to $n_1 = 0$.

n_1 are always fixed at zero. The results are reported in Figs. 37–46, where the evolution of the system is depicted for different values of m_3, while $n_3 = 0$, modeling a flux of immigrants from outside the group, and then for different values of n_3 with $m_3 = 0$, representing an emigration phenomenon.

In Fig. 37 the time history of the healthy individuals $S_1(t)$ is plotted for different values of m_3 while Fig. 38 reports the same situation for the infected subjects $I(t)$. Their behaviors are compatible with what could be expected: the number of individuals in the class of infected $I(t)$ grows as m_3 increases; at the same time, the uninfected people decrease due to the augmented probability of infection due to the larger $I(t)$ population. However, since the influence of the immigration acts directly on the dynamic of $I(t)$, the growth of $I(t)$ is more accentuated than the decrement of $S_1(t)$, the latter being a secondary effect.

This characteristic of the system behavior is confirmed looking at Figs. 39 and 40 in which the fraction of infected and healthy population are reported, respectively; the first increases while the second decreases. However, these variations have the combined effect of making the total population sensibly increase, as reported in Fig. 41, so having a society with higher number of individuals but composed of an even higher number of infected components.

The opposite effect of an emigration from the class of infected $I(t)$, modeled setting $m_3 = 0$ and choosing different values for n_3, can be analyzed making use of

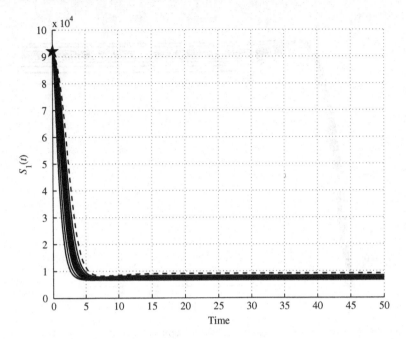

Fig. 37 Time history of $S_1(t)$ for different values of m_3. The *dotted line* corresponds to $m_3 = 0$.

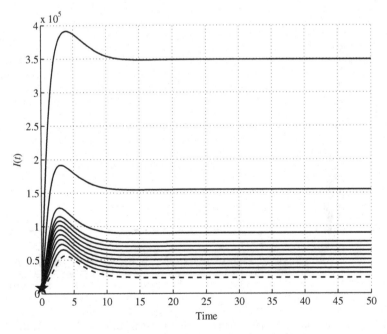

Fig. 38 Time history of $I(t)$ for different values of m_3. The *dotted line* corresponds to $m_3 = 0$.

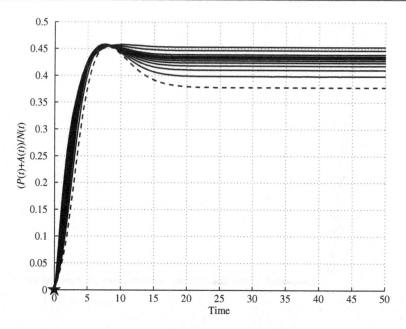

Fig. 39 Time evolution of the relative number of diagnosed patients $P(t) + A(t)$ w.r.t. the total population for different values of m_3. The *dotted line* corresponds to $m_3 = 0$.

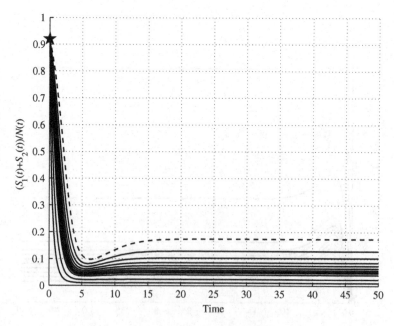

Fig. 40 Time evolution of the relative number of healthy individuals $S_1(t) + S_2(t)$ w.r.t. the total population for different values of m_3. The *dotted line* corresponds to $m_3 = 0$.

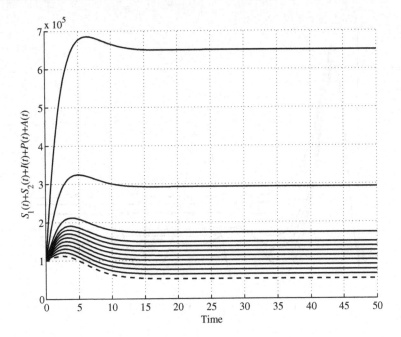

Fig. 41 Time evolution of all the population for different values of m_3. The *dotted line* corresponds to $m_3 = 0$.

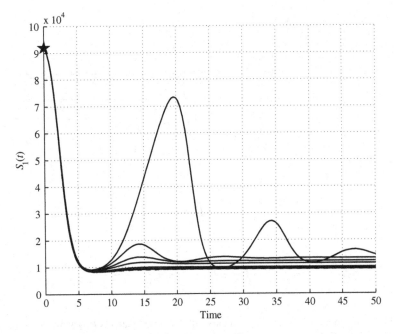

Fig. 42 Time history of $S_1(t)$ for different values of n_3. The *dotted line* corresponds to $n_3 = 0$.

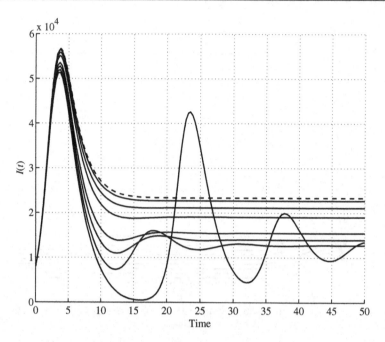

Fig. 43 Time history of $I(t)$ for different values of n_3. The *dotted line* corresponds to $n_3 = 0$.

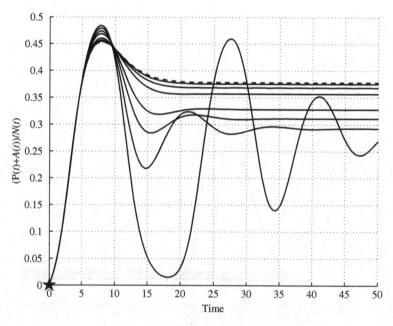

Fig. 44 Time evolution of the relative number of diagnosed patients $P(t) + A(t)$ w.r.t. the total population for different values of n_3. The *dotted line* corresponds to $n_3 = 0$.

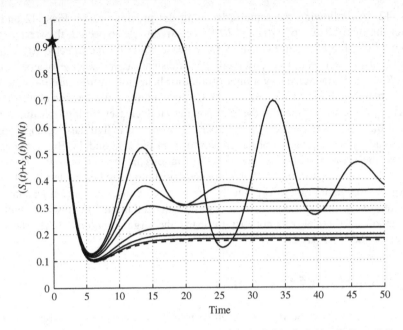

Fig. 45 Time evolution of the relative number of healthy individuals $S_1(t) + S_2(t)$ w.r.t. the total population for different values of n_3. The *dotted line* corresponds to $n_3 = 0$.

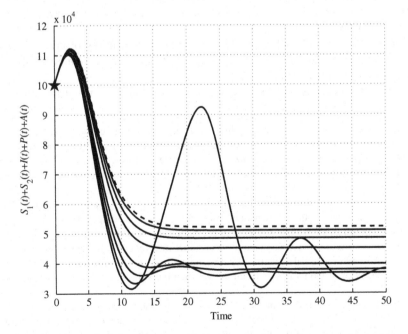

Fig. 46 Time evolution of all the population for different values of n_3. The *dotted line* corresponds to $n_3 = 0$.

Figs. 42–46. Following the same order as in the previous cases, Figs. 42 and 43 depict the time history of $S_1(t)$ and $I(t)$, respectively. As expected, the steady-state values of the uninfected individuals $S_1(t)$ increase while for the infected $I(t)$ a more sensible decrement could be observed. The transient, as in the case of emigration from $S_1(t)$, is characterized by a high oscillatory behavior, in this case more accentuated than that shown in Figs. 32 and 33. Correspondingly, the fraction of diagnosed infected people with respect to the total population decreases for higher values of n_3 and, at the same time, the relative number of uninfected individuals increases, once their steady-state behavior is observed as shown in Figs. 43 and 45. But in the same figures, the transients show a very large variations in the components. This fact proves the necessity to have a reliable mathematical model to be able to predict some unexpected behaviors or to explain the presence of values in some time intervals that, intuitively, are not obvious. This is well evidenced in Fig. 46, in which, without the support of a model, it is not easy to justify the large variations in the total population as the time passes.

8   Discussion of the results

A final discussion on the results presented in the previous sections, with particular reference to Sections 5 and 7, is shortly reported in this section. The approach followed in the present work aims to put in evidence the effects of incoming and/ or outgoing migrations, when they involve healthy individuals who do not put a great attention to the modalities of the spread of virus or infected persons not aware of being ill and contagious.

Among the goals of such an analysis, there is the possibility of previewing and understanding the characteristics of the behaviors of some classes of individuals under particular external contributions.

The main aspects that can be evidenced once the numerical results are compared and interpreted are as follows:

(i) An immigration involving healthy individuals not well informed on the risks of unwise behaviors produces an increment in people in all the classes of the population, since the interactions producing virus transmission increase. However, at steady state, the ratio of the healthy population and the one of infected individuals with respect to the total one is constant.

(ii) Emigration involving the same population as in Case (i) implies an evolution opposite to the previous one, with a decrement in people in all classes, leading, at steady state, again to constant ratios, but with the presence of oscillations in the transient. This peculiar behavior is interesting to be stressed since there are time intervals with increment in individuals, healthy and infected, despite the emigration phenomena. This fact does not characterizes the behavior of the dynamics when the vaccination, which acts on the same subjects, is applied.

(iii) When the immigration involves infected people unaware of their status, the effect is the expected one: the number of healthy people decreases while all the infected ones increase. This produces an increment in the total population, but it is due to a higher number of infected individuals.

(iv) The emigration among people of the same class as in Case (iii) at steady state shows the opposite behavior with respect to the previous case, as expected. However, the transient is characterized by an oscillatory behavior, as in the Case (ii); this fact implies that a correct interpretation of the population dynamics cannot be correctly performed over short-time period.

9 Conclusions and future developments

In this chapter, the study of the effects of the interactions among different groups is performed by modeling the whole population as the aggregate of each one and introducing terms in the dynamics which take into account the possible interconnections. These results are obtained starting from the development of the analysis for the single group dynamics modeling each interaction as additional inputs and/or changes in the parameter values. Such effects can be modeled as constant fluxes, which introduce additive constant terms in the dynamics, or can be supposed to be driven by some characteristic of the group behavior, such as a *higher* or *lower* level of healthy, for example. As case study, the epidemic spread considered is the HIV-AIDS, assuming a new model in which the classical scheme that includes susceptible people and three classes of infected ones (infected but not aware of their status, patients in the pre-AIDS and AIDS conditions) is enriched by splitting the class of susceptible individuals into those aware of the risks of this virus and those who adopt irresponsible acts. The proposed analysis, particularly suitable in describing, for the chosen model, the migration phenomena, is useful to predict unexpected behaviors especially in the transient period in which oscillatory behaviors for the classes of infected patients appear, as evidenced in the discussion.

Since all the results presented are highly dependent on the model structure and parameter values, future developments should involve real data analysis and model validation. Moreover, the explicit introduction of the detailed interactions between more than one population is mandatory for putting in evidence the migration fluxes. This is preparatory for the introduction of a control strategy for reducing the spread of virus between populations despite the globalization needs.

References

Behncke, H., 2000. Optimal control of deterministic epidemics. Optimal Control Appl. Methods 21 (2), 269–285.

Casagrandi, R., Bolzoni, L., Levin, S.A., Andreasen, V., 2006. The SIRC model and influenza A. Math. Biosci. 200 (2), 152–169.

Chalub, F., Souza, M.O., 2011. The SIR epidemic model from a PDE point of view. Math. Comput. Model. 58, 1568–1574.

Chang, H., Astolfi, A., 2009. Control of HIV infection dynamics. IEEE Control Syst. 213 (2), 28–39.

Cinati Jr., J., Hoever, G., Morgenstern, B., Preiser, W., Vogel, J.U., Hofmann, W.K., Bauer, G., Michaelis, M., Rabenau, H.F., Doerr, H.W., 2004. Infection of cultured intestinal epithelial cells with severe acute respiratory syndrome coronavirus. Cell. Mol. Life Sci. 61 (16), 2010–2012.

Crush, J., Williams, B., Gouws, E., Lurie, M., 2005. Migration and HIV/AIDS in South Africa. Dev. South Afr. 22 (3), 293–317.

Dadlani, A., Kumar, M.S., Murugan, S., Kim, K., 2014. System dynamics of a refined epidemic model for infection propagation over complex networks. IEEE Syst. J. 10 (4), 1316–1325.

Di Giamberardino, P., Iacoviello, D., 2017a. Optimal control of SIR epidemic model with state dependent switching cost index. Biomed. Signal Process. Control 31 (2), 377–380.

Di Giamberardino, P., Iacoviello, D., 2017b. An optimal control problem formulation for a state dependent resource allocation strategy. In: Proceedings of the 14th International Conference on Informatics in Control, Automation and Robotics, vol. 1, pp. 186–195.

Di Giamberardino, P., Iacoviello, D., 2018. LQ control design for the containment of the HIV/ AIDS diffusion. Control Eng. Pract. 77, 162–173.

Di Giamberardino, P., Compagnucci, L., De Giorgi, C., Iacoviello, D., 2019. Modeling the effects of prevention and early diagnosis on HIV/AIDS infection diffusion. IEEE Trans. Syst. Man Cybern. Syst. 49 (10), 2119–2130.

Iacoviello, D., Liuzzi, G., 2008a. Fixed/free final time SIR epidemic models with multiple controls. Int. J. Simul. Model. 7 (2), 81–92.

Iacoviello, D., Liuzzi, G., 2008. Optimal control for SIR epidemic model: a two treatments strategy. In: Proceedings of IEEE 16th Mediterranean Conference on Control and Automation, vol. 7, pp. 81–92.

Iacoviello, D., Stasio, N., 2013. Optimal control for SIRC epidemic outbreak. Comput. Methods Programs Biomed. 110 (3), 333–342.

Joshi, H.R., 2002. Optimal control of an HIV immunology model. Optimal Control Appl. Methods 23 (2), 199–213.

Kryftis, Y., Mastorakis, G., Mavromoustakis, C.X., Batalla, J.M., Rodrigues, J.J.P.C., Dobre, C., 2017. Resource usage prediction models for optimal multimedia content provision. IEEE Syst. J. 11 (4), 2852–2863.

Kuniya, T., Nakata, Y., 2012. Permanence and extinction for a nonautonomous SEIRS epidemic model. Appl. Math. Comput. 218 (2), 9321–9331.

Mascio, M., Ribeiro, R., Markowitz, M., Ho, D., Perelson, A., 2004. Modeling the long-term control of viremia in HIV-1 infected patients treated with antiretroviral therapy. Math. Biosci. 188 (25), 47–62.

Nagelkerke, N.J., Jha, P., de Vlas, S.J., Korenromp, E.L., Moses, S., Blanchard, J.F., Plummer, F.A., 2002. Modelling HIV/AIDS epidemics in Botswana and India: impact of interventions to prevent transmission. Bull World Health Organ. 80 (2), 89–96.

Naresh, R., Tripathi, A., Sharma, D., 2009. Modeling and analysis of the spread of AIDS epidemic with immigration of HIV infectives. Math. Comput. Model. 49 (25), 880–892.

Nowzari, C., Preciado, V.M., Pappas, G.J., 2016. Analysis and control of epidemics. A survey of spreading processes on complex networks. IEEE Control Syst. Mag. 80 (2), 24–26.

Pinto, C., Rocha, D., 2012. A new mathematical model for co-infection of malaria and HIV. In: 4th IEEE International Conference on Nonlinear Science and Complexity, vol. 49. pp. 33–39.

Tanaka, G., Urabe, C., 2014. Random and targeted interventions for epidemic control in metapopulation models. Sci. Rep. 4 (2), 1–8.

Wodarz, D., 2001. Helper-dependent vs. helper-independent CTL responses in HIV infection: implications for drug therapy and resistance. J. Theor. Biol. 213 (2), 447–459.

Wodarz, D., Nowak, M., 1999. Specific therapy regimes could lead to long-term immunological control of HIV. Proc. Natl Acad. Sci. USA 96 (25), 14464–14469.

Yan, X., Zou, Y., 2008. Optimal and sub-optimal quarantine and isolation control in SARS epidemics. Math. Comput. Model. 47 (2), 235–245.

Zhou, Y., Yang, K., Zhou, K., Wang, C., 2014. Optimal treatment strategies for HIV with antibody response. J. Appl. Math. 52 (2), 1–13.

Reinforcement learning-based control of drug dosing with applications to anesthesia and cancer therapy

Regina Padmanabhan[a], Nader Meskin[a], Wassim M. Haddad[b]
[a]Department of Electrical Engineering, Qatar University, Doha, Qatar, [b]School of Aerospace Engineering, Georgia Institute of Technology, Atlanta, GA, United States

1 Introduction

This chapter presents a general framework that utilizes reinforcement learning (RL)-based method to regulate multiple parameters during intravenous drug administration. First, the Q-learning algorithm which is an RL-based method is used to fine tune continuous infusion of the drug propofol for patients in the ICU. We control the infusion of propofol so as to keep the bispectral index (BIS) and mean arterial pressure (MAP) of the patient at a desired range. Next, the use of a similar Q-learning-based controller is discussed to regulate the drug titration while different drugs with synergistic interactive effects are administered simultaneously. Finally, an RL-based controller design strategy for cancer chemotherapy treatment is also presented.

1.1 Motivation

During the last few decades, the critical and complex task of anesthesia administration has been widely studied and discussed in the literature using clinical as well as in silico trials. Consequently, many recent reviews on the currently adopted strategies highlight several aspects of the problem that need further research attention (Ionescu et al., 2014; Van Den Berg et al., 2017). Moreover, common anesthetics, such as propofol and midazolam that are necessary for various medical procedures, have side effects that include cough, nausea, skin irritations, numbness, delirium, seizures, muscle pain, weak or shallow breathing, and hemodynamic instability. The overdosing of some anesthetic and analgesic drugs is known to cause death (Jacobi et al., 2002; Mehta et al., 2006). Typically, patients admitted to the intensive care units suffer from multiple illnesses which necessitate the use of many drugs for life support and treatment. When it comes to the combined administration of several drugs such as anesthetic, analgesic, neuromuscular blockades, and cardiac drugs, the fact that the mechanisms of action are complex, interlaced, and not yet completely understood makes the problem very challenging. In the case of the continuous and simultaneous

Control Applications for Biomedical Engineering Systems. https://doi.org/10.1016/B978-0-12-817461-6.00009-3

infusion of these drugs for long periods, it is evident that an appropriate closed-loop control approach can be used to improve patient safety (Absalom et al., 2011; Ionescu et al., 2014; Jacobi et al., 2002).

Cancer chemotherapy treatment is another important therapeutic approach that involves continuous infusion of intravenous drugs. Surveys conducted in the area of cancer diagnosis and treatment highlight that the relative survival rate for many types of cancer have improved significantly over the years (ACS, 2015; WHO, 2018). These reports suggest that proficiency in early diagnosis and improvement in treatment methods are the important factors that contributed to reducing the morbidity rate and mortality rate associated with cancer. Even though there has been obvious improvement in the overall prognosis, diagnosis, and treatment of cancer, a steady increase in the incidence of this disease is a matter of concern. Like any other drug-dosing application, there are several factors that determine the drug dose required for a patient to culminate certain desired response. In the case of cancer chemotherapy, these factors include the type and stage of cancer, age, and weight of the patient, immune response of the patient, and presence any other illness. Accordingly, the clinician chooses the type and amount of the drug to be given to a patient by following certain established treatment protocols and guidelines.

However, several clinical trials and scientific studies point out the limitations of this approach and highlight the need for optimal and patient-specific dosing of chemotherapeutic drugs (Chen et al., 2012; Sbeity and Younes, 2015). These literatures highlight the importance to conduct clinical and in silico trials to study the effectiveness and feasibility of novel chemotherapy, plan to improve the therapeutic benefits of the treatment (Sbeity and Younes, 2015). However, clinical trials are often tedious to conduct, require long trial time, and are expensive. On the other hand in silico trials are cost effective and provide flexible techniques to evaluate novel treatment plans.

Even though several control methodologies have been suggested in the literature for the closed-loop control of intravenous drug administration, very few findings have attracted the attention of clinicians. This is mainly due to the discrepancy between the actual clinical requirement and the one that is considered for study. An ideal closed-loop controller that can effectively facilitate the complex task of drug delivery should account for multiple clinical phenomena such as drug interaction, drug overdosing and underdosing, significant variabilities in the drug response(s) of different patients, nonlinearities and disturbances in the system, and major drug-induced side effects such as immunosuppression or hemodynamic instability.

1.2 Literature review

1.2.1 Drug-dosing control for anesthesia administration

Anesthesia is mainly used to facilitate invasive and painful clinical procedures such as endotracheal intubation, ventilation, suction, and hemodialysis. Too much or too little anesthetic can cause increased morbidity. Hence, the rate of infusion of anesthetic drugs is critical, requiring continuous monitoring and repeated adjustments (Haddad et al., 2010). Typically, open-loop drug infusion is facilitated by a medical practitioner or via target controlled infusion (TCI) pump (Absalom and Mason, 2017; Absalom et al., 2011; Masui et al., 2010). TCI pumps are programmed to derive the required drug dose for a patient by using a nominal model of the patient. However, recent investigations in the area of anesthetic and analgesic drug dosing have documented several positive outcomes of the closed-loop control approaches compared to open-loop ones (Absalom et al., 2011; Brogi et al., 2017; Kuizenga et al., 2016; Soltesz et al., 2013). Specific advantages of the closed-loop control approaches include improved patient safety, early recovery time, and reduced treatment cost. Moreover, closed-loop control relieves the clinicians from doing frequent mechanical adjustments which in turn allow them to indulge in more critical aspects of therapy to improve overall well being of the patient (Haddad et al., 2010).

Patients admitted to ICU often suffer from multiple illnesses or even organ system failure. Hence, it is necessary to evaluate the health of these patients using various physiological monitors and provide required assistance using life-supporting devices. Some of the life-supporting procedures such as mechanical ventilation involve invasive endotracheal tube insertion which leaves the patient in physical as well as mental distress. Moreover, due to anxiety and discomfort related to these procedures the patients are often restless and in an incoherent state of mind. Hence, in order to comfort the patients and to perform painful clinical procedures in a cooperative and safe manner, often these patients are kept in a state of moderate sedation for a long period of time. Apart from the complications in the normal physiological functioning of the body which arise due to an inherent illness, side effects of the drugs used for treatment can also have an adverse effect on the overall health of these patients. For instance, most of the sedatives and analgesics used these days are identified to impair cardiac and respiratory functions (Absalom et al., 2011; Jacobi et al., 2002; Minto et al., 2000; Robinson et al., 1997). Thus, the critically ill patients in the ICUs who are treated using multiple intravenous drugs for long periods also demand the regulation of multiple physiological variables such as MAP, heart rate, respiratory rate, level of unconsciousness and pain sensation, and other vital parameters within acceptable safe limits (Heusden et al., 2018; Jacobi et al., 2002).

Analyzing drug anesthetic effects requires pharmacokinetic (PK) models to account for the drug disposition and pharmacodynamic (PD) models to capture drug concentration effects. In order to formulate the mathematical equivalent of a human drug disposition system with a time-dependent drug dose as an input signal, several physiological and nonphysiological models have been proposed (Absalom et al., 2009; Haddad et al., 2010). Among these, deterministic PK models, represented by compartmental models, which involve single or multiple compartments to capture the drug distribution and metabolism have gained wide acceptance (Absalom et al., 2009; Masui et al., 2010). In the case of intravenous infusion of anesthetic drugs, the mechanism of drug disposition can be effectively represented using a three-compartmental model with an additional effect-site compartment to model the time-lag in the drug dynamics at the locus of the drug effect (Masui et al., 2010). It should be noted that underlying illness, drug interaction, and other clinical disturbances alter the drug requirements (Absalom et al., 2011; Jacobi et al., 2002; Minto et al., 2000; Robinson et al., 1997).

Advancements in the area of automation and control engineering have fostered human health care in many ways. There exist many control methods that have been successfully used to design controllers for applications that require tracking a certain desired response. However, the requirement for an accurate mathematical model that depicts human physiology and difficulty in measuring certain system parameters that are required for feedback are the two main hurdles that limit the utilization of such control methods in the area of drug dosing. Several clinical and in silico trials conducted to evaluate the efficacy of the fixed-gain and linear controllers for the closed-loop control of anesthesia administration have proved inadequate (Absalom et al., 2011; Bailey and Haddad, 2005; Haddad et al., 2013; Hahn et al., 2012; Soltesz et al., 2013). This set back is mainly due to the complexity and uncertainty involved in the intricate task of anesthesia administration.

Furutani et al. (2010) reported 79 clinical trials conducted to evaluate the performance of model predictive controllers (MPC) for the closed-loop control of anesthesia administration. This study marks improved performance of the closed-loop control approach over manual control in terms of the amount of drug used and tracking error in reference output (BIS). However, the performance of the MPC-based controller was not so good compared to the that reported by Morley et al. (2000), Absalom and Kenny (2003), Liu et al. (2006), and Struys et al. (2001). Even though optimal control methods can account for system state constraints and control constraints, as pointed out by Furutani et al. (2010) such methods demand more accurate mathematical model to improve the tracking ability and robustness of the closed-loop control system. Haddad et al. (2003) documented the improved performance of adaptive disturbance rejection controller in addressing the system uncertainties and system disturbances associated with anesthesia administration. However, adaptive controllers cannot embody optimality requirements of the system optimality. Thus, it is necessary to develop novel methods that are capable of addressing problems that arise due to the system disturbances and system uncertainties, while deriving at optimal control laws to enhance the applicability and safety of automated anesthesia administration.

1.2.2 Drug-dosing control for cancer chemotherapy

Most of the cancer chemotherapy control algorithms reported in the literature are implemented using optimization methods (Chen et al., 2012, 2014; Doloff and Waxman, 2015; Engelhart et al., 2011; Kiran et al., 2009; Noble et al., 2010; Swierniak et al., 2003). Chen et al. (2012) and Noble et al. (2010) discussed an MPC-based controller which uses a new state measurement at the end of each sampling period to update the model used for solving the optimization problem. Kiran et al. (2009) used a multiobjective optimization approach to regulate the use of therapeutic agents and derive optimal treatment schedule for immunotherapy and chemotherapy. Similarly, Engelhart et al. (2011) investigated the problem of deriving optimal treatment plan for immunotherapy, chemotherapy, or/and antiangiogenic therapy with respect to various objective functions.

Batmani and Khaloozadeh (2013) and Çimen (2010) resorted to state-dependent Riccati equation (SDRE)-based controller design approach for deriving treatment schedule for cancer chemotherapy. Specifically, Batmani and Khaloozadeh (2013) used a state observer to estimate the unavailable system states. In Babaei and Salamci (2015), a hybrid method that compounds SDRE and model reference adaptive controller design method is used to determine a personalized drug dose for cancer treatment. As mentioned earlier, the efficacy of the optimal control approaches depends on the accuracy of the mathematical used. However, it is often impossible to derive an ideal mathematical model which can accommodate all the complex dynamics involved in the tumor microenvironment (Pillis and Radunskaya, 2001; Sbeity and Younes, 2015; Swan, 1990). Typically, these dynamics include the tumor growth, immune response to tumor growth, changes in the vascular network that supply nutrients to the tumor, and the effect of the drug on various cell types in the tumor microenvironment to name some.

Evolutionary algorithms (EA)-based approaches have also been used to derive optimal drug-dosing schedules for chemotherapy (Tan et al., 2002; Tse et al., 2007). Even though the EA-based approaches exhibit competitive performance compared to the other existing chemotherapy optimization approaches, difficulty in the selection of the initial population and significant computation effort involved limits the acceptance of these methods (Sbeity and Younes, 2015).

1.2.3 RL-based algorithms

Even though several control methodologies have been suggested in the available literature for the closed-loop control of intravenous drug administration, very few findings have attracted the attention of clinicians. This is mainly due to the discrepancy between the actual clinical situation and the one that is considered for study. An ideal closed-loop controller that can effectively facilitate the complex task of drug delivery should account for multiple clinical phenomena such as drug interaction, drug overdosing and underdosing, significant variabilities in the drug response(s) of different patients, nonlinearities and disturbances in the system, and major drug-induced side effects such as immunosuppression or hemodynamic instability. RL-based control is a novel promising approaches for the control of intravenous drug administration to achieve multiple clinical objectives simultaneously.

RL-based methods have been used for many years to derive optimal control inputs in the presence of system disturbances and in the absence of knowledge of complete system dynamics (Bertsekas and Tsitsiklis, 1996; Sutton and Barto, 1998; Vrabie et al., 2013). RL algorithms arrive at an optimal solution by performing control policy updates based on a reward or performance index defined with respect to the controlled system. Such algorithms are based on dynamic programming and give optimal solutions when the iterations converge (Barto et al., 1983; Sutton, 1988; Sutton and Barto, 1998). Moreover, these are interactive algorithms which can account for time-varying system dynamics and performance requirements (Vrabie et al., 2013). RL methods rely on the speed, efficiency, and computational advantages of the digital computers to assess the impact of each possible control action on the system and derive the best control action in an uncertain noisy environment.

RL-based control strategies have exhibited satisfactory performance in the areas of aeronautics, robotics, and clinical pharmacology when used for the control, automation, motion planning, signal processing, and networking (Abbeel et al., 2007; Dadhich et al., 2016; Hong et al., 2016; Sedighizadeh and Rezazadeh, 2008). RL methods can derive an optimal controller by exploring the advantage of each possible action in driving the system to a target (goal) state. After training, the controller uses the learned optimal control policies to regulate the transience of the system under control from an arbitrary initial state to the goal state. RL is suitable for deriving optimal drug-dosing schedules mainly because this method does not require the model of the system and it can learn the best sequence of actions or the optimal control law using the response of the patient to the control input (drug infusion). In the context of RL, the term agent is used as a synonym of the term controller in the field of control theory (Sutton and Barto, 1998). Here, a policy can be either a function of system states, or a path or a plan to transition the system from an arbitrary initial state to the goal state, or it can even be rule-based such as "if in this state, then do this." A reward function is used to assess the advantage of an action with respect to system states.

Recently, RL-based control strategies have been used in the drug-dosing control scenario to optimize the dosing of erythropoietin during hemodialysis, develop dynamic treatment regime for patients with lung cancer, assist insulin regulation in diabetic patients, regulate heparin dosing, and administer anesthetic drugs to induce and maintain the desired sedation level (Daskalaki et al., 2013; Martin-Guerrero et al., 2009; Moore et al., 2014; Nemati et al., 2016; Zhao et al., 2011). Moore et al. (2010) discussed RL-based optimal controller for the regulation of hypnosis during surgery. Specifically, the authors derived optimal control solutions by penalizing the control actions that correspond to an increase or decrease in BIS value from the target BIS value and rewarding the control actions that maintain the BIS output of the patient at the target value of BIS. The authors used three control actions (drug dose) u such as 0, 20, or 40 mg to train the RL-based controller. Currently, bedside monitors that can provide a measure of the depth of anesthesia in terms of BIS index value are available (Johansen et al., 2000). Extending the RL-based controller presented by Moore et al. (2010, 2014) and conducted the first clinical trial to evaluate the closed-loop control of hypnosis using human volunteers. This RL-based controller showed a patient-specific control of hypnosis with an enhanced control accuracy with respect to other similar investigations in the literature.

The remaining chapter is organized as follows. In Section 2, a general framework is presented to formulate the intravenous drug-dosing control problem in a finite Markov decision process (MDP) framework and develop a Q-learning based controller to design a multiobjective controller to regulate anesthesia administration by accounting for important physiological parameters of the patients. In Section 3, an optimal drug-dosing profile is derived by accounting for PK and PD disturbances such as drug interaction in the human body under treatment. Finally, in Section 4, the Q-learning-based controller design approach presented in Section 2 is adapted to address specific cases in cancer chemotherapy treatment.

2 Control of BIS by accounting for MAP

The aim of this section is to develop a multiobjective controller that can regulate sedation by simultaneously accounting for drug-induced hemodynamic instability in a
patient. Administration of sedative drug such as propofol can have adverse effects
on the hemodynamic stability of the patient. Specifically, propofol causes vasodilation
leading to the decrease in MAP, drug overdose, and even cardiovascular collapse (Fan
et al., 2012). Consequently, during propofol infusion, along with regulating the
desired drug response (BIS index) it is important to maintain hemodynamics parameters (e.g., MAP) of the patient in an clinically acceptable and safe range (Haddad
et al., 2010; Rao and Bequette, 2000). Toward this end, first, a general framework
for the development of an RL-based controller for the control of nonlinear dynamical
systems is presented. Next, the PK and PD models of propofol related to patient
responses such as the BIS index and MAP are discussed. Here, this model serves
as a patient model which is then used to generate input-output data required to train
the RL agent or the RL-based controller.

2.1 Problem formulation

The problem of obtaining control solutions to follow a desired system trajectory often
requires sequential decision making and can be solved by representing it in a finite
MDP framework (Vrabie et al., 2013). During anesthesia administration, the aim is
to reach a defined goal state (desired BIS value) with decisions predicated on the best
sequence of control actions (propofol infusion) required to transition the system from
a given arbitrary initial condition to the desired goal state. Toward this end, consider
the nonlinear dynamical system given by

$$\dot{x}(t) = f(x(t), u(t)), \quad x(0) = x_0, \quad t \geq 0, \tag{1}$$

$$y(t) = h(x(t)), \tag{2}$$

where $x(t) \in \mathbb{R}^n$, $t \geq 0$, is a vector with n states of the system as the elements, $u(t) \in \mathbb{R}$,
$t \geq 0$, denotes the control input, $y(t) \in \mathbb{R}^l$, $t \geq 0$, represents l number of outputs or
responses of the system, $f \colon \mathbb{R}^n \times \mathbb{R} \to \mathbb{R}^n$ is locally Lipschitz continuous and
$h \colon \mathbb{R}^n \to \mathbb{R}^l$ is continuous. RL-based control approaches are suitable for problems that
require a goal-oriented decision making (Sutton and Barto, 1998).

A finite MDP can be defined using the four-tuple $(\mathcal{S}, \mathcal{A}, \mathcal{P}, \mathcal{R})$, where $\mathcal{S}$ is a finite
set of states of the system or environment, $\mathcal{A}$ is a finite set of possible actions when in
the states $s_k \in \mathcal{S}$, $\mathcal{P}$ represents a state transition probability matrix, $\mathcal{P}_{a_k}(s_k, s_{k+1})$ is the
probability that the state $s_k \in \mathcal{S}$ at k transits to the state s_{k+1} at $k+1$ with an action $a_k \in$
$\mathcal{A}$ at k, and $\mathcal{R}$ represents a reward that assesses the advantage of an action $a_k \in \mathcal{A}$ for all
$s_k \in \mathcal{S}$. Note that the transition probability matrix denoted as $\mathcal{P}$ represents the system
dynamics and is equivalent to the function $f(\cdot, \cdot)$ which is assumed to be unknown. The
discrete states in the finite set $\mathcal{S}$ are denoted as $(\mathcal{S}_i)_{i \in \mathbb{I}^+}$, where $\mathbb{I}^+ \triangleq \{1, 2, ..., q\}$ and q

represents the total number of states. Similarly, the discrete actions in the finite sequence $\mathcal{A}$ are denoted as $(\mathcal{A}_j)_{j \in \mathbb{J}^+}$, where $\mathbb{J}^+ \triangleq \{1, 2, ..., p\}$ and p represents the total number of actions.

RL-based methods, such as Q-learning (Watkins and Dayan, 1992), have gained significant attention in recent years. The main reason for the increased acceptance of RL-based methods is the fact that it does not rely on a model of the system for the design of the controller. Moreover, RL-based methods can account for changes in the system that happens during the learning phase. Fig. 1 shows the schematic diagram of an RL-based approach in which the agent or controller learn a useful policy or an action plan using the information on the action taken, reward observed, and a new state to which the system reached due to the current action. In other words, the Q-learning-based controller design method can train an RL agent to learn the best sequence of control actions to regulate the states s_k of the system without using the system state $x(t)$, $t \geq 0$. Instead it uses the information gathered at time steps $k \in \{1, 2, ...\}$, $kT \leq t < (k+1)T$ along the system trajectories. At every time step k, the RL agent identifies the current state s_k from the set $\mathcal{S}$ and then it chooses an action a_k from the defined action set $\mathcal{A}$. Consequently, the system stochastically transitions from the current state s_k to a new state s_{k+1} incurring a numerical reward $r_{k+1} \in \mathbb{R}$.

Since learning is predicated on the knowledge of the discrete states $s_k \in \mathcal{S}$ and which should be measurable at time step k. Hence, the states s_k of the RL environment are defined with respect to the system response given by $y(t)$, $t \geq 0$, as

$$s_k = g(y(t)), \quad kT \leq t < (k+1)T, \tag{3}$$

where $g: \mathbb{R}^l \to \mathcal{S} \subset \mathbb{R}$ is a mapping between the system response $y(t)$ and the state representation s_k, $k = 1, 2, ...$ Here, $T > 0$ is the sampling time.

The agent aims to maximize the reward it earns over an infinite horizon. This can be achieved by using different strategies (Watkins and Dayan, 1992). A straightforward approach is to choose each action a_k such that it maximizes the expected value of the discounted return (Moore et al., 2014; Watkins and Dayan, 1992). In this case, the objective function is given by

$$J(R_k) = \mathbb{E}\left[\sum_{i=1}^{\infty} \theta^{(i-1)} r_{i+k}\right], \tag{4}$$

Fig. 1 Reinforcement learning schematic (Padmanabhan et al. (2015)).

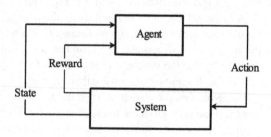

where $\mathbb{E}[\cdot]$ denotes expectation, R_k denotes the total discounted return, and $\theta \in [0, 1]$ is a *discount rate parameter* which represents the horizon of interest to the agent. For $\theta = 0$, $J(R_k) = r_k$, that is, for learning, the agent considers only the current reward. Alternatively, for θ approaching 1, the weight of the costs incurred in the future is increased.

2.2 Learning an optimal policy

RL-based control relies on learning an optimal control policy while interacting with the system. Information obtained while interacting with the system is used to enhance the agent's decision-making policy over time. Thus, the agent interacts with the system to learns the optimal policy starting from an initial arbitrary policy. In the case of linear systems, optimal control law pertaining to the certain defined objective function and system constraints can be derived by solving associated algebraic Riccati equation. However, deriving optimal control law for nonlinear systems is tedious and requires the solution of complex Hamilton-Jacobi-Bellman partial differential equation (Balashevich et al., 2002; Haddad and Chellaboina, 2008).

Watkin's Q-learning is an RL-based approach which uses each state transition to update each entry of a table Q which forms the control policy. The policy is stored in a table so that appropriate responses can be retrieved quickly with respect to the state of the system. The entry $Q(s_k, a_k)$ of the Q table for each pair of state s_k and action a_k, $k \in \{1, 2, \ldots\}$ shows the value of the state s_k when associated with action a_k. The controller or RL agent assess the measured variables, and implement control actions according to the learned optimal policy given by $Q(s_k, a_k): \mathcal{S} \times \mathcal{A} \to \mathbb{R}$ (see Fig. 1).

For every k and state s_k, the controller or agent selects the control action a_k as

$$a_k = \arg\max_{a \in \mathcal{A}} Q(s_k, a). \tag{5}$$

The numerical reward $r_k \in \mathbb{R}$ guides the agent to whether the action chosen at the time step k was "good" or "bad." After the transition $s_k \to s_{k+1}$, having taken an action a_k and received a reward r_{k+1}, the Q table is updated by

$$Q_k(s_k, a_k) \leftarrow Q_{k-1}(s_k, a_k) + \eta_k(s_k, a_k)[r_{k+1} + \theta \max_{a_{k+1}} Q_{k-1}(s_{k+1}, a_{k+1}) - Q_{k-1}(s_k, a_k)], \tag{6}$$

where $\eta_k(s_k, a_k) \in [0, 1)$ is the learning rate or step size parameter that is related to the size of adjustment after each experiment and θ denotes the discount rate parameter.

It has been shown by Bertsekas and Tsitsiklis (1996), Sutton and Barto (1998), and Watkins and Dayan (1992) that the Q-learning algorithm (6) converges to the optimal Q-function while maximizing Eq. (4) with probability one as long as

$$\sum_{k=1}^{\infty} \eta_k(s_k, a_k) = \infty, \quad \sum_{k=1}^{\infty} \eta_k^2(s_k, a_k) < \infty, \quad (s_k, a_k) \in \mathcal{S} \times \mathcal{A}. \tag{7}$$

Note that $\sum_{k=1}^{\infty} \eta_k(s_k, a_k) = \infty$ requires that all state-action pairs (s_k, a_k) are visited infinitely often, whereas $\sum_{k=1}^{\infty} \eta_k^2(s_k, a_k) < \infty, (s_k, a_k) \in \mathcal{S} \times \mathcal{A}$ is the condition required to ensure convergence of the algorithm with probability one. As mentioned earlier, the Q-learning algorithm starts with an initial arbitrary estimate of the unknown $Q(s, a)$ and then the algorithm iteratively updates the Q table until convergence is achieved, that is, until $\Delta Q = 0$, where ΔQ is the change in the Q table, or when the updates satisfy a minimum threshold $\Delta Q \leq \delta$, where δ is a prespecified tolerance parameter. Note that the optimal value of the Q table depends on the parameter values that are used for each iteration.

The framework introduced in this section is used to develop a Q-learning-based controller for the closed-loop regulation of the BIS and MAP by controlling the continuous infusion of propofol. The controller is designed to regulate the system output $y(t) \in \mathbb{R}^l, t \geq 0$, using the control input $u(t) \in \mathbb{R}, t \geq 0$. For our simulation, we use the simulated patient model as introduced in the following section to train the RL agent. Thus, the RL system or the environment shown in Fig. 1 is replaced by the nominal PK and PD model of the patient as shown in Fig. 2.

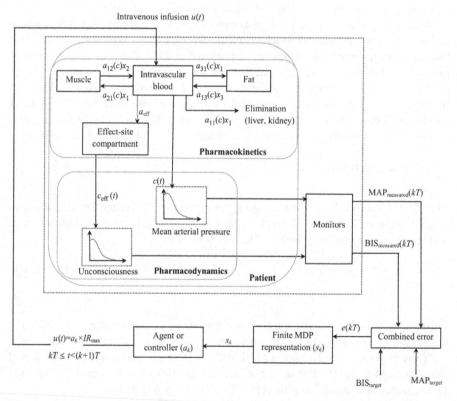

Fig. 2 Closed-loop control using RL showing the interaction between the agent and the simulated patient for simultaneous regulation of BIS and MAP management (Padmanabhan et al. (2015)).

2.3 Pharmacokinetic and pharmacodynamic patient model

As mentioned earlier, propofol has interlaced effects on the consciousness and hemodynamic stability of the patient. Specifically, most anesthetic drugs reduce sympathetic tone and alter arterial pressure of the patient by bringing about venodilation. Propofol infusion also decreases cardiac output and thereby reduces drug disposition. This reduction in disposition of the drug in body is compensated by titrating more drug which may lead to overdose. A nonlinear dynamical patient model given by the function $f(x(t), u(t))$ is used to represent the nonlinear PK and PD of the drug, where $u(t), t \geq 0$, denotes the intravenous infusion of the drug propofol and $x(t), t \geq 0$, is the system states.

As shown in Fig. 2, a nonlinear three-compartment model with an effect-site compartment is used to represent the patient dynamics. The control input is the continuous infusion of propofol to the central compartment. In this model, $x_1(t), t \geq 0$, represents the amount of the drug in the arteries and veins (intravascular blood). In addition to the intravascular blood, $x_1(t), t \geq 0$, also includes the mass of the drug in organs with very high blood supply such as the brain, heart, kidneys, and liver. The states $x_2(t), t \geq 0$, and $x_3(t), t \geq 0$, represent the rest of the drug in the body which is assumed to be in two peripheral compartments, comprised of muscle and fat, respectively. These two peripheral compartments receive less than 20% of the overall blood supply (cardiac output) in the body.

The three-compartment model is given by Haddad et al. (2010), Soltesz et al. (2013), and Absalom et al. (2009)

$$\dot{x}_1(t) = -[a_{11}(c(t)) + a_{21}(c(t)) + a_{31}(c(t))]x_1(t) + a_{12}(c(t))x_2(t) + u_1(t),$$
$$x_1(0) = x_{10}, \quad t \geq 0, \tag{8}$$

$$\dot{x}_2(t) = a_{21}(c(t))x_1(t) - a_{12}(c(t))x_2(t), \quad x_2(0) = x_{20}, \tag{9}$$

$$\dot{x}_3(t) = a_{31}(c(t))x_1(t) - a_{13}(c(t))x_3(t), \quad x_3(0) = x_{30}, \tag{10}$$

$$\dot{c}_{\text{eff}}(t) = a_{\text{eff}}(x_1(t)/V_c - c_{\text{eff}}(t)), \quad c_{\text{eff}}(0) = c_{\text{eff}0}, \tag{11}$$

where $c(t) = x_1(t)/V_c, t \geq 0$, denotes the drug concentration in the central compartment denoted by $x_1(t), t \geq 0$, and V_c represents the volume of the central compartment, $a_{ij}(c(t)) = A_{ij}(\text{AC}_{50}^{\gamma_a}/(\text{AC}_{50}^{\gamma_a} + (c(t))^{\gamma_a})), i, j = 1, 2, 3$, denote the nonnegative mass transfer coefficients between the jth and ith compartment, A_{ij} are positive constants, γ_a is a parameter that determines the steepness of the concentration-effect relationship, and AC_{50} is the drug concentration associated with a 50% decrease in the transfer coefficient. The relation between the amount of the drug in the system and its effect on the output variables such as BIS and MAP follow a nonlinear sigmoidal dynamics and thus, the function $h(\cdot)$ in Eq. (2) can be modeled using the Hill equation given by Haddad et al. (2010)

$$h(x(t)) = [\text{BIS}_{\text{measured}}(c_{\text{eff}}(t)), \text{MAP}_{\text{measured}}(c(t))]^T, \tag{12}$$

where $BIS_{measured}(c_{eff}(t))$ and $MAP_{measured}(c(t))$ are the drug effects captured by

$$BIS_{measured}(c_{eff}(t)) = BIS_0\left(1 - \frac{(c_{eff}(t))^\gamma}{(c_{eff}(t))^\gamma + (C_{50})^\gamma}\right), \tag{13}$$

$$MAP_{measured}(c(t)) = MAP_0\left(1 - \frac{(c(t))^\alpha}{(c(t))^\alpha + (MC_{50})^\alpha}\right), \tag{14}$$

where BIS_0 represents the baseline value, which is typically assigned a value of 100 to denote an awake state, C_{50} denotes the concentration of the drug related to the half-maximal effect of the BIS and models the patient's sensitivity to the drug, γ denotes the degree of nonlinearity, MAP_0 is the initial value of MAP of the patient before drug infusion, MC_{50} denotes the concentration of the drug related to the half-maximal effect of the MAP, and α denotes the degree of nonlinearity associated with MAP of the patient (Haddad et al., 2010).

2.4 Closed-loop control of BIS and MAP using RL

In this section, the Q-learning algorithm is used to develop a drug-dosing agent for the simultaneous regulation of anesthesia and hemodynamic status. The control variable $u(t)$, $t \geq 0$, in the dynamical system given by Eq. (1) is the continuous intravenous infusion of propofol. In the RL framework, since the agent interacts with the patient at discrete time steps the propofol infusion rate at each time step k is defined as

$$IR_k = a_k \times IR_{max}, \tag{15}$$

where $k \in \{1, 2, ...\}$, IR_{max} is the maximum allowable infusion rate, and a_k is a particular action from the action set $\mathcal{A}$ selected at the kth time step. Thus, between any two time steps k and $k + 1$, the infusion rate remains constant and is given by $u(t) = IR_k$, $kT \leq t < (k + 1)T$, where T is the time duration between any two time steps. The action $a_k \in \mathcal{A}$ at the kth time step can vary from 0 (no infusion) to 1 (maximum rate of infusion) within the finite action set $\mathcal{A} = \{0, 0.01, 0.02, 0.03, 0.04, 0.05, 0.08, 0.1, 0.15, 0.2, 0.25, 0.3, 0.35, 0.4, 0.5, 0.6, 0.7, 0.8, 0.9, 1\}$, where $\mathbb{J}^+ = \{1, 2, ..., 20\}$. Since IR_{max} is a configurable parameter, one of the benefits of the infusion rate scheme given by Eq. (15) is that it is easy to set its value according to the sedation requirements of the patient in the ICU.

In the RL framework, the controller or agent makes a decision about the action to be taken at each time step based on the current state of the system $s_k = g(y(t))$, $s_k \in \mathcal{S}$, $t \in [kT, (k+1)T)$. Hence, the state s_k of the system should be observable for decision making. Therefore, the states s_k of the RL system is defined based on the measurable parameters $BIS_{measured}(c_{eff}(t))$ and $MAP_{measured}(c(t))$, $kT \leq t < (k + 1)T$. The state s_k is defined based on the error $e(t)$, $kT \leq t < (k + 1)T$, given by

$$e(t) = \sqrt{\beta_w BIS_{error}^2(t) + MAP_{error}^2(t)}, \tag{16}$$

where $\beta_w > 0$ is a weighing factor, which can be used to weigh the importance of anesthesia control over hemodynamic control,

$$BIS_{error}(t) = \frac{BIS_{measured}(c_{eff}(t)) - BIS_{target}}{BIS_{target}} \times 100, \qquad (17)$$

and

$$MAP_{error}(t) = \frac{MAP_{measured}(c(t)) - MAP_{target}}{MAP_{target}} \times 100. \qquad (18)$$

For ICU sedation, the agent aims to learn the best sequence of infusion rates which minimize BIS_{error} and MAP_{error}. Hence, defining the system states denoted by s_k with respect to the error $e(t)$, $t \geq 0$, is reasonable. Moreover, using $e(t)$, $t \geq 0$, for training purposes has the advantage of involving single measurement given by Eq. (16) rather than two separate measurements of $BIS_{error}(t)$ and $MAP_{error}(t)$, $t \geq 0$. This decreases the complexity of the training algorithm. In this case, the action of the agent is predicated on the values of BIS and MAP. For our simulations, the parameters BIS and the MAP are calculated using the propofol concentration in the PD models (13), (14). In real time, both these variables can be measured in ICU using corresponding bedside monitors. To model possible measurement limitations in the BIS and MAP monitors, a sampling time $T = 6$ s is used. Thus, the agent interacts with the patient at every 6 s (Moore et al., 2014).

Here, the agent seeks to learn the best action sequence that will transition the system from given initial state to target states identified as $BIS_{target} = 65$ and $MAP_{target} = 80$. The range of output variables that are considered for our simulation are $BIS_{measured}(t) \in [0, 100]$ and $MAP_{measured}(t) \in [0, 120]$. Note that $BIS_{error}(t)$, $t \geq 0$, remains positive when $BIS_{measured}(t) \in (65, 100]$ and negative when $BIS_{measured}(t) \in [0, 65)$. However, this change in sign is not reflected in the value of $e(t)$ for $BIS_{measured}(t) \in (65, 100]$ and $BIS_{measured}(t) \in [30, 65)$. See that, as shown in Fig. 3, $e(t)$ when calculated using Eq. (16) gives the same value for $BIS_{measured}(t) \in (65, 100]$ and $BIS_{measured}(t) \in [30, 65)$. The agent should increase the infusion of the sedative drug when $BIS_{error}(t)$ is positive and decrease it when $BIS_{error}(t)$ is negative. In order to account for this, separate set of states is assigned for positive and negative values of $BIS_{error}(t)$, $t \geq 0$. Specifically, $s_k \in \{1, 2, \ldots, 13\}$ is assigned for $e(kT) \in [0, e_p(t)]$ and $s_k \in \{14, 15, \ldots, 20\}$ is assigned for $e(kT) \in [0, e_n(t)]$, where $e_p(t)$ and $e_n(t)$ denote the maximum error in the region of error $e(kT)$ where $BIS_{error}(t)$ is positive and negative, respectively. See Table 1 for the mapping between the error $e(kT)$ and state s_k. The entries in the Q table which corresponds to the states 1–13 for positive values of $BIS_{error}(t)$, $t \geq 0$, and 14–20 for negative values of $BIS_{error}(t)$, $t \geq 0$, are updated using Eq. (6).

It can be seen from Table 1 that there is a dense discretization of $e(kT)$ near the region where $e(kT) = 0$. Moreover, compared the case when $BIS_{error}(t)$ is negative, more number of states are assigned when $BIS_{error}(t)$ is positive. This is to account for the fact that when $BIS_{error}(t)$ is negative the patient is oversedated and hence the ideal infusion rate is zero as $e(kT)$ approaches the value 300.

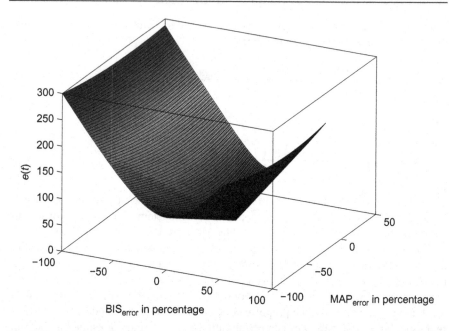

Fig. 3 Normalized percentage error of BIS and MAP versus combined error $e(t)$ (Padmanabhan et al. (2015)).

Table 1 State assignment based on $e(t)$ (Padmanabhan et al., 2015)

$BIS_{error} > 0$		$BIS_{error} < 0$	
State s_k	$e(kT)$	State s_k	$e(kT)$
1	[0, 2]	14	[0, 10]
2	[2, 4]	15	[10, 50]
3	[4, 10]	16	[50, 100]
4	[10, 15]	17	[100, 150]
5	[15, 25]	18	[150, 200]
6	[25, 35]	19	[200, 250]
7	[35, 45]	20	[250, 300]
8	[45, 60]		
9	[60, 80]		
10	[80, 100]		
11	[100, 120]		
12	[120, 140]		
13	[140, 165]		

On the other hand, when $BIS_{error}(t)$ is positive the patient is undersedated and the infusion rate should vary considerably according to how close is the patient from targeted BIS value.

Choosing an appropriate reward function is a very important step during the implementation of Q-learning-based algorithm. Note that the reward function is used to assess the advantage of each action in the action set. Reward function plays a key role in reinforcing the agent's decision-making policies and hence choosing reward function requires a careful consideration. For ICU sedation, it is apparent that the action that decreases the difference between the measured value of BIS and MAP denoted by $BIS_{measured}$ and $MAP_{measured}$ and the targeted value of BIS and MAP denoted by BIS_{target} and MAP_{target}, respectively, must incur more reward. An appropriate reward has to steer the agent to learn the optimal policy for the regulation of BIS and MAP responses toward the required target values. Hence, the reward r_{k+1} corresponding to action a_k at k is computed by

$$r_{k+1} = \begin{cases} \dfrac{e(kT) - e((k+1)T)}{e(kT)}, & e((k+1)T) < e(kT), \\ 0, & e((k+1)T) \geq e(kT). \end{cases} \tag{19}$$

For an error $e((k+1)T) \geq e(kT)$, the algorithm assigns $r_{k+1} = 0$. This means that if certain action imparted to the system at k could not reduce the error at the time step $k+1$ then that action is given a zero reward. This assignment penalizes bad control actions. On the other hand, if certain action imparted to the system at the current time step reduces the error at the next time step, then that action is given a reward proportional to the difference in error $(e(kT) - e((k+1)T))$ between two time steps. Note that the Q table is updated using Eqs. (6), (19). Here, for each state s_k, the action in the set $\mathcal{A}$ that results in maximum value for $e(kT) - e((k+1)T)$ is assigned the highest value of reward.

RL-based algorithms exploit the computational power of computers to execute all possible actions from each state to asses which action will steer the system closer toward the desired target state. The aim is to drive the system from a given initial state $s_0 \in \mathcal{S}$, $\mathcal{S} = \{1, 2, ..., 20\}$, to the desired target state 1 as $k \to \infty$. A *policy* is defined as the sequences of state actions which can steer the system from an initial arbitrary state to the target state. Among the possible policies, the *optimal policy* is the one which earns a maximum reward. Thus, successful training is achieved when, for each state $s_k \in \mathcal{S}$, the agent identifies the best action a_k^* among all possible actions $a_k \in \mathcal{A}$ resulting in a maximum reward. Maximizing the reward in turn implies that the action a_k^* will drive the system closer to the desired state 1 as compared to all other possible actions in the given action set. The learned optimal policy is unique for a given set of states and action set (Sutton and Barto, 1998).

First, the RL agent learns by experimenting with the system using the permissible actions and assessing the response (output) of the simulated patient as shown in Fig. 4. For our simulations, the patient model is assigned an arbitrary initial condition.

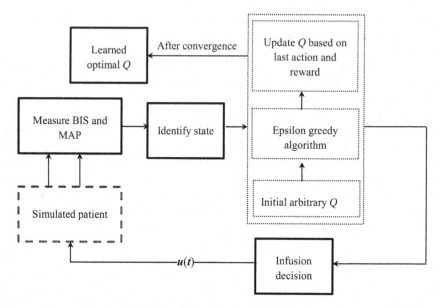

Fig. 4 Schematic representation of training sequence to obtain the optimal Q table (Padmanabhan et al. (2015)).

Note that, as shown in Fig. 2, the patient is replaced by a nominal population model which represents the PK and PD of the drug. The combined error $e(t)$, $t \geq 0$, is derived using the PD model for the response variables BIS and MAP given in Eqs. (13), (14), respectively. As shown in Fig. 4, using the value of $e(t)$, $t \geq 0$, the current state of the system is identified. Initially, the values of the Q table are set to zero. Using Eq. (5), in a Q table with all zero entries, the agent is always directed to choose the same action as a_k^*. In order to avoid this initial difficulty, and to facilitate learning, an ϵ-greedy policy can be used (Sutton and Barto, 1998). Using ϵ-greedy policy, the agent executes random actions with probability ϵ, where ϵ is a small positive number. These random actions help the agent to gather information according to the pharmacology of a patient. Toward this end, the agent infuses propofol at different rates defined in the action set and observes the response of the patient. After each experiment, the agent calculates the reward incurred and updates the corresponding state-action entry in the Q table using Eqs. (6), (19), respectively. This is done to associate each state with the best action in the action set $\mathcal{A}$.

Given the current state $s_k \in \mathcal{S}$, a "good" control action $a_k \in \mathcal{A}$ imparted by the controller results in a positive reward ($r_{k+1} > 0$) for $e((k+1)T) < e(kT)$. Similarly, a "bad" control action $a_k \in \mathcal{A}$ results in zero reward ($r_{k+1} = 0$), for $e((k+1)T) \geq e(kT)$. Note that r_{k+1} is used in Eq. (6) to update the Q table. To facilitate learning, the agent tries every possible action $a_k \in \mathcal{A}$ for all possible states $s_k \in \mathcal{S}$ and observes the utility of each action in earning a positive reward (Sutton and Barto, 1998). According to Eq. (7), to arrive at an optimal policy with respect to a defined set of states and actions, the controller or the RL agent should explore all states and actions and utilize the information pertaining to the previous trails that were useful or effective in incurring more reward (Sutton and Barto, 1998). With $k \rightarrow \infty$, all the defined states and actions in the Q table will be executed recurrently which enables the Q table to converge the optimal Q table. Another condition required to ensure convergence of the Q table and to learn the optimal policy is to reduce the learning rate $\eta_k(s_k, a_k)$ defined in Eq. (7) over time (Sutton and Barto, 1998).

2.5 Details of the simulation

In this section, simulation results are presented to illustrate the use of RL-based controller for the closed-loop control of BIS and MAP. Simulations are conducted by setting iteration number to 50,000 (arbitrarily high) scenarios, where a scenario represents the series of transitions from an arbitrary initial state to the required final state 1. Furthermore, initially $\eta_k(s_k, a_k) = 0.2$ (for scenarios 1–499) is assigned and subsequently halved $\eta_k(s_k, a_k)$ every 500th scenario. For each scenario, a new set of randomized initial states $x_1(0) \in [0, 0.084]$ g, $x_2(0) \in [0, 0.067]$ g, $x_3(0) \in [0, 0.039]$ g, and $c_{\mathrm{eff}}(0) \in [0, 0.005]$ g L^{-1} of propofol was assigned to the simulated patient model and then the learning phase was repeated until convergence and the performance goals were met; that is, keeping the BIS and the MAP values within the desired ranges. For our simulation, the Q table converged before reaching the maximum iteration. After convergence, for every state s_k, the agent chose an action $a_k = \arg\max_{a \in \mathcal{A}} Q(s_k, a)$.

After the training phase, that is, once the agent learned the optimal sequence of infusion rates required for each state $s_k \in \mathcal{S}$ to reach the desired goal state, the efficacy of the learned agent in a sequence of scenarios is evaluated over individual patients to check how well the agent can perform drug administration based on its optimal control policy during practical situations.

During anesthesia administration oversedation and undersedation is not acceptable. Hence, after training exploration or random actions are avoided to update the Q table, but used the optimal $Q(s_k, a_k)$ discussed in the previous section for making drug infusion decisions for the 30 simulated patients.

The value of $\mathrm{BIS}_{\mathrm{error}}(t)$, $t \geq 0$, and $\mathrm{MAP}_{\mathrm{error}}(t)$, $t \geq 0$, is in range the range of 0%–100%. Next, in order to prioritize the control of BIS over MAP a positive-weighing parameter β_w is used. A high value for β_w decreases the regulation of MAP, on the other hand choosing a small value for β_w will reduce the control of BIS. For our simulation $\beta_w = 8$ is used which is set by trial and error. The recommended dose of propofol given in ASHP guidelines (Jacobi et al., 2002) is an initial bolus (20 mg) followed by continuous infusion (5–80 μgkg^{-1}min^{-1}). The maximum amount of drug required for a 100-kg patient during the maintenance phase of anesthesia administration is 8 mg min^{-1}. Hence, for training the RL agent $IR_{\mathrm{max}} = 20$ mgmin^{-1} is used.

The evaluation of the performance of the Q-learning algorithm-based controller is conducted in 30 simulated patient models using hypnosis scenarios that lasted for 2 h.

Table 2 Perturbation values (Padmanabhan et al., 2015)

Parameter	Perturbation range
Concentration at half maximal effect of BIS, C_{50}	0.004 ± 0.001 g L^{-1}
Concentration at half maximal effect of MAP, MC_{50}	0.004 ± 0.001 g L^{-1}
Concentration at half maximal effect of a_{ij}, AC_{50}	0.004 ± 0.001 g L^{-1}
Degree of nonlinearity of BIS(c_{eff}), γ	3 ± 1
Degree of nonlinearity of MAP(c), α	3 ± 1
Degree of nonlinearity of a_{ij}, γ_a	3 ± 1
Time lag between $c_{\text{eff}}(t)$ and $c(t)$, a_{eff}	$\in [0.17, 1]$ (min^{-1})
Volume of central compartment, V_c	16 ± 1 L
Transfer coefficients, a_{ij}	$\pm 0.5\%$ (min^{-1})

The pharmacological parameter values of the 30 simulated patients are taken randomly from a predefined parameter range as given in Table 2. In addition, $A_{11} = 0.119 \text{min}^{-1}$, $A_{12} = 0.0550 \text{min}^{-1}$, $A_{21} = 0.112 \text{min}^{-1}$, $A_{31} = 0.0419 \text{min}^{-1}$, and $A_{13} = 0.0033 \text{min}^{-1}$ are used (Bailey and Haddad, 2005). Unlike surgery which requires deep sedation, often the procedures in ICU can be carried out with moderate sedation. Thus, for our simulation, the target values of output variables are set to $\text{BIS}_{\text{target}} = 65$ and $\text{MAP}_{\text{target}} = 80$. Simulation results showing the steady-state performance of the Q-learning-based anesthesia control approach for two case studies are presented. Statistical analysis pertaining to these simulation studies is also conducted.

2.6 Results and discussion

Fig. 5 shows the implementation of the closed-loop control strategy using the learned optimal policy. At each time step K, the RL agent chooses an infusion rate based on the learned optimal Q table. Note that, even though the training of the agent is done using a simulated patient model, performance evaluation is conducted on a population of 30 simulated patients. To study the efficacy of the trained RL agent in the closed-loop regulation of anesthesia, common performance matrices such as the root mean square error (RMSE), median performance error (MDPE), and median absolute performance error (MDAPE) are used (Moore et al., 2014). The performance error (PE) is defined as

$$\text{PE}_i(j) \triangleq \frac{\text{Measured value}_i(j) - \text{Target value}}{\text{Target value}} \times 100, \quad j = 1, \ldots, N, \qquad (20)$$

where $i \in \{1, \ldots, 30\}$ represents the ith patient, j represents the set of PE measurements for an individual, N is the number of measurements for each patient, and Measured value and Target value in Eq. (20) refer to BIS and MAP, respectively. Note that for the controlled variables BIS and MAP, the PE is the same as the $\text{BIS}_{\text{error}}(t)$ and

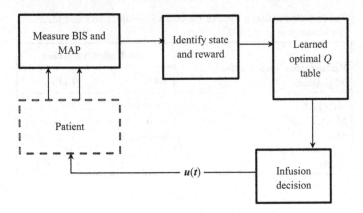

Fig. 5 RL-based optimal and robust closed-loop control of BIS and MAP (Padmanabhan et al. (2015)).

$MAP_{error}(t)$, $t \geq 0$, given by Eqs. (17), (18), respectively. The MDPE gives the control bias observed and is computed by

$$MDPE_i = \text{median}(PE_i(j)), \quad j = 1,...,N, \tag{21}$$

whereas

$$MDAPE_i = \text{median}(|PE_i(j)|), \quad j = 1,...,N, \tag{22}$$

and

$$RMSE_i = \sqrt{\frac{\sum_{j=1}^{N}(\text{Measured value}_i(j) - \text{Target value})^2}{N}}, \tag{23}$$

where $MDAPE_i$ denotes the median of the absolute value of PE and it reflects the accuracy of the trained RL agent in keeping the targeted values of the control variables BIS and MAP for each of the 30 simulated patients (Moore et al., 2014). $RMSE_i$ is the RMSE for each patient. Table 3 shows the performance metrics for the Q-learning-based agent for 30 simulated patients during the 2 h of hypnosis scenario considered. The amount of inaccuracy reflected in the values of the MDAPE metrics listed in Table 3 is in the acceptable clinical performance range (Moore et al., 2014).

To further elucidate the performance of the RL agent, the central tendency as well as the range of measured variables is evaluated for all of 30 simulated patients. Specifically, the amount of time that the outputs are within a desired band of the targeted values, that is, ± 5, and the percentage of all of the patients for which the outputs are within a predefined band is calculated. For the 2-h drug infusion period considered in our simulation, the measured value of the output variable BIS is within ± 5 of BIS_{target} for 90.41% of the time for all 30 simulated patients. Similarly, the measured value of

Table 3 Performance metrics for control variables BIS and MAP
(Padmanabhan et al., 2015)

Performance metrics (for 30 patients)	Controlled variables	
	BIS	MAP
MDPE (%)	3.97 ± 2.32	4.05 ± 2.50
MDAPE (%)	4.19 ± 6.43	5.31 ± 5.30
Min−max	66.43−68.25	75.52−89.46
Interquartile range	0.55	7.16
RMSE	2.12−3.30	2.30−9.50

the output variable MAP is within ±5 of MAP_{target} for 76.65% of the time for 60% of 30 simulated patients.

Table 3 shows the minimum and maximum values of BIS and MAP, respectively, during the maintenance period of drug administration. The time range $t \in [0, 10)$ and $t \in [10, 120]$ are considered as the induction period and maintenance period of anesthesia administration, respectively. This table also lists the variability or midspread of the controlled variables determined in terms of the interquartile range (IQR). IQR is the value of the middle of a data set arranged in ascending order. In order to obtain the IQR, the average value of the controlled variables BIS and MAP during $t \in [10, 120]$ for all of the simulated patients is used. The IQR of BIS variable is 0.55 and that of MAP is 7.16. Note that BIS has comparatively lesser variability than MAP.

Fig. 6 shows the closed-loop anesthesia control scenario for three randomly selected simulated patients from 30 simulated patients. These plots further elucidate the variations of the controlled variables BIS and MAP around the target MAP_{target} and BIS_{target} values with respect to RL-based control. Often postsurgical patients are kept in ICU under moderate sedation to facilitate treatment procedures. Patient 1 is assumed to be a postsurgical patient and hence a nonzero initial condition is assigned to indicate the presence of propofol in the patient's body. This is to model the residual quantity of anesthetic drugs in the patient's body that has been administered during surgery. The initial conditions of Patients 2 and 3 are set to zero. The RL-based controller is able to regulate the output BIS and MAP value close to the target values. The trained RL-based controller demonstrates acceptable performance with respect to the simultaneous control of BIS and MAP (Moore et al., 2014). Fig. 6 along with the performance evaluation metrics given in Table 3 demonstrates the significance of the β_w parameter in Eq. (16) for prioritizing the control of BIS relative to MAP.

Note that our simulations show similar performance compared to the clinical trial conducted by Moore et al. (2014) for the evaluation of RL-based closed-loop control of intraoperative hypnosis. With respect to this clinical trial, the authors report that the range of the percentage values of MDPE and MDAPE is −2.8 to 8.8 and 3.4−9.6, respectively. For the 15 patients considered for the experiment, the range of the value of RMSE is 3.3−6.5. These figures are comparable with our results given in Table 2 for

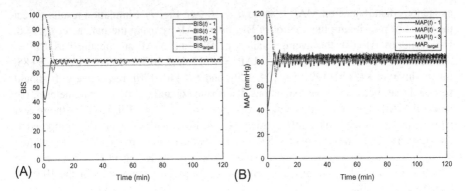

Fig. 6 Simulation results for three patients chosen randomly from the test set of 30 patients. (A) BIS index versus time for $BIS_{target} = 65$. (B) MAP versus time for $MAP_{target} = 80$ (Padmanabhan et al. (2015)).

the 30 simulated patients. Apart from the closed-loop regulation of the BIS, a methodology for the control of MAP is developed. In addition to that the RL-based controller does not rely on a system model and it demonstrates optimal and robust performance (see Fig. 6 and Table 3). This is an added advantage when it comes to the control of uncertain biological systems wherein developing accurate system models are very challenging.

In the context of simultaneous control of output variables such as BIS and MAP, one could debate that including an additional parameter (secondary variable(s)) can adversely affect the regulation of BIS (primary variable). However, instead of regulating sedation alone, simultaneous and balanced maintenance of the level of sedation along with other important requirements such as hemodynamic and respiratory system stability, pain management, muscle relaxation, etc. should be considered for improving patient safety. This is important as many of the sedative drugs (e.g., propofol) are known to induce significant changes in the heart rate, cardiac output, MAP, and respiratory rate of the patient. For our simulations, we consider MAP as the secondary control variable as propofol infusion reduces the sympathetic tone of the patient and induces venodilation. As a consequence, significant changes in the cardiac output and MAP is reported in the literature (Robinson et al., 1997).

Next, two case studies are presented to further elucidate the effect of simultaneous control of the BIS and MAP of a patient. First, a hemodynamic disturbance is simulated to account for the effect of hemorrhage on MAP by altering the MAP values by d units. For the second case study, irrespective of propofol infusion, the value of the secondary controlled variable (MAP) is held constant throughout the simulation period. This is to model the case of the intubated patients in the ICU who suffer from complications due to postaortic aneurysm repair. Another similar clinical situation in which the MAP becomes dangerously low is in the case of the septic patients.

First, to test the efficacy of the RL agent due to exogenous hemodynamic disturbance, a random patient from the population of 30 generated patients is simulated for: (i) $MAP(t)$, $t \geq 0$; (ii) $MAP(t) + d$, $t > 20$; and (iii) $MAP(t) - d$, $t > 40$. Here, the value of the exogenous disturbance on MAP is set to $d = 10$ (see Fig. 7). Simulation results reflect the effect of prioritizing the control of BIS over MAP by using the parameter $\beta_w = 8$. For scenarios (i)–(iii), the average values of BIS and MAP are obtained as BIS = 68.18, MAP = 81.82, BIS = 66.84, MAP = 90.17, and BIS = 66.86, MAP = 69.85, for the interval $t \in [10, 120]$, $t \in [20, 120]$, and $t \in [40, 120]$, respectively. It should be noted that, because of the exogenous disturbance quantified by the parameter d, the value of MAP_{error} is more and so the error signal is $e(t)$, $t \geq 0$. This explains the reason why the patient is sedated slightly more in the case of scenarios (ii) and (iii) compared to that of scenario (i). The control of BIS is affected by the increase in $e(t)$, $t \geq 0$, contributed mainly by the disturbance in the MAP value. However, the RL agent is able to keep the variation in BIS value within the acceptable range given by ± 5 units of the BIS_{target} (Moore et al., 2014).

During the second case study, irrespective of the propofol infusion, the value of $MAP(t)$, $t \geq 0$, is kept constant at the values 120, 100, 60, and 40 for all of the 30 simulated patients. The efficacy of the RL agent in regulating the value of BIS for these constant values of MAP is analyzed. Note that for all these simulation studies, the effect of propofol infusion on MAP is not considered. Instead, the value of the MAP is held constant. As shown in Table 4, for the cases with $MAP(t)$, $t \geq 0$, kept at values 100 and 60, the RL agent is able to keep the variation in BIS value within the acceptable range given by ± 5 units of the BIS_{target} (Moore et al., 2014). However, for the cases in which $MAP(t)$, $t \geq 0$, is kept constant at values 120 and 40, the variations in the value of BIS are in the range of $BIS_{target} \pm 10$. For such extreme scenarios, it is recommended to use an RL agent which is trained by setting a large value for the parameter β_w to improve the regulation of BIS.

Efficacy of RL methods is demonstrated in several real-time applications, however, for clinical scenarios, decision making predicated on online identification requires a careful consideration. This work is a preliminary study toward the

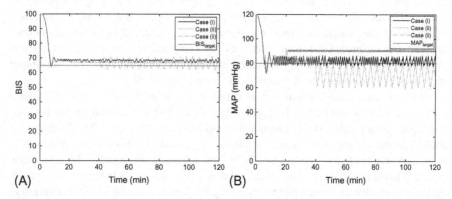

Fig. 7 Simulation results for a patient chosen randomly from the test set of 30 patients; Case (i) $MAP(t)$, $t \geq 0$; Case (ii) $MAP(t) + d$, $t > 20$; and Case (iii) $MAP(t) - d$, $t > 40$, where $d = 10$ units represents a disturbance in the hemodynamic system of patient. (A) BIS index versus time for $BIS_{target} = 65$. (B) MAP versus time for $MAP_{target} = 80$ (Padmanabhan et al. (2015)).

Table 4 Performance metrics for the control variable BIS by keeping MAP constant; for 30 simulated patients (Padmanabhan et al., 2015)

Performance metrics	MAP(t) = 120	MAP(t) = 100	MAP(t) = 60	MAP(t) = 40
MDPE (%)	9.43 ± 0.63	−0.84 ± 0.45	−0.87 ± 0.49	9.44 ± 0.62
MDAPE (%)	9.43 ± 0.95	1.56 ± 0.43	1.58 ± 0.45	9.44 ± 0.56
Min−max	69.18−74.28	62.51−66.50	62.48−66.64	69.14−74.31
Interquartile range	2.6	2.15	2.16	2.67

implementation of RL-based closed-loop control of anesthesia. Some of the factors that contribute to the interindividual variations in the pharmacological parameters within a patient population are the patient physiological features, age, and concurrent illness. Drug habituation due to the frequent use of certain drugs also affect the response of a patient to the drug. Note that compared to the nominal model used for training, the variation in the pharmacologic parameters of the patient under treatment will be reflected in the response of the patient. Consequently, the error signal $e(t)$, $t \geq 0$, varies accordingly and thus the state s_k. This implies that as the RL agent executes control actions with respect to the state s_k, it can indirectly address pharmacological variations in patient to a certain extent. However, if the drug habituation or any other clinical situation results in significant and nonlinear changes in the patient pharmacology, then adaptive decision making is essential. As mentioned, the amount of propofol required to result in certain desired sedation level changes with the gender, age, weight, and height of the patient. These patient features are reflected in the pharmacological model parameters such as a_{ij}, a_{eff}, C_{50}, MC_{50}, V_c, γ, and α. The RL agent is trained by setting $IR_k = a_k \times IR_{\max}$, where $IR_{\max} = 20\,\text{mg\,min}^{-1}$. Table 2 shows the range of the patient pharmacological features used to obtain the 30 simulated patients.

Figs. 6 and 7 and Tables 3 and 4 show that the RL-based controller demonstrates acceptable performance for the 30 simulated patients with a wide range of pharmacological features. However, patient pharmacological features will considerably vary between different patient populations such as elderly, adults, children, and infants. Accordingly, the value of $IR_{\max}$ should be fixed and the RL agent needs to be trained with the new value of $IR_{\max}$. Similarly, to address the drug-dosing requirements of each patient population with vivid pharmacological features, it is recommended to use a bank of RL agents in which each agent is trained by using an appropriate $IR_{\max}$.

Finally, even though the RL-based controller demonstrates good performance, one of the limitations of this approach is the use of discrete state space and action space. Continuous-time state space and action space can enhance the robust adaptation of the RL-based controller and thereby derive more patient-specific and optimal control solution. However, this improved performance comes at the cost of increased computational cost. The performance of the RL agent can be further enhanced by adjusting the value of the discount factor θ, learning rate $\eta(s_k, a_k)$, and by choosing a more appropriate reward function (Matignon et al., 2006).

3 Control of BIS by accounting for synergistic drug interaction

In this section, the use of an RL-based controller to fine tune the drug titration while different drugs with interactive effects are administered simultaneously is discussed. It is important to consider the interactive effects of the drugs to restrict the drug usage to the optimal level required to achieve certain therapeutic effects. In Section 2, a general framework is presented to formulate the problem of closed-loop control of intravenous drug administration using a finite MDP framework and the development of the Q-learning-based controller (Padmanabhan et al., 2015). In view of automated drug delivery for ICU sedation, another relevant factor that needs attention is the interactive effects of the drugs that are administered together.

In the following section, a Q-learning-based controller is developed to account for the synergistic effect during the combined administration of sedatives and analgesics. Then, the simulated patients who are used to train the RL agent and to conduct in silico trials are explained. The aim is to develop a controller to derive an optimal drug-dosing profile by accounting for the PK and PD disturbances in the human body under treatment.

3.1 Training the RL agent

The real-time system in this context is a dynamical system that represents the PK and PD of the multiple drugs that are administered together. The system description that follows is required to comprehend the input-output information needed to train the RL agent.

Consider the nonlinear dynamical system given by

$$\dot{x}(t) \;=\; f(x(t), u(t)), \quad x(0) \;=\; x_0, \; t \geq 0, \tag{24}$$

$$y(t) \;=\; h(x(t)), \tag{25}$$

where for $t \geq 0$, $x(t) \in \mathbb{R}^{(n+p)}$ is the state vector, n and p are the number of states used to represent the PK of the sedative agent and analgesic agent, respectively, $u(t) \in \mathbb{R}^{(m+r)}$ is the control input, m and r are the number of sedative agents and analgesic agent infused, $y(t) \in \mathbb{R}^l$ is the output (controlled variable) of the system, $f \colon \mathbb{R}^{(n+p)} \times \mathbb{R}^{(m+r)} \to \mathbb{R}^{(n+p)}$ is locally Lipschitz continuous, and $h \colon \mathbb{R}^{(n+p)} \to \mathbb{R}^l$ is continuous. Here, the controlled variable of interest is the sedation level of the patient.

The aim is to develop an RL-based agent for the closed-loop control of the primary drug during their combined administration with any other drugs with a synergistic interactive effect.

Toward this end, the equivalent finite MDP representation of the system presented in Section 2.1, which involves a finite set of states S of the system, a finite set of action $\mathcal{A}$ that is available for each state $s_k \in S$, a scalar reward $r_k \in \mathbb{R}$, and the transition probability matrix $\mathcal{P}$ that depends on the function $f(\cdot, \cdot)$ defined in Eq. (24) which is assumed to be unknown is used. With respect to the infusion of the sedative agent, the finite action set with p number of discrete actions defined as $(\mathcal{A}_j)_{j \in \mathbb{J}^+}$, $\mathbb{J}^+ \triangleq \{1, 2, ..., p\}$ is considered. As explained in Section 2.2, a Q-function is progressively updated as per Eq. (6) using the available information with respect to system (24), which involve current state, action taken, new state reached, and reward received for the state transition.

3.2 Simulated patient

In this section, the patient models used for our simulations are presented. A superscript S or A denote that the parameter is associated with a sedative or an analgesic drug, respectively. First, consider the dynamical system

$$\dot{x}(t) = Ax(t) + Bu(t), \quad x(0) = x_0, \quad t \geq 0, \tag{26}$$

where $A \in \mathbb{R}^{(n+p) \times (n+p)}$ is a compartmental matrix, $B \in \mathbb{R}^{(n+p) \times (m+r)}$ is an input matrix, $x(t) \in \mathbb{R}^{(n+p)}$, $t \geq 0$, is the state vector, and $u(t) \in \mathbb{R}^{(m+r)}$, $t \geq 0$, is given by $u(t) = [(u^S(t))^T, (u^A(t))^T]^T$, where $u^S(t) \in \mathbb{R}^m$, $t \geq 0$, and $u^A(t) \in \mathbb{R}^r$, $t \geq 0$, represent the sedative and analgesic drug infusion, respectively. Next, rewrite Eq. (26) as

$$\dot{x}(t) = Ax(t) + B\bar{u}^S(t) + d(t), \quad x(0) = x_0, \quad t \geq 0, \tag{27}$$

where $\bar{u}^S(t) = [(u^S(t))^T, 0]^T$, $d(t) = B\bar{u}^A(t)$, and $\bar{u}^A(t) = [0, (u_A(t))^T]^T$. For each drug, a three-compartment model with an effect-site compartment is used to represent the drug disposition in the human body. While infusing several drugs simultaneously, the mass distribution of each drug in these three compartments and the effect site can be represented using the respective system states for each drug.

For the simultaneous infusion of a sedative and an analgesic drug, we consider the state vector $x(t) = [x_1(t), x_2(t), x_3(t), c_{\text{eff}}^S(t), x_5(t), x_6(t), x_7(t), c_{\text{eff}}^A(t)]^T$, where $x_i(t), t \geq 0$, $i = 1, 2, 3$, and $x_i(t), t \geq 0, i = 5, 6, 7$, denote the masses of the sedative and analgesic in the ith compartment, respectively, and $c_{\text{eff}}^S(t), t \geq 0$, and $c_{\text{eff}}^A(t), t \geq 0$, are the effect-site concentrations of the sedative and analgesic, respectively. In particular,

$$\dot{x}_1(t) = -(a_{11}^S + a_{21}^S + a_{31}^S)x_1(t) + a_{12}^S x_2(t) + u^S(t), \quad x_1(0) = x_{10}, \quad t \geq 0, \tag{28}$$

$$\dot{x}_2(t) = a_{21}^S x_1(t) - a_{12}^S x_2(t), \quad x_2(0) = x_{20}, \tag{29}$$

$$\dot{x}_3(t) = a_{31}^S x_1(t) - a_{13}^S x_3(t), \quad x_3(0) = x_{30}, \tag{30}$$

$$\dot{c}_{\text{eff}}^S(t) = a_{\text{eff}}^S(x_1(t)/V_c - c_{\text{eff}}^S(t)), \quad c_{\text{eff}}^S(0) = c_{\text{eff0}}^S, \tag{31}$$

and

$$\dot{x}_5(t) = -(a_{11}^A + a_{21}^A + a_{31}^A)x_5(t) + a_{12}^A x_6(t) + u^A(t), \quad x_5(0) = x_{50}, \ t \geq 0, \tag{32}$$

$$\dot{x}_6(t) = a_{21}^A x_5(t) - a_{12}^A x_6(t), \quad x_6(0) = x_{60}, \tag{33}$$

$$\dot{x}_7(t) = a_{31}^A x_5(t) - a_{13}^A x_7(t), \quad x_7(0) = x_{70}, \tag{34}$$

$$\dot{c}_{\text{eff}}^A(t) = a_{\text{eff}}^A(x_5(t)/V_c - c_{\text{eff}}^A(t)), \quad c_{\text{eff}}^A(0) = c_{\text{eff0}}^A, \tag{35}$$

where a_{ij}^S and a_{ij}^A denote the rate of mass transfer between the jth and ith compartment for the sedative and analgesic drug, respectively, and V_c is the volume of the central compartment (blood).

When two drugs with interactive effects are administered simultaneously, their drug effect varies according to the ratio of the two drugs denoted as ϕ and their normalized drug concentration U. We use the common sedation assessment measure given by the BIS (Johansen et al., 2000) to assess the sedation level of the patient. The net sedative effect of an anesthetic drug when administered along with an analgesic drug which has synergistic interactive effect is given by

$$\text{BIS}_{\text{measured}}(t) = \text{BIS}_0 \left(1 - \frac{\left(\dfrac{U^S(t) + U^A(t)}{U_{50}(\phi)}\right)^{\gamma(\phi(t))}}{1 + \left(\dfrac{U^S(t) + U^A(t)}{U_{50}(\phi(t))}\right)^{\gamma(\phi)}} \right), \tag{36}$$

where $\phi(t) \triangleq \frac{U^S(t)}{U^S(t) + U^A(t)}$, $\gamma(\phi(t))$, $t \geq 0$, is the steepness of the concentration-response relation at ratio $\phi(t)$, and $U_{50}(\phi(t))$ is the number of units associated with 50% of maximum effect at ratio $\phi(t)$ (Minto et al., 2000). Furthermore, $U^S(t)$, $t \geq 0$, and $U^A(t)$, $t \geq 0$, are the normalized drug concentrations of the sedative and analgesic drugs and are given by $U^S(t) = \frac{c_{\text{eff}}^S(t)}{C_{50}^S}$ and $U^A(t) = \frac{c_{\text{eff}}^A(t)}{C_{50}^A}$, where C_{50}^S and C_{50}^A are the drug concentrations of the sedative and analgesic that cause 50% drug effects, respectively. The BIS value corresponding to fully conscious patient is denoted by BIS_0. For training the RL agent, $e(t) = \text{BIS}_{\text{error}}(t)$ is assigned, where $\text{BIS}_{\text{error}}(t)$ is given by Eq. (17).

3.3 Results and discussion

In this section, the efficacy of the RL-based controller in deriving optimal infusion rates of an anesthetic drug so as to achieve certain desired sedation level by simultaneously accounting for the infusion of an synergistic analgesic is discussed. For our simulation, the most widely used sedative and analgesic drugs, propofol and remifentanil, respectively, are used. These drugs have synergistic interactive effects (Mehta et al., 2006).

For our simulations, 25 simulated patients using clinically relevant patient parameters are used. The pain experienced by a patient during the clinical procedures such as surgery, tracheal tube insertion, or physiotherapy treatment varies considerably. In the case of analgesic drugs, the drug concentration that causes half-maximal effect (pain relief) denoted by C_{50}^A varies with the intensity of the pain associated. For instance, the C_{50}^S and C_{50}^A of patients with or without liver disorders varies considerably (Mehta et al., 2006). Hence, to account for such variations in the pharmacological parameter C_{50}^A with respect to different pain stimulus, the values in the range 0.025 ± 0.007 mg L^{-1} are used (Mehta et al., 2006). Table 5 summarizes the range of PK and PD parameters of the drugs propofol and remifentanil that are used to generate 25 simulated patients. For training the RL agent using a simulated patient, the PK parameter values $C_{50}^S = 5.6$ μg L^{-1} for propofol, and $C_{50}^A = 30$ ng L^{-1} for remifentanil are used. The response of the patient given by Eq. (36) are calculated using the relation $U_{50}(\phi) = 1 - \theta_B \phi + \theta_B \phi^2$, where $\theta_B = 0.22$ and $\gamma(\phi) = 0.85$ (Padmanabhan et al., 2014).

At each time step k, the Q-learning algorithm (6) requires the values of s_k, a_k, s_{k+1}, and r_{k+1} to progressively derive the optimal action set. Toward this end, the states s_k are defined based on the error $e(kT)$. The values $s = 10$ when BIS$_{error} < 0$ and $s_k \in \{1, 2, ..., 9\}$ when BIS$_{error} > 0$ are used. The range of values of the error $e(kT)$ for each $s_k \in \{1, 2, ..., 9\}$ is ([0, 1], (1, 3], (3, 8], (8, 12], (12, 18], (18, 25], (25, 35], (35, 45], (45, 54]), respectively. We use a finite action set $\mathcal{A} = \{0, 0.02, 0.04, 0.1, 0.25, 0.5, 0.7, 0.8, 0.9, 1\}$ for the RL agent. At each time step k, the agent imparts an infusion rate $u(t) = IR_k$, $IR_k = a_k \times IR_{max}$, where IR_{max} is the maximum allowable infusion rate for the sedative drug propofol. For our simulations, $IR_{max} = 25$ mg min^{-1} and BIS$_{target} = 65$ are used. The action a_k at the kth time step is chosen from the finite action set $\mathcal{A}$.

Table 5 Range of values used to generate 25 simulated patients (Padmanabhan et al., April, 2017a)

Parameter	Propofol	Remifentanil
C$_{50}$	0.004 ± 0.001 g L^{-1}	0.025 ± 0.007 mg L^{-1}
V_c (L)	16 ± 1	16 ± 1
a_{ij} (min^{-1})	$\pm 0.5\%$	$\pm 0.5\%$
a_{eff} (min^{-1})	$\pm 0.5\%$	$\pm 0.5\%$

Another factor to consider is that the patient PK and PD vary significantly according to the health condition of the patient. The recommended propofol infusion rate for a patient treated for ailments in renal, hepatic, or cardiac function is 2.8 ± 1.1 mg kg^{-1} h^{-1} of propofol. Likewise, for a patient with respiratory ailments, the recommended drug dose titration rate of propofol is 1.25 ± 0.87 mg kg^{-1} h^{-1}. The recommended remifentanil infusion rate for the combined administration of propofol and remifentanil is 0.6–15 µg kg^{-1} h^{-1} (Mehta et al., 2006). For an 80-kg patient, this range is equivalent to 0.008–0.02 mg min^{-1}. Hence, the efficacy of the Q-learning-based controller with respect to two different drug infusion rates; 0.05 and 0.1 mg min^{-1} are tested. Figs. 8 and 9 show the controlled variable (BIS) for the two different infusion rates of remifentanil. For both cases, first the drug propofol alone during the time interval $t \in [0, 60)$ min is administered and then

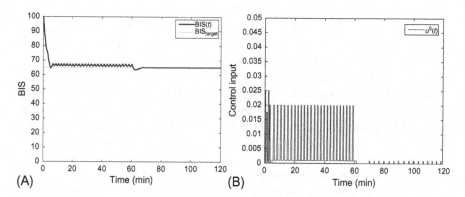

Fig. 8 Simulation results with remifentanil infusion rate of $u^{A}(t) = 0.1$ mg min^{-1} during $t = [60, 120]$. (A) BIS index versus time for BIS$_{\text{target}} = 65$. (B) Control input $u^{S}(t)$ versus time (Padmanabhan et al., April, 2017a).

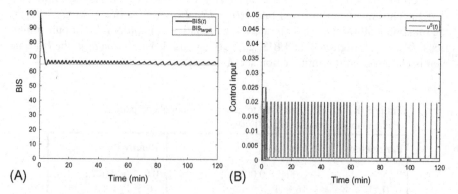

Fig. 9 Simulation results with remifentanil infusion rate of $u^{A}(t) = 0.05$ mg min^{-1} during $t = [60, 120]$. (A) BIS index versus time for BIS$_{\text{target}} = 65$. (B) Control input $u^{S}(t)$ versus time (Padmanabhan et al., April, 2017a).

Table 6 Performance metrics for 25 patients for the controlled variable BIS for $u^A(t) = 0.1 \,\mathrm{mg\,min^{-1}}$ (Padmanabhan et al., April, 2017a)

Performance metrics	Drugs	
	Propofol	Propofol and remifentanil
MPE (%)	1.4657 ± 0.3110	0.499 ± 4.865
MDPE (%)	1.4349 ± 0.3127	-1.678 to 0
MDAPE (%)	1.4349 ± 0.9479	$0-1.678$

Table 7 Performance metrics for 25 patients for the controlled variable BIS for $u^A(t) = 0.05 \,\mathrm{mg\,min^{-1}}$ (Padmanabhan et al., April, 2017a)

Performance metrics	Drugs	
	Propofol	Propofol and remifentanil
MPE (%)	1.2675 ± 0.3535	0.6111 ± 0.4006
MDPE (%)	1.4349 ± 0.3125	0.1182 ± 1.5518
MDAPE (%)	1.4349 ± 0.9479	0.2698 ± 1.4079

during $t \in [60, 120]$ remifentanil along with propofol is infused. Since the two drugs have synergistic interactive effects, the desired drug effect of $\mathrm{BIS_{target}} = 65$ can be achieved using lower doses of propofol when administered along with remifentanil. It can be seen from Figs. 8 and 9 that the target anesthetic effect is achieved and maintained using a lower dose of propofol when both drugs are administered together.

As mentioned earlier, for our simulations two different drug infusion rates of the analgesic drug remifentanil are considered. Tables 6 and 7 show the statistical performance indices such as mean performance error (MPE), MDPE, and MDAPE used to evaluate the RL-based controller for the two different values of u^A used (Moore et al., 2014; Padmanabhan et al., 2015). It can be seen from Figs. 8 and 9, and Tables 6 and 7 that the performance of the RL-based controller is within the acceptable clinical ranges (Moore et al., 2014).

Simulations are conducted to generate the 25 simulated patients using the pharmacological parameters given in Table 5. Our simulation results (see Tables 6 and 7) show comparable performance with the recent in silico trial conducted on 24 virtually generated patients using model-based predictive control algorithm for automatic induction and regulation of depth of anesthesia (Nascu et al., 2011). In this in silico trial, the range of the percentage value of PE is -2.12 ± 15.13, value of MDPE is 0.8664, and the value of MDAPE is 1.114 for the 24 patients.

4 Control of cancer chemotherapy treatment

In this section, an RL-based controller design approach for the closed-loop control of cancer chemotherapy is developed. Specifically, a multiobjective RL-based controller to eradicate tumor cells by simultaneously accounting for the damage of normal cells and immune cells in a patient is discussed. The Q-learning-based approach presented in this section follows the general framework discussed in Section 2.1 to implement a similar controller for the control drug dosing pertaining to chemotherapy. The efficacy of the RL-based controller is evaluated using a nonlinear model of cancer chemotherapy controller. In the following section, a pharmacological model for cancer chemotherapy treatment is presented.

4.1 Mathematical model of cancer chemotherapy

Recently, there have been considerable efforts to develop various mathematical models to depict cancer dynamics. This is mainly to support research activities associated with the prediction of cancer incidence, drug development and its validation, and evaluation of novel drug-dosing approaches (De Pillis and Radunskaya, 2003; Sbeity and Younes, 2015). A mathematical model of cancer chemotherapy essentially accounts for the growth, death, mutation, PK, and PD in the tumor microenvironment. Typically, tumor microenvironment involves many types of cells, extracellular matrix, proteins, blood vessels, lymph vessels, etc. However, for the studies related to cancer chemotherapy, the three main cell types identified are tumor cells, immune cells, and host (normal) cells. All these cell types share common habitat (tumor microenvironment) and resources (nutrition and oxygen), resulting in nonlinear and interdependent cell dynamics. In general, a tumor without blood vessels and with blood vessels are referred as benign cancer and malignant cancer, respectively. Malignant cancers are capable of spreading (metastasize) from a tumor microenvironment to a healthy new site (ACS, 2015). Metastatic cancer is potentially lethal and hence it is often recommended to eradicate the tumor in the initial stage itself to avoid metastases.

Similar to the previous section concerning the RL-based control of anesthesia administration, in this section, a nonlinear model representing the cancer dynamics as given by Batmani and Khaloozadeh (2013) and De Pillis and Radunskaya (2003) is used to train the RL agent. In this model, the cell dynamics involved in the tumor microenvironment is explained by using the number of tumor cells, immune cells, normal cells, and drug concentration denoted by $T(t), t \geq 0, I(t), t \geq 0, N(t), t \geq 0, C(t), t \geq 0$, respectively. The model is given by

$$\dot{x}_1(t) = r_2 x_1(t)[1 - b_2 x_1(t)] - c_4 x_1(t) x_2(t) - a_3 x_1(t) x_4(t), \quad x_1(0) = x_{10}, \quad t \geq 0,$$

$$(37)$$

$$\dot{x}_2(t) = r_1 x_2(t)[1 - b_1 x_2(t)] - c_2 x_3(t) x_2(t) - c_3 x_2(t) x_1(t) - a_2 x_2(t) x_4(t),$$
$$x_2(0) = x_{20},$$

$$(38)$$

$$\dot{x}_3(t) = s + \frac{\rho x_3(t) x_2(t)}{\alpha_c + x_2(t)} - c_1 x_3(t) x_2(t) - d_1 x_3(t) - a_1 x_3(t) x_4(t), \quad x_3(0) = x_{30},$$

$$(39)$$

$$\dot{x}_4(t) = -d_2 x_4(t) + u(t), \quad x_4(0) = x_{40}, \tag{40}$$

where $x_1(t) = N(t), t \geq 0, x_2(t) = T(t), t \geq 0, x_3(t) = I(t), t \geq 0$, and $x_4(t) = C(t), t \geq 0$, $u(t), t \geq 0$, is the drug infusion rate, s denotes the (constant) influx rate of immune cells to the site of the tumor, r_1 and r_2 represent the per capita growth rate of the tumor cells and normal cells, respectively, b_1 and b_2 represent the reciprocal carrying capacities of both the cells, d_1 is the death rate of immune cells, d_2 denotes the per capita decay rate of the injected drug, and a_1, a_2, and a_3 denote the fractional cell kill rates of the immune cells, tumor cells, and normal cells, respectively (Batmani and Khaloozadeh, 2013; De Pillis and Radunskaya, 2003; Pillis and Radunskaya, 2001).

The common response of the immune system of our body toward any identified harmful infection or disease is to increase the number of immune cells. This happens whenever the body's immunosurveillance identifies a tumor cell and is modeled in Eq. (39) using the term $\frac{\rho x_3(t) x_2(t)}{\alpha_c + x_2(t)}$, where ρ and α_c represent the immune response rate and immune threshold rate, respectively (De Pillis and Radunskaya, 2003; Pillis and Radunskaya, 2001). As mentioned earlier, since all the three main cell types share common habitat and resources the increase in the survival rate of one cell type adversely effects the existence of the other type of cell. These interaction between the cell types are modeled in Eqs. (37)–(39) using the terms $-c_1 x_3(t) x_2(t), -c_2 x_3(t) x_2(t), c_3 x_2(t) x_1(t)$, and $c_4 x_1(t) x_2(t)$, where $c_i, i = 1, 2, \ldots, 4$, represent the competition terms (De Pillis and Radunskaya, 2003). Similarly, the effect of the chemotherapeutic drug on all the three cell types is modeled in Eqs. (37)–(39) using the terms $a_i x_4(t), i = 1, 2, \ldots, 3$.

Note that, apart from annihilating the tumor cells, the chemotherapeutic agent can also adversely affect the proliferation and survival of the normal cells and immune cells. Other typical side effects of chemotherapeutic drugs include hair loss, nausea, frequent infections due to the reduction in immune cell number, neuropathy, anemia, and organ damage (ACS, 2015). Hence, the aim is to derive optimal control input $u(t), t \geq 0$, for the control of drug infusion during chemotherapy so that the desired drug effect is maximized and the drug-induced side effects are minimized.

4.2   RL-based optimal control for chemotherapic drug dosing

The model (37)–(40) is used along with the general framework discussed earlier in this chapter to develop an RL-based control approach for the closed-loop control of cancer chemotherapy. Similar to anesthesia administration discussed earlier, the problem of obtaining control solution for eradicating tumor cells using chemotherapeutic agents requires sequential decision making and can be solved using RL-based approaches. Here, the objective is to drive the system given by Eqs. (37)–(40) from a nonzero initial condition to the final goal state such as $x_2(t) = 0$ at some time t. This requires deriving the best sequence of actions in terms of the drug infusion rates so as to steer the cancer patient from the state $x_2(t) \geq 0, t \geq 0$, to the desired state $x_2(t) = 0$. Toward this end, first rewrite Eqs. (37)–(40) in the state space form given by

$$\dot{x}(t) = f(x(t), u(t)), \quad x(0) = x_0, \quad t \geq 0, \tag{41}$$

$$y(t) = h(x(t)), \tag{42}$$

where $f: \mathbb{R}^n \times \mathbb{R} \to \mathbb{R}^n$, $h: \mathbb{R}^n \to \mathbb{R}^l$, $x(t) \in \mathbb{R}^n$, $t \geq 0$, denotes the state vector, $u(t) \in \mathbb{R}$, $t \geq 0$, represents the control input, and $y(t) \in \mathbb{R}^l$, $t \geq 0$, denotes the system output.

In this control problem, the number of tumor cells, given by $x_2(t)$, $t \geq 0$, is the feedback parameter used to derive the optimal amount of drug to be titrated to the patient. In case of a peripheral tumor, the number of tumor cells can be measured employing a caliper. However, for externally inaccessible tumors, such as that in the brain or lungs, imaging techniques can be used to assess the volume of the tumor (Batmani and Khaloozadeh, 2013; Gholami et al., 2011; Huang et al., 2011; Suzuki et al., 2008).

Fig. 10 shows the schematic diagram of the training steps that is used to obtain the optimal Q function for cancer chemotherapy. Here, the objective is to derive the optimal control sequence that results in a minimum tumor volume, ideally $x_2(t) = 0$. As the training of the RL agent is predicated on the availability of the discrete states $s_k \in \mathcal{S}$, it is straightforward to define the state s_k of the patient in terms of measurable output $y(t)$, $t \geq 0$. Thus, $s_k = g(y(t))$, $kT \leq t < (k + 1)T$, is assigned, where $g: \mathbb{R}^l \to \mathcal{S} \subset \mathbb{R}$ (Pachmann et al., 2001; Pillis and Radunskaya, 2001).

Sections 2.1 and 2.2 explain the use of RL-based approaches for the development of algorithms for optimal decisions making (Padmanabhan et al., 2017; Sutton and

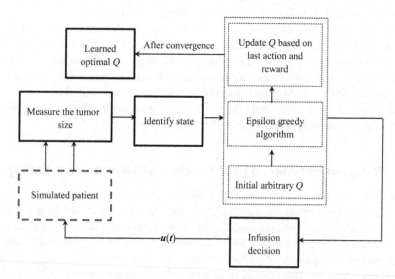

Fig. 10 Schematic diagram of training sequence to obtain optimal Q table (Padmanabhan et al., April, 2017b).

Barto, 1998; Vrabie et al., 2013). In this section, the RL framework discussed in Sections 2.1 and 2.2 is used to develop a closed-loop controller for regulating cancer chemotherapy treatment. As shown in Fig. 1, the main elements include an agent and a system. Note that here also, as the Q-learning algorithm does not use transition probability matrix denoted by $\mathcal{P}$ in order to derive the optimal control policy, it is assumed to be unknown. A controller or an agent is developed to maximize the reward it receives over an infinite horizon defined by Eq. (4). As explained in Section 2.2, a Q function is progressively updated as per Eq. (6) using the available information with respect to system (41) involving the current state, action taken, new state reached, and reward received for the state transition.

4.3 Results and discussion

This section details the numerical examples that demonstrate the performance of the Q-learning-based approach for the closed-loop control of drug dosing related to chemotherapy. To account for real-time situations, three different clinical scenarios were used to train the RL-based controller. Specifically, the case of an adult patient with cancer, a pregnant woman with cancer, and a critically ill elderly patient with cancer is considered. Here, different RL agents are developed to address the drug-dosing control in each of these cases. It is apparent that the ability of the human body to grow, repair, and defend disease is different for different age groups (Batmani and Khaloozadeh, 2013). The reason behind choosing these three case studies is to demonstrate the changes required in the RL algorithm to implement clinically relevant treatment strategies.

For instance, in case of a young cancer patient, the first preference of an oncologist will be to eradicate tumor cells to prevent metastasis. As young patients have a good growth ability, even if some of the normal cells are damaged as a side effect of chemotherapy that will be easily compensated by the body. However, this is not the case with an elderly patient who suffers from cancer as well as other diseases. For an elderly patient with cancer, the oncologist will try to eradicate cancer while preserving normal cells as well. Similarly, if the patient is suffering from brain cancer or cancer in any of vital organ, then also it is important to restrict damage of normal cells. These conditions are accounted for by selecting appropriate reward function (19). Moreover, for specific patient population like infants, children, and pregnant women, the oncologist needs to restrict the upper limits of the drug dose. This can be achieved by appropriately choosing the maximum value of the drug infusion rate $u_{\max}$ in $IR_k = a_k \times u_{\max}$ while training the RL agent.

The parameters listed in Table 8 are used in Eqs. (37)–(40) to generate simulated patients for the training and testing of the RL-based control algorithm. In the simulation, the maximum number of iteration is assigned as 50,000 scenarios. Here, a scenario is the series of state transitions from a random initial state to the desired final state s_k. The value of $\eta_k(s_k, a_k) = 0.2$ is assigned initially for the first 499 scenarios and then the value of $\eta_k(s_k, a_k)$ is subsequently halved after every 500th scenario. After convergence of the Q table to the optimal Q function, for every state s_k, the agent

Table 8 Parameter values used to generate simulated patient (Batmani and Khaloozadeh, 2013; De Pillis and Radunskaya, 2003) (Padmanabhan et al., April, 2017b)

Parameter	Parameter description	Value	Unit
a_1	Fractional immune cell kill rate	0.2	$\text{mg}^{-1}\text{L}\,\text{day}^{-1}$
a_2	Fractional tumor cell kill rate	0.3	$\text{mg}^{-1}\text{L}\,\text{day}^{-1}$
a_3	Fractional normal cell kill rate	0.1	$\text{mg}^{-1}\text{L}\,\text{day}^{-1}$
b_1	Reciprocal carrying capacity of tumor cells	1	cell^{-1}
b_2	Reciprocal carrying capacity of normal cells	1	cell^{-1}
c_1	Immune cell competition term (between T and I cells)	1	$\text{cell}^{-1}\text{day}^{-1}$
c_2	Tumor cell competition term (between T and I cells)	0.5	$\text{cell}^{-1}\text{day}^{-1}$
c_3	Tumor cell competition term (between N and T cells)	1	$\text{cell}^{-1}\text{day}^{-1}$
c_4	Normal cell competition term (between N and T cells)	1	$\text{cell}^{-1}\text{day}^{-1}$
d_1	Immune cell death rate	0.2	day^{-1}
d_2	Decay rate of injected drug	1	day^{-1}
r_1	Per unit growth rate of tumor cells	1.5	day^{-1}
r_2	Per unit growth rate of normal cells	1	day^{-1}
s	Immune cell influx rate	0.33	$\text{cell}\,\text{day}^{-1}$
α_c	Immune threshold rate	0.3	cell
ρ	Immune response rate	0.01	day^{-1}

chooses an action $a_k = \arg\max_{a \in \mathcal{A}} Q(s_k, a)$. Next, the changes required in the development of the RL-based agent for three clinical situations are discussed.

Case 1. First, the case of a young patient with cancer is considered. In this case, since the patient has a good growth ability, the patient's body can more easily compensate for the loss of normal cells and immune cells as the side effect of chemotherapy. In such a situation, the oncologist typically tries to eradicate the cancer cells $x_2(t)$, $t \geq 0$, completely. Thus, here the objective is to annihilate the tumor cells to attain the desired state $x_{2d} = 0$. Therefore, the error $e(t)$, $t \geq 0$, is defined as $e(t) = x_2(t) - x_{2d}$. The criteria used for the state assignment based on the error $e(t)$, $kT \leq t < (k+1)T$ is shown in Table 9. In this case, the reward r_{k+1} is calculated by setting $e(t) = x_2(t)$. For this case, an RL agent trained with $u_{\max} = 10\,\text{mg}\,\text{L}^{-1}\text{day}^{-1}$ is used.

Fig. 11 shows the response of the patient when a chemotherapeutic drug is administered using an RL-based controller and includes the plots of the number of normal cells, the number of tumor cells, the number of immune cells, and the concentration of chemotherapeutic drug in blood. The number of normal cells and tumor cells given in Fig. 11 are normalized values. Note that with treatment, the number of tumor cells have reduced and the normal cells have increased. However, see that initially the

Table 9 State assignment for Cases 1–3 based on $e(t)$ (Padmanabhan et al., April, 2017b)

Cases 1 and 2		Case 3	
State s_k	**$e(kT)$**	**State s_k**	**$e(kT)$**
1	[0, 0.0063]	1	[0, 0.03]
2	[0.0063, 0.0125]	2	[0.03, 0.1]
3	[0.0125, 0.025]	3	[0.1, 0.2]
4	[0.025, 0.01]	4	[0.2, 0.3]
5	[0.01, 0.05]	5	[0.3, 0.4]
6	[0.05, 0.1]	6	[0.4, 0.5]
7	[0.1, 0.2]	7	[0.5, 0.6]
8	[0.2, 0.25]	8	[0.6, 0.7]
9	[0.25, 0.3]	9	[0.7, 0.8]
10	[0.3, 0.35]	10	[0.8, 0.9]
11	[0.35, 0.4]	11	[0.9, 1]
12	[0.4, 0.45]	12	[1, 1.2]
13	[0.45, 0.5]	13	[1.2, 1.4]
14	[0.5, 0.55]	14	[1.4, 1.6]
15	[0.55, 0.6]	15	[1.6, 1.8]
16	[0.6, 0.65]	16	[1.8, 2]
17	[0.65, 0.7]	17	[2, 2.2]
18	[0.7, 0.8]	18	[2.2, 2.5]
19	[0.8, 0.9]	19	[2.5, 3]
20	[0.9, ∞]	20	[3, ∞]

Notes: Case 1: young cancer patient, Case 2: pregnant woman with cancer, and Case 3: an elderly patient who has cancer along with other critical illnesses.

number of immune cells decreases as an adverse effect of chemotherapy, and later their number improves. The amount of drug administered for Case 1 is shown in Fig. 12.

Case 2. For this case, a young pregnant woman with cancer is considered. Here, the aim is to keep the amount of the chemotherapeutic drug used to a minimum level and thus not harm the fetus. After child birth, the use of the chemotherapeutic drug can be increased to the required level to eradicate tumor. In similar situations, the oncologist often resorts to a two-stage chemotherapy. Here, for our simulations, it is assumed that the patient 7 months pregnant. In the first stage, the maximum amount of the drug infused is restricted by setting $u_{max} = 0.5\text{mgL}^{-1}\text{day}^{-1}$. However, after child birth, the maximum amount of the drug infused is increased to $u_{max} = 10\text{mgL}^{-1}\text{day}^{-1}$ (Batmani and Khaloozadeh, 2013).

For training the RL agent to derive drug infusion rates during the first stage, the value of u_{max} was set to 0.5 $\text{mgL}^{-1}\text{day}^{-1}$. Similarly, for the second state, the value of u_{max} was set to 10 $\text{mgL}^{-1}\text{day}^{-1}$. Figs. 13 and 14 show the simulation results for the two-stage chemotherapy for the young pregnant woman using RL-based controllers. Note that during the initial 90 days, the drug concentration in the plasma is

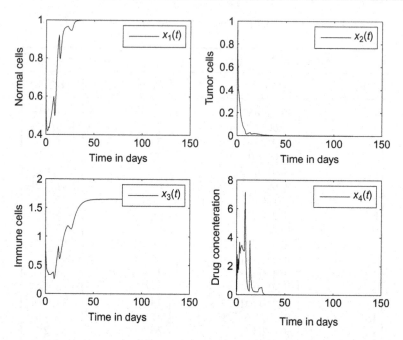

Fig. 11 Response of young patient with cancer (Case 1), $u_{max} = 10\,\text{mg L}^{-1}\text{day}^{-1}$ (Padmanabhan et al., 2017b).

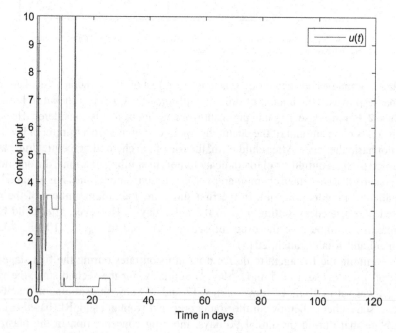

Fig. 12 Amount of drug administered (Case 1), $u_{max} = 10\,\text{mg L}^{-1}\text{day}^{-1}$ (Padmanabhan et al., 2017b).

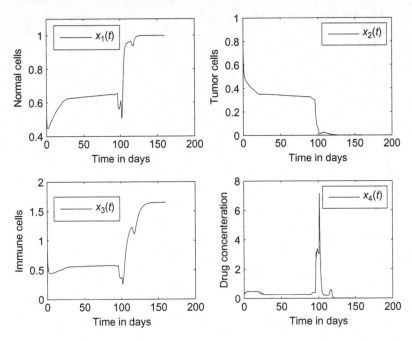

Fig. 13 Response of young pregnant woman with cancer (Case 2), $u_{max} = 0.5\,\text{mg}\,\text{L}^{-1}\text{day}^{-1}$ until delivery (90 days) and then $u_{max} = 10\,\text{mg}\,\text{L}^{-1}\text{day}^{-1}$ (Padmanabhan et al., 2017b).

restricted to 0.5 mgL^{-1}, however, after child birth the amount of drug used to eradicate the tumor completely has increased.

Case 3. For this case, an elderly patient with cancer and other illnesses is considered. This scenario represents the clinical situations wherein it is essential to minimize the damage to the normal cells while annihilating maximum number of tumor cells. In order to account for this requirement while training the RL agent, the parameter β_w which denotes a weighing factor is used to prioritize between the normal cells and cancer cells. The objective is to attain $x_{1d} = 1$ and $x_{2d} = 0$, where x_{1d} and x_{2d} represent the target values of $x_1(t)$, $t \geq 0$, and $x_2(t)$, $t \geq 0$, respectively. Here, an RL agent is trained using a reward function defined based on the deviation of the number of normal cells and tumor cells from the respective desired values. Specifically, the state s_k, $kT \leq t < (k+1)T$, is defined in terms of the error

$$e(t) = \beta_w x_2(t) + (1 - \beta_w)[1 - x_1(t)]. \tag{43}$$

The reward is calculated using the value of error $e(t)$, $kT \leq t < (k+1)T$ in Eq. (19). For our simulation, the parameter values used are $\theta = 0.7$, $\eta = 0.2$, and $\beta_w = 0.9$. Simulation results showing the response of the closed-loop control of the chemotherapeutic drug for Case 3 is shown in Figs. 15 and 16. Here the RL agent is trained with respect to the error defined by Eq. (43). Note that, compared to Case 1 that is shown in

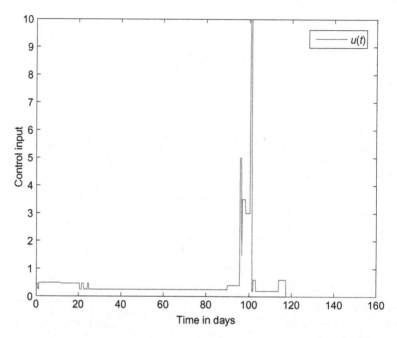

Fig. 14 Amount of drug administered (Case 2), $u_{max} = 0.5mg\,L^{-1}day^{-1}$ until delivery (90 days) and then $u_{max} = 10mg\,L^{-1}day^{-1}$ (Padmanabhan et al., 2017b).

Figs. 11 and 12, the amount of drug used in this case is lesser (see Figs. 15 and 16). This is to minimize the damage to the healthy normal cells.

Table 9 shows the criteria used for the state assignment for Cases 1–3. For Cases 1 and 3, and the second stage of Case 2, the finite action set $\mathcal{A} = \{0, 0.01, 0.02, 0.03, 0.04, 0.06, 0.08, 0.1, 0.15, 0.2, 0.25, 0.3, 0.35, 0.4, 0.5, 0.6, 0.7, 0.8, 0.9, 1\}$ is used. However, for the first stage of Case 2, with $u_{max} = 0.5mgL^{-1}day^{-1}$, the finite action set $\mathcal{A} = \{0.5, 0.55, 0.6, 0.65, 0.7, 0.75, 0.78, 0.80, 0.82, 0.85, 0.87, 0.9, 0.91, 0.92, 0.93, 0.94, 0.95, 0.97, 0.98, 1\}$ is used. Moreover, for Cases 1 and 3, and the second stage of Case 2 during the training of RL agent, the goal state as $s = 1$ is used to eradicate the tumor completely. However, for the first stage of Case 2, during the training of RL agent, the goal state is set as $s = 7$, which represents a limited tumor size.

In order to demonstrate the robustness of the controller, the trained optimal RL-based controller is used for the drug dosing of three different simulated patients. In Case (i), we consider the simulated patient with a nominal model generated using the parameters given in Table 8. In Cases (ii) and (iii), simulated patients with -10% and $+15\%$ parameter variations with respect to the values given in Table 8 are used. Figs. 17 and 18 show the corresponding simulation results. It can be seen that the controller is able to impart patient-specific infusion rates in accordance with the parameter variations. This is mainly due to the fact that the drug-dosing decision is made using the optimal Q table with respect to the state s_k. Recall that the state s_k is defined

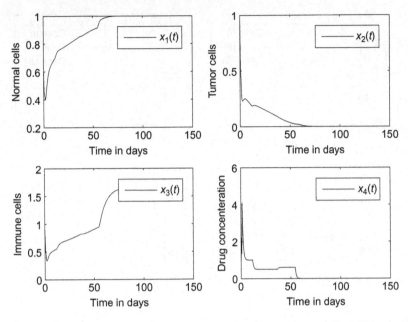

Fig. 15 Response of an elderly patient who has cancer along with other critical illnesses (Case 2), $u_{max} = 10\text{mg L}^{-1}\text{day}^{-1}$ (Padmanabhan et al., 2017b).

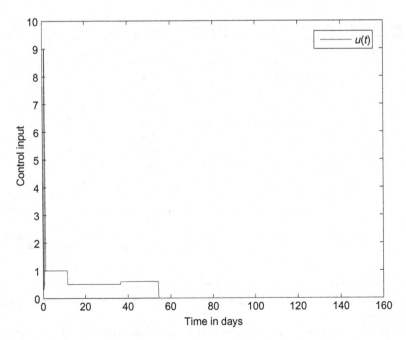

Fig. 16 Amount of drug administered (Case 2), $u_{max} = 10\text{mg L}^{-1}\text{day}^{-1}$ (Padmanabhan et al., 2017b).

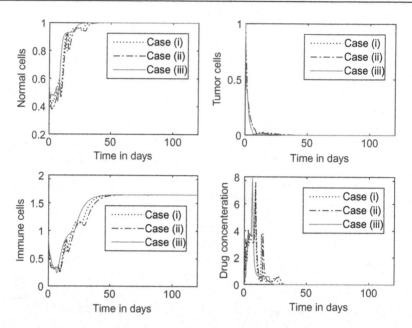

Fig. 17 Response for three different patient models: Case (i) with nominal model, Case (ii) with −10% parameter variation, and Case (iii) with +15% parameter variation (Padmanabhan et al., 2017b).

Table 10 Statistical analysis for 15 simulated patients (Padmanabhan et al., April, 2017b)

Parameter		N_{dev}	T_{per}
Percent value; before chemotherapy	Min	40	100
	Max	40	100
	Mean	40	100
Percent value; after 1 week of chemotherapy	Min	10.17	19.34
	Max	87.75	0.0096
	Mean	45.05	2.50
Percent value; after 4 weeks of chemotherapy	Min	0	0.5324
	Max	3.47	0
	Mean	0.4271	0.1708
Percent value; after 7 weeks of chemotherapy	Min	0	0.0634
	Max	0.0560	0
	Mean	0.0059	0.0064

based on the error $e(t)$, $t \geq 0$, which reflects the patient-specific response to drug intake. Thus, the value of the error $e(t)$, $t \geq 0$, varies according to the patient characteristics.

Table 10 shows the statistical results of the simulations performed on 15 simulated patients using the RL agent trained for Case 1. We generated 15 simulated patients

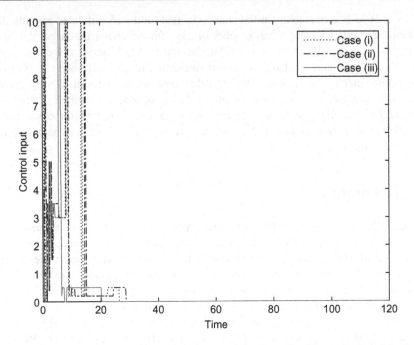

Fig. 18 Control input for three different patient models: Case (i) with nominal model, Case (ii) with -10% parameter variation, and Case (iii) with $+15\%$ parameter variation (Padmanabhan et al., April, 2017b).

with the parameter ranges of fraction cell kill a_i, $i = 1, 2, 3, 0 < a_i \leq 0.5, a_3 \leq a_1 \leq a_2$, carrying capacities $b_1^{-1} \leq b_2^{-1} = 1$, competition terms $0.3 \leq c_i \leq 1, i = 1, \dots, 4$, death rates $0.15 \leq d_1 \leq 0.3, d_2 = 1$, per unit growth rates, $1.2 \leq r_1 \leq 1.6, r_2 = 1$, immune source rate $0.3 \leq s \leq 0.5$, immune threshold rate $0.3 \leq \alpha \leq 0.5$, and immune response rate $0.01 \leq \alpha_c \leq 0.05$. See De Pillis and Radunskaya (2003) for further details on the parameter ranges of the cancer chemotherapy model.

The percent deviation of the number of normal cells from the target value ($x_{1d} = 1$) given in Table 10 is calculated as

$$N_{dev} = \frac{|\text{Measured value} - \text{Target value}|}{\text{Target value}} \times 100 = |x_1(t^*) - 1| \times 100,$$

where $t^* = 0, 1, 4,$ or 7 weeks. The percent value of the number of tumor cells with respect to the initial value is calculated as

$$T_{per} = \frac{\text{Measured value}}{\text{Initial value}} \times 100 = \frac{x_2(t^*)}{x_2(0)} \times 100.$$

It can be seen from Table 10 that by week 7, the percent deviation of the number of normal cells from the target value is 0.0059 and the percent value of the number of tumor cells with respect to the initial value is 0.0064 for the 15 simulated patients. The minimum, maximum, and mean number of days for achieving the target values of $x_1(t)$, $t \geq 0$, and $x_2(t)$, $t \geq 0$, are 13, 50, 28, and 6, 52, 27 days, respectively, for the 15 simulated patients. Comparing our simulation results with those by Batmani and Khaloozadeh (2013), it can be seen that both methods result in very similar responses. In both cases the tumor is eradicated using optimal chemotherapy drug dosing and the controllers are robust to parameter variations. However, the advantage of the RL-based method is that it does not require a model of the system in order to develop a controller.

5 Summary

First, in Section 2, an RL-based controller design approach for the simultaneous regulation of sedation and MAP using the controlled titration of the sedative drug propofol is detailed. Simulation studies conducted using 30 simulated patients with varying pharmacological parameters show that the RL-based controller design approach is promising in developing closed-loop controllers for ICU sedation while regulating multiple vital physiological parameters simultaneously.

Next, in Section 3, an RL-based controller that can account for the simultaneous administration of drugs with synergistic interactive effect is presented. The RL agent is trained using Q-learning algorithm with respect to the states defined in terms of the error associated with the desired output. A similar method can be used for the case of the drugs with inhibitive drug interactive effect. Moreover, our simulations demonstrate that the RL-based methods can be used to implement closed-loop control systems which are robust to system uncertainties. Further experiments are warranted to refine the optimal control agent and to extend it by including the simultaneous control of additional vital physiological parameters such as heart rate, cardiac output, and respiratory rate.

Finally, in Section 4, the efficacy of the RL-based method is investigated for different cases of cancer treatment. The method results in an optimal and robust controller. In order to preserve normal cells while eradicating tumor cells, a scaled value of the error is used in the reward function. The controller using the RL method can be extended to account for different constraints in cancer treatment by appropriately choosing the reward function. The main advantage of the RL-based control method is that the algorithm does not require knowledge of the system dynamics. However, different RL agents need to be trained to account for the patient characteristics of different patient groups.

It is apparent that the credibility of feedback information that is used to guide the controller is one of the main factors that determines the closed-loop performance of the controller. In case of biomedical applications, most of the monitoring systems has to account for various errors in the measured signal which may arise due to the movement of cables, electrodes, and patient, and also the noise and interferences from other devices. Hence, modern biomedical monitors include intricate filtering algorithms to improve signal-to-noise ratio of the measurement. This in turn increases the computation time and hence introduces time delay in measurement. Hence, studying the scope for improvement in performance of the proposed Q-learning-based controller design methodology by accounting for possible time delays during intravenous drug administration is desirable.

Acknowledgments

This publication was made possible by the GSRA grant no. GSRA1-1-1128-13016 from the Qatar National Research Fund (a member of Qatar Foundation). The findings achieved herein are solely the responsibility of the authors.

References

Abbeel, P., Coates, A., Quigley, M., Ng, A.Y., 2007. An application of reinforcement learning to aerobatic helicopter flight. In: Neural Information Processing Systems, vol. 19, pp. 1–8.

Absalom, A.R., Kenny, G.N.C., 2003. Closed-loop control of propofol anaesthesia using bispectral index: performance assessment in patients receiving computer-controlled propofol and manually controlled remifentanil infusions for minor surgery. Br. J. Anesth. 90 (6), 737–741.

Absalom, A.R., Mason, K.P., 2017. Total Intravenous Anesthesia and Target Controlled Infusions: A Comprehensive Global Anthology. Springer, Switzerland.

Absalom, A.R., Mani, V., De Smet, T., Struys, M.M., 2009. Pharmacokinetic models for propofol defining and illuminating the devil in the detail. Br. J. Anaesth. 103 (1), 26–37.

Absalom, A.R., De Keyser, R., Struys, M.M.R.F., 2011. Closed-loop anesthesia: are we getting close to finding the holy grail? Anesth. Analg. 112 (3), 516–518.

ACS, 2015. Cancer facts and figures 2015. American Cancer Society, Atlanta, Georgia. http://www.cancer.org/acs/groups/content/@editorial/documents/document/acspc-044552.pdf.

Babaei, N., Salamci, M.U., 2015. Personalized drug administration for cancer treatment using model reference adaptive control. J. Theor. Biol. 371, 24–44.

Bailey, J.M., Haddad, W.M., 2005. Drug dosing control in clinical pharmacology. IEEE Control Syst. Mag. 23 (2), 35–51.

Balashevich, N.V., Gabasov, R., Kalinin, A.I., Kirillova, F.M., 2002. Optimal control of nonlinear systems. Comput. Math. Math. Phys. 42 (7), 931–956.

Barto, A.G., Sutton, R.S., Anderson, C.W., 1983. Neuron like adaptive elements that can solve difficult learning control problems. IEEE Trans. Syst. Man Cybernet. 13, 834–846.

Batmani, Y., Khaloozadeh, H., 2013. Optimal chemotherapy in cancer treatment: state dependent Riccati equation control and extended Kalman filter. Optimal Control Appl. Methods 34 (5), 562–577.

Bertsekas, D.P., Tsitsiklis, J.N., 1996. Neuro-Dynamic Programming. Athena Scientific, Belmont, MA.

Brogi, E., Cyr, S., Kazan, R., Giunta, F., Hemmerling, T.M., 2017. Clinical performance and safety of closed-loop systems: a systematic review and meta-analysis of randomized controlled trials. Anesth. Analg. 124 (2), 446–455.

Chen, T., Kirkby, N.F., Jena, R., 2012. Optimal dosing of cancer chemotherapy using model predictive control and moving horizon state/parameter estimation. Comput. Methods Programs Biomed. 108 (3), 973–983.

Chen, C.S., Doloff, J.C., Waxman, D.J., 2014. Intermittent metronomic drug schedule is essential for activating antitumor innate immunity and tumor xenograft regression. Neoplasia 16 (1), 84–96.

Çimen, T., 2010. Systematic and effective design of nonlinear feedback controllers via the state-dependent Riccati equation (SDRE) method. Annu. Rev. Control 34 (1), 32–51.

Dadhich, S., Bodin, U., Sandin, F., Andersson, U., 2016. Machine learning approach to automatic bucket loading. In: 24th Mediterranean Conference on Control and Automation, pp. 1260–1265.

Daskalaki, E., Diem, P., Mougiakakou, S.G., 2013. Personalized tuning of a reinforcement learning control algorithm for glucose regulation. In: 35th Annual International Conference of the IEEE Engineering in Medicine and Biology Society, pp. 3487–3490.

De Pillis, L.G., Radunskaya, A., 2003. The dynamics of an optimally controlled tumor model: a case study. Math. Comput. Model. 37 (11), 1221–1244.

Doloff, J.C., Waxman, D.J., 2015. Transcriptional profiling provides insights into metronomic cyclophosphamide-activated, innate immune-dependent regression of brain tumor xenografts. BMC Cancer 15 (1), 375.

Engelhart, M., Lebiedz, D., Sager, S., 2011. Optimal control for selected cancer chemotherapy ODE models: a view on the potential of optimal schedules and choice of objective function. Math. Biosci. 229 (1), 123–134.

Fan, S.Z., Wei, Q., Shi, P.F., Chen, Y.J., Liu, Q., Shieh, J.S., 2012. A comparison of patient's heart rate variability and blood flow variability during surgery based on the Hilbert Huang transform. Biomed. Signal Process. Control 7 (5), 465–473.

Furutani, E., Tsuruoka, K., Kusudo, S., 2010. A hypnosis and analgesia control system using a model predictive controller in total intravenous anesthesia during day-case surgery. In: SICE Annual conference, Taipei, Taiwan, pp. 223–226.

Gholami, B., Agar, N.Y.R., Jolesz, F.A., Haddad, W.M., Tannenbaum, A.R., 2011. A compressive sensing approach for glioma margin delineation using mass spectrometry. In: Annual International Conference of the IEEE Engineering in Medicine and Biology Society. IEEE, pp. 5682–5685.

Haddad, W.M., Chellaboina, V., 2008. Nonlinear Dynamical Systems and Control: A Lyapunov-Based Approach. Princeton University Press, Princeton, NJ.

Haddad, W.M., Hayakawa, T., Bailey, J.M., 2003. Adaptive control for nonnegative and compartmental dynamical systems with applications to general anesthesia. Int. J. Adapt Control Signal Process. 17, 209–235.

Haddad, W.M., Chellaboina, V., Hui, Q., 2010. Nonnegative and Compartmental Dynamical Systems. Princeton University Press, Princeton, NJ.

Haddad, W.M., Bailey, J.M., Gholami, B., Tannenbaum, A.R., 2013. Clinical decision support and closed-loop control for intensive care unit sedation. Asian J. Control 15 (2), 317–339.

Hahn, J.O., Dumont, G.A., Ansermino, J.M., 2012. Robust closed-loop control of hypnosis with propofol using WAVCNS index as the controlled variable. Biomed. Signal Process. Control 7 (5), 517–524.

Heusden, K.V., Ansermino, J.M., Dumont, G.A., 2018. Robust MISO control of propofol-remifentanil anesthesia guided by the NeuroSENSE monitor. IEEE Trans. Control Syst. Technol. 26 (5), 1758–1770.

Hong, M., Razaviyayn, M., Luo, Z.Q., Pang, J.S., 2016. A unified algorithmic framework for block-structured optimization involving big data: with applications in machine learning and signal processing. IEEE Signal Process. Mag. 33 (1), 57–77.

Huang, J., Gholami, B., Agar, N.Y.R., Norton, I., Haddad, W.M., Tannenbaum, A.R., 2011. Classification of astrocytomas and oligodendrogliomas from mass spectrometry data using sparse kernel machines. In: Annual International Conference of the IEEE Engineering in Medicine and Biology Society, IEEE, pp. 7965–7968.

Ionescu, C.M., Nascu, I., De Keyser, R., 2014. Lessons learned from closed loops in engineering: towards a multivariable approach regulating depth of anaesthesia. Int. J. Clin. Monit. Comput. 28 (6), 537–546.

Jacobi, J., Fraser, G.L., Coursin, D.B., Riker, R.R., Fontaine, D., Wittbrodt, E.T., Chalfin, D.B., Masica, M.F., Bjerke, H.S., Coplin, W.M., Crippen, D.W., Fuchs, B.D., Kelleher, R.M., Marik, P.E., Nasraway, S.A., Murray, M.J., Peruzzi, W.T., Lumb, P.D., 2002. Clinical practice guidelines for the sustained use of sedatives and analgesics in the critically ill adult. Am. J. Health Syst. Pharm. 59, 150–178.

Johansen, J.W., Sebel, P.S., Smet, T.D., Struys, M.M., 2000. Development and clinical application of electroencephalographic bispectrum monitoring. Anesthesiology 93, 1336–1344.

Kiran, K.L., Jayachandran, D., Lakshminarayanan, S., 2009. Multi-objective optimization of cancer immuno-chemotherapy. In: 13th International Conference on Biomedical Engineering, pp. 1337–1340.

Kuizenga, M.H., Vereecke, H.E., Struys, M.M., 2016. Model-based drug administration: current status of target-controlled infusion and closed-loop control. Curr. Opin. Anesthesiol. 29 (4), 475–481.

Liu, N., Chazot, T., Genty, A., Landais, A., Restoux, A., McGee, K., Laloë, P.A., Trillat, B., Barvais, L., Fischler, M., 2006. Titration of propofol for anesthetic induction and maintenance guided by the bispectral index: closed-loop versus manual control: a prospective, randomized, multicenter study. J. Am. Soc. Anesthesiol. 104 (4), 686–695.

Martin-Guerrero, J.D., Gomez, F., Soria-Olivas, E., Schmidhuber, J., Climente-Marti, M., Jemenez-Torres, N.V., 2009. A reinforcement learning approach for individualizing erythropoietin dosages in hemodialysis patients. Expert Syst. Appl. 36, 9737–9742.

Masui, K., Upton, R.N., Doufas, A.G., Coetzee, J.F., Kazama, T., Mortier, E.P., Struys, M.M., 2010. The performance of compartmental and physiologically based recirculatory pharmacokinetic models for propofol: a comparison using bolus, continuous, and target-controlled infusion data. Anesth. Analg. 111, 368–379.

Matignon, L., Laurent, G.J., Fort-Piat, N.L., 2006. Reward function and initial values: better choices for accelerated goal-directed reinforcement learning. In: 16th International Conference on Artificial Neural Networks, Athens, Greece, pp. 840–849.

Mehta, S., Burry, L., Fischer, S., Motta, J.C.M., Hallet, D., Bowman, D., Wong, C., Meade, M.O., Stewart, T.E., Cook, D.J., 2006. Canadian survey of the use of sedatives, analgesics, and neuromuscular blocking agents in critically ill patients. Crit. Care Med. 34 (2), 374–380.

Minto, C., Schnider, T., Short, T., Gregg, K., Gentilini, A., Shafer, S., 2000. Response surface model for anesthetic drug interactions. Anesthesiology 92, 1603–1616.

Moore, B.L., Panousis, P., Kulkarni, V., Pyeatt, L.D., Doufas, A.G., 2010. Reinforcement learning for closed-loop propofol anesthesia. In: Proceedings of 22th Annual Conference on Innovative Applications of Artificial Intelligence, Atlanta, Georgia, USA, pp. 1807–1813.

Moore, B.L., Pyeatt, L.D., Kulkarni, V., Panousis, P., Kevin, Doufas, A.G., 2014. Reinforcement learning for closed-loop propofol anesthesia: a study in human volunteers. J. Mach. Learn. Res. 15, 655–696.

Morley, A., Derrick, J., Mainland, P., Lee, B.B., Short, T.G., 2000. Closed loop control of anaesthesia: an assessment of the bispectral index as the target of control. Anaesthesia 55 (10), 953–959.

Nascu, I., Ionescu, C.M., Nascu, I., De Keyser, R., 2011. Evaluation of three protocols for auto-matic DOA regulation using propofol and remifentanil. In: 9th IEEE International Conference on Control and Automation, pp. 573–578.

Nemati, S., Ghassemi, M.M., Clifford, G.D., 2016. Optimal medication dosing from suboptimal clinical examples: a deep reinforcement learning approach. In: 38th Annual International Conference of the IEEE Engineering in Medicine and Biology Society, pp. 2978–2981.

Noble, S.L., Sherer, E., Hannemann, R.E., Ramkrishna, D., Vik, T., Rundell, A.E., 2010. Using adaptive model predictive control to customize maintenance therapy chemotherapeutic dosing for childhood acute lymphoblastic leukemia. J. Theor. Biol. 264 (3), 990–1002.

Pachmann, K., Heiß, P., Demel, U., Tilz, G., 2001. Detection and quantification of small numbers of circulating tumour cells in peripheral blood using laser scanning cytometer (LSC®). Clin. Chem. Lab. Med. 39 (9), 811–817.

Padmanabhan, R., Meskin, N., Haddad, W.M., 2014. Direct adaptive disturbance rejection control for sedation and analgesia. In: Middle East Conference on Biomedical Engineering, Doha, Qatar, pp. 175–179.

Padmanabhan, R., Meskin, N., Haddad, W.M., 2015. Closed-loop control of anesthesia and mean arterial pressure using reinforcement learning. Biomed. Signal Process. Control 22, 54–64.

Padmanabhan, R., Meskin, N., Haddad, W.M., 2017. Reinforcement learning-based control for combined infusion of sedatives and analgesics. In: 4th International Conference on Control, Decision and Information Technologies, Barcelona, Spain, pp. 505–509.

Padmanabhan, R., Meskin, N., Haddad, W.M., 2017b. Reinforcement learning-based control of drug dosing for cancer chemotherapy treatment, Mathematical Biosciences 293, 11–20.

Pillis, L.G.D., Radunskaya, A., 2001. A mathematical tumor model with immune resistance and drug therapy: an optimal control approach. Comput. Math. Methods Med. 3 (2), 79–100.

Rao, R.R., Bequette, B.W., 2000. Simultaneous regulation of hemodynamic and anesthetic states: a simulation study. Ann. Biomed. Eng. 28 (1), 71–84.

Robinson, B.J., Ebert, T.J., Brien, T.J.O., Colinco, M.D., Muzi, M., 1997. Mechanisms whereby propofol mediates peripheral vasodilation in humans: sympathoinhibition or direct vascular relaxation? Anesthesiology 86, 64–72.

Sbeity, H., Younes, R., 2015. Review of optimization methods for cancer chemotherapy treatment planning. J. Comput. Sci. Syst. Biol. 8, 74–95.

Sedighizadeh, M., Rezazadeh, A., 2008. Adaptive PID controller based on reinforcement learning for wind turbine control. World Acad. Sci. Eng. Technol. 2, 1–23.

Soltesz, K., Hahn, J.O., Hagglund, T., Dumont, G.A., Ansermino, J.M., 2013. Individualized closed-loop control of propofol anesthesia: a preliminary study. Biomed. Signal Process. Control 8 (6), 500–508.

Struys, M.M., De Smet, T., Versichelen, L.F., Van de Velde, S., Van den Broecke, R., Mortier, E.P., 2001. Comparison of closed-loop controlled administration of propofol using bispectral index as the controlled variable versus "standard practice" controlled administration. Anesthesiology 95 (1), 6–17.

Sutton, R.S., 1988. Learning to predict by the methods of temporal difference. Mach. Learn. 3, 9–44.

Sutton, R.S., Barto, A.G., 1998. Reinforcement Learning: An Introduction. MIT Press, Cambridge, MA.

Suzuki, C., Jacobsson, H., Hatschek, T., Torkzad, M.R., Boden, K., Eriksson-Alm, Y., Berg, E., Fujii, H., Kubo, A., Blomqvist, L., 2008. Radiologic measurements of tumor response to treatment: practical approaches and limitations. Radiographics 28 (2), 329–344. https://doi.org/10.1148/rg.282075068.

Swan, G.W., 1990. Role of optimal control theory in cancer chemotherapy. Math. Biosci. 101 (2), 237–284.

Swierniak, A., Lezewicz, U., Schattler, H., 2003. Optimal control for a class of compartmental models in cancer chemotherapy. Int. J. Appl. Math. Comput. Sci. 13 (3), 357–368.

Tan, K.C., Khor, E.F., Cai, J., Heng, C.M., Lee, T.H., 2002. Automating the drug scheduling of cancer chemotherapy via evolutionary computation. Artif. Intell. Med. 25 (2), 169–185.

Tse, S.-M., Liang, Y., Leung, K.-S., Lee, K.-H., Mok, T.S.-K., 2007. A memetic algorithm for multiple-drug cancer chemotherapy schedule optimization. IEEE Trans. Syst. Man Cybern. B Cybern. 37 (1), 84–91.

Van Den Berg, J.P., Vereecke, H.E.M., Proost, J.H., Eleveld, D.J., Wietasch, J.K.G., Absalom, A.R., Struys, M.M., 2017. Pharmacokinetic and pharmacodynamic interactions in anaesthesia. A review of current knowledge and how it can be used to optimize anaesthetic drug administration. Br. J. Anaesth. 118 (1), 44.

Vrabie, D., Vamvoudakis, K.G., Lewis, F.L., 2013. Optimal Adaptive Control and Differential Games by Reinforcement Learning Principle. Institution of Engineering and Technology, London.

Watkins, C.J.C.H., Dayan, P., 1992. Q-learning. Mach. Learn. J. 8 (3), 279–292.

WHO, 2018. Fact Sheets. Available from: http://www.who.int/mediacentre/factsheets/fs297/en/.

Zhao, Y., Zeng, D., Socinski, M.A., Kosorok, M.R., 2011. Reinforcement learning strategies for clinical trials in nonsmall cell lung cancer. Biometrics 67 (4), 1422–1433.

Control strategies in general anesthesia administration[*]

Adriana Savoca, Davide Manca
PSE-Lab, Process Systems Engineering Laboratory, Dipartimento di Chimica, Materiali e Ingegneria Chimica "Giulio Natta," Politecnico di Milano, Milano, Italy

List of abbreviations

ADME	absorption distribution metabolism elimination
AP	arterial pressure
AUC	area under the curve
BIS	bispectral index
BM	body mass
BSA	body surface area
CPU	central processing unit
DOA	depth of anesthesia
EMA	European Medicines Agency
FDA	Food and Drug Administration
GICS	gastrointestinal circulatory system
HO	highly perfused organs
IR	infusion rate
LOC	loss of consciousness
MAP	mean arterial pressure
MPC	model predictive control
(PB)PK	(physiologically based) pharmacokinetic
PCLCS	physiological closed-loop controlled systems
PD	pharmacodynamic
PF	penalty function
P(ID)	proportional (integral derivative)
PT	poorly perfused tissues
TCI	target controlled infusion
TIVA	total intravenous anesthesia
WMA	World Medical Association

[*] No parts of this paper may be reproduced or elsewhere used without the prior written permission of the authors.

Control Applications for Biomedical Engineering Systems. https://doi.org/10.1016/B978-0-12-817461-6.00010-X

List of symbols

C_i	concentration referred to compartment i
k_{10}	central compartment elimination constant
k_{12}	central-rapid compartment transfer constant
k_{13}	central-slow compartment transfer constant
k_{21}	rapid-central compartment transfer constant
k_{31}	slow-central compartment transfer constant
k_{e0}	plasma-effect-site equilibration constant
E_0	baseline value of the effect
E	effect
EC_{50}	concentration at 50% of the maximum effect
h_c	control horizon
h_c	prediction horizon
γ	hill coefficient
δ	model mismatch
e_y	error of the controlled variable respect to desired setpoint
y	model prediction of the controlled variable
y_{sp}	set point of the controlled variable
y_r	real measured value of the controlled variable
y_{min}	lower-bound value of the controlled variable
y_{max}	upper-bound value of the controlled variable
C_p	plasma concentration
$C_{p\ max}$	upper-bound value of the plasma concentration
w	weight in objective function

1 Introduction

Section 1 introduces current approach to intravenous anesthesia administration, discussing its limitations and comparing it to automated anesthesia delivery (Sections 1.1 and 1.2). Section 1.3 provides a short dissertation on the differences among the most diffused control strategies applied to anesthesia delivery, i.e., feedback and model-based control.

1.1 Anesthesia delivery today

"The practice of medicine is an art based on science" said the father of modern medicine Sir William Osler (1849–1919). This mindset can be partially interpreted as one of the reasons contributing to slowness by biomedicine and pharmacology to integrate computer-aided modeling/simulation and control systems, compared to other fields such as transports and electronics (Leil and Bertz, 2014). Additional factors are the prejudice that biological systems are too complicated to model, and the difficulty of creating multidisciplinary teams that would work in this direction. Nonetheless, in recent years, technological advances in biomedicine and pharmacology have allowed making considerable steps forward (Leil and Bertz, 2014).

One of the most interesting emerging applications of control systems in biomedicine is automated anesthesia delivery. At present, 230 million anesthetic procedures are carried out yearly all over the world (Schiff and Wagner, 2016). Anesthesia is today considered rather safe, and its associated risks represent a small fraction of the total risk of surgical procedures. Yet, with such a high number of surgical or other painful procedures taking place every year, those risks cannot be overlooked. Although a number of studies claim an ongoing decrease in the overall global anesthesia-associated mortality, an equivalent number of studies claim the opposite and highlight discrepancies and differences in the statistics (Schiff and Wagner, 2016). In any case, researchers agree that the events leading to anesthesia-related death are often associated to medication errors, which result into overdose and critical side effects, with particular impact on the respiratory and cardiovascular systems (Schiff and Wagner, 2016). Conversely, awareness episodes related to underdosing are not that rare among patients. Indeed, selection of the optimal anesthetic dose is not an easy task. In fact, both anesthetic and analgesic agents feature narrow therapeutic windows, which mean a limited range between (i) minimum concentration levels to induce desired clinical effects and (ii) maximum levels to avoid undesired dangerous ones. The optimal dose depends not only on the patient's physical characteristics, age, possible diseases, and genetics, but also on the type of surgical operation (Absalom and Struys, 2007). The aim of the anesthetist is on one hand to achieve the desired unconsciousness and pain relief levels and on the other to accomplish a proper postanesthesia recovery.

A balanced anesthesia results from the combination of three aspects, i.e., three types of drugs: (i) a hypnotic agent causing unconsciousness, (ii) an analgesic agent producing pain relief, and (frequently) (iii) a muscle relaxant, to avoid undesired reflex activity that would interfere with surgery (essential in abdominal and cardiac surgery). Today, basing on the patient's features and their own experience, the anesthetist selects an initial dose for induction of the desired depth of anesthesia (DOA). Actually, this level of unconsciousness is not uniquely defined among anesthetists, and neither standard methods nor certified technologies are available for quantitative evaluation. Thus, the anesthetist observes the patient response and manually adjusts the initial dose until the desired DOA is reached. This phase is defined as the induction of anesthesia. After surgery starts, intraoperative surgical stimuli are likely to produce disturbances of the anesthetic state. The role of the anesthetist is to maintain DOA against such disturbances, by making further adjustments based on patient's vital parameters and electroencephalographic trace monitoring (i.e., maintenance phase of anesthesia) (Absalom et al., 2011).

The required drugs can be either administered by means of (i) standard syringe pumps (with desired dose as input) or (ii) target-controlled infusion (TCI) pumps, first proposed in 1983 (Absalom and Struys, 2007). In this case, rather than the dose, the anesthetist selects a target (i.e., desired) concentration either in plasma or in the drug site of action (aka effect site), and the pump evaluates the corresponding infusion rate (IR) to be administered to the patient. This calculation is typically based on a three-compartment pharmacokinetic (PK) model, which correlates the drug IR to the expected concentration. Section 1.2 provides further details on the working principles and limitations of TCI pumps, while Section 2.3 digresses on the features of compartmental PK models.

1.2   Open-loop or closed-loop anesthesia?

Despite their name, TCI pumps work rather differently than conventional control systems. Inputs to these pumps are main physical characteristics of the patients (e.g., gender, body mass, height, and age), and a desired value of the drug concentration. Calculation of the corresponding dose is based on the analytical/numerical solution of a 3-equation ordinary differential system, which describes the drug concentration evolution in the body. As explained better in Section 2.3, classical PK models have empirical foundation and their structure is not strictly related to the patient's anatomy and physiology. Mostly, the models implemented in commercial TCI pumps were identified at the end of the '90s [e.g., Schnider model for propofol (Schnider et al., 1998), Minto model for remifentanil (Minto et al., 1997)] grounding on PK studies on a limited number of healthy volunteers. As a matter of fact, several authors highlighted the necessity to identify specific three-compartment models in case of different populations, e.g., obese, children, and patients with some sort of disease (Cortínez et al., 2010; Marsh et al., 1991; Rigouzzo et al., 2010). Thus, interindividual variability is likely to represent a serious concern for the anesthetist used to commercial TCI pumps.

The most critical and limiting feature of TCI pumps is that they do not adjust IR after any feedback measure(s) of the patient's anesthetic state (contrary to any real feedback control system). In fact, in case of total intravenous anesthesia (TIVA), it is not possible to measure the drug concentration in real time during the surgical operation. Consequently, the anesthesiologist can never be certain that the expected target concentration is reached within the body of the patient. In this sense, TCI pumps deliver an *open-loop* configuration, where the anesthetist acts as a human controller, as they physically close the *loop* by personally monitoring the vital parameters of the patient and appropriately regulating the target concentration. The same occurs in case of manual syringe pumps. Only, in this case, the anesthetist directly regulates the drug IR or chooses to administer intermittent boluses, instead of modifying the target concentration. In some cases, this regulation occurs before any disturbance (i.e., surgical stimuli) manifestation. Indeed, before a particularly stimulating procedure, the anesthetist may decide to increase the target concentration/drug IR to avoid possible alterations of the DOA and analgesia level of the patient, relying on their experience.

It is evident that human factors play a paramount role in TIVA procedure, in terms of continuous monitoring, training and experience, communication with the operating room team, level of fatigue and stress. Because of this way of proceeding, the *control action* on the anesthetic state of the patient may result irregular and intermittent, characterized by rather "random" changes of the drugs IRs (Absalom et al., 2011). Automated anesthesia delivery systems are designed with the purpose of supporting medical doctors and improving their control action of the patient's state.

Indeed, to guarantee a safer and more stable induction and maintenance of anesthesia, this alternative approach was proposed for the first time by Mayo et al. (1950). Specifically, they experimented automated delivery of anesthesia with ether in patients subject to abdominal surgery, based on the electroencephalogram trace monitoring. They applied this methodology successfully to 50 patients and proved its potentiality. Since then, several studies have designed, developed, and tested *closed-loop* controllers of anesthesia using different control strategies, types of inputs, and anesthetic/analgesic agents (El-Nagar and El-Bardini, 2014; Gentilini et al., 2002; Liu et al., 2011; Nascu et al., 2015; West et al., 2013; Zhusubaliyev et al., 2015). For instance, West et al. (2013) and Zhusubaliyev et al. (2015) opted for a proportional integral derivative (PID) control strategy, whereas Nascu et al. (2015) proposed several model-based strategies for anesthesia control with propofol. All of these papers rely on classical PK modeling (see Section 2.3) for either evaluation of the PID controllers or model-based control. Different strategies were also proposed to address interindividual variability issues. In a preliminary in silico study, Savoca and Manca (2019a) proposed physiologically based model predictive control (MPC) of both analgesia and anesthesia components. To face interindividual variability, El-Nagar and El-Bardini (2014) proposed a fuzzy neural network-based controller, which was tested over a wide range of patient parameters. A Monte Carlo approach was instead used by Soltesz et al. (2013) to individualize the patient model of the dose-response relation via system identification during induction.

The basic principle of *closed-loop* controlled anesthesia is to regulate the drugs IRs automatically with limited human intervention grounding on measured indexes of the patient's DOA. The main advantage is that these systems are not distractible, and their application allows the anesthetists to focus on the patient's state, by reducing their workload. Indeed, the goal of these systems is not to replace the anesthetist in their task and experience, but rather to support them and reduce human error incidence. In fact, over the years, control systems resulted in costs decrease and efficiency increase in a wide variety of applications (Dumont, 2014).

1.3 Classic feedback vs model-predictive control of anesthesia

The strategy of a controller for anesthesia can be either (i) model-based or (ii) model-free. In case the control strategy is based on a model, the difficulty of its development and the assessment of its reliability may arise concerns in medical doctors. Conversely, model-free design is simpler and may be based on clinical guidelines embedded in the controller hardware (e.g., expert system), which may result in easier understanding by clinicians. However, in both cases (i.e., model-based or model-free), a reliable model of the patient is needed whenever one wants to assess in silico the controller performance and/or tune the parameters with the goal of obtaining a reasonable tradeoff between robustness and responsiveness of the control action (Parvinian et al., 2018).

Feedback control grounds on a model-free strategy and is the most employed in the industry. The reason consists in the combination of simplicity and efficacy (Sha'aban et al., 2013). Feedback control does not rely on any type of model to regulate the manipulated variable(s), but only on a proportional (P), proportional integral (PI), or proportional integral and differential (PID) action respect to the error between the *set point* (i.e., desired value of the controlled variable(s)) and the measured value of the controlled variable(s). Thus, on the plus side, the performance is independent of any error or uncertainty in the modeling, and the simplicity of the working principle can enhance acceptance and understanding by clinicians. However, it is less efficient in managing multivariable problems with strong interactions among controlled variables. In addition, its intrinsic nature does not allow anticipating any corrective action before a disturbance has produced its effects on the system (this is indeed due to the feedback feature).

Conversely, MPC is certainly more complex than feedback control, but has been applied successfully for decades in industrial processes (Forbes et al., 2015). MPC is a model-based strategy and therefore features some uncertainty issues, which are unavoidable. Indeed, a model of the system is used to predict its future evolution and optimize the control actions. The real-time experimental measure of the process is used to correct the so-called model mismatch (between the real system and the modeled one). One of the main advantages of MPC is the ability to tackle multivariable problems (Dumont, 2014). As already discussed, general anesthesia is a delicate balance of multiple drugs administration. In addition, the anesthetists rely on different parameters and measures to assess the patient's state; thus, the problem cannot be fully treated by focusing on a single aspect. In addition, MPC ability to tackle both linear and nonlinear constraints is extremely valuable. The objective function (which is at the heart of the MPC mathematical formulation and allows identifying the optimal trajectory of the manipulated variables) can be suitably designed to account for therapeutic windows of drugs and critical ranges of the physiological variables (e.g., dangerous cerebral activity, hypotension, and hypertension). In addition, oscillations of the manipulated variables (i.e., drugs IRs) can be limited by means of an intelligent tuning of the weights that contribute to the objective function. The pure in silico assessment of the MPC performance and reliability (before any in vivo implementation and clinical application) calls for the implementation of a different model of the patient than the one embedded in the controller, for the sake of recreating the model *mismatch* intrinsic to real applications.

2 Case study: Model-predictive control of anesthesia with propofol and remifentanil

Section 2 focuses on an in silico application of MPC to anesthesia using two largely diffused drugs, i.e., propofol and remifentanil. After a short introduction on the characteristics of propofol and remifentanil (see Sections 2.1 and 2.2), Section 2.3 presents the main features of classic and physiologically based pharmacokinetic (PBPK) models, as both are applied to the following case study. Lastly, Sections 2.4 and 2.5 describe the working principles of MPC (developed for anesthesia with propofol and remifentanil) and show some results of the simulations of induction and maintenance phases.

2.1 Propofol

Propofol is an intravenous hypnotic drug used for induction and maintenance of general anesthesia. Its popularity is related not only to the resulting fast induction of unconsciousness (about 40 s from the start of administration, due to rapid distribution to central nervous system (CNS)) but also to the reduced postrecovery side effects in comparison with volatile anesthetics (e.g., thiopental, methohexital, and etomidate). Propofol is chiefly eliminated by hepatic conjugation to inactive metabolites which are excreted by the kidneys (Knibbe et al., 1999). Among side effects, respiratory depression is the most critical.

2.2 Remifentanil

Remifentanil was approved by Food and Drug Administration (FDA) in 1996. It was developed and designed purposely to provide a rapid and predictable dissipation of the effects in the recovery phase of the patient, once administration is stopped. In fact, the peculiar configuration of remifentanil (presence of alkyl-ester bonds on the piperidine ring) makes it susceptible to metabolism by a specific esterase in blood and tissues (Dershwitz et al., 1996; Navapurkar et al., 1998; Pitsiu et al., 2004), resulting in a shorter time of elimination than other analgesic opioids. Remifentanil produces dose-dependent increases in analgesic effect. Based on its ability to provide analgesia in volunteers after a single bolus dose administration, remifentanil is 20–30 times more potent than other analgesic opioids (Egan et al., 1993) and shares the side effects on the respiratory system (Glass et al., 1999). Remifentanil is often used in combination with propofol.

2.3 Classic and physiologically based pharmacokinetic-pharmacodynamic models

Classic PK models are based on an empirical compartmental approach, i.e., the model compartments do not have any actual connections with the anatomy and physiology of mammals. They usually involve 1–3 compartments, depending on the features of the drug experimental pharmacokinetic curve. Despite of the simplistic approach, classic PK models are especially useful for experimental data analyses and derivation of meaningful PK parameters such as the maximum concentration, C_{max}, the area under the curve, AUC, and the volume of distribution, V_d. They are also used for prediction purposes.

While the one-compartment model approximates the whole body as a single compartment by assuming that the drug concentration is uniformly distributed and is eliminated by a first-order process, the three-compartment model features (i) a central compartment model that corresponds to plasma, (ii) a rapidly-equilibrating tissues compartment, and (iii) a slowly-equilibrating tissues compartment. Its mathematical description corresponds to a set of three ordinary differential equations that describe the drug transfers among the three compartments (Eqs. 1–3).

Three-compartment models commonly describe the pharmacokinetics of both intravenous anesthetic and analgesic drugs. Minto and coauthors (Minto et al., 1997) introduced the main reference model of remifentanil pharmacokinetics. Conversely, a number of authors proposed three-compartment models of propofol pharmacokinetics. In our application of MPC to in silico anesthesia, we consider the model of Schnider et al. (1998), which is also widely used.

$$\frac{dC_1}{dt} = -(k_{10} + k_{12} + k_{13})C_1 + k_{21}C_2 + k_{31}C_3 + \frac{IR(t)}{V_1} \tag{1}$$

$$\frac{dC_2}{dt} = k_{12}C_1 - k_{21}C_2 \tag{2}$$

$$\frac{dC_3}{dt} = k_{13}C_1 - k_{31}C_3 \tag{3}$$

As far as PBPK models are concerned, Teorell proposed for the first time in 1937 a different approach that combines real anatomy and physiology with PK modeling (Teorell, 1937). Indeed, PBPK modeling associates compartments to the organs and tissues of the body, and the drug transfer among them follows the anatomical and physiological pathways. Thus, such models are powerful tools to investigate, describe, and predict the processes of absorption, distribution, metabolism, and elimination (ADME) of drugs. Although the paper by Teorell (1937) was only theoretical, as the mathematical tools to solve such complex systems of equations were not available at that time, PBPK modeling has had a huge development since then. Today PBPK modeling has primary importance in several fields of pharmacology. Despite their higher level of mathematical complexity and parameters number, PBPK models are used in a number of applications such as interspecies extrapolation, toxicology studies, and comparison of different routes of administration (Jones et al., 2006, 2011; Savoca and Manca, 2019b).

The reference PBPK model used in this book chapter comes from Abbiati et al. (2016a). In case of IV drugs, the counterdiffusion of the drug from the gastrointestinal circulatory system (GICS) to both small and large intestinal lumina is considered negligible. Thus, the model features only 5 compartments, i.e., plasma, GICS, liver, highly perfused organs (HO, i.e., lumping of kidneys, lungs, brain, and spleen), and poorly perfused tissues (PT, i.e., lumping of fat, muscles, bones, and skin). The comprehensive mathematical description is reported elsewhere (Abbiati et al., 2016b, 2018; Abbiati and Manca, 2017). The model equations correspond to drug material balances on these compartments, and the featured parameters can be divided into three categories: (i) individualized, (ii) assigned, and (iii) adaptive. Individualized parameters (i.e., compartment volumes and blood flowrates among them) are calculated via correlations available in the scientific literature. Particularly, compartment volumes are based on the organ/tissue mass fractions respect to the total body, whereas flowrates are calculated as fractions of the cardiac output that reach the organs/tissues, with cardiac output depending on the patient body surface area (BSA) (calculated as a function of patient's mass, height, and gender). Assigned parameters depend on the drug properties, e.g., protein binding, and are available in the pharma/med literature. Finally, adaptive parameters are mainly mass transfer coefficients representing diffusive and convective transport processes and metabolic constants, which are identified via a nonlinear regression procedure of PK experimental data. In particular, we used adaptive parameters identified for remifentanil with PK data available in Egan et al. (1998), and validated the model with PK data from a number of studies (Glass et al., 1993; Pitsiu et al., 2004;

Westmoreland et al., 1993). Equally, we regressed the adaptive parameters for propofol with experimental data from Gepts et al. (1987) and validated the results with Schnider et al. (1998) and Wiczling et al. (2016).

PK models describe the drug concentration evolution in the body and need to be combined with pharmacodynamic (PD) models for a comprehensive characterization of the dynamic pharmacological effects. Several drugs explicate their effect in a specific organ or tissue (i.e., site of action) to which they are conveyed by systemic circulation. In case of analgesic opioids and anesthetic agents, this is the brain. Hence, there is an experimental delay between the time course of the plasma drug concentration and the drug pharmacological effect. A simple approach to deal with this hysteresis is the effect-site compartment approximation. An additional equation represents the material balance in the effect-site (Eq. (4)).

$$\frac{dC_e}{dt} = k_{e0}C_p - k_{e0}C_e \tag{4}$$

This effect site is a virtual compartment that does not affect the PK and allows accounting for the aforementioned hysteresis. Either the Hill equation or suitable similar forms (see Eq. (5)) allow describing the pharmacological effect of these drugs.

$$E = E_0 - (E_0 - E_{max})\frac{C_e^\gamma}{EC_{50}^\gamma + C_e^\gamma} \tag{5}$$

where E_0 is the baseline value (i.e., assigned and corresponding to the awake state), E_{max} is the maximum drug effect, EC_{50} is the concentration that corresponds to 50% of the maximum effect, and γ is a fitting parameter, also called Hill coefficient. In our work, the effects of propofol and remifentanil are identified via BIS, the bispectral index, which is an index of unconsciousness state, and MAP, the mean arterial pressure, which is one of the hemodynamic parameters whose changes are related to the analgesia level, respectively. PD parameters are obtained via a nonlinear regression procedure with BIS and MAP data from the literature (Liu et al., 2015; Thompson et al., 1998).

2.4 Model predictive control principles

A model predictive controller architecture consists of: (i) the set-point (desired) trajectories of the controlled variables, (ii) the optimizer, and (iii) the model implemented in the controller. The MPC interacts both in input and output directions with the real process (see Fig. 1). At each control action (which occurs every Δt), the controller receives a new measured value of the real controlled variable(s). The process model is used to predict the possible future evolution of the real process over the prediction horizon $h_p \Delta t$. This horizon is discretized into multiple time steps $k, k+1, ..., k+h_p$ with a Δt time interval. At each time step k, the controller optimizes a set of h_c control actions (i.e., inputs of the process) based on the model predictions. h_c is the "control horizon" and its value is lower than h_p. The optimizer minimizes the

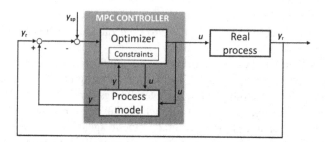

Fig. 1 Classic MPC loop components scheme: (i) the set point specifications, (ii) the optimizer, (iii) the model implemented in the controller, and (iv) the real process to be controlled. Set points y_{sp} are the desired set point values of the controlled variables, u are the input manipulated variables to the process, while y_r are the real measured values of the controlled variables. The optimizer periodically calls the process model to evaluate the predictions y as a response to future control actions u.

distance of the model predictions (adjusted according to the real measured values) from the desired trajectories. The MPC supervisor engineer can add user-defined constraints to describe the expected behavior of the real process better. Only the first control action of the control horizon h_c is then implemented in the real process at the following time step $k+1$. This control procedure is then iterated at every control action (with a Δt sampling time).

In case of automated anesthesia delivery, the real process corresponds to the body of the patient, whose inputs are the drugs IRs. The outputs are the corresponding pharmacological effects, in particular the anesthetic and analgesic levels, evaluated in practice by means of the monitored vital parameters in the operating room, e.g., heart rate, arterial pressure, bispectral index. A PK/PD model describes the patient body dose-effect relation in the controller. The degree of complexity of the model affects the computational time taken by the optimizer to solve the numerical problem and identify the optimal IRs. The anesthetist role is to set and adjust the desired anesthetic state in terms of the controlled variable(s) (set point) and the aforementioned constraints related to drugs therapeutic indexes and the patient's characteristics.

For the sake of exhaustiveness, it is worth observing that MPC can also solve rectangular problems where the number of controlled variables differs from the number of manipulated variables. Indeed, the pairing feature between manipulated and controlled variables, which is intrinsic to feedback control loops (e.g., P(ID) loops), loses its meaning in case of MPC. Actually, MPC tackles the control problem as a whole, where the set of manipulated variables acts on the set of controlled variables to achieve the controllability of the real process optimally. As already mentioned, another enhancing feature of MPC over classic feedback control is the potential of implementing lower and/or upper bounds for the controlled variables, which can play the role of asking the MPC to keep the real process within those safe bands instead of, or in addition to, approaching a user-defined setpoint. The advantages of MPC over feedback control are further epitomized by a dual approach to safe bands of manipulated variables with their customized lower and upper bounds (to preserve the physical consistency of control actions) together with possible recommended maximum changes of the manipulated variables between successive control actions according to a *bumpless* implementation philosophy. Finally, depending on the specific control problem and its layout, the MPC setup may also feature some recommended/expected values for the manipulated variables (or a subset of them), which are advisable to approach once the other higher-priority constraints have been already implemented and met. For the sake of clarity, it may happen that the MPC user knows some optimal values of the manipulated variables that would make the real process smoother and/or safer once the controllability of the system has been reached. In this case, such expected values of the manipulated variables are configured in the whole objective function of the MPC problem to be optimized by the numerical procedure, which is the pulsating heart of the optimizer (see Fig. 1). The MPC user can balance the importance of the aforementioned contributions by weighing their relative importance. These considerations can be translated into the mathematical formulation of the problem, as in Eqs. (6)–(11).

In Eqs. (6)–(11), u represents the manipulated variables, i.e., the drugs IRs, and y the controlled variables, i.e., in our in silico simulation BIS and MAP. Eq. (6) displays the objective function of the optimization problem. The first term (also explicated in Eq. 7) evaluates the distance from the desired anesthetist-defined DOA in terms of BIS and MAP. PF stands for penalty function and allows taking into account constraints to the drugs plasma concentration and BIS and MAP, to avoid exiting the therapeutic ranges and fall into critical situations, e.g., hypotension, hypertension, and excessively deep anesthesia levels. Boundary values should not to be intended as fixed and rigid constraints, but more like indications, because they are subject to interpatient variability and conditions (e.g., type of premedication, combination with other drugs, and patient physical state). The second term in Eq. (6) allows avoiding sharp changes of the IR, by inducing a *bumpless* action. In fact, excessive oscillations of the IR may cause dangerous plasma peaks and consequent adverse effects on the patient's state. The term $\delta_y(k)$ in Eqs. (7), (8) allows correcting the model prediction y by means of *real* collected measured data (y_r) at each sampling time k. w_i are weights that sort the priority of the objective function terms.

$$\min{}_{IR(k),IR(k+1),...,IR(k+h_c-1)} \left\{ \sum_{j=k+1}^{k+h_p} \left[w_y\, e_y^2(j) + PF_y(j) + PF_{C_p}(j) \right] + \sum_{i=k}^{k+h_p-1} \left[w_u\, \Delta u^2(i) \right] \right\} \tag{6}$$

$$e_y(j) = \frac{\left[y(j) + \delta_y(k) \right] - y_{sp}(j)}{y_{sp}(j)} \tag{7}$$

$$\delta_y(k) = y_r(k) - y(k) \tag{8}$$

$$\Delta u(i) = \frac{u(i) - u(i-1)}{u(i-1)} \tag{9}$$

$$PF_{C_p}(j) = w_{C_p}^{PF} \left\{ \mathrm{Max} \left[0, \frac{C_p(j) - C_{p\max}}{C_{p\max}} \right] \right\}^2 \tag{10}$$

$$PF_y(j) = w_y^{PF} \left\{ \left\{ \text{Max} \left[0, \frac{y(j) - y_{\max}}{y_{\max}} \right] \right\}^2 + \left\{ \text{Min} \left[0, \frac{y(j) - y_{\min}}{y_{\min}} \right] \right\}^2 \right\} \qquad (11)$$

As our case study on anesthesia is simulated in silico (see Section 2.5), the in silico patient is represented by modeling its pharmacokinetics and pharmacodynamics. A model mismatch is thus required between the model implemented in the controller and the one representing the actual patient dose-effect relation (process model and real model, respectively, in Fig. 1). To do this, we simulate the real patient with a PBPK/PD model, while the one implemented in the controller is a simpler three-compartment PK/PD model, which is also more CPU-time efficient (see Fig. 2).

2.5 Simulation of surgical operations in different patients

An optimal induction of anesthesia leads the patient safely and as fast as possible from the initial awake state to the desired DOA level. In engineering terms, this problem is called "servo problem." In practical terms, the controller must find a suitable compromise between velocity of the control action and safety of the (in silico) patient. In terms of performance assessment, this should result in an acceptable rise time (i.e., time required to reach the set point level for the first time) and limited overshoot (i.e., maximum exceeded distance from the set point). In clinical practice, it is desirable to accomplish induction within 10–15 min. Deep anesthesia levels correspond to BIS in the range 40–60 [−]; thus, in our simulations, setpoint is set to BIS = 50 [−]. Although adequate analgesia is more difficult to quantify, it is usually associated to hemodynamic stability throughout the procedure. For normal values of baseline MAP (associated to healthy patients before the start of the procedure), a 20%–25% drop is expected during induction with propofol and remifentanil (Batra et al., 2004; Hall et al., 2000; Thompson et al., 1998), with consequent postinduction values of

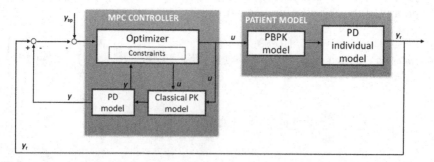

Fig. 2 Scheme of the in silico MPC-loop of anesthesia. The PBPK/PD model, closer to the reality of patient's anatomy and physiology, represents the dose-effect relation of the real patient. The model implemented in the controller is a classical three-compartment PK/PD model. Set point y_{sp} is the vector of desired BIS and MAP values, u is the vector of propofol and remifentanil infusion rates, while y_r is the output of the real patient (in our case simulated in silico), i.e., values of the measured variables BIS and MAP.

MAP around 70 mmHg. In our simulations, this value is the MAP set point to be maintained.

Fig. 3 shows the induction phase of anesthesia for nine in silico adult patients (see Table 1 for demographic features) and different PD characteristics that allow simulating interindividual variability.

Top panel of Fig. 3 shows MAP (left) and BIS (right) dynamics, while bottom panel shows corresponding remifentanil and propofol IRs. Desired set point values are indicated by horizontal dashed black lines for both MAP and BIS. Despite PD variability of the patients, all the curves remain within the safety clinical limits, i.e., MAP above the lower limit of 60 mmHg (red dotted line) and BIS in the 40–60 [−] range (shaded red area). In fact, BIS values higher than 60 may lead to awareness episodes and below 40 may lead to critical recovery of the patients.

BIS is slower to stabilize at the desired set point (i.e., BIS = 50 [−]) because of a slower velocity of re-equilibration between plasma and effect site of propofol

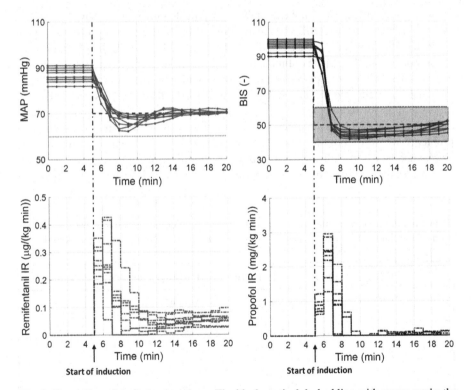

Fig. 3 Simulation of the induction phase. The black vertical dashed line with arrow marks the start of induction. Left panel refers to analgesia, right panel to hypnosis. (Top) Dynamics of controlled variables MAP and BIS. (Bottom) Dynamics of manipulated variables, i.e., remifentanil and propofol IRs. The straight horizontal red line (gray line in print version) corresponding to MAP = 60 mmHg shows the lower bound before hypotension occurs. The shaded area in the BIS plot displays the recommended clinical range, BIS = 40–60 [−].

Table 1 Demographics of the simulated patients

N°	BM (kg)	H (cm)	Sex (−)
1	75.5	180	M
2	71.3	173	F
3	94.5	195	M
4	90.3	181	F
5	50.5	159	F
6	57.7	162	F
7	56.7	164	M
8	95	203	M
9	85	185	M
Mean	75.2	178	4F/5M
St. Dev.	17.2	15.1	−

compared to remifentanil. On the modeling side, this feature manifests in a slower k_{e0} value. For both MAP and BIS, the mean values of the nine in silico patients are quite above the minimum thresholds. This value can be suitably modulated by changing the tuning weights of the objective function. The simulated control action (see Table 2) is capable of achieving acceptable rise times of 2–3 min for MAP and BIS. In addition, the controlled trajectories are stable within 15–20 min, which is an acceptable response time for induction.

In engineering terms, maintenance of anesthesia is equivalent to a "regulator problem" where a controller must maintain the desired set point (i.e., in our case DOA) against external disturbances. Surgical stimuli are likely to cause disturbances of the anesthetic state during surgery. In fact, they can lead to alteration of the

Table 2 Performance evaluation of the controller. First column lists patients' number, second and third columns report rise times for MAP and BIS, fourth and fifth columns list minimum values of MAP and BIS

#	$t_{rise, MAP}$ (min)	$t_{rise, BIS}$ (min)	Min_{MAP} (mmHg)	Min_{BIS} (−)
1	1.893	1.963	64.9	43.9
2	2.521	1.822	64.8	44.2
3	1.519	2.861	67.7	46.7
4	1.789	2.013	63.2	45.9
5	2.002	1.867	61.7	41.8
6	2.347	2.049	68.2	47.0
7	3.449	2.065	69.4	45.0
8	2.874	1.925	68.1	43.2
9	2.733	2.029	67.3	42.6
Mean	1.917	2.669	66.1	44.5
St. Dev.	0.344	0.709	2.6	1.8

monitored physiological variables, for instance in terms of cardiovascular and electroencephalographic changes. Some stimuli may be so strong to produce pain and patient arousal. Therefore, the anesthesia controller must guarantee a fast response against critical changes of the controlled physiological variables. Fig. 4 shows the results of the simulation of the maintenance phase for the nine in silico patients. Black vertical lines point at a series of events, i.e., (0) induction of anesthesia, (1) intubation (performed 4 min after the start of infusion), (2) incision, (3) and (4) arousal episodes manifested by an increase of BIS, and (5) end of infusion. We assumed that analgesia level is sufficient to prevent peripheral noxious stimuli from reaching the brain. Consequently, we suppressed BIS reaction to intubation, because according to Nakayama et al. (2003), intubation is mediated at the subcortical level, and therefore may be unrelated to BIS, which is an indicator of cerebral cortical activity. For this reason,

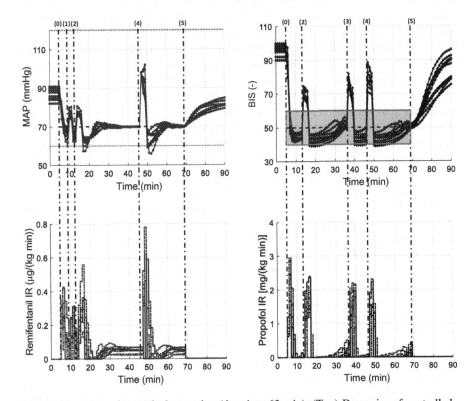

Fig. 4 Simulation of a surgical operation (duration: 63 min). (Top) Dynamics of controlled variables MAP (left) and BIS (right). (Bottom) Dynamics of manipulated variables, IRs of remifentanil (left) and propofol (right). Black vertical dashed lines mark specific events: (0) induction of anesthesia, (1) intubation (4 min after start of induction), (2) incision, (3) and (4) arousal episodes, and (5) end of infusion. Horizontal red lines (gray lines in print version) in the MAP diagram mark the bounds of hypo- and hypertension regions (i.e., MAP < 60 mmHg and MAP > 120 mmHg). The shaded area in the BIS plot denotes the recommended clinical range, i.e., BIS = 40–60 [−].

in our simulations, only incision induces BIS changes. MPC reacts promptly with steep and short increases of IRs to keep the patient within the safe boundaries. Maximum values of the IRs are never out of the ranges indicated in the dosing guidelines of propofol and remifentanil.

It is worth observing that few minutes are required to re-establish targeted depths of anesthesia and analgesia after disturbances, and only one patient experiences a MAP value visibly below the assigned lower bound of 60 mmHg, although for a rather short time interval. Time for awakening after the end of infusion, identified with the time required to restore BIS higher than 90, is in line with experimental studies. In fact, according to Frost (2014), time for awakening from modern anesthetic agent is rather short. Although it varies depending on the patient's physical characteristics, administered drug(s), and surgical procedure type and duration, all patients are expected to be responsive within 60 min from the end of infusion, and most of them are fully conscious after 15 min. According to our simulations, the mean time for awakening over the nine simulated patients is 18 min.

Fig. 5 portrays the predicted drugs pharmacokinetics in different parts of the body after the IRs trajectories of Fig. 3, assuming lack of disturbances and stimulation. Since measuring drug concentrations in the organs and tissues of human subjects is too invasive, it is not possible to validate the PK simulations in body compartments other than plasma. However, such simulations provide useful and complementary information to the anesthetist in terms of decision making and situational awareness. This is especially true as they are produced by adapting the PBPK model not only to the patients' physical characteristics, but also to the specific properties of the drug, for instance, in terms of metabolic/elimination pathways. For instance, remifentanil is not metabolized via hepatic route. Therefore, the elimination pathways of remifentanil PBPK model account for plasma and tissue esterases elimination (see Fig. 5, showing extrahepatic metabolized amount for remifentanil). Conversely, for propofol, both tissue and hepatic metabolisms are considered, although the latter plays the main role. As expected for the IV route, the highly perfused organs and liver levels are high (in fact, no first-pass hepatic effect occurs, as it would happen in case of oral administration). It is worth observing that remifentanil peak of blood concentration falls within the recommended range for induction of adequate analgesia [3–8 ng/mL, (Absalom and Struys, 2007)]. The propofol blood concentration of Fig. 5 tends to a maintenance value that is consistent with medical guidelines (Absalom and Struys, 2007), i.e., target values of plasma concentration amid 4-5 µg/mL for unpremedicated patients and in presence of an analgesic component.

An advantage of using the PBPK model consists also in the possibility of adapting to different populations, with suitable modifications in the description of the ADME processes and anatomical/physiological features. For instance, we consider the simulation of controlled anesthesia in pediatric patients. Age is a most relevant factor affecting both pharmacokinetics and pharmacodynamics of several drugs. In fact, children should not be simply considered as *small* adults; indeed, allometric scaling is not always sufficient to account for the differences between children and adults subject to drug administration (Mahmood, 2014). ADME processes of drugs undergo significant changes depending on the development and growth stage. Several studies

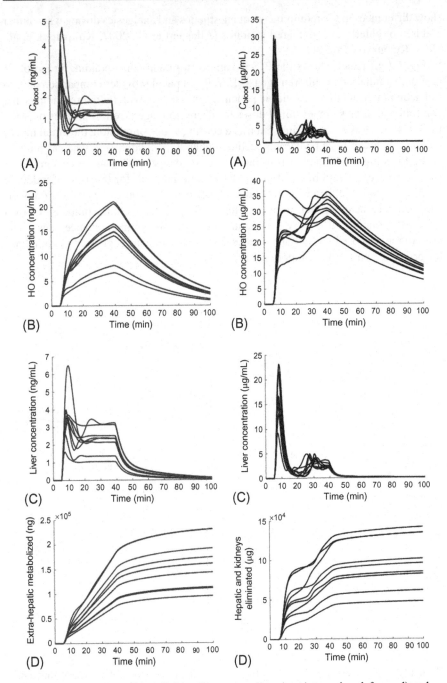

Fig. 5 Simulation of remifentanil (blue lines—gray lines in print version, left panel) and propofol (black lines, right panel) body compartments pharmacokinetics of nine in silico patients. A–C panels show concentration dynamics in blood, highly perfused organs (HO), and liver. D panels show the eliminated amount via extrahepatic routes (remifentanil case) and hepatic metabolism and renal elimination (propofol case).

show differences in the requirements of anesthetics and analgesics dosing when either children or elderly are compared to adults (Allegaert et al., 2007; Kirkpatrick et al., 1988; Rigouzzo et al., 2008).

Fig. 6 (left panel) shows the BIS dynamics (top) and corresponding propofol IR (bottom) simulation in children aged 3–15 y. Right panel provides comparison of control action in the adult and pediatric patients. Consistently with literature findings, the controller simulates higher initial doses for decreasing age (see IRs profiles over the first 10 minutes after start of induction). According to our simulation, the initial induction dose is lower for a pediatric patient, compared to an adult one, whereas maintenance IR is higher. Induction is in fact the most "dangerous" phase, as a significant amount of drug is administered in a reduced time interval, for the sake of inducing unconsciousness within a reasonable time. Since clinical ranges are narrower for pediatric patients, the controller proposes a lower IR during induction, compensated by a stable IR (nonnull, differently than adults) in the following maintenance phase. The level of the proposed IR for maintenance is similar to the one suggested in guidelines for pediatric anesthesia (Gregory, 2012).

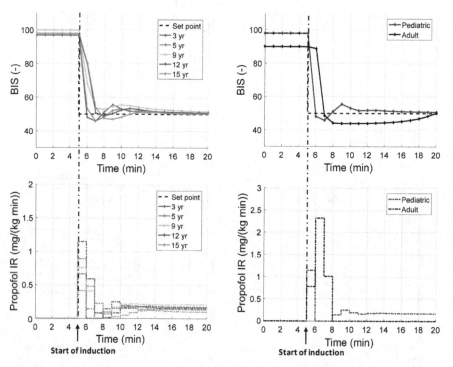

Fig. 6 (Top) BIS and (bottom) propofol IR dynamics during induction. (Left panel) Simulation of induction of anesthesia in pediatric patients (3–15 y). (Right panel) Comparison of adult (black line) versus 5-year patient (blue line—gray line in print version) simulation of induction of anesthesia.

These preliminary results can be considered as a starting point for the assessment of the MPC performance. Standardized performance indexes and evaluation methods are still to be established for encouraging wide acceptance and application of controllers to anesthesia. In this respect, Section 3 introduces some related ethical concerns and clinical consequences.

3 Ethical concerns and clinical outcomes of closed-loop controlled anesthesia

After considering the advantages and technological aspects of automated delivery of anesthesia, the goal of Section 3 is to stimulate the interested reader to consider the controversial ethical and clinical outcomes.

The introduction of any new invention in biomedicine is expected to raise ethical concerns. With respect to the implementation of new surgical technologies and techniques, the authors of Strong et al. (2014) identify six key ethical considerations that can be suitably transferred to the context of automated anesthesia delivery. We reformulate these six points into Sections 3.1–3.5. Finally, Section 3.6 discusses potential clinical impact of control systems adoption in anesthesia.

3.1 Guarantee of the safety of new technology and management of timing and process for implementation

Regulatory agencies are in charge of managing the guarantee of safety of new technologies. The most well-known are USA FDA and EMA (European Medicines Agency). With similar regulatory processes, their common mission is to ensure safety and efficacy of new drugs and medical devices, while guaranteeing a rapid introduction of innovative therapies (Van Norman, 2016). This task is complicated by the fact that efforts of increasing safety often result into an increase in the costs and times for approval. It is worth underlining that there are neither universal regulatory frameworks nor processes. A very clear instance in the field of anesthesia is the case of TCI pumps. The authors of Absalom et al. (2016) identified and questioned commercial companies who were actively manufacturing and distributing TCI devices between 2004 and 2013. Despite claiming that more than 40,000 devices were sold in those years, such devices are still not approved/sold in all the countries. For instance, officially, no TCI pumps have ever been sold in the USA.

A related issue is the timing and process for implementation of new technologies. In fact, there is always a conflict between different realities: (i) the national/international framework and (ii) the local institution. Although the national organizations provide guidelines on these matters, the final effective decision on the use of new technologies is prerogative of the local institution, e.g., the hospital, the medical staff, the academic department head (Sachdeva and Russell, 2007).

3.2 Patient's informed consent

Necessary condition for human experimentation to be both legal and ethical is the patient's informed consent. According to the Declaration of Helsinki developed by the World Medical Association (WMA) as a statement of ethical principles for medical research involving human subjects, *"each potential subject must be adequately informed of the aims, methods, sources of funding, any possible conflicts of interest, institutional affiliations of the researcher, the anticipated benefits and potential risks of the study and the discomfort it may entail, post-study provisions and any other relevant aspects of the study"* (World Medical Association, 2013). This means that the medical doctor must engage the patients in a discussion aimed at not only informing them but also educating, understanding, and listening to potential doubts and questions. In case of anesthesia, this is particularly important. Since the drugs involved in the procedure may have critical adverse effects and affect not only the outcome of the surgical operation, but also the postrecovery phase, the patient must be put in the condition of complete trust in the anesthetist and their capacity of judgement and use of any tools involved in the procedure.

3.3 Training and credentialing physicians in new technology or technique

Often neglected, training is a critical issue regarding new medical devices. In the specific case of anesthesia, an interesting example is again the case of TCI pumps. According to Absalom et al. (2016), there are no specific rules on who can or cannot use TCIs. In addition, there is no regulated training for the use of TCI systems. This delicate point is related strictly to the clinical consequences discussed in Section 3.6. In fact, an exhaustive and appropriate training, ideally provided by multidisciplinary experts (e.g., clinicians and engineers), is essential for a responsible and effective use of automated devices for anesthesia delivery. This may help avoiding on one hand excessive trust in the instrument and on the other loss of situational awareness of the anesthetist.

3.4 Track and assessment of new technology outcomes

During the time interval between approval and clinical adoption/application, data on the new technology are still limited. Thus, there is an ethical obligation for early adopters to track outcomes of the new technology. Once again, rules on how to manage this pharmacovigilance phase should be issued by regulatory agencies. Indeed, it is crucial to lead to a wide acceptance of the new device. In the context of PCLC devices (i.e., physiological closed-loop controllers), it is especially important to establish common performance indexes for both their development and evaluation after clinical application. The authors of Parvinian et al. (2018) refer to the consensus standard IEC 60601-1-10 as working in this direction. In fact, this collateral standard specifies requirements for the development (i.e., not only analysis and design, but also verification and validation) of a PCLCS in medical electrical equipment. Automated anesthesia delivery systems can be included within the PCLCS class.

3.5 Balancing responsibilities to patients and society

In relation to the cost and value of new surgical technologies, the authors of Strong et al. (2014) point out the possible conflict of responsibility of the medical doctors, i.e., on one side toward the patients' well-being and on the other toward society (e.g., from the market and the industry) for potential pressure to introduce innovative solutions. They cite the American Board of Internal Medicine principle according to which the main priority must always consist of the health and benefit of patients. We agree that this principle should absolutely be extended to closed-loop control of anesthesia, as the aim is to support anesthetists and increase the safety and stability of patients' state during anesthesia.

3.6 Clinical impact and risks

In the current concept of automated anesthesia, the anesthetist becomes a supervisor (i.e., a decision maker) of the procedure and can focus on monitoring the patient's DOA status. As discussed in Parvinian et al. (2018), depending on the degree of automation, consequences can lead to loss of situational awareness, complacency, and skill degradation. Loss of situational awareness may occur because of the anesthetist that evolves from being a human manual controller to a supervisor and decision maker of the automated control action. The natural consequence is the reduction of the time and actual procedures during which the anesthetist is actively involved in the care of the patient's anesthetic state. Additionally, if the control system proves exceptionally reliable and efficient (as it hopefully should be to guarantee a safe application), the anesthetist is exposed to the risk of overtrusting the tool. A related consequence is progressive degradation of skills. With the medical doctor less and less involved in the conscious procedure about dosing, future anesthetists may exhibit skill decay, as their skills would not be used as often as in the past, in case of high degree of automation.

A series of issues related to engineering aspects will have to be carefully assessed and analyzed in the future, such as the presence of fail-safe mechanisms, modeling uncertainty, and algorithms robustness.

These considerations are common hazards introduced by the implementation of automated systems. However, the active involvement of clinicians in the development and testing of control systems for anesthesia delivery, coordinated with the introduction of adequate methods for the evaluation and standardization of training represent the key milestones to limit likely negative impacts on clinical practice.

4 Conclusions

This chapter provided a discussion on the application of advanced control systems in anesthesia. Firstly, we discussed the limitations of the current manual approach to anesthesia delivery, mainly related to the impact of human factors on the choice of the optimal dosing, and the limited use of data on the monitored vital parameters (i.e., only for subjective interpretation of the DOA level). Afterwards, we compared the main control strategies that are being studied and presented a case study on MPC of anesthesia with propofol and remifentanil. In silico results were presented and critically analyzed, along with the potential of application of modern PBPK modeling to individualize the prediction of the dose-effect relation respect to different individuals (e.g., adult vs pediatric). Finally, some ethical and clinical implications were discussed to stimulate the reader to make their own mind on the topic and propose solutions to the remaining open problems.

We find that the "engineering" challenges of control systems applications in anesthesia can be summarized into two specific points: (i) interindividual variability of the response to drugs and surgical stimuli and (ii) the complexity of the process that involves multiple drugs, and thus several effects that manifest their mechanism of actions and interactions. For what concerns the first issue, we believe that population PK/PD models are hardly adequate. Future work should focus on efforts to develop models able to individualize the prediction. Taking into account physiological and anatomical differences in the individuals, related to age, body mass and height, presence of disease, genetics, can make the difference. In addition, use of the online information of cardiovascular changes can help improve the prediction of drugs disposition within the body with consequent changes in the manifestation of their effects. PBPK models use blood flows as parameters, and thus can take into account these changes. The literature reports that PD variability is higher than PK variability among individuals. Adaptive control techniques can be evaluated to face this matter, so that the predicted PD response is adapted to the specific patients undergoing anesthesia. As for the second issue, it is no mystery that anesthesia is complex and some underlying mechanisms are still unclear. With advancing knowledge on drug-drug interactions mechanisms, and improved tools and methods to manage multivariable problems, researchers should address the issue by increasing the quantity of data that the controller can process and use, so that drug-drug interactions and several pharmacodynamic effects can be embedded into the control problem.

Despite difficulties related to the lack of standardization and universal regulatory frameworks, nowadays TCI pumps are a mature technology (Absalom et al., 2016). This opens an encouraging perspective for control systems application in anesthesia based on real-time monitored measures of the patient's anesthetic and analgesic state. Such systems have the potential to not only reduce anesthesia-associated risks, human error incidence, and anesthetist's workload for a more efficient monitoring and focus on the patient's state, but also standardize anesthesia procedure, by reducing subjectivity of the optimal dose selection. Obviously, clinical consequences must be carefully assessed and engineering aspects validated before any clinical adoption and application.

It is our opinion that the only road for successful application of control systems in anesthesia calls for engineers and clinicians working together. In fact, on one hand, engineers are essential for an effective design and development of the patient's model and control system. On the other hand, clinicians must guide engineers by providing information on crucial aspects such as the key indexes for an optimal and safe assessment of the control system performance and the essential data required to quantify the DOA level. In addition, clinicians can provide useful recommendations and specifications for the optimal design of the graphical user interface of the automated delivery tool.

References

Abbiati, R.A., Depetri, V., Scotti, F., Manca, D., 2016a. A new approach for pharmacokinetic model application towards personalized medicine. In: Zdravko, K., Miloš, B. (Eds.), Computer Aided Chemical Engineering. In: vol. 38. Elsevier, pp. 1611–1616.

Abbiati, R.A., Lamberti, G., Grassi, M., Trotta, F., Manca, D., 2016b. Definition and validation of a patient-individualized physiologically-based pharmacokinetic model. Comput. Chem. Eng. 84, 394–408.

Abbiati, R.A., Manca, D., 2017. Enterohepatic circulation effect in physiologically based pharmacokinetic models: the Sorafenib case. Ind. Eng. Chem. Res. 56, 3156–3166.

Abbiati, R.A., Savoca, A., Manca, D., 2018. (Chapter 2). An engineering oriented approach to physiologically based pharmacokinetic and pharmacodynamic modeling. In: Manca, D. (Ed.), Computer Aided Chemical Engineering. In: vol. 42. Elsevier, pp. 37–63.

Absalom, A., Struys, M., 2007. An Overview of TCI and TIVA. Academia Press.

Absalom, A.R., De Keyser, R., Struys, M.M., 2011. Closed loop anesthesia: are we getting close to finding the holy grail? Anesth. Analg. 112, 516–518.

Absalom, A.R., Glen, J.I., Zwart, G.J., Schnider, T.W., Struys, M.M., 2016. Target-controlled infusion: a mature technology. Anesth. Analg. 122, 70–78.

Allegaert, K., de Hoon, J., Verbesselt, R., Naulaers, G., Murat, I., 2007. Maturational pharmacokinetics of single intravenous bolus of propofol. Paediatr. Anaesth. 17, 1028–1034.

Batra, Y.K., Al Qattan, A.R., Ali, S.S., Qureshi, M.I., Kuriakose, D., Migahed, A., 2004. Assessment of tracheal intubating conditions in children using remifentanil and propofol without muscle relaxant. Paediatr. Anaesth. 14, 452–456.

Cortínez, L.I., Anderson, B.J., Penna, A., Olivares, L., Muñoz, H.R., Holford, N.H.G., Struys, M.M.R.F., Sepulveda, P., 2010. Influence of obesity on propofol pharmacokinetics: derivation of a pharmacokinetic model. Br. J. Anaesth. 105, 448–456.

Dershwitz, M., Hoke, J.F., Rosow, C.E., Michalowski, P., Connors, P.M., Muir, K.T., Dienstag, J.L., 1996. Pharmacokinetics and pharmacodynamics of remifentanil in volunteer subjects with severe liver disease. Anesthesiology 84, 812–820.

Dumont, G.A., 2014. Feedback control for clinicians. J. Clin. Monit. Comput. 28, 5–11.

Egan, T.D., Huizinga, B., Gupta, S.K., Jaarsma, R.L., Sperry, R.J., Yee, J.B., Muir, K.T., 1998. Remifentanil pharmacokinetics in obese versus lean patients. Anesthesiology 89, 562–573.

Egan, T.D., Lemmens, H.J., Fiset, P., Hermann, D.J., Muir, K.T., Stanski, D.R., Shafer, S.L., 1993. The pharmacokinetics of the new short-acting opioid remifentanil (GI87084B) in healthy adult male volunteers. Anesthesiology 79, 881–892.

El-Nagar, A.M., El-Bardini, M., 2014. Interval type-2 fuzzy neural network controller for a multivariable anesthesia system based on a hardware-in-the-loop simulation. Artif. Intell. Med. 61, 1–10.

Forbes, M.G., Patwardhan, R.S., Hamadah, H., Gopaluni, R.B., 2015. Model predictive control in industry: challenges and opportunities. IFAC-Pap. OnLine 48, 531–538.

Frost, E.A., 2014. Differential diagnosis of delayed awakening from general anesthesia: a review. Middle East J. Anaesthesiol. 22, 537–548.

Gentilini, A., Schaniel, C., Morari, M., Bieniok, C., Wymann, R., Schnider, T., 2002. A new paradigm for the closed-loop intraoperative administration of analgesics in humans. IEEE Trans. Biomed. Eng. 49, 289–299.

Gepts, E., Camu, F., Cockshott, I.D., Douglas, E.J., 1987. Disposition of propofol administered as constant rate intravenous infusions in humans. Anesth. Analg. 66, 1256–1263.

Glass, P.S., Gan, T.J., Howell, S., 1999. A review of the pharmacokinetics and pharmacodynamics of remifentanil. Anesth. Analg. 89, S7–14.

Glass, P.S., Hardman, D., Kamiyama, Y., Quill, T.J., Marton, G., Donn, K.H., Grosse, C.M., Hermann, D., 1993. Preliminary pharmacokinetics and pharmacodynamics of an ultrashort-acting opioid: remifentanil (GI87084B). Anesth. Analg. 77, 1031–1040.

Gregory, G., 2012. Anesthesia for Premature Infants. Gregory's Pediatric Anesthesia.

Hall, A.P., Thompson, J.P., Leslie, N.A., Fox, A.J., Kumar, N., Rowbotham, D.J., 2000. Comparison of different doses of remifentanil on the cardiovascular response to laryngoscopy and tracheal intubation. Br. J. Anaesth. 84, 100–102.

Jones, H.M., Gardner, I.B., Collard, W.T., Stanley, P.J., Oxley, P., Hosea, N.A., Plowchalk, D., Gernhardt, S., Lin, J., Dickins, M., Rahavendran, S.R., Jones, B.C., Watson, K.J., Pertinez, H., Kumar, V., Cole, S., 2011. Simulation of human intravenous and oral pharmacokinetics of 21 diverse compounds using physiologically based pharmacokinetic modelling. Clin. Pharmacokinet. 50, 331–347.

Jones, H.M., Parrott, N., Jorga, K., Lave, T., 2006. A novel strategy for physiologically based predictions of human pharmacokinetics. Clin. Pharmacokinet. 45, 511–542.

Kirkpatrick, T., Cockshott, I.D., Douglas, E.J., Nimmo, W.S., 1988. Pharmacokinetics of propofol (diprivan) in elderly patients. Br. J. Anaesth. 60, 146–150.

Knibbe, C.A.J., Voortman, H.-J., Aarts, L.P.H.J., Kuks, P.F.M., Lange, R., Langemeijer, H.J.M., Danhof, M., 1999. Pharmacokinetics, induction of anaesthesia and safety characteristics of Propofol 6% SAZN vs Propofol 1% SAZN and Diprivan®-10 after bolus injection. Br. J. Clin. Pharmacol. 47, 653–660.

Leil, T.A., Bertz, R., 2014. Quantitative Systems Pharmacology can reduce attrition and improve productivity in pharmaceutical research and development. Front. Pharmacol. 5, 247.

Liu, N., Chazot, T., Hamada, S., Landais, A., Boichut, N., Dussaussoy, C., Trillat, B., Beydon, L., Samain, E., Sessler, D.I., Fischler, M., 2011. Closed-loop coadministration of propofol and remifentanil guided by bispectral index: a randomized multicenter study. Anesth. Analg. 112, 546–557.

Liu, N., Lory, C., Assenzo, V., Cocard, V., Chazot, T., Le Guen, M., Sessler, D.I., Journois, D., Fischler, M., 2015. Feasibility of closed-loop co-administration of propofol and remifentanil guided by the bispectral index in obese patients: a prospective cohort comparison†. BJA: Brit. J. Anaesth. 114, 605–614.

Mahmood, I., 2014. Dosing in children: a critical review of the pharmacokinetic allometric scaling and modelling approaches in paediatric drug development and clinical settings. Clin. Pharmacokinet. 53, 327–346.

Marsh, B., White, M., Morton, N., Kenny, G.N.C., 1991. Pharmacokinetic model driven infusion of propofol in children. Br. J. Anaesth. 67, 41–48.

Mayo, C.W., Bickford, R.G., Faulconer Jr., A., 1950. Electroencephalographically controlled anesthesia in abdominal surgery. J. Am. Med. Assoc. 144, 1081–1083.

Minto, C.F., Schnider, T.W., Egan, T.D., Youngs, E., Lemmens, H.J., Gambus, P.L., Billard, V., Hoke, J.F., Moore, K.H., Hermann, D.J., Muir, K.T., Mandema, J.W., Shafer, S.L., 1997. Influence of age and gender on the pharmacokinetics and pharmacodynamics of remifentanil. I. Model development. Anesthesiology 86, 10–23.

Nakayama, M., Kanaya, N., Edanaga, M., Namiki, A., 2003. Hemodynamic and bispectral index responses to tracheal intubation during isoflurane or sevoflurane anesthesia. J. Anesth. 17, 223–226.

Nascu, I., Krieger, A., Ionescu, C.M., Pistikopoulos, E.N., 2015. Advanced model-based control studies for the induction and maintenance of intravenous anaesthesia. IEEE Trans. Biomed. Eng. 62, 832–841.

Navapurkar, V.U., Archer, S., Gupta, S.K., Muir, K.T., Frazer, N., Park, G.R., 1998. Metabolism of remifentanil during liver transplantation. Br. J. Anaesth. 81, 881–886.

Parvinian, B., Scully, C., Wiyor, H., Kumar, A., Weininger, S., 2018. Regulatory considerations for physiological closed-loop controlled medical devices used for automated critical care: Food and Drug Administration workshop discussion topics. Anesth. Analg. 126, 1916–1925.

Pitsiu, M., Wilmer, A., Bodenham, A., Breen, D., Bach, V., Bonde, J., Kessler, P., Albrecht, S., Fisher, G., Kirkham, A., 2004. Pharmacokinetics of remifentanil and its major metabolite, remifentanil acid, in ICU patients with renal impairment. Br. J. Anaesth. 92, 493–503.

Rigouzzo, A., Girault, L., Louvet, N., Servin, F., De-Smet, T., Piat, V., Seeman, R., Murat, I., Constant, I., 2008. The relationship between bispectral index and propofol during target-controlled infusion anesthesia: a comparative study between children and young adults. Anesth. Analg. 106, 1109–1116 (table of contents).

Rigouzzo, M.D.A., Servin, M.D.P.D.F., Constant, M.D.P.D.I., 2010. Pharmacokinetic-pharmacodynamic modeling of propofol in children. Anesthesiology 113, 343–352.

Sachdeva, A.K., Russell, T.R., 2007. Safe introduction of new procedures and emerging technologies in surgery: education, credentialing, and privileging. Surg. Clin. North Am. 87 (853-866), vi–vii.

Savoca, A., Manca, D., 2019a. A physiologically-based approach to model-predictive control of anesthesia and analgesia. Biomed. Signal Process. Control 53, 101553, 1–12.

Savoca, A., Manca, D., 2019b. Physiologically-based pharmacokinetic simulations in pharmacotherapy: selection of the optimal administration route for exogenous melatonin. ADMET DMPK 7, 44–59.

Schiff, J.H., Wagner, S., 2016. Anesthesia related mortality? A national and international overview. Trend Anaesth. Crit. Care 9, 43–48.

Schnider, T.W., Minto, C.F., Gambus, P.L., Andresen, C., Goodale, D.B., Shafer, S.L., Youngs, E.J., 1998. The influence of method of administration and covariates on the pharmacokinetics of propofol in adult volunteers. Anesthesiology 88, 1170–1182.

Sha'aban, Y.A., Lennox, B., Laurí, D., 2013. PID versus MPC Performance for SISO Dead-time Dominant Processes. IFAC Proc. Vol. 46, 241–246.

Soltesz, K., Hahn, J.-O., Hägglund, T., Dumont, G.A., Ansermino, J.M., 2013. Individualized closed-loop control of propofol anesthesia: a preliminary study. Biomed. Signal Process. Control 8, 500–508.

Strong, V.E., Forde, K.A., MacFadyen, B.V., Mellinger, J.D., Crookes, P.F., Sillin, L.F., Shadduck, P.P., 2014. Ethical considerations regarding the implementation of new technologies and techniques in surgery. Surg. Endosc. 28, 2272–2276.

Teorell, T., 1937. Kinetics of distribution of substances administered to the body. I. The extravascular modes of administration. Arch. Int. Pharmacodyn. Ther. 57, 205–225.

Thompson, J.P., Hall, A.P., Russell, J., Cagney, B., Rowbotham, D.J., 1998. Effect of remifentanil on the haemodynamic response to orotracheal intubation. Br. J. Anaesth. 80, 467–469.

Van Norman, G.A., 2016. Drugs and devices: comparison of European and U.S. approval processes. JACC: Basic Transl. Sci. 1, 399–412.

West, N., Dumont, G.A., van Heusden, K., Petersen, C.L., Khosravi, S., Soltesz, K., Umedaly, A., Reimer, E., Ansermino, J.M., 2013. Robust closed-loop control of induction and maintenance of propofol anesthesia in children. Pediatr. Anesth. 23, 712–719.

Westmoreland, C.L., Hoke, J.F., Sebel, P.S., Hug Jr., C.C., Muir, K.T., 1993. Pharmacokinetics of remifentanil (GI87084B) and its major metabolite (GI90291) in patients undergoing elective inpatient surgery. Anesthesiology 79, 893–903.

Wiczling, P., Bieda, K., Przybylowski, K., Hartmann-Sobczynska, R., Borsuk, A., Matysiak, J., Kokot, Z.J., Sobczynski, P., Grzeskowiak, E., Bienert, A., 2016. Pharmacokinetics and pharmacodynamics of propofol and fentanyl in patients undergoing abdominal aortic surgery—a study of pharmacodynamic drug-drug interactions. Biopharm. Drug Dispos. 37, 252–263.

World Medical Association, 2013. World Medical Association Declaration of Helsinki: ethical principles for medical research involving human subjects. JAMA 310, 2191–2194.

Zhusubaliyev, Z.T., Medvedev, A., Silva, M.M., 2015. Bifurcation analysis of PID-controlled neuromuscular blockade in closed-loop anesthesia. J. Process Control 25, 152–163.

Computational modeling of the control mechanisms involved in the respiratory system

Alejandro Talaminos-Barroso[a], Javier Reina-Tosina[a], Laura María Roa-Romero[a], Francisco Ortega-Ruiz[b], Eduardo Márquez-Martín[b]
[a]Biomedical Engineering Group, Universidad de Sevilla, Seville, Spain, [b]Medical-Surgical Unit of Respiratory Diseases, Instituto de Biomedicina de Sevilla (IBiS), University Hospital Virgen del Rocío, Seville, Spain

1 Introduction

The respiratory system comprises a set of anatomical structures that work synergistically to perform a specific task: transport atmospheric oxygen (O_2) to cells and release carbon dioxide (CO_2) produced by metabolism. To achieve this purpose, a set of active and passive processes of diverse nature (mechanical, physical, chemical, or electrical) are perfectly coordinated in order to introduce air into the lungs, diffuse O_2 into the bloodstream, transport gases to the tissues, and remove CO_2 produced by following the reverse path. The control mechanisms of the respiratory system regulate all these processes and adapt them to internal or external disturbances or physiopathological situations.

Respiratory control mechanisms have historically been studied from two separate perspectives, although in nature they are integrated: respiratory mechanics and control, respectively. Respiratory physiology was already studied in Galen's writings (Fitting, 2015) about the anatomy of the diaphragm and other respiratory muscles. Leonardo Da Vinci (West, 2017) was also interested in the study of respiratory mechanics and described the intercostal muscles and the movement of the rib cage, as well as the physiology of the trachea and other airways. Later on, Robert Boyle (Walmsley, 2007) built an artificial air pump, demonstrating that air was an essential component for breathing and life. At the same time, Robert Hooke (West, 2014) carried out different experiments with animals and proved with two bellows that breathing only required a continuous flow of fresh air and it was not necessary to move the thorax and the rest of the muscles. From this discovery, John Locke (West, 2014) concluded that the mixing of blood and air occurred in lungs. Subsequently, Joseph Priestley (Priestley, 1776) discovered and isolated oxygen, although he did not correctly interpret its chemical behavior. Antoine Lavoisier laid the foundations of modern chemistry and with the collaboration of Pierre Simon Laplace, demonstrated the role of oxygen in respiration, performing different experiments to measure carbon dioxide produced both at rest and under physical activity. At the beginning of the 20th century,

Control Applications for Biomedical Engineering Systems. https://doi.org/10.1016/B978-0-12-817461-6.00011-1

experimentation was still the only possible method for the study of phenomena originated in nature to reproduce behaviors under particular conditions. With the advent of the digital computer and advances in respiratory physiology, the first techniques employed for the study of the integration of the respiratory control system and the associated mechanical action began to be introduced. The functional and structural characteristics of the respiratory system, as well as the diversity of associated neurodynamic behaviors, made it an ideal system for modeling and describing the fundamental principles of control theory at a physiological level from an integrative perspective. For this reason, the first computational models emerged in the mid-20th century and attempted to integrate the two aspects of respiratory physiology. Since then and up to the present day, computational models of respiratory physiology have grown in quantity, diversity, and complexity.

This chapter analyzes the dynamic behaviors associated with respiratory control in terms of how they have been historically reproduced through mathematical modeling and simulation techniques (computational modeling). The chapter is structured as follows: after this introduction, in Section 2, an overview of the respiratory system is presented and structured into two subsections. The first one introduces fundamental concepts of respiratory mechanics, transport, and exchange of gases. On the other hand, the second part exposes the mechanisms involved in the control of respiration, from a biomedical engineering perspective that allows to understand the computational models that will be presented later. Next, the third section of the chapter is focused on computational modeling as a tool for the prediction and analysis of dynamic behaviors of different variables in physiological and clinical applications. In the fourth section, computational models of the respiratory system are analyzed, from the most classical ones that emerged in the 1960s to the most modern ones, highlighting the importance of computational modeling as a low-cost technique for simulating a wide range of physiopathological conditions and as a complement to animal research and experimentation. Finally, the conclusions of the chapter present an overview of all the aspects addressed in the different sections.

2 Control mechanisms in the respiratory system

2.1 Gas exchange at the pulmonary and capillary levels

Breathing is a rhythmic mechanical process and centrally regulated that controls the movement of gas into and out of the basic unit of ventilation (alveoli) by contraction and relaxation of the skeletal muscles of the diaphragm, abdomen, and rib cage (Staub, 1991). This concept should not be confused with respiration, which is a chemical process that provides energy to tissue cells by transforming oxygen and nutrients into carbon dioxide and water. In this process of respiration, multicellular organisms use the external environment, on the one hand, to capture the oxygen and deliver it to tissue cells, and on the other hand, to remove the waste substances that are released from metabolism. Mammals have two systems to carry out this task: the respiratory and circulatory systems, usually considered as a global system (cardiorespiratory system).

The respiratory system acts as a gas exchanger at the level of the pulmonary alveoli between oxygen and carbon dioxide, while the circulatory system is responsible for transporting the gases dissolved in the blood that are consumed and produced during cellular respiration.

Breathing consists mainly of two phases: inspiration and exhalation, rhythmic movements that cause air to enter and leave the lungs. With regard to the generation of respiratory movements (control of breathing) during inspiration/expiration, lungs are considered as passive organs that depend on the external forces generated by the respiratory muscles (respiratory mechanic) (Ratnovsky et al., 2008). As it is well known, inspired and exhaled air is a mixture of gases governed by Dalton's law of partial pressures Eq. (1), which states that the total pressure of a mixture of gases is the sum of the partial pressures of the component gases. This law applies to all body compartments where there is a mixture of gases.

$$P_{total} = \sum_{i_1}^{n} p_i = p_{O_2} + p_{CO_2} + p_{N_2} + p_{H_2O} + \dots + p_n \tag{1}$$

A simplified diagram of the pulmonary system considered as a two-dimensional extensible container is shown in Fig. 1. During inspiration, the diaphragm (d) and other ventilatory muscles cause an expansion of the thoracic cavity (x) with respect to its size at rest (x_0). Respiratory muscles depend on pulmonary compliance, which is defined as the ratio of the change of lung volume related to the change of the pressure difference. This relationship can be modeled as an elastic spring dependent on an elastic constant (k_s) associated with respiratory tissue, as well as the frictions between the gas flow through the airways and on the alveolar surface. Expansion of the rib cage causes a fall in pleural pressure (P_L) and alveolar pressure (P_a), allowing the gas (q) to flow into the lungs from the external medium due to the pressure gradient with respect to the atmospheric pressure (P_e). When the diaphragm relaxes, an elastic recoil of the thoracic cavity is produced and the air is expelled toward the outside due to the difference between the pressure in the alveolus versus the atmospheric pressure.

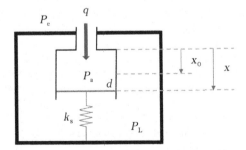

Fig. 1 Simplified diagram of the pulmonary system as a two-dimensional extensible container in one direction.

In addition, certain diseases (Milioli et al., 2016) can lead to a pathophysiological condition associated with low pulmonary compliance in which a higher pressure is required to produce volume changes. In this case, the air entering the lungs (ventilation) is usually lower and the control mechanisms of the respiratory system compensate for this by increasing the respiratory rate. On the other hand, the exhalation process may be incomplete when pulmonary compliance is high due to the reduction of the elastic recoil of the lungs, associated with abnormalities of pulmonary gas exchange (Tran et al., 2018). High pulmonary compliance may lead to muscle fatigue (Ioannidis et al., 2015), which increases the development of ventilatory insufficiency or oxygenation. These effects are shown in Fig. 2, presenting the relationship between transpulmonary pressure (difference between airway pressure and pleural pressure) and the variation of lung volume depending on the pulmonary compliance.

Pulmonary compliance is also associated with lung static volumes, widely used in clinical practice. In this sense, pulmonary compliance is highest at normal lung volumes and lower at volumes that are very low or very high (Esquinas, 2015). The main lung volumes are: total lung capacity (TLC), residual volume (RV), residual functional capacity (FRC), and tidal volume (V_t). In particular, TLC is the volume of gas in the lungs at the end of a forced maximum inspiration (maximum capacity of a healthy lung). On the other hand, RV is the volume of gas in the lungs at the end of a maximum forced exhalation. The difference between TLC and RV is known as vital capacity (VC), while FRC is the lung volume at the end of a normal breath and V_t is the circulating volume between a normal inspiration and exhalation without additional effort. Finally, other volumes used in clinical practice include inspiratory reserve volume (ERV), which is defined as the difference between FRC and RV, and inspiratory capacity (IC), which is the difference between TLC and FRC. The lung volumes already introduced are presented in Fig. 3 and their magnitudes depend on factors such as gender, age, height, weight, race, or ethnicity (Talaminos-Barroso et al., 2018). In clinical practice, spirometry or plethysmography tests are used to estimate these lung volumes.

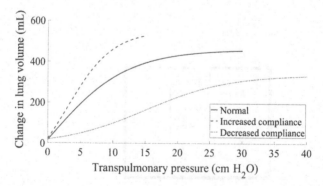

Fig. 2 Representation of pulmonary compliance as the relationship between transpulmonary pressure and the variation of lung volume.

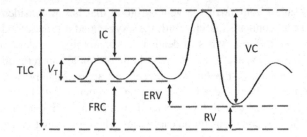

Fig. 3 Relationship between the main lung volumes used in clinical practice.

Most of the inspired air is distributed to the middle and lower lung area where the gas exchange of O_2 and CO_2 takes place between the pulmonary alveoli and the capillaries. The efficiency of this process is measured by the $\dot{V}_a/\dot{Q}$ ratio, i.e., the relationship between ventilation and the blood flow reaching the alveoli through the capillaries (perfusion). The respiratory control system optimizes the efficiency of this process with the least possible effort, including the work associated with respiratory muscle strength and pumping of the heart right ventricle.

The surface area where gas exchange occurs is the alveolar-capillary barrier or membrane. However, there is a portion of gas that does not take part in gas exchange, because it remains in the airways or reaches alveoli that are poorly perfused. This volume is known as dead space and is shown in Fig. 4, including the derivation of the dead

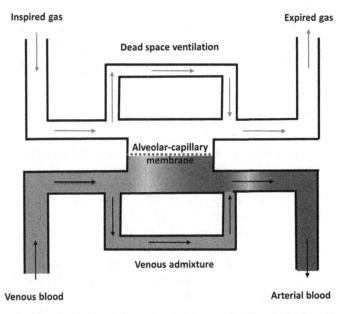

Fig. 4 Schematic representation of gas exchange between alveoli and blood capillaries in the alveolar-capillary membrane.

space, as well as of content of the venous blood that is not considered in the gas exchange and that connects directly with the oxygenated arterial blood via a shunt.

There are three different types of dead space: anatomic, alveolar, and physiological. Anatomical dead space refers to the volume of inhaled air remaining in the airways that does not reach the lungs, representing approximately 30% of the tidal volume. In contrast, alveolar dead space is often associated with pathological conditions resulting from inadequate blood perfusion in the alveoli, due to decreased cardiac output or obstruction of the blood capillaries. The sum of anatomical and alveolar dead space is known as physiological dead space.

In clinical practice, the efficiency of gas exchange is quantified by using Bohr's equation (Crossman et al., 1970), which calculates the ratio of physiological dead space (V_d) and tidal volume (V_t). This relationship is presented in Eq. (2), where P_aCO_2 is the partial pressure of CO_2 in arterial blood and P_eCO_2 is the partial pressure of CO_2 in exhaled air.

$$\frac{V_d}{V_t} = \frac{P_aCO_2 - P_eCO_2}{P_aCO_2} \tag{2}$$

The movement of O_2 and CO_2 molecules between alveoli and capillaries is called passive diffusion because the exchange of gases produces a gradient of partial pressures on each side of the membrane. In particular, this process is governed by Fick's law (Stoller, 2015), which states that the net rate of diffusion of a gas through a semipermeable membrane is proportional to the area of the membrane and the difference between pressures on both sides of the membrane, and inversely proportional to the thickness. This law is simplified to involve a single dependent variable (time), considering that the exchange of gases occurs in one main direction. In general, Eq. (3) is used in practice, where A is the surface of the semipermeable membrane, d its thickness, $P_1 - P_2$ the pressure gradient on both sides of the membrane, and D the diffusion constant.

$$\dot{V} = \frac{D \cdot A}{d}(P_1 - P_2) \tag{3}$$

The exchange of gases occurs not only in the lungs during breathing, but also between the blood capillaries and the mitochondrial organelle during the breathing process in cells. In this sense, partial pressures are defined as the alveolar partial pressure of O_2 (PAO_2) and CO_2 ($PACO_2$) in lungs, the arterial partial pressures of O_2 (P_aO_2) and CO_2 (P_aCO_2), and the partial venous pressures of O_2 (P_vO_2) and CO_2 (P_vCO_2), considering O_2 diffusing from the alveoli into the arterial blood and CO_2 from the venous blood into the alveoli. All these pressures and its normal values are presented in Fig. 5.

After gas exchange, respiratory gases are transported in the blood to the tissue capillaries, where O_2 is consumed as a consequence of the metabolic processes of cells, while CO_2 is produced to be transported back to the lungs (Hall, 2015). In relation to transport, Henry's law states that the concentration of gas in a fluid is a proportional

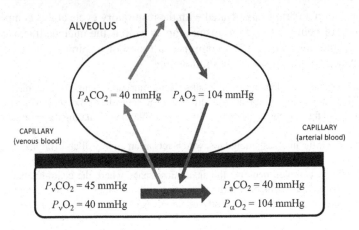

Fig. 5 Exchange of O_2 and CO_2 between alveoli and blood.

relationship between the partial pressure of the gas (P) and the coefficient of solubility of the gas in the fluid (α), as stated in Eq. (4) for O_2.

$$[O_2] = \alpha_{O_2} \cdot P_{O_2} \tag{4}$$

The O_2 dissolved in blood is unable to meet the metabolic needs of the cells considering its low solubility in water and the magnitude of PAO_2. Therefore, nature had to develop a protein to allow the transport of O_2 through the bloodstream (Scholander, 1960). Thus, O_2 is transported almost in its entirety tied to the hemoglobin protein, while the rest (about 3%) is dissolved in plasma. For the same reason, CO_2 is transported in the blood in three different ways: dissolved in blood plasma, in the form of bicarbonate (70% approximately) and in combination with hemoglobin (Marieb and Hoehn, 2012).

The affinity of CO_2 and O_2 is different in arterial blood and venous blood. This is because the affinity of CO_2 increases when O_2 dissociates from hemoglobin (venous blood) at the tissue level (Haldane effect), while the affinity of O_2 decreases when CO_2 bonds with hemoglobin (Bohr's effect).

2.2   Control of ventilation

Respiratory control mechanisms are a combination of physical and chemical phenomena that occur automatically and can be partially overridden by voluntary control and temporarily interrupted by swallowing food or by involuntary acts such as sneezing, coughing, or hiccupping (Katz, 2018). Respiration is regulated by control systems with great flexibility to adapt to change under different environmental and pathophysiological conditions, including, for example, posture, speech, physical exercise, climate, or altitude.

Neural connections in the respiratory control system are present both inside and outside the spinal cord, mainly divided into two neural groups: the dorsal respiratory group (DRG) and the ventral respiratory group (VRG). The DRG mainly concerns the synchronization and times associated with the respiratory cycle and is composed predominantly of neurons related to inspiration as well as the nucleus tractus solitarius (NTS). On the other hand, VRG comprises a group of respiratory neurons (Horner and Malhotra, 2016), including the following:

- Caudal VRG: the nucleus retroambigualis (Hemmings and Egan, 2013), which fundamentally integrates upper motor neurons in charge of controlling the contralateral expiratory muscles, and the nucleus para-ambigualis, responsible for stimulating the contralateral inspiratory muscles.
- Rostral VRG: the nucleus ambigualis, which acts as an airway dilator (larynx, pharynx, and tongue) (Lumb, 2016) and provides coordination of the respiratory muscle activity.
- Pre-Bötzinger complex neurons: the anatomical space where the central pattern generator (CPG) is located, responsible for the generation and transmission of the respiratory rhythm (Muñoz-Ortiz et al., 2018). Currently under investigation, its exact role is not clear yet (Baertsch et al., 2018).
- Bötzinger complex neurons: play a key role in control of ventilation and hypoxia response.

Unlike the pacemaker that initiates the electrical impulse in the heart electrical conduction system, the respiratory rhythm does not have a single activating cell and does not produce sound usually (Ikeda et al., 2017). In particular, the pacemaker responsible for initiating respiration is associated with groups of neurons concentrated in the pre-Bötzinger complex, although the respiratory function is controlled by at least six groups of neurons that are not located at the same point, but extend throughout the medulla. It comprises the following groups of neurons: preinspiratory (pre-I), early inspiratory (early-I), inspiratory augmenting (Iaug), late inspiratory (late-I), postinspiratory (post-I), and expiratory augmenting (Eaug) (Passino et al., 2017).

The resulting respiratory cycle is divided into three phases (Fig. 6):

- Inspiration phase: early-I neurons are suddenly activated at the beginning of this phase. However, their intensity gradually declines and, at the same time, the excitatory activity increases for Iaug neurons, responsible for the stimulation of the inspiratory muscles, including pharyngeal and laryngeal dilators (Ratnovsky et al., 2008).

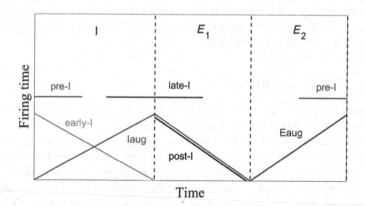

Fig. 6 Phases of the respiratory cycle and activation/deactivation of the different groups of neurons responsible for respiratory rhythm (explanations in the text).

- First phase of expiration: begins with the decrease of the excitation of Iaug neurons and its effect on the inspiratory muscles. Post-I neurons show a decreasing frequency throughout this phase, while during the rest of the cycle are silenced. Therefore, this phase of expiration is characterized by passive exhalation, a progressive fall in inspiratory muscle activity, and a decrease in the flow of gas into the lungs.
- Second phase of expiration: at the beginning of this phase, all inspiratory muscles are inactive, at which point Eaug neurons increase excitatory activity with the aim of gradually increasing expiratory muscle activity.

The main sensors within the respiratory control system are chemoreceptors. These sensory receptor cells are responsible for converting a chemical signal into an electrical signal. Chemoreceptors are stimulated mainly by P_aO_2, P_aCO_2, and the concentration of hydrogen ions ($[H^+]$), and depending on their location, can be classified into central or peripheral.

Central chemoreceptors are located in the central nervous system, specifically in the ventrolateral medullary surface, which is immersed in the cerebrospinal fluid (CSF) and separated from the blood by the blood-brain barrier (Fig. 7).

When blood P_aCO_2 increases, CO_2 diffuses through the blood-brain barrier and combines with water to form $[H^+]$, that stimulate chemoreceptors, as indicated by the reversible chemical reaction presented in Eq. (5).

$$CO_2 + H_2O \leftrightarrow H_2CO_3 \leftrightarrow H^+ + HCO_3^- \tag{5}$$

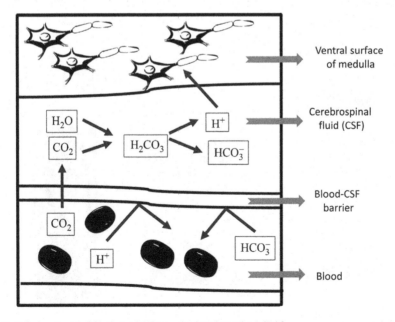

Fig. 7 Transport and diffusion of CO_2 to the cerebrospinal fluid.

The excitation of chemoreceptors generates a rise in alveolar ventilation, increasing approximately from 2 to 3 L/min per mmHg of P_aCO_2. In the long term, P_aCO_2 is controlled by reabsorption of bicarbonate ($[HCO_3^-]$) in the renal system, represented by the reaction (5), in inverse direction to the stimulation of chemoreceptors. In this way, there is a normalization of the blood $[H^+]$ and CSF within hours.

On the other hand, peripheral chemoreceptors, located at the bifurcation of the carotid arteries and the aortic arch, are not only sensitive to the concentration of $[H^+]$, but also the P_aO_2 (Patel and Majmundar, 2018). In general, peripheral chemoreceptors are sensitive to arterial hypoxia when P_aO_2 falls below 60 mmHg.

A simplified diagram of the peripheral and central chemoreceptors for controlling P_aO_2, P_aO_2 pressures and $[H^+]$ concentration is shown in Fig. 8, demonstrating the fundamental role of feedback for respiratory control at the physiological level.

Due to the low values of $[H^+]$ concentration in blood (in normal conditions between 4.46×10^{-8} and 3.54×10^{-8} moles/L), the concept of pH is used to quantify acid-base balance and is defined as the logarithm of the inverse of $[H^+]$ in a fluid (6).

$$pH = \log \frac{1}{[H^+]} \tag{6}$$

Under normal conditions, the range for pH is between 7.35 and 7.45. Its value is calculated using Eq. (5), considering equilibrium conditions and obtaining the $[H^+]$ concentration of the reaction. Thus, pH measurement is determined by Eq. (7), where the numerator refers to the concentration of $[HCO_3^-]$, while the denominator is associated

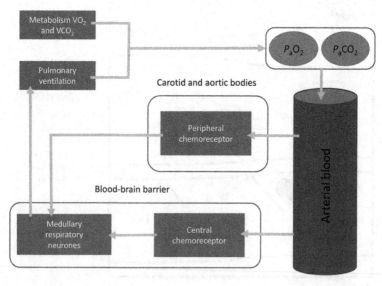

Fig. 8 Diagram of the peripheral and central chemoreceptors, as well as the metabolic aspects of O_2 consumption and CO_2 production.

with the concentration of CO_2, expressed as a function of its diffusion coefficient and the PCO_2.

$$pH = -\log \left(\frac{HCO_3^-}{0.03 \cdot PCO_2} \right) \tag{7}$$

In this way, when [H^+] concentration rises, pH decreases, and on the other hand, when [H^+] concentration falls, pH increases. pH levels determine blood acid-base status and thus there are two pH measurements: one referring to arterial blood (pH_a) and one associated to venous blood (pH_v), due to the different concentrations of CO_2.

3 Computational modeling as a tool for diagnosis and therapy

The diversity of behaviors of a biological system can be studied in two ways: by laboratory experimentation or by modeling and simulation. Modeling can be understood as the construction process of a model, this meaning a conception or simplified object that represents a certain real system. Digital computer simulation provides a common basis of techniques for the study and investigation of different situations and a large number of problems incorrectly defined, but resolvable. The heart of the simulation model is a precise description (mathematical model) of the system to be studied. Our interest focuses on mathematical models, which are models that express their relationships through mathematical equations. Mathematical models of continuous-time dynamic systems, where analysis is conducted by means of a digital computer, are called computational models (Thagard, 2018).

Computational modeling techniques of biological systems allow a methodology for the design, development, analysis, and evolution of a specific system. This allows reproducing the dynamic behaviors of many different variables that are difficult to measure in the clinical practice, but can be important for the diagnosis, evolution, and treatment of diseases. The use of computational modeling as an alternative and/or complement to animal experimentation has been highlighted by the Virtual Physiological Human (VPH) institute (Nendza et al., 2018), with an approach aimed at the so called 3R principles (replacement, refinement, and reduction of animal testing).

Due to the biological complexity, during an experimental study, it is often not possible to carry out direct measurements (in vivo) of physiological variables that may be of interest. Among the reasons are the inherent limitations imposed on the measurement process in biology and medicine, including technical and ethical difficulties.

In particular, the use of computational models helps to understand the system described, test hypotheses, teach, design experiments, predict behaviors, and integrate information with other heterogeneous data in order to generate clinical knowledge. In scientific terms, models can be used to describe, interpret, or predict phenomena that occur in nature (Haefner, 2005), being the predictive process the most interesting from a research perspective.

On the other hand, the development of computational models requires intuition, imagination, skill, knowledge, and common sense at the same time. In addition, for the construction of computational models, it is necessary to evaluate a set of considerations, which are presented as:

- Nature of system: it is related to which aspects of the real system will be considered in the model, as well as to the evolution of the system state over time. The characteristics considered will depend mainly on the objectives that the model is intended to achieve.
- Nature of behavior: it defines the qualitative or quantitative nature of the system variables. If the description is quantitative, it will be necessary to check whether the variables will be related to deterministic or stochastic behavior.
- Nature of model: it defines the methodology used for the construction of the model from the available knowledge of the system. In this sense, a distinction can be made between functional and structural models. The former try to find a mathematical expression between an input and an output. On the other hand, structural models are more based on the constituent elements of the system to be modelled. Thus, the simplified diagrams of functional and structural modeling methodology applied to biological systems are presented in Figs. 9 and 10. In particular, Fig. 10 shows how the underlying physiological mechanisms are explicitly represented at an appropriate level of approximation and resolution considering a priori knowledge available, as well as the assumptions about the behavior of the system.
- Quantification: it describes the relationships established in the model by means of mathematical equations. Quantification is normally based on the consideration of the system dynamics and transfer, storage, or loss of energy or mass.
- Implementation: it includes the codification of the mathematical equations to be implemented in a digital computer using a high-level programming language.
- Validation: the problem of this stage is closely linked to the elements that have been considered in the construction of the model (Gross and MacLeod, 2017). The process can vary from the search for relationships between input/output data or the establishment of a predictive model that incorporates relevant physiological structures that are close to the system to be reproduced. A computational model is considered valid if it describes all aspects of the

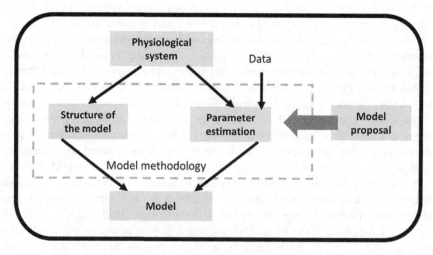

Fig. 9 Methodological diagram of functional modeling.

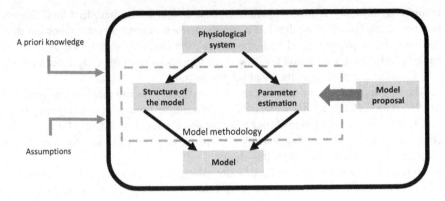

Fig. 10 Methodological diagram of structural modeling.

model structure considered in the initial hypotheses, together with a correct prediction of the relevant system behaviors for which experimental data are available. However, in many cases, it is not possible to have all the necessary data and theories about the different mechanisms involved and the computational model cannot be validated.

- Experimentation or usage of the model: this phase is aimed at reproducing different physiological or pathological situations that help to generate clinical knowledge.

4 Computational models for different control mechanisms of the respiratory system

The respiratory system is made up of a large number of elements that act in a coordinated manner to achieve a purpose: exchange O_2 and CO_2 between the external environment and the cell level. For this purpose, the respiratory system integrates the necessary mechanics that allow the introduction of an air flow into the lungs by means of a pressure gradient generated by the inspiratory/expiratory muscles. However, this is not enough, as the respiratory system requires other mechanisms that include the transport of gases in the blood with the heart acting as a pump, as well as passive phenomena of diffusion or perfusion, chemical processes such as the detection of acid-base disorders and, finally, the generation of electrical stimuli from the nervous system to excite the respiratory musculature and generate the respiratory rhythm. These physical and chemical processes, which are strongly coupled and are of different nature, made the respiratory system become the first system that awoke the interest of scientists in the beginnings of the development of computational models.

At the beginning of the 20th century, the idea that pulmonary ventilation was regulated as a closed loop to try to keep the P_aCO_2 within narrow limits was already considered (Haldane and Priestley, 1905). However, thanks to the advances in control theory in the 1940s and 1950s, it was decades later when the first mathematical models of respiratory control appeared following a description in stationary state.

The first models of the respiratory system made assumptions and simplifications of a large number of physiological processes, with the aim of avoiding interactions that complicate the analysis and computational cost. In this sense, these first models were functional and therefore neither had a clearly defined internal structure, nor did they rest on a strong physiological basis. In general, the feedback signal used by the controller to adjust the ventilation level was typically taken from the CO_2 concentration, which varied depending on the inspired gas mixture.

The first approach to the quantification of ventilatory control mechanisms emerged in 1946 (Gray, 1946). In this work, a theory of multiple factors influencing the chemical regulation of ventilation through the additive effects of P_aCO_2, P_aO_2 and pH (Fig. 11) was presented. The narrow control range of P_aCO_2 and the ventilation response was already evident in this study.

Gray modeled these relationships by describing the pulmonary gas exchange system and blood buffering system, considering three inputs: P_aCO_2, P_aO_2, and $[H^+]$. The relationships between these inputs and ventilation were established by four equations derived from mass-balance under stationary conditions. One of these equations is presented in Eq. (8), in particular that which relates PACO₂ with PAO₂,

$$PAO_2 = \frac{(B - 47 - PACO_2)(RQ \cdot FO_2 + FCO_2) - PACO_2(1 - FO_2[1 - RQ])}{RQ + FCO_2(1 - RQ)}$$

(8)

where B is the barometric pressure, RQ is the respiratory exchange ratio, and FO_2 and FCO_2 are the volumetric fractions of O_2 and CO_2 in the inspired air. Gray equated the alveolar partial pressures (PA) to arterial pressures (P_a) to simplify the model. Finally, alveolar ventilation was calculated by means of the sum of the partial effects of the inputs considered (P_aCO_2, P_aO_2, and $[H^+]$), as presented in Eq. (9). The coefficients were empirically adjusted to reproduce ventilation changes from clinical data.

$$\dot{V}_a = 0.262 \cdot P_aCO_2 + \frac{105}{10^{0.038 \cdot P_aO_2}} + 0.22 \cdot [H^+] - 18.0$$

(9)

Gray's model was improved by Grodins (1950), by adding a new factor to include the response to physical exercise. However, computational limitations restricted the development to simplified static models, preventing the increase of the range of applications.

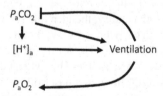

Fig. 11 Functional diagram of Gray's model.

These approaches of Gray and Grodins were in contrast to the observations of Nielsen and Smith (1952), who argued that the interaction of P_aO_2 and P_aCO_2 with ventilation was not additive, but multiplicative.

The first dynamic model was presented by Grodins et al. (1954). This model included some dynamic properties of the lungs and tissues in the exchange of gases and reproduced the ventilatory response to different concentrations of CO_2 in the inspired air. To achieve this purpose, the physiological system was considered as a regulator designed to control the level of CO_2 concentration at tissue level. Fig. 12 shows the simplified model proposed by Grodins, with the lung and tissue reservoirs connected by the circulatory system.

The air flow between the two reservoirs was formulated by following the law of mass conservation. The calculation of the volume of CO_2 in the pulmonary reservoir (V) is presented in Eq. (10),

$$V = K_7 \cdot \dot{V}_a - \dot{q}_1 - \dot{q}_2 + \dot{q}_3 \tag{10}$$

where K_7 is the fraction of CO_2 in the inspired air, $\dot{V}_a$ is the ventilation, $\dot{q}_1$ is the flow of CO_2 in the exhaled gas, $\dot{q}_2$ is the flow of CO_2 in the arterial blood, and $\dot{q}_3$ is the flow of CO_2 in the venous blood.

Defares et al. (1960) extended Grodins model to include two distinct compartments (brain and tissues), considering also the blood flow as a function of P_aCO_2. However, the effects of O_2 as a controller of ventilation and the delays in gas transport from lungs to peripheral tissues were not considered.

With the advances in computing and control theory, more dynamic aspects were included in the computational models to extend the range of application and utility, enabling the simulation of more physiological conditions. In addition to these works, Gray's model was improved by Horgan and Lance (1963) to introduce transport delays and allowing to reproduce the Cheyne-Stokes respiration. The following works by Milhorn and Guyton (1965) and Longobardo et al. (1966) also focused on the study of Cheyne–Stokes respiration, although Milhorn only incorporated CO_2 into the equations while Longobardo considered O_2 and CO_2. Both studies agreed that

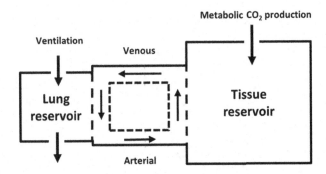

Fig. 12 Simplified model of the respiratory system proposed by Grodins.

Cheyne-Stokes respiration was due to the high sensitivity to CO_2, as well as the delay in circulation (Laura-María, 1998).

During those years, Lloyd and Cunningham (1963) also presented a mathematical model that described the multiplicative interaction of ventilatory response to changes in P_aCO_2 and P_aO_2. Two years later, one of the most detailed models of the respiratory system to date was presented (Milhorn et al., 1965). In this work, alveolar ventilation was obtained from P_aCO_2, P_aO_2 and concentration of hydrogen ions ([H^+]), as presented in Eqs. (11)–(13).

$$\left(\dot{V}_a\right)_{CO_2} = a_1 \cdot P_aCO_2 - b_1 \tag{11}$$

$$\left(\dot{V}_a\right)_{O_2} = d_1(m_1 - P_aO_2)^n \geq 0 \tag{12}$$

$$\left(\dot{V}_a\right)_{[H^+]} = c_1 \cdot [H^+] \tag{13}$$

where a_1, b_1, c_1, d_1, m_1, and n are constants obtained empirically, while the terms on the left-hand side are the partial alveolar ventilation that were used to calculate the total alveolar ventilation (14).

$$\dot{V}_a = \left(\dot{V}_a\right)_{P_aCO_2} + \left(\dot{V}_a\right)_{P_aO_2} + \left(\dot{V}_a\right)_{[H^+]} \tag{14}$$

Fig. 13 shows the results of calculation proposed by Milhorn to obtain the alveolar ventilation as independent functions of P_aCO_2, P_aO_2, and [H^+].

In the next year, Lloyd (1966) separated the effects of PCO_2 considering the central pressure of CO_2 (P_cCO_2) and P_aCO_2. Thus, the effect on ventilation is considered to be the sum of the central chemoreflective effects (V_c) dependent on P_cCO_2, peripheral chemoreflexes (V_p) dependent on P_aCO_2, and a basal term (V_b), as presented in Eq. (15).

$$\dot{V}_a = V_c + V_p + V_b \tag{15}$$

Thus, V_c is proportional to P_cCO_2 for a given threshold (T_c), as shown in Eq. (16),

$$V_c = \{S_c(P_cCO_2 - T_c) \rightarrow P_cCO_2 > T_c, 0 \rightarrow P_cCO_2 < T_c\} \tag{16}$$

where S_c is a constant that represents the sensitivity to P_cCO_2. In addition, V_p is proportional to P_aCO_2 for a certain threshold (T_p) and dependent on P_aO_2, as it is expressed in Eqs. (17), (18),

$$V_p = \begin{cases} S_p(P_aCO_2 - T_p) \rightarrow P_aCO_2 > T_p \\ 0 \rightarrow P_aCO_2 < T_p \end{cases} \tag{17}$$

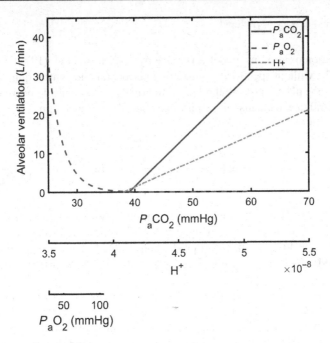

Fig. 13 Effect on alveolar ventilation as independent functions of P_aCO_2, P_aO_2, and $[H^+]$. (Figure adapted from Milhorn et al., 1965.)

$$S_p = S_{p_{min}} + \frac{A}{P_aO_2 - P_aO_{2_{min}}} \tag{18}$$

where S_p is a hyperbolic rectangular function that represents the sensitivity to P_aO_2 (18), S_{pmin} is a minimum reference threshold, A is a constant, and P_aO_{2min} is a reference value for P_aO_2.

Later, Grodins et al. (1967) proposed a detailed model, which brought together much of the knowledge accumulated to date with respect to previous models, including modeling of gas transport, peripheral and cerebral blood flow, concentration of respiratory gases in brain and CSF, chemoreceptors, cardiac output, and transport delays. Unlike the aforementioned work of Grodins, computational limitations could be overcome with the use of a digital computer.

Shortly after, Olszowka presented a model (Olszowka and Farhi, 1968) that described additional mechanisms involved in the transport of blood gases, integrating some recent advances obtained in respiratory physiology using the Doppler technique with respect to cerebral blood flow (Poulin and Robbins, 1996). In this sense, the concentration of bicarbonate ions ($[HCO_3^-]$) was calculated using pH and following the Henderson-Hasselbalck equation (19).

$$\left[HCO_3^-\right] = [CO_2] \cdot 10^{pH-6.1} \tag{19}$$

A modification of Grodins model was made by Loeschcke et al. (1970) and Mitchell et al. (1963), without specifically separating peripheral and central quimioreceptors. In this case, the pH of extracellular brain fluid (pH_{LCR}) was considered as an activator of the central mechanisms of respiration control. Its calculation is expressed in Eq. (20),

$$pH_{LCR} = pH_{LCR}(0) - \log\left(\frac{\dot{V}_a}{1+\dot{V}_a}\left(\frac{FICO_2}{VCO_2}\right)\right) - \log\left(\dot{V}_a(0) + \frac{FICO_2}{VCO_2}\right) \tag{20}$$

where $\dot{V}_a(0)$ and $pH_{LCR}(0)$ are normal values for the alveolar ventilation and pH in the extracellular brain fluid, respectively, $FICO_2$ is the fraction of carbonic gas in the inspired air, and VCO_2 is the CO_2 production. Fig. 14 shows the relation between $PACO_2$ and alveolar ventilation, considering a VCO_2 constant under normal conditions.

Years later, Milhorn published a state-of-the-art of mathematical models (Milhorn, 1976) of the respiratory control system to date. However, although all the models reviewed so far reproduced a large number of pathophysiological conditions, the respiratory cycle was ignored in most cases and the lungs were considered as a ventilated rigid container. This trend began to change in the following years with the increase of knowledge about the neural factors associated with the generation of the respiratory rhythm (Bradley et al., 1975; von Euler, 1986). As a consequence, the first models trying to describe neural interactions with the chemical components of the respiratory system began to be introduced. The objective was to include the respiratory neural activation from signals collected

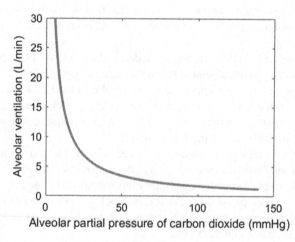

Fig. 14 Relation between $PACO_2$ and alveolar ventilation.

by peripheral and central chemoreceptors, and to generate the periodic output to the respiratory musculature, in charge of producing the variations in frequency and tidal volume.

In 1980, Saunders et al. (1980) presented a modification of Grodins model to incorporate the effects of the cyclic respiratory activity under normal conditions and with variations in tidal volume and dead space. Lung volume was modeled as a sine wave where frequency and amplitude were modulated by the control action. The mathematical expression for the lung volume (V_L) in the model proposed by Saunders is presented in Eq. (21),

$$V_L = FRC + \frac{V_t}{2}[1 - \cos(2 \cdot \pi \cdot f \cdot t + \theta)] \tag{21}$$

where FRC is the residual functional capacity, V_t is the tidal volume, f is the respiratory rate, t is the time, and θ denotes an offset of the sinusoidal function. In addition, the pulmonary flow ($\dot{V}_L$) was also considered in the Saunders model, as expressed in Eq. (22),

$$\dot{V}_L = [\pi \cdot f \cdot V_t \cdot \sin(2 \cdot \pi \cdot f \cdot t + \vartheta)] + (\dot{T}_{O_2} - \dot{T}_{CO_2}) \tag{22}$$

$$\dot{T}_{O_2} = \dot{Q}(C_aO_2 - C_vO_2)\left[\frac{863}{PB - 47}\right] \tag{23}$$

$$\dot{T}_{CO_2} = \dot{Q}(C_vCO_2 - C_aCO_2)\left[\frac{863}{PB - 47}\right] \tag{24}$$

where $\dot{T}_{O_2}$ and $\dot{T}_{CO_2}$ are the transfer rates of O_2 and CO_2 (Eqs. 23, 24) respectively, $\dot{Q}$ is the cardiac output, C_aO_2 is the O_2 concentration in the arterial blood, C_aCO_2 is the CO_2 in the arterial blood, C_vO_2 is the O_2 concentration in the venous blood, C_vCO_2 is the CO_2 concentration in venous blood, PB is the atmospheric pressure, and 47 mmHg is the partial pressure of water.

Subsequently, Fincham proposed a similar model in 1983 (Fincham and Tehrani, 1983), where a controller was described with respiratory pattern parameters that were calculated in each respiratory cycle, as it happens physiologically. Volume changes at the pulmonary level were regarded as a sine wave to simulate both inspiration and expiration intervals, as it is expressed in Eq. (25),

$$\frac{dv}{dt} = \dot{V}_a \sin(2 \cdot \pi \cdot f \cdot t) \tag{25}$$

where $\dot{V}_a$ is the alveolar ventilation, f is the respiratory rate, and t is the time. Also, $\dot{V}_a$ and f were optimized in each breath to minimize the work of breathing. Later, Guyton (1986) incorporated the Hering-Breuer reflex into this model, allowing the controller to modify the amplitude and respiratory rate at the end of the respiratory cycle. Finally, this model was improved by Tehrani (1993) to include changes in some parameters to adjust the ventilatory response to newborn infants.

In parallel, Patil et al. (1989) considered the structure of Saunders model, assuming the dead space as a constant, and proposed a double compartment for the central and peripheral components. In this model and others similar, the difficulty of incorporating tidal breathing into models of the chemical regulation of breathing was evident. In fact, until then, the action of the muscles in respiratory mechanics was still ignored.

Despite these limitations, these early models could regulate the ventilatory behavior in response to stimuli from chemoreceptors and provided an adequate explanation for phenomena such as inhalation of CO_2 or hypoxia. Other pathophysiological events such as exercise hypernea were more limited due to the still insufficient knowledge of the mechanisms associated with respiratory control (Grodins, 1981; Herczynski, 1988). As knowledge progressed, different approaches to the study of specific physiopathologies by means of computational modelling were proposed, including panic attacks (Clark et al., 1985), sleep-disordered breathing in children (Nugent and Finley, 1987), peripheral chemoreflex response to CO_2 (Khoo and Marmarelis, 1989), homeostasis of CO_2 (Rapoport et al., 1993), ventilatory response to isocapnic hypoxia (Painter et al., 1993), or hypoxic ventilatory depression (Takahashi and Doi, 1993). However, most of these models had limitations regarding the inadequate description of the neural control mechanisms and the impact of metabolic factors. As a result, these issues began to be addressed.

The first mathematical models related to neural center mechanisms associated with CPG emerged in the early 1990s (Botros and Bruce, 1990; Gottschalk et al., 1992). In particular, Botros/Bruce already considered a model of the respiratory cycle made up of three phases: inspiration, postinspiration, and expiration, including the five respiratory neural groups (inspiratory, late inspiratory, postinspiratory, expiratory, and early inspiratory neurons). The neural activity patterns (X_i) in the Botros/Bruce model were expressed by Eq. (26),

$$\frac{dX_i}{dt} = -\left(\frac{1}{\tau_i}\right)X_i - \sum W_{ji} \cdot S\left(X_j\right) - W_{ii} \cdot S(X_i) + B_i \tag{26}$$

where τ_i is the time constant, W_{ji} represents a term a self-limitation for each of the neurons, $S(X_j)$ is a sigmoid function that relates the neuron activation frequency to its activity, and B_i is the excitatory tonic input due to chemical and neural effects. Another model related with CPG was presented later (Matsugu et al., 1998), using a neural network oscillator driven by constant and periodic inputs of variable amplitudes, frequencies, and phases. The model was a simplification of Botros/Bruce model with only two neurons (one inspiratory and one expiratory) that were mutually inhibitory, as well as a "virtual interneuron" to provide stability. However, the model incorporated excitatory phasic inputs to complement the excitatory tonic inputs. Fig. 15 depicts the interactions between the inspiratory neuron (I), expiratory neuron (E), and "virtual interneuron" (V), considering the different tonic and phase excitatory inputs.

Tonic and phasic excitatory inputs

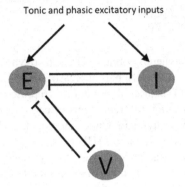

Fig. 15 Interaction diagram between the inspiratory neuron (I), expiratory neuron (E), and "virtual interneuronal" (V) in the RPG model proposed by Matsugu et al. (1998).

Shortly thereafter, a related work divided into three parts was presented (Butera et al., 1999a, b; Del Negro et al., 2001) where a model of the respiratory rhythm generation in the pre-Bötzinger complex of mammals was proposed considering a heterogeneous population of pacemaker neurons coupled by synaptic excitation. In addition, this same work also reported an exhaustive state-of-the-art of models of this nature to date.

In 2002 a computational model describing the mechanisms of the integrated responses of both the respiratory and cardiovascular systems during sleep was presented (Fan and Khoo, 2002). This work simulated disorders like obstructive sleep apnea and its relationship with heart rate, changes in pleural pressure or O_2 saturation, among other factors. Subsequently, this model was improved (Ivanova and K Khoo, 2004) to integrate a more detailed description of the motion of the heart and the circulatory and pulmonary components.

In the same year, one of the classic models related to ventilatory response to hypoxia and hypercapnia was presented (Topor et al., 2004). This model extended Grodins model to include a more precise description of chemoreceptors, considered as a sum of central and peripheral components. The model was validated with simulations performed for different pathophysiological conditions, specifically the response to hypoxia and the development of periodic breathing during sleep. In the following year, Longobardo et al. (2005) proposed an adaptation of the Botros/Bruce and Matsugu models associated with the RPG to be integrated into neurochemical control model of ventilation.

From another perspective, other models and works have focused on the analysis of the changes in respiratory response for O_2 decreased, or inhalation of gases such as carbon monoxide (CO), carbon dioxide, or other highly toxic gases. For example, Stuhmiller model (Stuhmiller and Stuhmiller, 2005) analyzes the effects of CO inhalation on ventilatory response. In this model, the union of CO and hemoglobin (carboxyhemoglobin) is calculated from the relationship between P_aCO and P_aO_2 (Haldane's second assumption), as expressed in Eq. (27),

$$\text{COHb} = O_2\text{Hb} \cdot M_{\text{CO}} \frac{P_a\text{CO}}{P_a O_2} \tag{27}$$

where COHb is the carboxyhemoglobin, O_2Hb is the oxyhemoglobin, M_{CO} is the Haldane coefficient for CO, P_aCO is the partial pressure of CO in arterial blood, and $P_a O_2$ is the partial pressure of O_2 in arterial blood. Fig. 16 presents the oxygen hemoglobin dissociation curve considering different levels of carboxyhemoglobin.

Other models related with hypoxia (Qutub and Popel, 2006) include an understanding of how hypoxia transcription factors (HIFs) work to detect oxygen and respond to hypoxia. A similar model was presented in the following year (Wolf and Garner, 2007) for the analysis of changes in ventilation, blood gas levels, and other critical variables during conditions of hypocapnia, hypercapnia, and the combination of these states with hypoxia, focusing specifically on ventilation changes due to high altitude.

In parallel to these works and at the beginning of the century, some models of chemoreflexive ventilatory control were also presented, both in steady state (Duffin, 2005; Mohan et al., 1998; Stephenson, 2004) and dynamic (Topor et al., 2007; Ursino and Magosso, 2004).

Later, Ben-Tal and Smith (2008) presented a model that integrated researches on RPG from previous years, both at the physiological and modeling level, and proposed a reduced description of mechanisms of neural control associated with gas exchange and transport. This model was based on previous work by the same authors (Ben-Tal, 2006) and described the pre-Bötzinger oscillator with an activity level and a persistent sodium channel activation, both formulated according to different variables that were defined by feedback functions, as expressed in Eqs. (28), (29),

$$\frac{dA}{dt} = \alpha(1 - A) - \beta \cdot A + \gamma \tag{28}$$

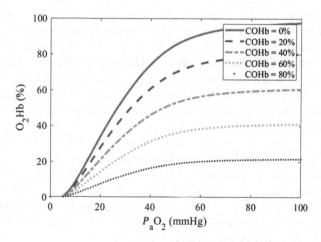

Fig. 16 Relation between $P_a O_2$ and O_2Hb for different levels of COHb.

$$\frac{dh_p}{dt} = \alpha_{hp}\left(1 - h_p\right) - \beta_{hp} \cdot h_p \tag{29}$$

where $\alpha(1 - A)$ represents the excitatory component of A (force of excitation), $\beta \cdot A$ is the inhibitory component of A, and γ is an external force that may be excitatory or inhibitory. Likewise, $\alpha_{hp}(1 - h_p)$ and $\beta_{hp} \cdot h_p$ factors represent the excitatory and inhibitory impulses for the persistent sodium channel (h_p).

Ben-Tal model comprised not only the effects of diaphragm, but also of other abdominal muscles, which were modeled as springs. However, the model presented a simplified description of neural control, not including feedback mechanisms from the lungs. Consequently, this model was extended years later (Molkov et al., 2014) to incorporate different mechanical and chemical feedbacks for the simulation and analysis of breathing during hypercapnia. To the best of authors' knowledge, it was the first closed-loop model of neural control of respiration in mammals related on active exhalation under hypercapnia conditions, including an exhaustive description of neural control, as well as respiratory mechanics, gas exchange, and transport. The main contribution of this work was to describe the response properties of the respiratory system to two types of time delays: the one associated with circulatory transport and the other focused on neural feedback. In this way, the model reproduced multiple experiments associated with central mechanoreceptors and chemoreceptors related with neural control.

Currently, neural control continues to raise much interest among researchers. In this sense, different computational models have been presented in recent years (Bacak et al., 2016; Fietkiewicz et al., 2016; Jasinski et al., 2013; Lü et al., 2017) with the aim of analyzing the behaviors derived from the rhythmic rupture in the respiratory excitation and the generation of respiratory rhythm, particularly in the pre-Bötzinger complex. To achieve this purpose, several works try to model the possible roles of persistent sodium and calcium currents involved in rhythmic respiratory under different initial conditions. The models are generally implemented using a hierarchical neuron network, in the style of the Hodgkin-Huxley model (Ausborn et al., 2018; Gaiteri and Rubin, 2011; Jasinski et al., 2013; Rybak et al., 2004). For example, Jasinski's model (Jasinski et al., 2013), based on other works (Molkov et al., 2010; Smith et al., 2007) uses the definition of neural membrane potential (V) as a set of membrane ionic currents (I), as expressed in Eq. (30),

$$C\frac{dV}{dt} = -I_{\mathrm{Na}} - I_{\mathrm{NaP}} - I_{\mathrm{K}} - I_{\mathrm{Ca}} - I_{\mathrm{CAN}} - I_{\mathrm{Pump}} - I_{\mathrm{L}} - I_{\mathrm{SynE}} \tag{30}$$

where I_{Na}, I_{NaP}, I_{K}, I_{Ca}, I_{CAN}, I_{Pump}, I_{L}, and I_{SynE} are the currents of sodium, persistent sodium, potassium, calcium, cationic, sodium/potassium pump, exhaust and synaptic excitatory, respectively. The objective of these models is to analyze the interaction between the neurons in the Bötzinger and pre-Bötzinger complexes, as well as their relationship with the retrotrapezoidal nucleus (RTN) and parafacial respiratory region (pFRG). These BötC/pre-BötC complexes are thought to be responsible for generating oscillations to provide a force to the diaphragm during inspiration, while the RTN/pFRG regions contain the central chemoreceptors.

Other area of research that has been explored in recent years in the context of computational modeling of the respiratory system includes the integration of different levels of physiological complexity. Thus, some models (Ben-Tal and Smith, 2008; Molkov et al., 2014, 2017; Ursino et al., 2001) try to integrate different physiological mechanisms but related to the respiratory system, including neural control, respiratory mechanics, transport, and gas exchange. In relation to the interaction between the respiratory system and cardiovascular system (Elstad et al., 2018), several models have also been presented from this perspective (Bai et al., 1997; Ben-Tal and Smith, 2010; Cheng et al., 2010; Liang and Liu, 2006; Lu et al., 2001, 2003, 2004; Magosso and Ursino, 2001; Ursino and Magosso, 2000), with results that include simulations of volumes, flows and pressures of both systems under normal conditions and under different physiopathological conditions (generally hypoxia and sleep disorders). Perhaps the most complete model in this sense was presented in 2016 (Cheng et al., 2016), with more than 250 parameters and variables describing the cardiorespiratory system through a set of modules characterized by multiple inputs and outputs that are modeled through differential equations.

Currently the topic of computational modeling in neural control is providing added value to pharmacological manipulation and animal experimentation to try to understand the importance of the pre-Bötzinger complex in neural control of the respiratory rhythm and its possible involvement in neurological diseases (Muñoz-Ortiz et al., 2018). A related state-of-the-art approach was recently presented in the literature (Molkov et al., 2017).

With respect to its clinical applicability in recent years, modeling of respiratory control mechanisms has been approached from different points of view: models focused on the response of hypercapnia, hypocapnia, and hypoxia; models that try to explain hyperventilation during physical exercise; and models that study the occurrence of periodic breathing, sleep apnea, and breathing stability. In general, most models describe the neurological and biomechanical mechanisms responsible for maintaining stable ventilation and provide oxygen to tissues and brain, which is the main objective of the respiratory system. In this sense, a review of neural network models and their integration in the cardiorespiratory response to hypoxia has recently been presented (Lindsey et al., 2018).

On the other hand, models that analyze the ventilatory response to different levels of exercise intensity also have interest in literature (Serna et al., 2018). The objective of these models is to reproduce the complex feedback system responsible for providing enough O_2 for metabolism and the removal of CO_2 produced, maintaining homeostasis of arterial blood. To achieve this purpose, the control of ventilation increases respiration in order to maintain the balance of gas with metabolism, so that the transfer of O_2 and CO_2 in lungs is equal to O_2 consumption and CO_2 production, respectively, in tissues and brain. One of the most researched aspects is the study of the mechanisms that adapt the respiratory control system in physical exercise for controlling ventilation and by maintaining arterial and cerebral pressures unchanged. One accepted theory is the so-called neurohumoral theory (Turner, 1991), where the dynamic response from the resting state to moderate exercise is characterized by three phases: the first one, determined by a sudden increase generally attributed to neurogenic mechanisms; the second one, by a gradual and exponential increase of ventilation; and finally, the third one, characterized by a stabilization to a singular steady state. However, there is still controversy about the feedback mechanisms involved (Serna et al., 2018).

5 Other research lines in computer modeling

As it has been demonstrated previously, the heterogeneity of strongly coupled components is an indisputable characteristic of biomedical systems in general, and of biological systems in particular. Therefore, computational modeling is presented as a powerful tool to integrate all this knowledge and offer a broader context that allows a global analysis of the various mechanisms that govern a system. In order to achieve this objective in the context of computational modeling, complexity is abstracted by modularity, including the separation of functionalities and the definition of independent but closely connected components. In this way, there is a large number of methodological strategies to address the modeling of different biological components (Bardini et al., 2017).

However, methodological approaches can vary substantially when the biological system which is the object of study is to be modeled from a multiscale perspective (Zohdi, 2017). In this sense, it is necessary to consider that the underlying structures of the system analyzed are not only defined by their functionality and nature, but can also be characterized by different scales, both temporal and spatial. This further highlights the diversity of behaviors and how the interactions of each of the constituent elements of a biological system become even more complex (Bardini et al., 2017), with an interaction probability that varies according to the proximity between components and stochastic events (Wilkinson, 2011).

The emergence of multiscale modeling is a direct consequence of the increasing availability of biological data, which is evolving the design patterns of computational models toward a multiscale scenario, and also, of multiple levels (Dallon, 2010). However, although multilevel and multiscale approaches have been used interchangeably in the literature, these terms do not represent the same concept (Uhrmacher et al., 2005). There is no rigorous notion in this sense, but in general, multilevel modeling can consider dynamic processes at multiple levels (at the level of tissue, cell, organ, etc.), whereas multiscale modeling incorporates multiple and different temporal and spatial scales into a single model, regardless the model has multiple levels or not. Therefore, a multilevel model is not necessarily a multiscale model and vice versa. However, usually multiple levels are in accordance with multiple spatial and temporal scales and, in addition to all these, a model can also be multidimensional (Liu et al., 2014).

On the other hand, the different spatial scales of biological systems have been categorized in the literature considering various methodological approaches (Gosak et al., 2018). In general, there seems to be consensus in establishing a certain organization of biological levels that range from genes, to proteins, individual cells, tissues, organs, organ systems, and finally the organism, including the interaction with the environment. In the context of this spatial organization (Dada and Mendes, 2011), it is necessary to consider the different time scales of biological processes, which are also strongly coupled. Dynamic behaviors over time can vary substantially from, for example, tenths of a second in genetic regulation, to phenomena such as the progressive shortening of telomeres, a process that extends throughout the life of an organism.

The use and application of computational models, validated by experimental data, can help to understand these spatial/temporal difficulties, as well as the relationships that are established to interact with external environment. A first approach suggested to address this problem and enable a transfer of knowledge between models at different scales is to use a serial or parallel methodology (Ayton et al., 2007). In addition to this communication between models, other aspects that should also be considered in multiscale modeling include redundancy of components (e.g., cells constituting a tissue), variation and evolution of components (e.g., mutation), hierarchical organization (e.g., cells sharing common functions), mobility (e.g., cell transport), replication (e.g., cell division), and destruction (e.g., cell death) (Liu and Heiner, 2013).

However, the quantitative success of multiscale modeling is currently still limited (Dada and Mendes, 2011) and its impact with respect to clinical applicability is insufficient. Despite this, the progress in computing technologies and development of new techniques and methodologies suggest a promising future for computational modeling of complex biological systems at the multiscale level. In this sense, it seems reasonable to consider that this tool will become fundamental in clinical practice and will revolutionize various domains of knowledge currently contextualized at a biological level (for example, synthetic biology in the study of microorganisms). The scope of different disciplines will need to be expanded to include other spatial levels of organization characterized by coupled hierarchical biological structures. These new paradigms will require a multiscale computational modeling approach based on a consistent management and cohesion. Finally, and due to underlying complexity, these new challenges will have to be overcome from a multidisciplinary perspective that brings together the efforts of professionals with very diverse skills, including physics, mathematics, biology, medicine, and engineering (Walpole et al., 2013).

6 Conclusion

The respiratory system comprises two control systems: the control system of respiratory mechanics and the control system of ventilation. Since the beginning of the 20th century, both have aroused much interest in the scientific community. However, at present, there are still different aspects associated with the control of respiration and more specifically the generation of respiratory rhythm, which remain controversial.

During the evolution of the knowledge about the mechanisms involved in respiratory control, computational modeling has facilitated the understanding of behaviors that have already been observed from animal experimentation. Although the early computational models made important simplifications and assumptions, with the increase in knowledge about respiratory system and performance of computers, more complex computational models emerged to include aspects not only related to respiratory control, but also to gas exchange, transport, or respiratory mechanics. In short, respiratory modeling has been approached from different perspectives throughout history depending on the objective to be achieved.

This chapter presents physiological aspects of the respiratory system with a greater consensus in the literature and that have already been incorporated into computational modeling works. Likewise, the importance of computational modeling has been highlighted in order to integrate knowledge derived from different heterogeneous sources and also as a tool to analyze certain mechanisms of physiological systems. In this way, simulation and reproduction of physiological and pathophysiological situations can help to better understand the diagnosis of diseases and their possible treatment. From the perspective of computational modeling and in the context of control mechanisms involved in the respiratory system, this work has presented a historical evolution, from a chronological point of view, of different computational models that, in the opinion of the authors, have had high scientific impact considering the advances in knowledge of respiratory physiology. For a better understanding, a brief introduction to basic physiology of the respiratory control mechanisms has been previously exposed, including the different constituent components of the respiratory system.

Current research lines involve the study of the generation of respiratory rhythm at neural level and its influence on the respiratory system considering different physiopathological states. Under a computational modeling perspective, the study of the respiratory control system has a promising future. The great diversity of behaviors associated with respiratory physiology and control of ventilation will require an integration of knowledge at different scales in order to incorporate new discoveries. In this sense, computational models will become an indispensable tool to integrate these new advances and the associated complexity.

Acknowledgments

This work was supported in part by the "Fondo de Investigaciones Sanitarias" (Instituto de Salud Carlos III, Spain) under Grants PI15/00306 and DTS15/00195, in part by the "Fundación Progreso y Salud" (Government of Andalucía, Spain) under Grant PI-0010-2013, PI-0041-2014 and PIN-0394-2017, in part by "Fundación Mutua Madrileña" under grant VÍA-RENAL.

References

Ausborn, J., Koizumi, H., Barnett, W.H., John, T.T., Zhang, R., Molkov, Y.I., Smith, J.C., Rybak, I.A., 2018. Organization of the core respiratory network: insights from optogenetic and modeling studies. PLoS Comput. Biol. 14. e1006148.
Ayton, G.S., Noid, W.G., Voth, G.A., 2007. Multiscale modeling of biomolecular systems: in serial and in parallel. Curr. Opin. Struct. Biol. Theory Ssimul. Macromol. Assemblage 17, 192–198.

Bacak, B.J., Segaran, J., Molkov, Y.I., 2016. Modeling the effects of extracellular potassium on bursting properties in pre-Bötzinger complex neurons. J. Comput. Neurosci. 40, 231–245.

Baertsch, N.A., Baertsch, H.C., Ramirez, J.M., 2018. The interdependence of excitation and inhibition for the control of dynamic breathing rhythms. Nat. Commun. 9, 843.

Bai, J., Lu, H., Zhang, J., Zhou, X., 1997. Simulation study of the interaction between respiration and the cardiovascular system. Methods Inf. Med. 36, 261–263.

Bardini, R., Politano, G., Benso, A., Di Carlo, S., 2017. Multi-level and hybrid modelling approaches for systems biology. Comput. Struct. Biotechnol. J. 15, 396–402.

Lloyd, B.B., Cunningham, D.J.C., 1963. The regulation of human respiration. In: Proceedings of the J. S. Haldane Centenary Symposium, Oxford, p. 331.

Ben-Tal, A., 2006. Simplified models for gas exchange in the human lungs. J. Theor. Biol. 238, 474–495.

Ben-Tal, A., Smith, J.C., 2010. Control of breathing: two types of delays studied in an integrated model of the respiratory system. Respir. Physiol. Neurobiol. 170, 103–112.

Ben-Tal, A., Smith, J.C., 2008. A model for control of breathing in mammals: coupling neural dynamics to peripheral gas exchange and transport. J. Theor. Biol. 251, 480–497.

Botros, S.M., Bruce, E.N., 1990. Neural network implementation of a three-phase model of respiratory rhythm generation. Biol. Cybern. 63, 143–153.

Bradley, G.W., von Euler, C., Marttila, I., Roos, B., 1975. A model of the central and reflex inhibition of inspiration in the cat. Biol. Cybern. 19, 105–116.

Butera, R.J., Rinzel, J., Smith, J.C., 1999a. Models of respiratory rhythm generation in the pre-Bötzinger complex. II. Populations of coupled pacemaker neurons. J. Neurophysiol. 82, 398–415.

Butera, R.J., Rinzel, J., Smith, J.C., 1999b. Models of respiratory rhythm generation in the pre-Bötzinger complex. I. Bursting pacemaker neurons. J. Neurophysiol. 82, 382–397.

Cheng, L., Albanese, A., Ursino, M., Chbat, N.W., 2016. An integrated mathematical model of the human cardiopulmonary system: model validation under hypercapnia and hypoxia. Am. J. Physiol. Heart Circ. Physiol. 310, H922–H937.

Cheng, L., Ivanova, O., Fan, H.-H., Khoo, M.C.K., 2010. An integrative model of respiratory and cardiovascular control in sleep-disordered breathing. Respir. Physiol. Neurobiol. Central Cardiorespir. Regul.: Physiol. Pathol. 174, 4–28.

Clark, D.M., Salkovskis, P.M., Chalkley, A.J., 1985. Respiratory control as a treatment for panic attacks. J. Behav. Ther. Exp. Psychiatry 16, 23–30.

Crossman, P.F., Bushnell, L.S., Hedley-Whyte, J., 1970. Dead space during artificial ventilation: gas compression and mechanical dead space. J. Appl. Physiol. 28, 94–97.

von Euler, C., 1986. Brain stem mechanisms for generation an control of breathing pattern. In: Cherniack, N.S., Widdicombe, J.G. (Eds.), Handbook of Physiology. Section 3. The Respiratory System.

Dallon, J.C., 2010. Multiscale modeling of cellular systems in biology. Curr. Opin. Colloid Interface Sci. 15, 24–31.

Defares, J.G., Derksen, H.E., Duyff, J.W., 1960. Cerebral blood flow in the regulation of respiration. Acta Physiol. Pharmacol. 9, 327–360.

Del Negro, C.A., Johnson, S.M., Butera, R.J., Smith, J.C., 2001. Models of respiratory rhythm generation in the pre-Bötzinger complex. III. Experimental tests of model predictions. J. Neurophysiol. 86, 59–74.

Duffin, J., 2005. Role of acid-base balance in the chemoreflex control of breathing. J. Appl. Physiol. 99, 2255–2265.

Elstad, M., O'Callaghan, E.L., Smith, A.J., Ben-Tal, A., Ramchandra, R., 2018. Cardiorespiratory interactions in humans and animals: rhythms for life. Am. J. Phys. Heart Circ. Phys. 315, H6–H17.

Esquinas, A.M., 2015. Noninvasive Mechanical Ventilation: Theory, Equipment, and Clinical Applications. Springer.

Fan, H.-H., Khoo, M.C.K., 2002. PNEUMA—a comprehensive cardiorespiratory model. In: Proceedings of the Second Joint 24th Annual Conference and the Annual Fall Meeting of the Biomedical Engineering Society. Engineering in Medicine and Biology. vol. 2, pp. 1533–1534.

Fietkiewicz, C., Shafer, G.O., Platt, E.A., Wilson, C.G., 2016. Variability in respiratory rhythm generation: In vitro and in silico models. Commun. Nonlinear Sci. Numer. Simul. 32, 158–168.

Fincham, W.F., Tehrani, F.T., 1983. A mathematical model of the human respiratory system. J. Biomed. Eng. 5, 125–133.

Fitting, J.-W., 2015. From breathing to respiration. RES 89, 82–87.

Gaiteri, C., Rubin, J.E., 2011. The interaction of intrinsic dynamics and network topology in determining network burst synchrony. Front. Comput. Neurosci. 5.

Gosak, M., Markovic, R., Dolensek, J., Slak Rupnik, M., Marhl, M., Stozer, A., Perc, M., 2018. Network science of biological systems at different scales: a review. Phys Life Rev 24, 118–135.

Gottschalk, A., Geitz, K.A., Richter, D.W., Ogilvie, M.D., Pack, A.I., 1992. Nonlinear dynamics of a model of the central respiratory pattern generator. In: Control of Breathing and Its Modeling Perspective. Springer, Boston, MA, pp. 51–55.

Gray, J.S., 1946. The multiple factor theory of the control of respiratory ventilation. Science 103, 739–744.

Grodins, F.S., 1981. Exercise hyperpnea. The ultra secret. In: Hutás, I., Debreczeni, L.A. (Eds.), Respiration, Pergamon Policy Studies on International Development. Pergamon, pp. 243–251.

Grodins, F.S., 1950. Analysis of factors concerned in regulation of breathing in exercise. Physiol. Rev. 30, 220–239.

Grodins, F.S., Buell, J., Bart, A.J., 1967. Mathematical analysis and digital simulation of the respiratory control system. J. Appl. Physiol. 22, 260–276.

Grodins, F.S., Gray, J.S., Schroeder, K.R., Norins, A.L., Jones, R.W., 1954. Respiratory responses to CO_2 inhalation; a theoretical study of a nonlinear biological regulator. J. Appl. Physiol. 7, 283–308.

Gross, F., MacLeod, M., 2017. Prospects and problems for standardizing model validation in systems biology. Prog. Biophys. Mol. Biol., Validat. Comput. Model. 129, 3–12.

Guyton, A.C., 1986. Textbook of Medical Physiology 7th edition by Guyton, Arthur C. (1986) Hardcover, seventh ed. Saunders College Publishing/Harcourt Brace.

Haefner, J.W., 2005. Modeling Biological Systems: Principles and Applications. Springer Science & Business Media.

Haldane, J.S., Priestley, J.G., 1905. The regulation of the lung-ventilation. J. Physiol. 32, 225–266.

Hall, J.E., 2015. Guyton and Hall Textbook of Medical Physiology, 13th ed. Saunders, Philadelphia, PA.

Hemmings, H.C., Egan, T.D., 2013. Pharmacology and Physiology for Anesthesia: Foundations and Clinical Application: Expert Consult—Online and Print. Elsevier Health Sciences.

Herczynski, R., 1988. The Mathematical Study of Respiratory Phenomena. The Respiratory System. Croom Helm, London.

Horgan, J.D., Lance, R.L., 1963. Digital computer simulation of the human respiratory system. Proc. IEEE 51, 534.

Horner, R.L., Malhotra, A., 2016. Control of breathing and upper airways during sleep. In: Broaddus, V.C., Mason, R.J., Ernst, J.D., King, T.E., Lazarus, S.C., Murray, J.F., ... Gotway, M.B. (Eds.), Murray and Nadel's Textbook of Respiratory Medicine. sixth ed. Saunders, Philadelphia. pp. 1511–1526. e1.

Ikeda, K., Kawakami, K., Onimaru, H., Okada, Y., Yokota, S., Koshiya, N., Oku, Y., Iizuka, M., Koizumi, H., 2017. The respiratory control mechanisms in the brainstem and spinal cord: integrative views of the neuroanatomy and neurophysiology. J. Physiol. Sci. 67, 45–62.

Ioannidis, G., Lazaridis, G., Baka, S., Mpoukovinas, I., Karavasilis, V., Lampaki, S., Kioumis, I., Pitsiou, G., Papaiwannou, A., Karavergou, A., Katsikogiannis, N., Sarika, E., Tsakiridis, K., Korantzis, I., Zarogoulidis, K., Zarogoulidis, P., 2015. Barotrauma and pneumothorax. J. Thorac. Dis. 7, S38–S43.

Ivanova, O., K Khoo, M., 2004. Simulation of spontaneous cardiovascular variability using PNEUMA. Conf. Proc. IEEE Eng. Med. Biol. Soc. 6, 3901–3904.

Jasinski, P.E., Molkov, Y.I., Shevtsova, N.A., Smith, J.C., Rybak, I.A., 2013. Sodium and calcium mechanisms of rhythmic bursting in excitatory neural networks of the pre-Bötzinger complex: a computational modelling study. Eur. J. Neurosci. 37, 212–230.

Katz, E.S., 2018. Disorders of central respiratory control. In: Pulmonary Complications of Non-Pulmonary Pediatric Disorders, Respiratory Medicine. Humana Press, Cham, pp. 163–175.

Khoo, M.C., Marmarelis, V.Z., 1989. Estimation of peripheral chemoreflex gain from spontaneous sigh responses. Ann. Biomed. Eng. 17, 557–570.

Laura-María, R., 1998. Modelado Matematico del Control Respiratorio. In: Neurobiología de Las Funciones Vegetativas. Universidad de Sevilla, Sevilla, pp. 231–255.

Liang, F., Liu, H., 2006. Simulation of hemodynamic responses to the valsalva maneuver: an integrative computational model of the cardiovascular system and the autonomic nervous system. J. Physiol. Sci. 56, 45–65.

Lindsey, B.G., Nuding, S.C., Segers, L.S., Morris, K.F., 2018. Carotid bodies and the integrated cardiorespiratory response to hypoxia. Physiology 33, 281–297.

Liu, F., Blätke, M.-A., Heiner, M., Yang, M., 2014. Modelling and simulating reaction-diffusion systems using coloured Petri nets. Comput. Biol. Med. 53, 297–308.

Liu, F., Heiner, M., 2013. Multiscale modelling of coupled Ca^{2+} channels using coloured stochastic Petri nets. IET Syst. Biol. 7, 106–113.

Lloyd, B.B., 1966. The interactions between hypoxia and other ventilatory stimuli. In: Cardiovascular and Respiratory Effects of Hypoxia, pp. 146–165.

Loeschcke, H.H., De Lattre, J., Schläfke, M.E., Trouth, C.O., 1970. Effects on respiration and circulation of electrically stimulating the ventral surface of the medulla oblongata. Respir. Physiol. 10, 184–197.

Longobardo, G., Evangelisti, C.J., Cherniack, N.S., 2005. Introduction of respiratory pattern generators into models of respiratory control. Respir. Physiol. Neurobiol. 148, 285–301.

Longobardo, G.S., Cherniack, N.S., Fishman, A.P., 1966. Cheyne-Stokes breathing produced by a model of the human respiratory system. J. Appl. Physiol. 21, 1839–1846.

Lu, K., Clark, J.W., Ghorbel, F.H., Robertson, C.S., Ware, D.L., Zwischenberger, J.B., Bidani, A., 2004. Cerebral autoregulation and gas exchange studied using a human cardiopulmonary model. Am. J. Phys. Heart Circ. Phys. 286, H584–H601.

Lu, K., Clark, J.W., Ghorbel, F.H., Ware, D.L., Bidani, A., 2001. A human cardiopulmonary system model applied to the analysis of the Valsalva maneuver. Am. J. Phys. Heart Circ. Phys. 281, H2661–H2679.

Lu, K., Clark, J.W.J., Ghorbel, F.H., Ware, D.L., Zwischenberger, J.B., Bidani, A., 2003. Whole-body gas exchange in human predicted by a cardiopulmonary model. Cardiovasc. Eng. 3, 1–19.

Lü, Z., Zhao, C., Zhang, B., Duan, L., 2017. Multitime scale study of bursting activities in the pre-Bötzinger complex. Int. J. Bifurcat. Chaos. 27. 1750172.

Lumb, A.B., 2016. Nunn's Applied Respiratory Physiology, eighth ed. Elsevier, Edinburgh; New York.

Magosso, E., Ursino, M., 2001. A mathematical model of CO_2 effect on cardiovascular regulation. Am. J. Physiol. Heart Circ. Physiol. 281, H2036–H2052.

Marieb, E.N., Hoehn, K.N., 2012. Human Anatomy & Physiology, nineth ed. Pearson.

Matsugu, M., Duffin, J., Poon, C.S., 1998. Entrainment, instability, quasi-periodicity, and chaos in a compound neural oscillator. J. Comput. Neurosci. 5, 35–51.

Milhorn, H.T., 1976. Simulation of the respiratory control system. Simulation 27, 169–172.

Milhorn, H.T., Benton, R., Ross, R., Guyton, A.C., 1965. A mathematical model of the human respiratory control system. Biophys. J. 5, 27–46.

Milhorn, H.T., Guyton, A.C., 1965. An analog computer analysis of Cheyne-Stokes breathing. J. Appl. Physiol. 20, 328–333.

Milioli, G., Bosi, M., Poletti, V., Tomassetti, S., Grassi, A., Riccardi, S., Terzano, M.G., Parrino, L., 2016. Sleep and respiratory sleep disorders in idiopathic pulmonary fibrosis. Sleep Med. Rev. 26, 57–63.

Mitchell, R.A., Loeschcke, H.H., Severinghaus, J.W., Richardson, B.W., Massion, W.H., 1963. Regions of respiratory chemosensitivity on the surface of the medulla. Ann. N. Y. Acad. Sci. 109, 661–681.

Mohan, R.M., Amara, C.E., Vasiliou, P., Corriveau, E.P., Cunningham, D.A., Duffin, J., 1998. Chemoreflex model parameters measurement. In: Advances in Modeling and Control of Ventilation, Advances in Experimental Medicine and Biology. Springer, Boston, MA, pp. 185–193.

Molkov, Y.I., Abdala, A.P.L., Bacak, B.J., Smith, J.C., Paton, J.F.R., Rybak, I.A., 2010. Late-expiratory activity: emergence and interactions with the respiratory CPG. J. Neurophysiol. 104, 2713–2729.

Molkov, Y.I., Rubin, J.E., Rybak, I.A., Smith, J.C., 2017. Computational models of the neural control of breathing. Wiley Interdiscip. Rev. Syst. Biol. Med. 9e1371.

Molkov, Y.I., Shevtsova, N.A., Park, C., Ben-Tal, A., Smith, J.C., Rubin, J.E., Rybak, I.A., 2014. A closed-loop model of the respiratory system: focus on hypercapnia and active expiration. PLoS One. 9. e109894.

Muñoz-Ortiz, J., Muñoz-Ortiz, E., López-Meraz, L., Beltran-Parrazal, L., Morgado-Valle, C., 2018. The pre-Bötzinger complex: generation and modulation of respiratory rhythm. Neurol. (Engl. Ed.).

Nendza, M., Kühne, R., Lombardo, A., Strempel, S., Schüürmann, G., 2018. PBT assessment under REACH: Screening for low aquatic bioaccumulation with QSAR classifications based on physicochemical properties to replace BCF in vivo testing on fish. Sci. Total Environ. 616–617, 97–106.

Nielsen, M., Smith, H., 1952. Studies on the regulation of respiration in acute hypoxia; with a appendix on respiratory control during prolonged hypoxia. Acta Physiol. Scand. 24, 293–313.

Nugent, S.T., Finley, J.P., 1987. Periodic breathing in infants: a model study. IEEE Trans. Biomed. Eng. BME-34, 482–485.

Dada, O., Mendes, P., 2011. Multi-scale modelling and simulation in systems biology. Integr. Biol. 3, 86–96.

Olszowka, A.J., Farhi, L.E., 1968. A system of digital computer subroutines for blood gas calculations. Respir. Physiol. 4, 270–280.

Painter, R., Khamnei, S., Robbins, P., 1993. A mathematical model of the human ventilatory response to isocapnic hypoxia. J. Appl. Physiol. 74, 2007–2015.

Passino, C., Cacace, E., Caratozzolo, D., Rossari, F., Saccaro, L.F., 2017. Mechanics and chemistry of respiration in health. In: The Breathless Heart. Springer, Cham, pp. 11–33.

Patel, S., Majmundar, S.H., 2018. Physiology, carbon dioxide retention. In: StatPearls. StatPearls Publishing, Treasure Island (FL).

Patil, C.P., Saunders, K.B., Sayers, B.M., 1989. Modelling the breath by breath variability in respiratory data. In: Swanson, G.D., Grodins, F.S., Hughson, R.L. (Eds.), Respiratory Control: A Modeling Perspective. Springer US, Boston, MA, pp. 343–352.

Poulin, M.J., Robbins, P.A., 1996. Indexes of flow and cross-sectional area of the middle cerebral artery using Doppler ultrasound during hypoxia and hypercapnia in humans. Stroke 27, 2244–2250.

Priestley, J., 1776. observations on respiration, and the use of the blood. Philos. Trans. R. Soc. Lond. 66, 226–248.

Qutub, A.A., Popel, A.S., 2006. A computational model of intracellular oxygen sensing by hypoxia-inducible factor HIF1α. J. Cell Sci. 119, 3467–3480.

Rapoport, D.M., Norman, R.G., Goldring, R.M., 1993. CO_2 homeostasis during periodic breathing: predictions from a computer model. J. Appl. Physiol. 75, 2302–2309.

Ratnovsky, A., Elad, D., Halpern, P., 2008. Mechanics of respiratory muscles. Respir. Physiol. Neurobiol. Respir. Biomech. 163, 82–89.

Rybak, I.A., Shevtsova, N.A., Paton, J.F.R., Dick, T.E., St. John, W.M., Mörschel, M., Dutschmann, M., 2004. Modeling the ponto-medullary respiratory network. Respir. Physiol. Neurobiol. Pontine Influences Breath. 143, 307–319.

Saunders, K.B., Bali, H.N., Carson, E.R., 1980. A breathing model of the respiratory system; the controlled system. J. Theor. Biol. 84, 135–161.

Scholander, P.F., 1960. Oxygen transport through hemoglobin solutions. Science 131, 585–590.

Serna, L.Y., Mañanas, M.A., Hernández, A.M., Rabinovich, R.A., 2018. An improved dynamic model for the respiratory response to exercise. Front. Physiol. 9, 69.

Smith, J.C., Abdala, A.P.L., Koizumi, H., Rybak, I.A., Paton, J.F.R., 2007. Spatial and functional architecture of the mammalian brain stem respiratory network: a hierarchy of three oscillatory mechanisms. J. Neurophysiol. 98, 3370–3387.

Staub, N.C., 1991. Basic Respiratory Physiology. Churchill Livingstone, New York.

Stephenson, R., 2004. A theoretical study of the effect of circadian rhythms on sleep-induced periodic breathing and apnoea. Respir. Physiol. Neurobiol. 139, 303–319.

Stoller, J.K., 2015. Murray & Nadel's textbook of respiratory medicine, 6th edition. Ann. ATS 12, 1257–1258.

Stuhmiller, J.H., Stuhmiller, L.M., 2005. A mathematical model of ventilation response to inhaled carbon monoxide. J. Appl. Physiol. 98, 2033–2044.

Takahashi, E., Doi, K., 1993. Destabilization of the respiratory control by hypoxic ventilatory depressions: a model analysis. Jpn. J. Physiol. 43, 599–612.

Talaminos-Barroso, A., Márquez Martín, E., Roa Romero, L.M., Ortega Ruiz, F., 2018. Factores que afectan a la función pulmonar: una revisión bibliográfica. Arch. Bronconeumol. 54, 327–332.

Tehrani, F.T., 1993. Mathematical analysis and computer simulation of the respiratory system in the newborn infant. IEEE Trans. Biomed. Eng. 40, 475–481.

Thagard, P., 2018. Computational Models in Science and Philosophy. In: Hansson, S.O., Hendricks, V.F. (Eds.), Introduction to Formal Philosophy, Springer Undergraduate Texts in Philosophy. Springer International Publishing, Cham, pp. 457–467.

Topor, Z.L., Pawlicki, M., Remmers, J.E., 2004. A computational model of the human respiratory control system: responses to hypoxia and hypercapnia. Ann. Biomed. Eng. 32, 1530–1545.

Topor, Z.L., Vasilakos, K., Younes, M., Remmers, J.E., 2007. Model based analysis of sleep disordered breathing in congestive heart failure. Respir. Physiol. Neurobiol. 155, 82–92.

Tran, D., Rajwani, K., Berlin, D.A., 2018. Pulmonary effects of aging. Curr. Opin. Anesthesiol. 31, 19.

Turner, D.L., 1991. Cardiovascular and respiratory control mechanisms during exercise: an integrated view. J. Exp. Biol. 160, 309–340.

Uhrmacher, A.M., Degenring, D., Zeigler, B., 2005. Discrete event multi-level models for systems biology. In: Transactions on Computational Systems Biology. I. Lecture Notes in Computer Science. Springer, Berlin, Heidelberg, pp. 66–89.

Ursino, M., Magosso, E., 2004. Interaction among humoral and neurogenic mechanisms in ventilation control during exercise. Ann. Biomed. Eng. 32, 1286–1299.

Ursino, M., Magosso, E., 2000. Acute cardiovascular response to isocapnic hypoxia. I. A mathematical model. Am. J. Physiol. Heart Circ. Physiol. 279, H149–H165.

Ursino, M., Magosso, E., Avanzolini, G., 2001. An integrated model of the human ventilatory control system: the response to hypoxia. Clin. Physiol. 21, 465–477.

Walmsley, J., 2007. John Locke on Respiration. Med. Hist. 51, 453–476.

Walpole, J., Papin, J.A., Peirce, S.M., 2013. Multiscale computational models of complex biological systems. Annu. Rev. Biomed. Eng. 15, 137–154.

West, J.B., 2017. Leonardo da Vinci: engineer, bioengineer, anatomist, and artist. Am. J. Phys. Lung Cell. Mol. Phys. 312, L392–L397.

West, J.B., 2014. Robert Hooke: early respiratory physiologist, polymath, and mechanical genius. Physiology 29, 222–233.

Wilkinson, D.J., 2011. Stochastic Modelling for Systems Biology, second ed. CRC Press, Boca Raton.

Wolf, M.B., Garner, R.P., 2007. A mathematical model of human respiration at altitude. Ann. Biomed. Eng. 35, 2003–2022.

Zohdi, T.I., 2017. Homogenization methods and multiscale modeling. In: Encyclopedia of Computational Mechanics. second ed. American Cancer Society, pp. 1–24.

Further reading

Cook-Snyder, D.R., Miller, J.R., Navarrete-Opazo, A.A., Callison, J.J., Peterson, R.C., Hopp, F.A., Stuth, E.A.E., Zuperku, E.J., Stucke, A.G., 2019. The contribution of endogenous glutamatergic input in the ventral respiratory column to respiratory rhythm. Respir. Physiol. Neurobiol. 260, 37–52.

Intelligent decision support for lung ventilation

Fleur T. Tehrani
California State University, Fullerton, CA, United States

1 Introduction

Artificial intelligence (AI) is gaining increasing significance in many areas of technology. The rapid expansion of the storage and processing capabilities of computers make them valuable tools as large information banks as well as effective decision-making aids when large amounts of data need to be analyzed and complex processes are involved. One of the areas where such systems can have significant applications is medicine. As various new tests become available and the amount of data for each patient increases, computers become practical tools for analyzing patient data.

Expert systems are computerized tools that mostly operate on the basis of the medical knowledge stored in them rather than any acquired knowledge through learning. Expert medical systems can be individualized based on patient specific data such as gender, age, height, weight, and other similar information and can be used for diagnosis and treatment. These systems have been presented for numerous applications in medicine to date such as diagnostic imaging (Suzuki et al., 2005), treatment of spinal deformities (Chalmers et al., 2015), detection of atrial fibrillations (Lin et al., 2010), multiple sclerosis (Gaspari et al., 2009), chemotherapy for cancer treatments (Dua et al., 2008; Pefani et al., 2013), control of anesthesia (Nascu and Pistikopoulos, 2017), and insulin delivery for type I diabetes (Dua and Pistikopoulos, 2005).

Self-learning AI systems as well as knowledge-based expert systems can be used as open-loop advisory tools, while some systems can also be used for provision of automatic treatment to patients via closed-loop control techniques.

In this chapter, an overview of the general structure of computerized decision-support systems (CDSSs) in medicine is provided. The chapter is particularly focused on the application of various CDSSs in mechanical ventilation, which is an essential treatment in the intensive care unit (ICU) settings of hospitals. The chapter briefly describes different technologies that are used in mechanical ventilation and provides the overviews of several CDSSs for this treatment technology.

Control Applications for Biomedical Engineering Systems. https://doi.org/10.1016/B978-0-12-817461-6.00012-3

2 General structure of CDSSs in medicine

2.1 Type I. Advisory (open-loop) systems

Fig. 1 shows a general block diagram of a CDSS designed as an advisory aid to physicians.

As seen in this figure, the physician provides inputs to this system that can include specific data/characteristics of the patient such as name, address, age, gender, height, weight, type of illness(es), etc. The patient monitoring system consists of sensors and monitors that measure the system's required patient data. The data depend on what the CDSS is designed for. For example, in a CDSS for use in pulmonary and critical care, the monitored data include cardiovascular and respiratory data, while in a system for endocrine treatment the data may include the blood glucose and insulin levels. The patient monitoring system may provide the data automatically or the monitored data can be measured intermittently and provided manually by the medical personnel via a keyboard.

The input data analyzer receives the inputs from the patient monitoring system and/ or the physician and processes the data for accuracy. This unit deletes erroneous data due to artifact and normally uses some smoothing processes to make the data usable by the processing and control unit that comes next and receives the validated data.

The function of the processing and control unit depends not only on the intended treatment but also on the nature of the CDSS. If the CDSS is an AI system designed to develop many of its rules based on learning, the control algorithms may involve the use of Fuzzy systems (Zadeh, 1983) and artificial neural networks. However, if the system is knowledge-based, the control algorithms will be based on treatment protocols and knowledge from the literature and expert physicians (Shortliffe, 1986).

Both deterministic and fuzzy control techniques are used in the design of the processing algorithms and the choice depends on the number of key monitored data, overlapping conditions, and the existence of clear treatment directions under different circumstances. The processing and control unit serves as the brain of the CDSS for

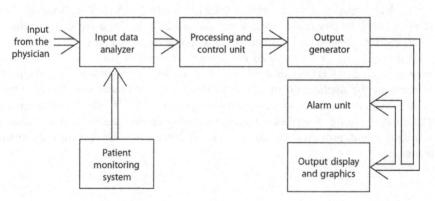

Fig. 1 A block diagram of an open-loop (advisory) CDSS.

processing numerous patient data, diagnosis, reviewing treatment options, and determining and recommending the preferred choices of treatment.

The output generator provides its outputs based on the inputs received from the processing and control unit. Those outputs can include instructions/recommendations for the next treatment options as well as warning messages to the clinician. If hazardous patient conditions are detected, alarm signals are generated by the alarm unit that receives its inputs from the output generator. The output signals are provided to the output display and graphics system that provides a user-friendly set of displays and graphs in relation to the patient treatment and conditions to the physician. An open-loop CDSS is activated and run by the clinician at certain intervals (Shortliffe, 1986; Hobbs et al., 1996; Montgomery et al., 2000; Walton et al., 1999).

2.2 Type II. CDSSs for use as both advisory and closed-loop systems

Apart from providing treatment advice, CDSSs can also be designed for closed-loop automatic treatment control. Fig. 2 shows a general block diagram of a CDSS adapted for closed-loop control of patients' treatments.

The controller in this system receives specific instructions/inputs from a physician, while the patient's monitored data are measured by sensor/monitors automatically and provided to the controller via feedback loops. The CDSS controller determines the next treatment based on the settings from the physician and the patient data from the sensor/monitors. In general, the controller has data smoothing and error detection algorithms embedded in its structure to delete erroneous data, detect artifacts, and process and smooth data to prevent abrupt changes in the system's outputs. The control algorithms may be based on fuzzy techniques to handle uncertainties or can use more deterministic strategies for choosing treatment options. If the controller detects hazardous patient data/conditions, it activates the Alarms unit. The outputs of the controller are constantly provided to a display/graphics unit to keep the medical personnel

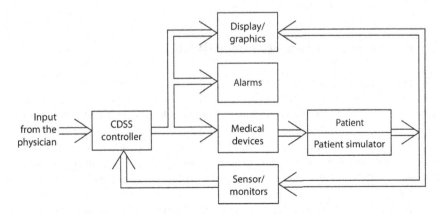

Fig. 2 Block diagram of a CDSS adapted for closed-loop control of patient's treatments.

informed of the provided treatments and the patient's conditions. Unlike the advisory systems, the signals generated by the CDSS controller are provided automatically to medical device(s) that are used to deliver the treatment. These devices can be implanted in the patient's body or can be external devices for provision of gas or IV anesthetics (Nascu and Pistikopoulos, 2017; Ritchie et al., 1987; Vishoni and Roy, 1991; Ting et al., 2004; Mahfouf, 2005; Dua et al., 2010), delivery of insulin (Campos-Delgado et al., 2006; Bequette, 2012; Breton et al., 2012; Dassau et al., 2013; Elleri et al., 2013; Hovorka et al., 2014), or mechanical ventilation (Dojat et al., 2000; Tehrani and Roum, 2008a, b; Tehrani and Abbasi, 2009; Tehrani, 2011) as examples.

The medical device(s) deliver the updated treatment to the patient or in some systems, also to a patient simulator. The patient simulator is a patient model that predicts the treatment results and provides its simulation data to the system. A model used in a CDSS is normally based on physiological characteristics of the part of the patient's body targeted by the treatment. If the CDSS is used to control drug delivery, the model should include the pharmacokinetics as well as pharmacodynamics of the delivered drug (Velliou et al., 2018). The pharmacokinetics describe and simulate the effects of the drug in the patient's body by focusing on the drug's rates of absorption, distribution to body organs, diffusion, and elimination. The pharmacodynamics simulations on the other hand describe the responses of the body to the delivered drug.

If a patient simulator is used by the CDSS, the simulator's predictions are provided to the display/graphics unit. However, in a closed-loop CDSS, the feedback data used by the controller are the actual measured patient data provided by the sensor/monitors unit and not the simulated data by the model.

Depending on whether the controlled medical device(s) are external or implanted in the patient's body and the location of the CDSS controller, the links between different compartments of Fig. 2 can be wired or wireless. The closed-loop control systems are gaining widespread applications at various levels of patients' treatments from the operating rooms and intensive and critical care units in hospitals (Nascu and Pistikopoulos, 2017; Tehrani and Roum, 2008b; Tehrani, 2011) to patient wards and long-term home treatments (Doyle et al., 2014).

3 CDSSs for mechanical ventilation

Mechanical ventilation is an essential life-saving ICU technology. In the early 20th century, negative pressure mechanical ventilation was used to treat patients who could not breathe on their own. By using this technology, the patient's body up to his neck was enclosed in an airtight tank and the volume of the tank was expanded so the patient's lungs would inflate as a result of the negative pressure applied to his thoracic area. Smaller versions of negative pressure ventilators that did not use a tank and applied negative pressure only to the patient's chest were later developed and known as jacket ventilators or cuirasses.

As the polio epidemic struck the western world in the mid-20th century, large numbers of patients who were affected by paralysis needed the mechanical ventilation treatment. At that time, a different kind of technology referred to as positive pressure mechanical ventilation was adapted to respond to the medical needs of the communities. In this technology, the patient's lungs are inflated by application of positive pressure to his airways. This treatment can be provided invasively by using an endotracheal tube or tracheostomy or noninvasively by use of a facial mask or a mouthpiece.

Different technologies are employed in today's positive pressure mechanical ventilation. In high-frequency jet ventilation (HFJV) and high-frequency oscillatory ventilation (HFOV), the oxygenated air is delivered at high frequencies to the patient's airways, which results in vibrating the alveoli at high rates. These treatments are provided to prevent lung collapse and injury to the alveolar tissue.

However, the most commonly used technique of positive pressure mechanical ventilation is referred to as conventional mechanical ventilation. In conventional mechanical ventilation, many ventilation methods and modalities are used to treat patients (Chatburn et al., 2014). By application of various modalities of ventilation, partial or full respiratory support can be given to patients and delivery of a certain volume of gas (volume controlled) or a certain gas pressure (pressure controlled) can be guaranteed. Depending on the modality of ventilation, either all of the ventilation parameters are set by the clinician (open-loop control) or the parameters are controlled automatically based on the patient's monitored conditions (closed-loop control) (Chatburn and Mireles-Cabodevila, 2011; Tehrani, 2008a, b).

In the ICU setting, the amount of patient data is large and the clinician's time to analyze such data and make crucial decisions is short. Under such conditions, an effective and user-friendly CDSS can be a valuable tool for the clinician to choose the appropriate ventilation treatment for his patients. In view of this fact, different CDSSs have been developed by many researchers for mechanical ventilation in the past few decades. Some of these CDSSs are designed as open-loop advisory systems to provide treatment advice to clinicians, while some other systems can also be used as closed-loop systems and control patients' ventilations automatically. In the next section, different design methodologies of CDSSs for mechanical ventilation are described.

4 Design methodologies

An important difference between CDSSs for mechanical ventilation is related to the design objective of the system: whether the system is going to be used as an advisory tool or if it will be capable of closed-loop automatic control of ventilation. Fig. 3 shows a block diagram of a CDSS for mechanical ventilation.

As shown in Fig. 3, the system includes an input data analyzer that takes its inputs from the clinician and the monitors. The measured patient data generally consist of measured data by the sensors and monitors connected to the patient. These data may typically include the following:

- The patient's blood gas information measured by using blood gas analyzers or by noninvasive sensors such as pulse oximeters and capnometers.
- The respiratory mechanics data and their rates of change.

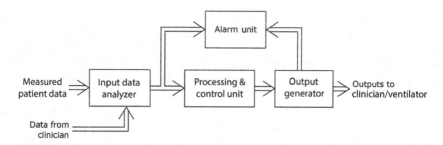

Fig. 3 Block diagram of a CDSS for mechanical ventilation.

- Measured ventilatory parameters such as tidal volume, minute ventilation, total respiratory rate, positive end-expiratory pressure (PEEP), the spontaneous breathing rate, the peak inspiratory pressure, and the rates of change of these data.
- The patient's blood pressure and heart rate information.

The listed data are either provided automatically by the monitors or are measured intermittently and entered to the system via a keyboard, with the latter method used mainly in the open-loop advisory systems.

The data from clinician may typically include the following:

- Patient's medical problems.
- Patient's age, gender, height, weight, body temperature, and ideal body weight.
- Set ventilation parameters such as tidal volume, minute ventilation, respiratory rate, the inspiratory to expiratory time ratio (I:E), inspiration time, pause time, the peak inspiratory pressure, fraction of inspired oxygen (F_{IO2}), and PEEP.
- Set alarm levels on the ventilator such as the maximum allowed pressure and volume levels.
- Other parameters used and required by the CDSS.

The input data analyzer in Fig. 3 receives the input data and analyzes the information before sending it to the processing and control unit. The input data analyzer uses validation algorithms to remove noise, eliminate erroneous data due to artifacts, and smooth data to prevent abrupt changes in the treatments provided by the system. This unit also detects hazardous patient conditions and activates appropriate alarms when the validated data fall in the predefined unsafe ranges or if the measured data are flawed due to disconnections or artifacts.

The processing and control unit receives the validated data from the input data analyzer and uses that information to determine the best treatment option at the next step. The methodologies used by this unit are quite different depending on the applied control mechanisms. The methodologies can be divided to three major categories: (a) rule-based systems, (b) model-based systems, and (c) rule-based + model-based systems.

In rule-based systems, the determinations of the processing and control unit are based on clinical and experimental protocols. Expert knowledge-based systems are the examples where clinical protocols are used to determine the next optimal treatment for the patient. In addition, neurofuzzy systems can also be considered as rule-based where the rules are derived from the inputs in combination with the expert clinical knowledge.

Table 1 Main categories of CDSSs for mechanical ventilation

Main features		Alternatives	
Type of technology	Open-loop (advisory)	Closed-loop (automatic)	Open-loop + Closed-loop
Basic structure	Rule-based	Model-based	Rule-based + Model-based
Ventilation modes	Single mode (e.g., PS)	Multiple modes	
Ventilation stage	Management phase	Weaning phase	Management & weaning phases
Patient groups	Adults	Neonates	Adults & neonates
Disease conditions	ARDS	COPD	Various diseases

In model-based systems, the treatment methods are optimized by using a simulation physiological model of the patient.

In rule-based + model-based systems, the rules to determine the optimal treatments are based on a combination of the patient's physiological model results as well as clinical protocols and guidelines.

The processing and control unit determines the next parameters of mechanical ventilation. The determined parameters depend on the design objectives of the CDSS. Those objectives may involve the applicable modes and stages of ventilation, the targetted patient population, and the specific respiratory disease conditions that the system is designed for. The stage of ventilation can be the management phase, the weaning phase, or both. The applicable modes of ventilation may be volume control (VC), pressure control (PC), pressure support (PS), or synchronized intermittent mandatory ventilation (SIMV) as some examples.

CDSSs can be designed for treatment of certain respiratory diseases such as the acute respiratory distress syndrome (ARDS) or chronic obstructive pulmonary disease (COPD), or may be usable in a wider range of applications. CDSSs can also be used for specific patient populations such as neonates or adults or may be applicable to both groups.

Table 1 shows various categories of CDSSs for conventional mechanical ventilation. Several examples of these systems can be found in (Strickland and Hassan, 1993; Shahsavar et al., 1995; Nemoto et al., 1999; Kwok et al., 2003; Rees et al., 2006; Lozano-Zahonero et al., 2011) and the descriptions of many major CDSSs can be found in some review papers on this subject (Tehrani and Roum, 2008b; Tehrani, 2011).

In the following sections, the overviews of several CDSSs for mechanical ventilation are provided.

5 A model-based CDSS for mechanical ventilation

In this section, the focus will be on a model-based CDSS for mechanical ventilation that can be used as an advisory system and is based on physiological models of the

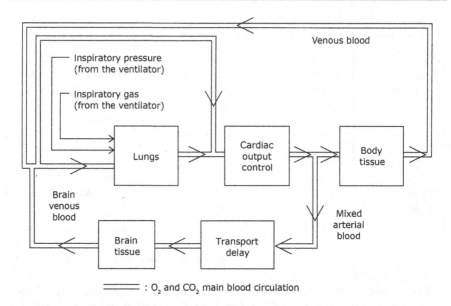

Fig. 4 A schematic block diagram of a model-based CDSS (Tehrani and Abbasi, 2012).

patient. The details of this CDSS were first introduced in 2012 (Tehrani and Abbasi, 2012). In this system, the modified plants of two previously published physiological models of the human respiratory system for adults (Fincham and Tehrani, 1983a) and infants (Tehrani, 1993) were used. Fig. 4 shows the block diagram of this system.

In the system shown in Fig. 4, the ispiratory gas is delivered to the patient's lungs by the ventilator. The ventilatory parameters are set by the clinician and input to the system. The Lungs are ventilated at the rate and the depth specified by the clinician and PEEP and F_{IO2} levels are also adjusted based on the corresponding input data by the clinician. The oxygenated blood leaves the lungs compartment and the cardiac output control unit adjusts the blood perfusion rates to the body and the brain tissues based on the arterial pressure levels of oxygen and carbon dioxide. The arterial blood perfuses the body tissue, is transported to the brain by the transport delay compartment, and perfuses the brain tissue. Then the venous blood from the brain and the body tissues mix before entering the lung compartment and the cycle is repeated.

By using this system, the clinician is provided with the system's predictions of any therapy and the best option can be chosen by the physician for the next stage of the patient's treatment. The equations describing different compartments of this system are provided in Tehrani and Abbasi (2012) and are not given here because they are too long. The predictions of this system were compared with the clinical data for adult and infant patient groups and were found to be in close agreement with those data.

6 Application of a model-based CDSS in differential lung ventilation

Differential lung ventilation treatment is a ventilation technique provided to patients with different lungs' characteristics. Independent lung ventilation (ILV) is a mechanical ventilation treatment provided in differential lung disease. This technically demanding treatment is used when conventional mechanical ventilation fails or is not appropriate. The 1st cases of ILV in the intensive care settings were reported in the 1970s (Glass et al., 1976; Anantham et al., 2005). The ILV treatment is given by using two types of techniques: (a) anatomical lung separation and (b) physiological lung separation.

Anatomical lung separation is considered a short-term rescue treatment while physiological lung separation is applied when the lungs have different characteristics such as in unilateral lung disease (Cinnella et al., 2001), or after single-lung transplant operations (Anantham et al., 2005; Ost and Corbridge, 1996).

In physiological lung separation, different treatment methods can be applied to each lung. In synchronous treatments, the frequency of respiration is the same for both lungs while the other ventilatory parameters can be independently assigned for each lung. In asynchronous treatments, the ventilation mode and all ventilatory parameters, including respiratory rate can be individually set for each lung.

Application of ILV can be a demanding technique for the intensivists. An effective CDSS can be a valuable aid to the physician to predict the results of various ventilation techniques for each lung. The next section provides a brief description of a CDSS for ILV treatments.

6.1 Methods

6.1.1 Description of the plant

The respiratory system's plant used in the CDSS is based on physiological models of adult and infant respiratory systems (Fincham and Tehrani, 1983a, b; Tehrani, 1993) that have been used by many researchers since their development (Hernandez et al., 2008). The CDSS has two lung compartments. A block diagram of the system's plant is shown in Fig. 5.

As shown in Fig. 5, the lung compartments can be ventilated by two separate ventilators and the ventilation parameters for each lung are set independently. The inspiratory gas from the ventilators comes into contact with the blood in each lung compartment. Then the oxygenated blood from the compartments is distributed to the body tissue and the brain. The cardiac output control regulates the rates of blood perfusion to the body and the brain based on the oxygen and carbon dioxide levels of the arterial blood. The arterial transport delay controls the transportation delay caused by the arteries to the brain. The circulating arterial blood flows through the body tissue and the brain tissue and the venous blood from these compartments mix before entering the lungs at the beginning of the next cycle. The details of different blocks of Fig. 5 are described in Tehrani (2019).

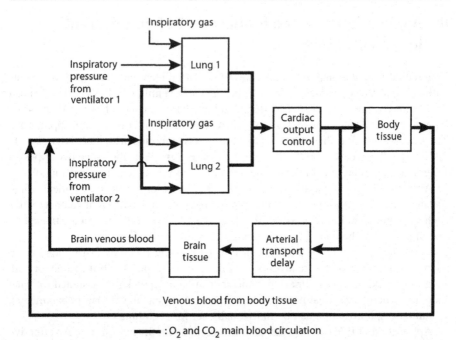

Fig. 5 A block diagram of a system's plant for ILV treatment.

6.2 *Equations of the plant*

The equations for the lung compartments are as follows:

For the alveolar space in lung 1:

For CO_2:

$$(C_{VTCO2} - C_{a1CO2})Q_T + (C_{VBCO2} - C_{a1CO2})Q_B$$
$$= [v_1/(P_b - 47)]dP_{A1CO2}/dt + FACT11 \tag{1}$$

For O_2:

$$(C_{a1O2} - C_{VTO2})Q_T + (C_{a1O2} - C_{VBO2})Q_B = [-v_1/(P_b - 47)]dP_{A1O2}/dt$$
$$+ FACT12 \tag{2}$$

where in inspiration that $dv_1/dt \geq 0$:

$$FACT11 = [(P_{A1CO2} - P_{I1CO2})/(P_b - 47)]dv_1/dt$$

and: $FACT12 = [(P_{I1O2} - P_{A1O2})/(P_b - 47)]dv_1/dt$

and during expiration that $dv_1/dt < 0$: $FACT11 = FACT12 = 0$.

In these equations, C_{a1} represents the arterial gas volume concentration in lung 1, C_{VT} is the gas volume concentration in the body tissue venous blood, C_{VB} represents the gas volume concentration in the brain venous blood, P_{A1} is the alveolar gas partial pressure in lung 1, and the second subscripts O_2 and CO_2 specify the gas concentrations to be for oxygen or carbon dioxide, respectively. P_{I1O2} and P_{I1CO2} represent the partial pressures of oxygen and carbon dioxide in the inspiratory gas in lung 1, respectively, v_1 is the alveolar volume in lung 1, P_b represents the barometric pressure, and Q_T and Q_B are the blood flow rates in the lumped body tissue and the brain tissue, respectively.

For the alveolar space in lung 2:

For CO_2:

$$(C_{VTCO2} - C_{a2CO2})Q_T + (C_{VBCO2} - C_{a2CO2})Q_B$$

$$= [v_2/(P_b\text{-}47)]dP_{A2CO2}/dt + FACT21 \tag{3}$$

For O_2:

$$(C_{a2O2} - C_{VTO2})Q_T + (C_{a2O2} - C_{VBO2})Q_B$$

$$= [-v_2/(P_b\text{-}47)]dP_{A2O2}/dt + FACT22 \tag{4}$$

where in inspiration that $dv_2/dt \geq 0$:

$$FACT21 = [(P_{A2CO2} - P_{I2CO2})/(P_b\text{-}47)]dv_2/dt$$

and: $FACT22 = [(P_{I2O2} - P_{A2O2})/(P_b\text{-}47)]dv_2/dt$

and during expiration that $dv_2/dt < 0$: $FACT21 = FACT22 = 0$

In Eqs. (3), (4), v_2 is the alveolar volume in lung 2, C_{a2} represents the arterial gas volume concentration in lung 2, P_{A2} is the alveolar gas partial pressure in lung 2, P_{I2} represents the inspiratory gas partial pressure in lung 2, with the second subscripts O_2 and CO_2 specifying the gas to be either oxygen or carbon dioxide, respectively.

In these equations, the alveolar and arterial partial pressures of CO_2 are assumed to be equal:

$$P_{A1CO2} = P_{a1CO2} \tag{5}$$

$$P_{A2CO2} = P_{a2CO2} \tag{6}$$

and for oxygen:

$$P_{A1O2} = P_{a1O2} + K_1 \tag{7}$$

$$P_{A2O2} = P_{a2O2} + K_2 \tag{8}$$

where P_{a1} and P_{a2} represent the arterial partial pressures of gas in lungs 1 and 2, respectively, with the second subscripts O_2 and CO_2 specifying the gas to be either oxygen or carbon dioxide, and K_1 and K_2 are the alveolar-arterial oxygen differences in lungs 1 and 2, respectively.

The blood O_2 and CO_2 gas concentrations are related to blood gas pressures according to the following equations:

$$C_{O2} = K_3 (1 - \exp(-K_4 P_{O2}))^2 \tag{9}$$

$$C_{CO2} = K_5 P_{CO2} \tag{10}$$

where C_{O2} and C_{CO2} represent blood gas volume concentrations of oxygen and carbon dioxide, respectively, and P_{O2} and P_{CO2} are the corresponding blood gas partial pressures of these gases. The blood gas dissociation constants for oxygen are K_3 and K_4 with typical values of 0.2 and 0.046 $mmHg^{-1}$, respectively, and K_5 is the blood gas constant for carbon dioxide with a typical value of 0.016 $mmHg^{-1}$.

By assuming homogeneous mixing of arterial blood from the two lung compartments:

$$C_{aCO2} = (C_{a1CO2} + C_{a2CO2})/2 \tag{11}$$

$$C_{aO2} = (C_{a1O2} + C_{a2O2})/2 \tag{12}$$

where C_{aCO2} and C_{aO2} represent the volume concentrations of carbon dioxide and oxygen in the arterial blood, respectively. The arterial oxygen saturation, S_{aO2}, is related to C_{aO2} as: $S_{aO2} = C_{aO2}/K_3$.

The equations used for the lumped body tissue and the brain compartments are the same as those in Fincham and Tehrani (1983a). Also, the dynamic changes in Q_T and Q_B that represent the blood flow rates in the lumped body tissue and the brain tissue are functions of the arterial partial pressure of CO_2 (P_{aCO2}) and the arterial partial pressure of oxygen (P_{aO2}), respectively. The equations for Q_T and Q_B as functions of arterial O_2 and CO_2 pressures are described elsewhere (Fincham and Tehrani, 1983a, 1983b). Those equations are not repeated here for brevity.

6.3 Application of ILV

In application of ILV, two separate ventilators can be used for the lungs. The lungs' volumes are affected by the levels of PEEP applied to the lungs:

$$\Delta FRC1 = C_{dyn1} PEEP1 \tag{13}$$

$$\Delta FRC2 = C_{dyn2} PEEP2 \tag{14}$$

where $\Delta FRC1$ and $\Delta FRC2$ are the changes in the functional residual capacities of lungs 1 and 2, respectively, C_{dyn1} is the dynamic compliance of lung 1, C_{dyn2} is the dynamic compliance of lung 2, and PEEP1 and PEEP2 are the PEEP levels applied to lungs 1 and 2, respectively.

The fraction of inspired oxygen in lung 1 ($F_{IO2}(1)$) and lung 2 ($F_{IO2}(2)$) are found as:

$$F_{IO2}(1) = P_{I1O2}/(P_b\text{-}47) \tag{15}$$

$$F_{IO2}(2) = P_{I2O2}/(P_b\text{-}47) \tag{16}$$

In the volume-controlled mode, any waveform can be applied. Regardless of the particular waveform, a sinusoidal airflow can be considered as the main harmonic of any airflow pattern as:

$$dv/dt = \pi V_A{}^{\bullet} \sin(2\pi ft) \tag{17}$$

where $V_A{}^{\bullet}$ is the alveolar ventilation in that lung in lit/s as:

$$V_A{}^{\bullet} = f(V_t - V_D) \tag{18}$$

where f is the respiration rate in breaths/s, V_D is the dead space, and V_t is the tidal volume in that lung. The dead space volume in normal lungs for adults can be estimated from the ideal body weight (Weight) as:

$$V_{D(\text{both lungs})} = 0.0026 \,(\text{Weight}) \tag{19}$$

In normal lungs, this value of V_D can be used to find an estimation for the dead space in each lung. However, in a diseased lung, its value can be adjusted based on the lung's conditions.

In the pressure-controlled mode of ventilation, the following equations are used for the lungs:

$$\Delta P_1 = K_1' v_1' + K_1'' dv_1/dt \tag{20}$$

$$\Delta P_2 = K_2' v_2' + K_2'' dv_2/dt \tag{21}$$

In these equations, ΔP_1 is the inspiratory pressure above PEEP1 applied to lung 1, K_1' is the respiratory elastance in lung 1, v_1' is the change in the volume of lung 1 above the functional residual capacity in lung 1, FRC1, and K_1'' is the airway resistance in lung 1. Similarly for lung 2, ΔP_2 is the inspiratory pressure above PEEP2 applied to lung 2, K_2' is the respiratory elastance in lung 2, K_2'' is the airway resistance in lung 2, and v_2' is the change in the volume of lung 2 above the functional residual capacity in lung 2, FRC2.

Table 2 The details of the ILV treatment provided to a patient (Al Herz, 2014)

Lung	Ventilation mode	F_{IO2}	PEEP (cm H_2O)	Peak pressure (cm H_2O)	Respiratory rate (breaths/min)	I/E
Right lung	Pressure-controlled	0.8	0	32	12	1:6.2
Left lung	Pressure-controlled	0.8	5	23	20	1:3.5

Table 3 The results of an ILV treatment provided to a patient as posted in Al Herz (2014) in comparison with a CDSS-predicted outcome (Tehrani, 2019)

Measured tidal volumes (mL)	Measured S_{aO2}	Measured P_{aCO2} (mmHg)	CDSS-predicted tidal volumes (mL)	CDSS-predicted S_{aO2}	CDSS-predicted P_{aCO2} (mmHg)
Right Lung:100 Left Lung:300	100%	34.3	Right Lung:110 Left Lung:300	100%	33

6.4 An example of the application of the system

The results of an asynchronous ILV treatment to a 69-year-old patient with COPD and air leak in the right lung are described in Al Herz (2014). The details of the ILV treatment are shown in Table 2.

The tidal volumes in the right and the left lungs of this patient were measured to be 100 and 300 mL, respectively. The arterial oxygen saturation, S_{aO2}, of the patient was 100% and his arterial partial pressure of carbon dioxide, P_{aCO2}, was measured at 34.3 mmHg. The CDSS was used to simulate this ILV treatment. Table 3 shows the clinically measured results in comparison with the CDSS predictions. As seen, the CDSS predicted the tidal volume of the right lung to be 110 mL and for the left lung, its prediction for the tidal volume was 300 mL. These results are in comparison with the measured tidal volumes of 100 and 300 mL in the right and the left lungs, respectively. The CDSS results for S_{aO2} and P_{aCO2} were 100% and 33 mmHg, respectively. The CDSS predictions are seen to be close to the measured values. Fig. 6 shows the simulation dynamic changes in P_{aCO2} for this patient as generated by the CDSS. As seen in this figure, according to the CDSS, the P_{aCO2} value reaches to about 33 mmHg after around 2.5 h of ILV treatment.

This CDSS is the first system designed to simulate ILV treatments. It can be used in both synchronous and asynchronous types of ventilations and all ventilatory parameters can be assigned independently to each lung. Further applications of this CDSS for different ILV patients in the future can improve the understanding of the effectiveness of the system under various disease conditions.

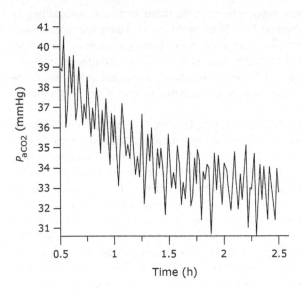

Fig. 6 A CDSS predictions of P_{aCO2} for a patient treated with ILV.

7 Examples of CDSSs used in commercial ventilators

In the preceding sections, the overviews of some model-based advisory CDSSs for mechanical ventilation were provided. In this section, the focus will be on CDSSs that have been used in commercial mechanical ventilators. One of those systems was first introduced in 1992 (Dojat et al., 1992). This system used three patient parameters as inputs: (a) the respiratory rate, (b) the tidal volume, and (c) the end-tidal partial pressure of carbon dioxide measured by using capnometry. This system was designed to be used in the PS mode and during the weaning phase of ventilation. By using this technique, if any of the input parameters was outside a predefined "comfort zone," the level of support by the machine was increased incrementally. If all three parameters were within the "comfort zone," the support level was reduced incrementally until the patient was ready for extubation. This system that underwent further clinical evaluations (Lellouche et al., 2006) was rule-based. The system's rules were fixed for all patients and were based on clinical protocols. A closed-loop version of this system has been marketed by Drager Medical and is known as SmartCare.

In 1991 another closed-loop control system for mechanical ventilation was introduced (Tehrani, 1991a, b). By using this system, the amount of ventilation and the rate of respiration were automatically controlled on the basis of the patient's bodily requirements. The respiratory mechanics data were measured and used to adjust the frequency of respiration on a breath-by-breath basis to minimize the respiratory work rate. The rationale was to provide a natural pattern of breathing to patients to stimulate spontaneous breathing and expedite weaning. For passive patients the ventilator delivered mandatory breaths at optimal depth and rate, while for active patients the machine provided additional support to meet the calculated tidal volume target. This closed-loop system was marketed by Hamilton Medical as adaptive support ventilation (ASV) from the late 1990s and an open-loop advisory version of the system was marketed in the US in the mid 2000s. This system used the continuously monitored respiratory mechanics data and the levels of the arterial oxygen and carbon dioxide pressures of the patient measured by using pulse oximetry and capnometry, respectively.

This closed-loop system that could also be implemented as an open-loop advisory tool was later modified and enhanced to provide automatic and optimal control of several other ventilatory parameters, including PEEP, F_{IO2}, and the level of support by the ventilator during the weaning phase of the treatment. The system used knowledge-based as well as model-based rules to determine the outputs. The enhanced system (Tehrani and Roum, 2008a) did not need any simulation physiological model for oxygen and carbon dioxide transport. However, many of the system rules were adaptive and derived, based on physiological models and hypotheses. As a result, the rules used by the system were derived for each individual patient and the same rules were not applied to all patients. This system could be used as an open-loop advisory tool or as a closed-loop system to control ventilatory parameters. This system was patented and could be used for adults as well as neonates (Tehrani and Abbasi, 2009). The commercial-mode ASV marketed by Hamilton Medical has also been enhanced to control PEEP, F_{IO2}, and the level of support during weaning in recent years.

8 An overview of a CDSS used in closed-loop control of mechanical ventilation

An overview of the last CDSS that was mentioned earlier is provided in this section. Fig. 7 shows a block diagram of this CDSS.

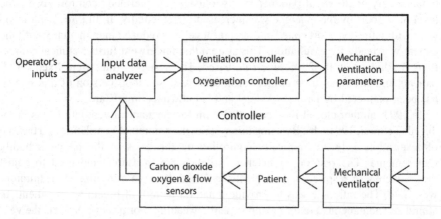

Fig. 7 Block diagram of a closed-loop control system for mechanical ventilation (Tehrani and Roum, 2008a; Tehrani and Abbasi, 2009).

The system shown in Fig. 7 can be used as an advisory open-loop or as an automatic closed-loop system for mechanical ventilation (Tehrani and Roum, 2008a; Tehrani and Abbasi, 2009). As was mentioned, mechanical ventilation treatment can be given noninvasively or invasively to provide partial or full respiratory support to patients. During this treatment, the ventilation and oxygenation levels provided to patients need to be in accordance to their bodily requirements and the pattern and rate of the breathing given by the machine need to be in synchrony with the patient's natural breathing. Nonoptimal treatment provided by the ventilator can lead to unnecessary prolongation of the treatment and many untoward effects on patients' health, and can cause many morbidities and mortalities. The system shown in Fig. 7 and described in Tehrani and Roum (2008a) and Tehrani and Abbasi (2009) is designed to provide optimal and synchronous breathing to patients on mechanical ventilation during the management and weaning phases of the treatment.

As shown in Fig. 7, the controller includes a ventilation controller and an oxygenation controller. The data provided to the controller include inputs from the physician operator such as the patient's illness, gender, weight, and height. The controller also receives data from the sensors and monitors that measure the patient's blood carbon dioxide and oxygen levels by using noninvasive techniques of capnometry and oximetry, respectively, as well as respiratory mechanics data whose values are found from the measurements of a flow sensor. The controller further receives data about the ventilatory parameters as well as the patient's spontaneous breathing activity such as the rate of the spontaneous breaths and their depth.

The input data analyzer receives the inputs and deletes erroneous data due to artifacts. If any patient data are found outside the safe range, the controller activates the alarms of the system accordingly.

The ventilation controller takes the input data and determines the required levels of ventilation, MV, the tidal volume of breaths, V_t, the frequency of respiration, f, the inspiration and expiration times, TI, and TE, and adjusts all the determined values in accordance to the safety rules embedded in the unit. The controller determines the level of ventilation based on the patient's bodily requirements and adjusts the rate and depth of respiration to minimize the respiratory work rate. The controller also determines whether the weaning procedure should be started or stopped and if weaning is already in progress, the controller decides if the ventilation support should be increased or decreased and if so, to what extent the ventilation support should change.

The oxygenation controller determines the concentration of oxygen in the patient's inspiratory gas, F_{IO2}, and the level of PEEP which are two important mechanical ventilation parameters affecting the patient's oxygenation. Based on the mode of ventilation, the controller can also determine the inspiratory pressure for the patient's next breath within safe limits. The determinations of the controller are used to set the outputs of the Mechanical Ventilation Parameters unit (e.g., MV, V_t, f, TI, TE, F_{IO2}, PEEP, the inspiratory pressure, and the weaning level).

If the system is used as an advisory aid, the links between the controller and the mechanical ventilator in Fig. 7 are not provided and instead, the controller's outputs are communicated to the medical personnel. In the advisory mode, the physician restarts or resets the system when the patient's status is reviewed and the ventilatory parameters need to be adjusted. In the advisory mode, a patient's record window is

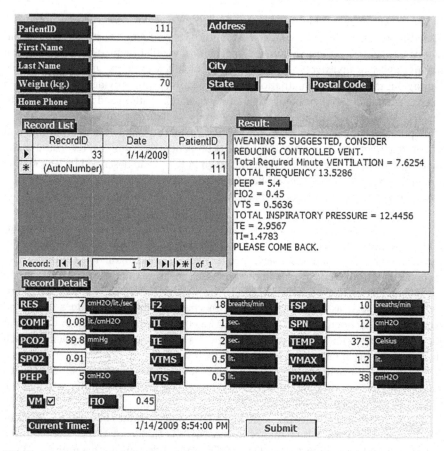

PatientID	111		Address	
First Name				
Last Name			City	
Weight (kg.)	70		State		Postal Code	
Home Phone				

Record List

	RecordID	Date	PatientID
▶	33	1/14/2009	111
✳	(AutoNumber)		111

Record: ◀◀ ◀ | 1 | ▶ | ▶◀ | ▶✳ | of 1

Result:

WEANING IS SUGGESTED, CONSIDER
REDUCING CONTROLLED VENT.
Total Required Minute VENTILATION = 7.6254
TOTAL FREQUENCY 13.5286
PEEP = 5.4
FIO2 = 0.45
VTS = 0.5636
TOTAL INSPIRATORY PRESSURE = 12.4456
TE = 2.9567
TI=1.4783
PLEASE COME BACK.

Record Details

RES	7 cmH2O/lit./sec	F2	18 breaths/min	FSP	10 breaths/min
COMP	0.08 lit./cmH2O	TI	1 sec.	SPN	12 cmH2O
PCO2	39.8 mmHg	TE	2 sec.	TEMP	37.5 Celsius
SPO2	0.91	VTMS	0.5 lit.	VMAX	1.2 lit.
PEEP	5 cmH2O	VTS	0.5 lit.	PMAX	38 cmH2O

VM ☑   FIO   0.45

Current Time: 1/14/2009 8:54:00 PM   Submit

Fig. 8 A patient's record window of the CDSS of Fig. 7 when the system is used in the advisory mode.

opened when the physician starts the procedure. Fig. 8 shows an example of a patient's record window. The data needed from the physician are the following:

- The patient's ID, name, telephone number, etc.
- The patient's ideal body weight (Weight) and his/her body temperature (TEMP).
- The patient's end-tidal carbon dioxide pressure (PCO2).
- The patient's arterial blood oxygen saturation level measured by using pulse oximetry (SPO2).
- The respiratory mechanics data (i.e., respiratory airway resistance (RES) and dynamic compliance (COMP)).
- Ventilatory parameters, including the set tidal volume (VTS), total measured respiratory rate (F2), PEEP, the inspiratory to expiratory time ratio (TI:TE), the fraction of inspired oxygen (FIO which is the same as F_{IO2}), and the ventilation mode, VM (VM = 0 in the volume control/assist modes and is set to "1" in the pressure control/assist modes).

- Measured ventilation parameters, including the measured tidal volume (VTMS), the spontaneous breathing rate (FSP), and the peak inspiratory pressure (SPN)
- The maximum alarm limits for pressure (PMAX) and volume (VMAX).

After the physician enters the required information and submits it for processing, the system comes back with its recommendations as shown in the "Results" box in Fig. 8. The recommendations include suggestions in relation to weaning, total required ventilation, the optimal respiration rate, the values of PEEP and F_{IO2}, the required tidal volume, the required inspiratory pressure, and the updated values for TI and TE as shown in the figure.

When the system of Fig. 7 is used as a closed-loop control system, the outputs of the Controller are provided to the Mechanical Ventilator to control and adjust the outputs of the machine automatically. The patient is provided with ventilation and oxygenation by using the new updated parameters of the system automatically and his/her blood gas values and respiratory mechanics data are measured and provided to the input data analyzer through automatic feedback lines as depicted in Fig. 7.

The system has been tested in adult and infant patient groups. Its detailed algorithms, mathematical equations, and descriptions can be found elsewhere (Tehrani and Roum, 2008a; Tehrani and Abbasi, 2009) and are not repeated here for brevity. This system is covered by US and UK patents (Tehrani, 1991b, 2010, 2008c, 2014) with counterparts in several other countries and marketed commercially as adaptive support ventilation (ASV).

9 Conclusion and future directions

Computerized decision-support systems have many applications in medicine. Those applications include diagnostic imaging, anesthesia, drug and insulin delivery, and mechanical ventilation. If an advisory CDSS is effective and user-friendly, there can be no doubt that it can be a helpful aid to physicians and other medical personnel in treating their patients.

The open-loop applications of many CDSSs can be regarded as the step prior to closed-loop treatment control in medicine. In fact, a number of closed-loop systems have already been commercialized and are known as advanced life-saving technologies, which are used in the operating rooms and the ICUs of hospitals. Examples are the closed-loop anesthesia delivery techniques and automatic ventilation and weaning systems as discussed above.

With the rapid development of technology, it can be anticipated that the trend of automation in medicine will continue and may even accelerate in the future. An effective and robust closed-loop control system, which in many cases can be derived and developed from an open-loop CDSS, not only can provide more optimal treatment to patients but can help prevent medical errors that cause large numbers of patients' morbidities and mortalities every year all over the world.

With regard to mechanical ventilation, with the increasing amount of patient data for analysis and the availability of more robust and reliable monitors and sensors to measure patients' data noninvasively, the applications of CDSSs for mechanical ventilation are on the rise. Among various CDSSs, those that can provide optimal treatments and can be used more easily by the clinicians are likely to find more practical applications.

Furthermore, the trend of mechanical ventilation is toward full automation and closed-loop systems for automatic adjustment of the ventilator parameters have already been marketed. Therefore, CDSSs that can be used both as advisory aids and as closed-loop control systems are likely to be preferred in many applications in the years to come.

Model-based CDSSs can be used to predict the outcomes of various ventilatory treatments as open-loop advisory aids to physicians, particularly in the clinical situations (such as in ILV) that open-loop techniques may be preferred. These systems can also be combined with other CDSSs to set the parameters of ventilation as open-loop or closed-loop systems.

In brief, technological advancements have led to numerous discoveries and revolutionized healthcare in the past several decades. It can be anticipated that the trend will continue and more helpful open-loop CDSSs and more robust and effective closed-loop control technologies and systems will be developed to improve healthcare and become available in the years to come. These systems that are developed in many areas of medicine, including diagnostic imaging, drug delivery, and mechanical ventilation can help physicians provide more optimal treatments to their patients, reduce medical errors, reduce the costs of treatment by reducing the operators' frequent interventions, and provide physicians and other medical personnel with more time to spend on clinical tasks that require their close attentions and personal interventions.

References

Al Herz, R.T., 2014. Double Lung Ventilation, Independent Lung Ventilation. Available at: https://www.slideshare.net/RawanHerz/double-lung-ventilation (last accessed 18 January 2019).

Anantham, D., Jagadesan, R., Tiew, P.E.C., 2005. Clinical review: independent lung ventilation in critical care. Crit. Care 9 (6), 594–600.

Bequette, B.W., 2012. Challenges and recent progress in the development of a closed-loop artificial pancreas. Annu. Rev. Control. 36, 255–266.

Breton, M., Farret, A., Bruttomesso, D., Anderson, S., Magni, L., Patek, S., Man, C.D., Place, J., Demartini, S., Del Favero, S., Toffanin, C., Hughes-Karvetski, C., Dassau, E., Zisser, H., Doyle, F.J., De Nicolao, G., Avogaro, A., Cobelli, C., Renard, E., Kovatchev, B., on behalf of the International Artificial Pancreas (iAP) Study Group, 2012. Fully integrated artificial pancreas in type 1 diabetes. Modular closed-loop glucose control maintains near normoglycemia. Diabetes 61, 2230–2237.

Campos-Delgado, D.U., Hernandez-Ordonez, M., Femat, R., Gordillo-Moscoso, A., 2006. Fuzzy-based controller for glucose regulation in type 1 diabetic patients by subcutaneous route. IEEE Trans. Biomed. Eng. 53 (11), 2201–2210.

Chalmers, E., Pedrycz, W., Lou, E., 2015. Human experts' and a fuzzy model's predictions of outcomes of scoliosis treatment: a comparative analysis. IEEE Trans. Biomed. Eng. 62 (3), 1001–1007.

Chatburn, R., El-Khatib, M., Mireles-Cabodevila, E., 2014. A taxonomy for mechanical ventilation: 10 fundamental maxims. Respir. Care 59 (11), 1747–1763.

Chatburn, R.L., Mireles-Cabodevila, E., 2011. Closed-loop control of mechanical ventilation: description and classification of targeting schemes. Respir. Care 56 (1), 85–102.

Cinnella, G., Dambrosio, M., Brienza, N., Guiliani, R., Bruno, F., Fiore, T., Brienza, A., 2001. Independent lung ventilation in patients with unilateral pulmonary contusion, Monitoring with compliance and $EtCO_2$. Intensive Care Med. 27 (12), 1860–1867.

Dassau, E., Zisser, H., Harvey, R.A., Percival, M.W., Grosman, B., Bevier, W., Atlas, E., Miller, S., Nimri, R., Jovanovic, L., Doyle, F.J., 2013. Clinical evaluation of a personalized artificial pancreas. Diabetes Care 36, 801–809.

Dojat, M., Harf, A., Touchard, D., Lemaire, F., Brochard, L., 2000. Clinical evaluation of a computer-controlled pressure support mode. Am. J. Respir. Crit. Care Med. 161, 1161–1166.

Dojat, M., Brochard, L., Lemaire, F., Harf, A., 1992. A knowledge-based system for assisted ventilation of patients in intensive care units. Int. J. Clin. Monit. Comput. 9 (4), 239–250.

Doyle III, F.J., Huyett, L.M., Lee, J.B., Zisser, H.C., Dassau, E., 2014. Closed-loop artificial pancreas systems: engineering the algorithms. Diabetes Care 37, 1191–1197.

Dua, P., Dua, V., Pistikopoulos, E.N., 2008. Optimal delivery of chemotherapeutic agents in cancer. Comput. Chem. Eng. 32, 99–107.

Dua, P., Pistikopoulos, E.N., 2005. Modelling and control of drug delivery systems. Comput. Chem. Eng. 29, 2290–2296.

Dua, P., Dua, V., Pistikopoulos, E.N., 2010. Modelling and multi-parametric control for delivery of anaesthetic agents. Med. Biol. Eng. Comput. 48, 543–553.

Elleri, D., Allen, J.M., Kumareswaran, K., Leelarathna, L., Nodale, M., Caldwell, K., Cheng, P., Kollman, C., Haidar, A., Murphy, H.R., Wilinska, M.E., Acerini, C.L., Dunger, D.B., Hovorka, R., 2013. Closed-loop basal insulin delivery over 36 hours in adolescents with type 1 diabetes. Diabetes Care 36, 838–844.

Fincham, W.F., Tehrani, F.T., 1983b. On the regulation of cardiac output and cerebral blood flow. J. Biomed. Eng. 5 (1), 73–75.

Fincham, W.F., Tehrani, F.T., 1983a. A mathematical model of the human respiratory system. J. Biomed. Eng. 5 (2), 125–133.

Gaspari, M., Saletti, D., Scandellari, C., Stecchi, S., 2009. Refining an automatic EDSS scoring expert system for routine clinical use in multiple sclerosis. IEEE Trans. Inf. Technol. Biomed. 13 (4), 501–511.

Glass, D.D., Tonnesen, A.S., Gabel, G.C., Arens, J.F., 1976. Therapy of unilateral pulmonary insufficiency with a double lumen endotracheal tube. Crit. Care Med. 4 (6), 323–326.

Hernandez, A.M., Mananas, M.A., Costa-Castello, R., 2008. Learning respiratory system function in BME studies by means of a virtual laboratory: respilab. IEEE Trans. Educ. 51 (1), 24–34.

Hobbs, F.D.R., Delaney, B.C., Carson, A., Kenkre, J.E., 1996. A prospective controlled trial of computerized decision support for lipid management in primary care. Fam. Pract. 13, 133–137.

Hovorka, R., Elleri, D., Thabit, H., Allen, J.M., Leelarathna, L., El-Khairi, R., Kumareswaran, K., Caldwell, K., Calhoun, P., Kollman, C., Murphy, H.R., Acerini, C.L., Wilinska, M.E., Nodale, M., Dunger, D.B., 2014. Overnight closed-loop insulin delivery in young people with type 1 diabetes: a free-living randomized clinical trial. Diabetes Care 37, 1204–1211.

Kwok, H.F., Linkens, D.A., Mahfouf, M., Mills, G.H., 2003. Rule-based derivation for intensive care ventilator control using ANFIS. Artif. Intell. Med. 29, 185–201.

Lellouche, F., Mancebo, J., Jolliet, P., Roeseler, J., Schortgen, F., Dojat, M., Cabello, B., Bouadma, L., Rodriguez, P., Maggiore, S., Reynaert, M., Mersmann, S., Brochard, L.,

2006. A multicenter randomized trial of computer driven protocolized weaning from mechanical ventilation. Am. J. Respir. Crit. Care Med. 174, 894–900.

Lin, C.T., Chang, K.C., Lin, C.L., Chiang, C.C., Lu, S.W., Chang, S.S., Lin, B.S., Liang, H.Y., Chen, R.J., Lee, Y.T., Ko, L.W., 2010. An intelligent telecardiology system using a wearable and wireless ECG to detect atrial fibrillation. IEEE Trans. Inf. Technol. Biomed. 14 (3), 723–733.

Lozano-Zahonero, S., Gottlieb, D., Haberthur, C., Guttmann, J., Moller, K., 2011. Automated mechanical ventilation: adapting decision making to different disease states. Med. Biol. Eng. Comput. 49 (3), 349–358.

Mahfouf, M., 2005. Constrained closed-loop control of depth of anaesthesia in the operating theatre during surgery. Int. J. Adapt. Control Signal Process. 19, 339–364.

Montgomery, A.A., Fahey, T., Peters, T.J., MacIntosh, C., Sharp, D.J., 2000. Evaluation of computer based clinical decision support system and risk chart for management of hypertension in primary care: randomized controlled trial. Br. Med. J. 320, 686–690.

Nascu, I., Pistikopoulos, E.N., 2017. Modeling, estimation and control of the anaesthesia process. Comput. Chem. Eng. 107, 318–332.

Nemoto, T., Hatzakis, G.E., Thorpe, C.W., Olivenstein, R., Dial, S., Bates, J.H.T., 1999. Automatic control of pressure support mechanical ventilation using fuzzy logic. Am. J. Respir. Crit. Care Med. 160, 550–556.

Ost, D., Corbridge, T., 1996. Independent lung ventilation. Clin. Chest Med. 17 (3), 591–601, 1996.

Rees, S.E., Allerod, C., Murley, D., Zhao, Y., Smith, B.W., Kjargaard, S., Thorgaard, P., Andreassen, S., 2006. Using physiological models and decision theory for selecting appropriate ventilator settings. J. Clin. Monit. Comput. 20 (6), 421–429.

Pefani, E., Panoskaltsis, N., Mantalaris, A., Georgiadis, M.C., Pistikopoulos, E.N., 2013. Design of optimal patient-specific chemotherapy protocols for the treatment of acute myeloid leukemia (AML). Comput. Chem. Eng. 57, 187–195.

Ritchie, R.G., Ernst, E.A., Late, B.L., Pearson, J.D., Sheppard, L.C., 1987. Closed-loop control of an anesthesia delivery system. Development and animal testing. IEEE Trans. Biomed. Eng. 34, 437–443.

Shahsavar, N., Ludwigs, U., Blomqvist, H., Gill, H., Wigertz, O., Matell, G., 1995. Evaluation of a knowledge-based decision-support system for ventilator therapy management. Artif. Intell. Med. 7, 37–52.

Shortliffe, E., 1986. Medical expert systems knowledge tools for physicians. West. J. Med. 145, 830–839.

Strickland Jr., J.H., Hassan, J.H., 1993. A computer-controlled ventilator weaning system. Chest 103, 1220–1226.

Suzuki, K., Li, F., Sone, S., Doi, K., 2005. Computer-aided diagnostic scheme for distinction between benign and malignant nodules in thoracic low-dose CT by use of massive training artificial neural networks. IEEE Trans. Med. Imaging 24 (9), 1138–1150.

Tehrani, F.T., Roum, J.H., 2008a. FLEX: a new computerized system for mechanical ventilation. J. Clin. Monit. Comput. 22 (2), 121–130.

Tehrani, F.T., Roum, J.H., 2008b. Intelligent decision support systems for mechanical ventilation. Artif. Intell. Med. 44 (3), 171–182.

Tehrani, F.T., Abbasi, S., 2009. Evaluation of a computerized system for mechanical ventilation of infants. J. Clin. Monit. Comput. 23 (2), 93–104.

Tehrani, F.T., 2011. Computerized decision support systems for mechanical ventilation. In: Jao, C. (Ed.), Efficient Decision Support Systems: Practice and Challenges in Biomedical Related Domain. INTECH Publishers, London, UK, pp. 227–238.

Tehrani, F.T., 2008a. Automatic control of mechanical ventilation. Part 1. Theory and history of the technology. J. Clin. Monit. Comput. 22 (6), 409–415.

Tehrani, F.T., 2008b. Automatic control of mechanical ventilation. Part 2. The existing techniques and future trends. J. Clin. Monit. Comput. 22 (6), 417–424.

Tehrani, F.T., Method and Apparatus for Controlling a Ventilator, U.K. Patent No. GB2423721, issued October 14, 2008c.

Tehrani, F.T., Abbasi, S., 2012. A model-based decision support system for critiquing mechanical ventilation treatments. J. Clin. Monit. Comput. 26 (3), 207–215.

Tehrani, F.T., 1991a. Automatic control of an artificial respirator. In: Proceedings of the 13th Annual International Conference of IEEE Engineering in Medicine and Biology, pp. 1738–1739.

Tehrani, F. T., Method and Apparatus for Controlling an Artificial Respirator, US Patent No. 4,986,268, issued January 22, (1991b).

Tehrani, F.T., Method and Apparatus for Controlling a Ventilator, U.S. Patent No. 7,802,571, issued September 28, 2010.

Tehrani, F.T., Weaning and Decision Support System for Mechanical Ventilation, U.S. Patent No. 8,695,593, issued April 15, 2014.

Tehrani, F.T., 2019. Computerised decision support for differential lung ventilation. Healthcare Technol. Lett. 6 (2), 37–41.

Tehrani, F.T., 1993. Mathematical analysis and computer simulation of the respiratory system in the newborn infant. IEEE Trans. Biomed. Eng. 40 (5), 475–481.

Ting, C.H., Arnott, R.H., Linkens, D.A., Angel, A., 2004. Migrating from target-controlled infusion to closed-loop control in general anaesthesia. Comput. Methods Prog. Biomed. 75, 127–139.

Velliou, E.G., Nascu, I., Zavitsanou, S., Pefani, E., Krieger, A., Giorgiadis, M.C., Pistikopoulos, E.N., 2018. Framework and tools: a framework for modelling, optimization and control of biomedical systems. In: Pistikopoulos, E.N., Nascu, I., Velliou, E.G. (Eds.), Modelling Optimization and Control of Biomedical Systems. John Wiley & Sons Ltd, NJ, USA, pp. 3–11.

Vishoni, R., Roy, R.J., 1991. Adaptive control of closed-circuit anesthesia. IEEE Trans. Biomed. Eng. 38 (1), 39–47.

Walton, R., Dovey, S., Harvey, E., Freemantle, N., 1999. Computer support for determining drug dose: systemic review and meta-analysis. Br. Med. J. 318, 984–990.

Zadeh, L.A., 1983. The role of fuzzy logic in management of uncertainty in expert systems. Fuzzy Sets Syst. 11, 199–227.

Customized modeling and optimal control of superovulation stage in in vitro fertilization (IVF) treatment

13

Urmila Diwekar[a], Apoorva Nisal[b], Kirti Yenkie[c], Vibha Bhalerao[d]
[a]Vishwamitra Research Institute, Stochastic Research Technologies LLC, The University of Illinois at Chicago, Chicago, IL, United States, [b]The University of Illinois, Chicago, IL, United States, [c]Chemical Engineering, Rowan University, Glassboro, NJ, United States, [d]Jijamata Hospital and IVF Center, Nanded, India

1   Introduction

A survey conducted by the World Health Organization (WHO) in 2010 using data from 190 countries over a period of 20 years found that around 2% women suffer from primary infertility and 10% women suffer from secondary infertility. Primary infertility is the inability to conceive a first live birth and secondary infertility is the inability to conceive after a prior live birth. Certain regions of Eastern Europe, North Africa, Middle East, Oceania, and Sub Saharan Africa showed greater prevalence of infertility (Mascarenhas et al., 2012). In the United States itself, data collected by the Center for Disease Control (CDC) over a 4-year span showed 6.7% of married women to be suffering from infertility (FastStats, 2018).

In vitro fertilization (IVF) process is one of the most commonly recommended treatments in Assisted Reproductive Technologies (ART). 1.7% of infants were born through ART in the United States in 2015 (Sunderam et al., 2018). IVF is a process by which oocytes or egg cells are fertilized by a sperm outside the body in a laboratory simulating similar conditions in the body, and then the fertilized eggs or embryos are implanted back in the uterus for a full-term pregnancy. It has four basic stages (Fritz and Speroff, 2010): superovulation, egg retrieval, insemination/fertilization, and embryo transfer as shown in Fig. 1.

IVF is an expensive treatment, and the out-of-pocket costs per cycle tend to be around $10,000–$15,000. This cost varies and increases with multiple factors such as unsuccessful IVF cycles, multiple births, low-birthweight infants, and preterm births occurring from IVF cycles (Sunderam et al., 2018). The cost of IVF depends upon the cost of superovulation. Currently, this step is executed using almost daily monitoring of the follicular development using ultrasound and blood test. The daily dosage of hormones is customized for each patient based on these tests. Conventionally, doses are prescribed based on empirical data instead of randomized control trials

Control Applications for Biomedical Engineering Systems. https://doi.org/10.1016/B978-0-12-817461-6.00013-5

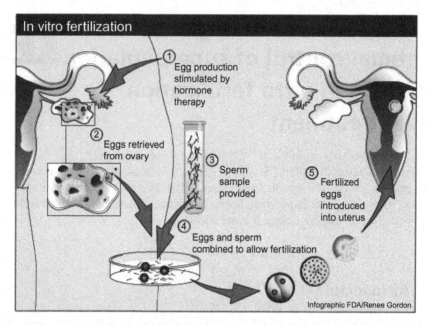

Fig. 1 Schematic diagram of the in vitro fertilization procedure (Gordon, 2012).
From Sage K. (2016) Reproductive Decisions. In: Child A. (eds) Diagnosis and Management of Marfan Syndrome. Springer, London.

and start at 150 or 225 IU. Devroey and team employed initial low-dose FSH (follicle-stimulating hormone) (100 IU) on a relatively young age group and recorded a high number of retrieved oocytes (Devroey et al., 1998). Prescribed minimum dosages start from 150 to 300 IU for younger patients and reach the absolute maximum at 450 IU for poor responders (Jungheim et al., 2015; Rombauts, 2007; Dorn, 2005). Certain factors, which come into play when choosing an FSH dose for a patient, are usually female age, anamnesis, clinical criteria, and ovarian markers such as AFC (antral follicle count) and AMH (anti-Mullerian hormone) (La Marca and Sunkara, 2014). FSH starting dose based on AFC was found to be less than 225 IU for most patients under the age of 35 years (La Marca et al., 2013). Although there are general guidelines for the dosage limits, the dose is not optimized for each patient. IVF procedure can have side effects such as the ovarian hyper stimulation syndrome (OHSS) (Alper et al., 2009), and the remedial actions are still unidentified. Around 1%–2% of women undergoing IVF suffer from a serious case of OHSS (Klemetti et al., 2005). Patients suffering from polycystic ovarian syndrome (PCOS) are found to be the ones most susceptible to OHSS. However, many patients who do not suffer from PCOS may also develop OHSS after stimulation. Protocols based on factors like age, AMH, AFC, FSH, BMI (body mass index) levels and smoking history predict optimal protocols with highest follicle yield and reduced occurrence of OHSS (Yovich et al., 2016). However, all the existing protocols are based on patient history, testing and monitoring, and professional judgment of the physician. The complications such as overstimulation or unsuccessful superovulation do occur. The cost associated with patient monitoring and testing as well as the hormonal drugs make the superovulation stage very expensive. The evidence is building in support of personalized IVF treatment (Nyboe Andersen et al., 2017, Sighinolfi et al., 2017; Simopoulou et al., 2018) and tools that can suggest optimal patient-specific drug-dosage profiles to reduce hyperstimulation, cost of treatment, improve the oocyte quality, and quantity to increase the overall success rate of IVF, resulting in successful pregnancies and live birth. Trials have shown that mild stimulation individualized protocol for poor response patients with low antral follicular count (AFC < 11) yield similar pregnancy rates with largely reduced dosages per woman (van Tilborg et al., 2017; Youssef et al., 2016). A report evaluating merits of mild stimulation protocol for poor responders suggests that dosage of ≤150 IU/day yields high-quality oocytes and similar pregnancy rates compared to conventional protocols (Practice Committee of the American Society for Reproductive Medicine, 2018). A nomogram prediction model based on age, day 3 FSH and AMH to select appropriate starting FSH dose proposed by La Marca and team was validated through a randomized trial and showed an increased number of patients exhibiting optimal oocyte retrieval response (Allegra et al., 2017; La Marca et al., 2013). Moon and team also employed nomogram to predict the number of oocytes retrieved from IVF cycles from different univariate and multivariate models based on age, serum FSH, AFC, AMH levels (Moon et al., 2016). While both AFC and AMH are good predictors of ovarian reserve and oocyte yield, AFC poses advantages such noninvasive measurement, ease of testing, and smaller testing time over AMH (Fleming et al., 2015; Kotanidis et al., 2016). Recently, a modeling approach and a computerized algorithm to generate customized hormonal dosing policies for enhanced superovulation results, and reduced cost and decreased testing were presented by our group (Yenkie et al., 2013; Yenkie and Diwekar, 2014; Nisal et al., 2019). There are four commonly used protocols for IVF.

The four protocols (Scoccia, 2017), which are generally used: (1) Long Lupron agonist protocol, (2) microflare agonist protocol, (3) three-stop Lupron agonist protocol, and (4) Flexible GnRH antagonist (Ganerelix or Cetrorelix) protocol with NEA. This approach is presented for the first of the four protocols in this chapter. The validation of the procedure is carried out using clinical data from patients who have previously undergone IVF cycles. Initial two-day data for each patient are used to obtain parameters of the model for that patient. The model is used then to predict FSD for the remaining days of the cycle. This procedure was conducted for 49 patients. The results of the customized models are found to be closely matching with the observed FSD on the successive days of the IVF superovulation cycle. This customized model is then used to optimize the dosage for this patient. The FSD at the end of the cycle was determined using the model and the optimized dosages. A small clinical trial was also conducted in India. This was a double-blinded trial. The results show that the dosage predicted by using the model is 40% less than that suggested by the IVF doctors. It also shows that the number of mature follicles obtained at the end of the cycle using the dosage predicted by the model is significantly higher than that of physician suggested dosage. These results were consistent with all patients in this clinical trial. The testing requirements for these patients with optimized drug dosage are also reduced by 72%.

The rest of the chapter is arranged as follows. Section 2 describes the customized modeling approach for each patient, followed by Section 3 on the optimal control for the determination of hormonal dosage profiles for each patient. These two sections provide results for clinical data from 49 patient cycles obtained from Jijamata Hospital, Nanded, India. The overall approach is summarized in Section 4. This section also presents results from the small clinical trial conducted at the Jijamata hospital in India. Section 5 summarizes the overall approach and provides insights into our ongoing work and future directions.

2 Modeling of in vitro fertilization

In the superovulation stage of IVF, multiple follicles enter into the growth phase and increase in size due to externally injected hormones (Baird, 1987). There are significant similarities between the superovulation stage of IVF and the particulate process of batch crystallization (Hill, 2005; Yenkie and Diwekar, 2012), which has been well studied in the chemical engineering discipline. We used these similarities to develop a model for the daily follicle size distribution in IVF (Table 1).

The properties of a particulate system can be represented by moments of its particle size distribution (Randolph, 1988). The moment model for follicle number and size was adapted from the concept of batch crystallization (Hill, 2005) based on the analogy between batch crystallization and superovulation presented in Table 1. The superovulation follicle growth model, in general, resembles greatly to the growth of seeded batch crystals (Hu et al., 2005). The aim of seeded batch crystallization is to allow the seeds added to the solution to grow to desired shape and size and truncate the process of nucleation by maintaining certain process conditions. The numbers of seed added to the solution are constant and hence the zeroth moment of seeded batch crystals, which corresponds to its number, is constant. Similarly, when we look at superovulation, the

Table 1 Analogy between batch crystallization and IVF superovulation stage

Batch crystallization	Superovulation (IVF stage I)
• Production of multiple crystals	• Production of multiple oocytes or eggs
• Crystal quality is determined in terms of size distribution and purity	• Oocyte quality is determined in terms of no abnormalities, similar size
• The rate of crystallization or crystal growth varies with time and process conditions	• The rate of ovulation or oocyte growth varies with time and drug interactions
• The process is affected by external variables like agitation, and process operating variables like temperature, pressure, etc.	• The process is affected by externally administered drugs and body conditions of the patient undergoing the process

number of follicles activated during an IVF cycle is constant. Thus, the moment model for both the processes can be similar; the growth term, which is a function of process variables like temperature and supersaturation in seeded batch crystallization, becomes a function of hormonal dosage in case of superovulation process.

2.1 Data organization and moment calculation

Due to ovarian stimulation using externally injected hormones, the number of follicles entering the ovulation stage is more in number as compared to a single follicle in a normal menstrual cycle. The superovulation cycle data obtained from Jijamata Hospital, India, had measurements of follicle size and number along with the amounts of hormone administration. The data for example for patient 1 are reorganized as shown in Table 2 in terms of bin sizes of various diameters of follicles.

The data represented in Table 2 can be converted to moments using the general expression shown in Eq. (1).

$$\mu_i = \sum n_j(r, t) r_j^i \Delta r_j \tag{1}$$

Here, μ_i is the ith moment, $n_j(r, t)$ is the number of follicles in bin "j" of mean radius "r" at time "t," r_j is the mean radius of jth bin, and Δr is the range of radii variation in each bin. Here, the follicle sizes are divided into 6 bins; thus for efficient process modeling, it is essential to consider at least first 6 orders of moments along with the zeroth moment (Flood, 2002). Table 3 shows the moment values evaluated using Eq. (1).

2.2 Model equations

The moment-based model for predicting follicle size and number will involve the follicle growth rate and moment equations. It is assumed that the follicle growth is dependent on the amount of FSH administered since it the most influential hormone in

Table 2 Variation of follicle size (diameter) with time and FSH dose in patient

↓ Size bins (mm)	Number of follicles			
Time →	Day 1	Day 5	Day 7	Day 9
0–4	4	0	0	0
4–8	12	0	0	0
8–12	8	17	1	0
12–16	2	6	15	3
16–20	0	3	10	15
20–24	0	0	0	8
FSH dose (IU/ml)	150	75	75	75

IU—International units used for hormonal dosage measurement.

Table 3 Moments evaluated for patient 1

Sr. No	Time (day)	μ_0	μ_1	μ_2	μ_3	μ_4	μ_5	μ_6	FSH (IU/mL)
1.	1	52	188	820	4028	21556	123,068	738,100	150
2.	5	52	308	1924	12740	89428	662,228	5,131,684	75
3.	7	52	400	3140	25120	204500	1,691,440	14,189,540	75
4.	9	52	488	4660	45224	445492	4,449,128	44,994,100	75

follicular dynamics. Thus, the growth term is written as a function of FSH dose; as shown in Eq. (2). Here, G is the follicle growth term, k is the rate constant, ΔC_{fsh} is the amount of FSH injected, and α is the rate exponent.

$$G(t) = k\Delta C_{fsh}(t)^{\alpha} \tag{2}$$

The number of follicles activated for growth is assumed to be a constant due to the literature suggested by Baird (1987) and clinical data from the Jijamata hospital; hence, zeroth moment is constant. The 1st to 6th order moments are used for efficient recovery of the size distributions. The moment equations for the follicle dynamics are Eqs. (3)–(9). Here, $G(t)$ is the follicle growth term and μ_i is the ith moment. It can be clearly seen that the $(n + 1)$th moment is dependent upon the nth moment.

$$\mu_0 = \text{constant} \tag{3}$$

$$\frac{d\mu_1}{dt} = G(t)\mu_0(t) \tag{4}$$

$$\frac{d\mu_2}{dt} = 2G(t)\mu_1(t) \tag{5}$$

$$\frac{d\mu_3}{dt} = 3G(t)\mu_2(t) \tag{6}$$

$$\frac{d\mu_4}{dt} = 4G(t)\mu_3(t) \tag{7}$$

$$\frac{d\mu_5}{dt} = 5G(t)\mu_4(t) \tag{8}$$

$$\frac{d\mu_6}{dt} = 6G(t)\mu_5(t) \tag{9}$$

Patient parameters μ_0, k, and α for the models are determined from fitting the results of Eqs. (1)–(9) to the moment data at different times as shown in Table 2. In this protocol, in the clinical (experimental) settings, the attending physician determines the initial dosage for the patient based on various patient factors. The first-day ultrasound and blood test provides the baseline total number of follicles. The same dose is continued for the first four days and then testing with ultrasound and blood test starts. Depending on the tests, each day dose is determined based on the follicular distribution seen in ultrasound and the blood test results. Since there are three parameters involved, we need a minimum of two days of data for the model. The validity of the model can be evaluated by comparing the FSD predicted by the model from 5th day onwards with that of observed or experimental values.

2.3   FSD evaluation

Eqs. (2)–(9) predict the moment values. However, the desired output is required in the form of FSD; thus, the approach to obtain FSD from moment values is illustrated later. The distribution is approximated using the inversion matrix (**A**) shown in Table 4 along with a nonlinear constrained optimization technique (Flood, 2002; Yenkie et al., 2013). The method suggested is represented in Eq. (10), which can be rewritten to keep the unknown variable (n) on the L.H.S. as Eq. (12).

$$\mu = \mathbf{A}\,n \tag{10}$$

$$n = \mathbf{A}^{-1}\mu \tag{11}$$

Here, n is the vector of a number of follicles in all the size bins at the ith day in the cycle, μ is the moment vector for the ith day, and **A** is the inversion matrix. For the current bin size of 2 mm (radii) and the number of bins as 6, the inversion matrix **A** is shown in Table 4.

2.4   Follicle number prediction algorithm

To predict the number of follicles in a particular size bin on a particular day in the FSH dosage regime, a constrained optimization algorithm is applied. The optimization variables in the suggested algorithm are the number of follicles (n) per day in the superovulation cycle. Using this follicle number prediction algorithm, the moment model for follicle growth can be validated for a given patient.

> **Step #1**: Assign some initial values for n (number of follicles/day) within the different size bins used in the model.
> **Step #2**: Obtain the moment values by multiplying the matrix **A** with the initially assumed n values.
> **Step #3**: Introduce the constraint for the total number of follicles. It is assumed that a constant number of follicles ($\mu_0/2$) enter the growth stage in the IVF cycle for a particular patient. Hence, for each day, the number of follicles must sum up to the assumed constant value.
> **Step #4**: Restrict the values of n to be either positive or zero since the number of follicles can never be negative.

Table 4 The inversion matrix A (6 × 6) to recover size distribution from moments

A =	2	6	10	14	18	22
	2	18	50	98	162	242
	2	54	250	686	1458	2662
	2	162	1250	4802	13,122	29,282
	2	486	6250	33,614	118,098	322,102
	2	1458	31,250	235,298	1,062,882	3,543,122

Step #5: Write the objective function Eq. (12) to minimize the sum of the square of errors between the model predicted moments and the ones obtained from Eq. (10).

$$\text{Min } F(nk, j) = \sum_{i=1;j=1}^{i=m;j=N_{bins}} \left(\frac{\mu_{i,j}^{eval} - \mu_{i,j}^{model}}{\mu_{i,j}^{eval}} \right)^2 \tag{12}$$

Here, $n_{k,j}$ is the number of follicles on the kth day in j bins, i is the order of the moment (1–6), j is the number of bins, $\mu_{i,j}^{eval}$ is obtained by using Eq. (10), and $\mu_{i,j}^{model}$—obtained from moment model using optimum k and α.

Step #6: Use a constrained nonlinear optimization method to obtain the values of n.
Step #7: Compare the optimum values of n obtained from this constrained optimization method to the actual data observed for the patient.

2.5 Model validation

The model described here uses the data collected on the first and fifth days to calibrate the model. However, if all-day data are used, the model fit is going to be much better. Fig. 2 shows the FSD for various days observed in real practice (denoted as experimental values (E)) compared to the model predictions (denoted as (M)) considering data from considering all-day data (Fig. 2A) and considering only the two-day data (Fig. 2B) for patient 2. Similarly, the results are presented for patient 3 in Fig. 3A and 3B, respectively. This shows that the model performs very well for these two patients irrespective of either two-day or all-day data. The results of these two patients are selected as they represent two ends of the different age spectrum.

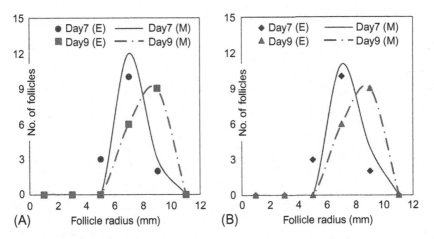

Fig. 2 Comparison of observed (E) follicular distribution with the follicle size distribution predicted by customized model (M) for various days for patient 2: (A) patient parameters estimated from all-day data and (B) patient parameters estimated using two-day data.

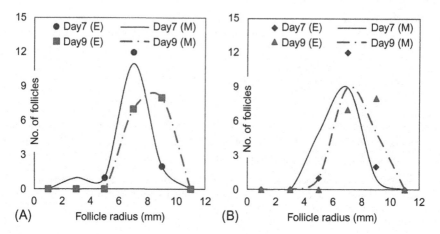

Fig. 3 Comparison of observed (*E*) follicular distribution with the follicular distribution predicted by customized model (*M*) for various days for patient 3: (A) patient parameters estimated from all-day data and (B) patient parameters estimated using two-day data.

As stated earlier, data have been gathered for 49 patients from Jijamata Hospital, India. The data are used to study the predictive capability of the model for the final day of stimulation. Fig. 4 presents the histogram of the ratio of final-day mature follicles predicted by the model ($n_{mature,\ M}$) to final-day mature follicles observed experimentally ($n_{mature,\ E}$) in real practice. Fig. 4A presents the prediction from all-day data and Fig. 4B presents predictions from the two-day data. For most of the patients (more than 90% of the patients), the model shows a good fit for all-day data versus 70%

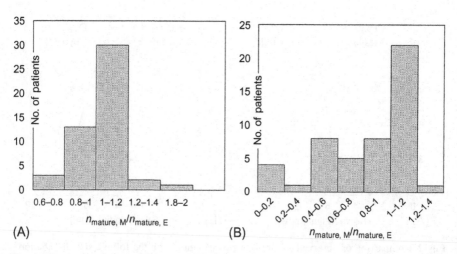

Fig. 4 Histogram of $n_{mature,M}/n_{mature,E}$ for 49 patients: (A) model fitted using all-day data and (B) model fitted using two-day data.

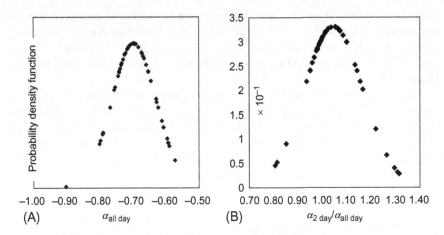

Fig. 5 Probability density function for the patient parameter α: (A) distribution of $\alpha_{\text{all-day}}$ and (B) distribution of error $\alpha_{2\text{-day}}/\alpha_{\text{all day}}$.

for two-day data predictions. Although the model predictions are not that good for 30% of the patients for the two-day data, it is important to find out whether the optimal control profile predicted using the 2-day data can be still used for these patients. This is studied in next section.

2.6 Results from parameter estimation

It has been observed that the model parameter k (follicle growth rate constant) is the same for all patients and is found to be 22. Therefore, it can be concluded that k is patient independent. However, α (follicle growth rate exponent) changes for each patient. The probability distribution for $\alpha_{\text{all-day}}$ and the distribution of error in $\alpha_{2\text{-day}}$ compared to $\alpha_{\text{all-day}}$ is shown in Fig. 5A and B, respectively. The analysis of the outliers from the histograms is presented in Fig. 4A. It is observed that two patients are outliers. Further analysis of these two patients revealed that $\alpha_{2\ \text{day}}$ value for both is above -0.92. Thus, it can be said that the model is the best fit for values of alpha ranging from "-0.5" to "-0.9."

3 Optimal control for customized optimal dosage determination

Optimal control is a method for evaluating the time-varying values of certain process variables, also known as the control variable, which aid in achieving the desired outcome. It falls under a special category of optimization problems in which the optimization variable is a time-dependent vector instead of a single value. It has a wide range of applications in industrial processes, unit operations, and biomedicine.

In biomedical field, optimal control has been used for predicting cancer chemotherapy and tumor degradation (Castiglione and Piccoli, 2007; Czako et al., 2017), drug scheduling in HIV infection treatment (Khalili and Armaou, 2008), and for blood glucose regulation in insulin-dependent diabetes patients (Ulas Acikgoz and Diwekar, 2010).

There are various methods for solving optimal control problems such as calculus of variations, dynamic programming, maximum principle, and nonlinear programming discussed in detail by Diwekar (2008). In IVF, the control variable is the value of hormonal doses per day of the treatment cycle. The objective of superovulation is to obtain a high number (maximum possible) of uniformly sized (18- to 22-mm diameter) follicles on the last day of FSH administration.

3.1 Mathematical formulation

The data on superovulation cycles from the collaborators indicate that after the initial 4–5 days of FSH administration the follicle size and number plots tend to follow a Gaussian/Normal distribution and as time progresses this distribution continues to follow a Gaussian trend with a shift in the mean and variance. This distribution is used to define the objective function in terms of the moments. The moment model for FSD prediction and the method for deriving normal distribution parameters (John et al., 2007) have been used as the basis for deriving expressions for the mean and coefficient of variation.

Since the data clearly reflect a normal distribution, it is quite reasonable to assume it as an a priori distribution for follicles and the following mean (Eq. 13) and coefficient of variation (Eq. 14) expressions can be derived in terms of moments. Here, $\bar{x}$ is the mean follicle size, CV is the coefficient of variation.

$$\bar{x} = \frac{\mu_1}{\mu_0} \tag{13}$$

$$CV = \sqrt{\frac{\mu_2 \mu_0}{\mu_1^2} - 1} \tag{14}$$

Thus, the objective of superovulation in the mathematical form can be stated as to minimize the coefficient of variation on the last day of FSH administration ($CV(t_f)$) and the control variable shall be the dosage of FSH with time ($C_{fsh}(t)$). Here t_f is the final day of the cycle.

To customize the model for each patient, the parameters are evaluated using the initial two-day observations of the follicle size and counts along with the FSH administered. The optimal dosage prediction for the desired superovulation outcome is represented as Eq. (15).

$$\underset{C_{fsh}}{\text{Min}} \, CV(t_f) \tag{15}$$

Subject to:

(i) Follicle growth term and moment model
(ii) Additional equations for the coefficient of variation (CV) and mean ($\bar{x}$)

$$\frac{dCV}{dt} = \frac{G\mu_0}{C_v\mu_1}\left\{1 - \frac{\mu_0\mu_2}{\mu_1^2}\right\} \tag{16}$$

$$\frac{d\bar{x}}{dt} = G \tag{17}$$

(iii) The final size of the follicles must not exceed 22 mm in diameter.

3.2 Solution by maximum principle

The control problem has 9 state variables resulting in 9 state equations. For simplicity of notations "y_i" is used to denote the ith state variable. In maximum principle, one adjoint variable is introduced corresponding to each state variable. Thus, there are 9 adjoint variables resulting in 9 additional equations. Let z_i be the ith adjoint variable. The objective function is then converted to the Hamiltonian form (H), which on expansion involves the state as well as adjoint variables. The optimality condition for the problem is given by Eq. (22). A tolerance level is fixed for the derivative of the Hamiltonian with respect to the control variable (dH/dC_{fsh}) and can be written in a more realistic manner as the condition "$abs[dH/dC_{fsh}|t] < tolerance$." Initial values are available for the state variables, whereas final values are available for the adjoint variables. This results in a two-point boundary value problem.

Let $y_i = [\mu_0\mu_1\mu_2\mu_3\mu_4\mu_5\mu_6 CV\bar{x}]$ then;

$$\text{Max}_{C_{fsh}(t)}\{-y_8(t_f)\} \tag{18}$$

$$\frac{dy_i}{dt} = f(y_i, t, C_{fsh}) \tag{19}$$

$$\frac{dz_i}{dt} = \sum_{j=1}^{9} z_j\frac{\partial f(y_i, t, C_{fsh})}{\partial y_i} = f(y_i, z_i, t, C_{fsh}) \tag{20}$$

$$H = \sum_{i=1}^{9} z_i f(y_i, t, C_{fsh}) \tag{21}$$

$$\left|\frac{dH}{dC_{fsh}}\right|_t = 0 \tag{22}$$

The system of equations are solved stepwise, beginning with the state equations, which are integrated in the forward direction from starting time t_0 till the end of

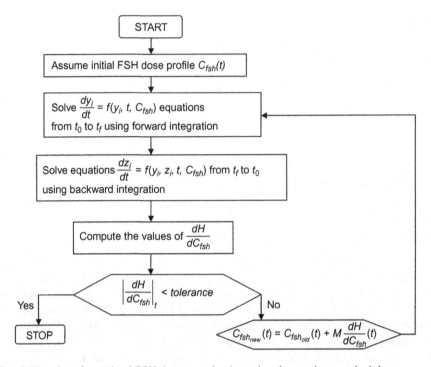

Fig. 6 Flowchart for optimal FSH dosage evaluation using the maximum principle.

the process or final time t_f. After this, the adjoint equations are integrated in the backward direction. Also, the optimality condition (Eq. 22) needs to be satisfied at every time point. The details of the calculation procedure are shown in the flowchart in Fig. 6.

3.3 Results from optimal control

As stated earlier, for days 1–4 the same dose is used, and no testing is done till 5th day. The optimal control method is applied to find dosage from 5th day onwards using the maximum principle. The patient parameters estimated using the two-day data are used, and the maximum principle method is applied to determine dosage from 5th day onwards. The optimal drug dosages for each patient are calculated, based on the starting dose, cycle days, and the initial FSD observed in each patient. The final-day mature follicle count using optimal control is then compared with observed mature follicles using the dosage specified by the attending physician for these 49 patients. Since the parameters from all-day data are more accurate than two-day data, those parameters are used with optimal control profile predicted by the two-day data for comparison. Fig. 7 shows the mature follicle distribution optimal versus experimental in Fig. 7A and the optimal dosage versus experimental dosage in Fig. 7B. The cumulative dose for this patient is found to be 2662.5 IU compared to

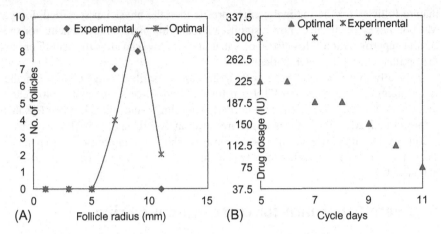

Fig. 7 (A) Follicular distribution for patient 3 predicted by optimal control with two-day parameters for the next day of the cycle versus observed follicle distribution from experiments for the last day of the cycle. (b) Optimal dosage for patient 3 predicted by optimal control with two-day parameters versus experimental dosage prescribed by doctor.

doctor prescribed dose of 3600 IU. The results serve as an example of the significant reduction in a dosage, which consequently reduces the costs to the patient.

The optimal control profile was calculated and customized for each patient for the clinical data available on 49 patient cycles. The histograms of the results are presented in Fig. 8. Fig. 8A shows the histogram of the ratio of optimal mature follicles to mature follicles observed using physician suggested dosage. Fig. 8B shows the % of reduction in dosage for each patient. It has been found that 98% of the patients show higher

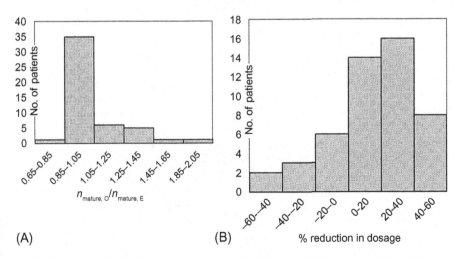

Fig. 8 Histogram of all patients: (A) Ratio of optimal mature follicles to experimental mature follicles. (B) % reduction in dosage.

mature follicles for the optimal control profile than the physician specified dosage. Most of these patients also show a significant reduction in dose requirements for successful superovulation. This also shows that the two-day data are sufficient to predict the optimal dosage for each patient.

Typically, older patients (age > 35 years) are prescribed dosages on the higher side ranging from 300 to 450 IU. Even for the higher age for patient 3 (40 years), the results show that the actual dose needed to get similar outcomes is much less than as prescribed. Also, the starting doses are lower at 300 IU and 225 IU, thus corroborating the idea that lower starting doses can also achieve similar responses in patients. This study found no correlation between the age of patients and higher doses of 300–450 IU.

4 Overall approach for customized medicine

The model and optimal control methods are implemented in integrated software for clinical trials. This software is called OPTIVF (Diwekar, 2018). The software uses the initial two-day data from the patient, i.e., their FSD and hormone dosage, as an input to the model. Optimization-based parameter estimation (iterative) of the moment model described in Section 2 is carried out to customize the model for each patient. The parameters then are used along with the iterative optimal control capability to find optimal drug-dosing profile for the remaining days of the cycle. Fig. 9 shows a schematic of this procedure. Thus, daily tests are avoided, and a reduced amount of drugs can be used to obtain significantly better outcomes.

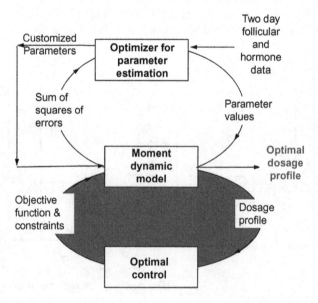

Fig. 9 Schematic of the overall approach and the steps in the OPTIVF software package.

4.1 Clinical trial using the software

Recently, the first clinical trial was conducted in Jijamata Hospital, Nanded. The trial involved 10 patients and was a double-blinded trial. Half of the patients were given dosage by the attending physician, and the other half were given the dosage predicted using this new approach using OPTIVF. Table 5 and Fig. 10 show the outcome for one of the patients in the clinical trial. Using the model and the optimized dosage, the follicular distribution at the end of the cycle in a clinical trial for this patient, it has been observed that the dosage predicted by using the model is 40 % less than that suggested by the IVF doctors (Fig. 10 and Table 5). Table 5 also shows that the number of mature follicles obtained at the end of the cycle using the model predicted dosage is significantly higher than that of physician suggested dosage. Percentage of good-quality eggs were similar from both the procedures. These results

Table 5 Comparison of optimal dosage profile with actual profile for patient 6

Patient 6	FSH Dr. Recom.	FSH optimal	No. of mature follicles on day 11 $9 \leq r \leq 12$		% of reduction in FSH due to optimization	% of reduction in testing due to optimization
Base follicle count -11	1950	1162.5	5 Dr. reco	11 (opt) actual follicles obtained	40.4%	72%

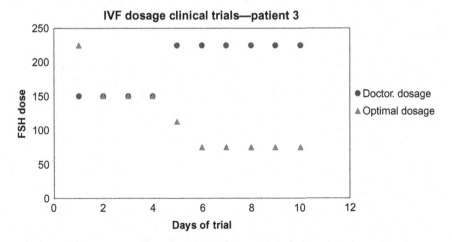

Fig. 10 Clinical trial patient 3, customized dosage comparison.

were consistent with all patients in this clinical trial. The testing requirement for patients using the optimized drug-dosage policy predicted by OPTIVF is reduced by 72%, and the number of follicles obtained was more than twice the number obtained by physician predicted dosage.

5   Summary and future work

IVF is the most common technique in assisted reproductive technology. Superovulation is a drug-induced method to enable multiple ovulation per menstrual cycle. The success of IVF depends upon successful superovulation, defined by the number and the uniformly high-quality of eggs retrieved in a cycle. Currently, this step is executed using almost daily monitoring of the follicular development using ultrasound and blood test. The daily dosage of hormones is customized for each patient based on these tests. Although there are general guidelines for the dosage, the dose is not optimized for each patient. The cost of testing and drugs makes this stage very expensive. To overcome the shortcoming of this system, a computer-assisted approach was presented for customized medicine for IVF. The approach uses customized models for each patient based on initial two-day data from each patient to determine the outcomes. Optimal control methods are then used on these customized models to obtain drug-dosage profiles for each patient. It has been found that this procedure provides better outcomes in terms of a higher number of mature follicles, reduced dosage, and reduced testing. This can reduce the side effects of the drugs significantly. A small clinical trial supports these theoretical findings. A user-friendly software was developed, which can provide a customized model of this stage for each patient, which would provide a basis for predicting the possible outcome based on the optimal drug-dosage profiles predicted by the optimal control functionality in the software. Further work is being carried out with the doctors in the United States to extend this approach to other protocols and for patients in the United States.

References

Allegra, A., Marino, A., Volpes, A., Coffaro, F., Scaglione, P., Gullo, S., La Marca, A., 2017. A randomized controlled trial investigating the use of a predictive nomogram for the selection of the FSH starting dose in IVF/ICSI cycles. Reprod. Biomed. Online 34, 429–438. https://doi.org/10.1016/j.rbmo.2017.01.012.

Alper, M.M., Smith, L.P., Sills, E.S., 2009. Ovarian hyperstimulation syndrome: current views on pathophysiology, risk factors, prevention, and management. J. Exp. Clin. Assist. Reprod. 6.

Baird, D.T., 1987. A model for follicular selection and ovulation: lessons from superovulation. J. Steroid Biochem. 27, 15–23.

Castiglione, F., Piccoli, B., 2007. Cancer immunotherapy, mathematical modeling and optimal control. J. Theor. Biol. 247, 723–732. https://doi.org/10.1016/j.jtbi.2007.04.003.

Czako, B., Sápi, J., Kovács, L., 2017. Model-based optimal control method for cancer treatment using model predictive control and robust fixed point method. In: 2017 IEEE 21st International Conference on Intelligent Engineering Systems (INES). Presented at the 2017 IEEE 21st International Conference on Intelligent Engineering Systems (INES), pp. 000271–000276. https://doi.org/10.1109/INES.2017.8118569.

Devroey, P., Tournaye, H., Van Steirteghem, A., Hendrix, P., Out, H.J., 1998. The use of a 100 IU starting dose of recombinant follicle stimulating hormone (Puregon) in in vitro fertilization. Hum. Reprod. 13, 565–566. https://doi.org/10.1093/humrep/13.3.565.

Diwekar, U., 2008. Introduction to applied optimization. In: Springer Optimization and Its Applications, second ed. Springer, New York.

Diwekar, U., 2018. OPTIVF: A User-Friendly Software for Optimization of IVF Cycles. Stochastic Research Technologies, LLC.

Dorn, C., 2005. FSH: what is the highest dose for ovarian stimulation that makes sense on an evidence-based level? Reprod. Biomed. Online 11, 555–561. https://doi.org/10.1016/S1472-6483(10)61163-7.

FastStats, 2018. https:/www.cdc.gov/nchs/fastats/infertility.htm ((Accessed 3 March 2019)).

Fleming, R., Seifer, D.B., Frattarelli, J.L., Ruman, J., 2015. Assessing ovarian response: antral follicle count versus anti-Müllerian hormone. Reprod. Biomed. Online 31, 486–496. https://doi.org/10.1016/j.rbmo.2015.06.015.

Flood, A.E., 2002. Thoughts on recovering particle size distributions from the moment form of the population balance. Dev. Chem. Eng. Miner. Process. 10, 501–519. https://doi.org/10.1002/apj.5500100605.

Fritz, M.A., Speroff, L., 2010. Clinical Gynecologic Endocrinology and Infertility, eighth ed. LWW, Philadelphia.

Gordon, R., 2012. In Vitro Fertilization: ivf Process: And Embryo Transfer (IVF – ET). https://infofru.com/health/pregnancy/infertility-management-recent-advancement/in-vitro-fertilization-and-embryotransfer-ivf-et/ (Accessed 25 July 2019).

Hill, P., 2005. Batch crystallization. In: Korovessi, E., Linninger, A.A. (Eds.), Batch Processes. Taylor and Francis, CRC Press, New York, USA.

Hu, Q., Rohani, S., Jutan, A., 2005. Modelling and optimization of seeded batch crystallizers. Comput. Chem. Eng. Control Multiscale Distrib. Process Syst. 29, 911–918. https://doi.org/10.1016/j.compchemeng.2004.09.011.

John, V., Angelov, I., Öncül, A.A., Thévenin, D., 2007. Techniques for the reconstruction of a distribution from a finite number of its moments. Chem. Eng. Sci. 62, 2890–2904. https://doi.org/10.1016/j.ces.2007.02.041.

Jungheim, E.S., Meyer, M.F., Broughton, D.E., 2015. Best practices for controlled ovarian stimulation in in vitro fertilization. Semin. Reprod. Med. 33, 77–82. https://doi.org/10.1055/s-0035-1546424.

Khalili, S., Armaou, A., 2008. Sensitivity analysis of HIV infection response to treatment via stochastic modeling. Chem. Eng. Sci., Control Particul. Process. 63, 1330–1341. https://doi.org/10.1016/j.ces.2007.07.072.

Klemetti, R., Sevón, T., Gissler, M., Hemminki, E., 2005. Complications of IVF and ovulation induction. Hum. Reprod. 20, 3293–3300. https://doi.org/10.1093/humrep/dei253.

Kotanidis, L., Nikolettos, K., Petousis, S., Asimakopoulos, B., Chatzimitrou, E., Kolios, G., Nikolettos, N., 2016. The use of serum anti-Mullerian hormone (AMH) levels and antral follicle count (AFC) to predict the number of oocytes collected and availability of embryos for cryopreservation in IVF. J. Endocrinol. Invest. 39, 1459–1464. https://doi.org/10.1007/s40618-016-0521-x.

La Marca, A., Grisendi, V., Giulini, S., Argento, C., Tirelli, A., Dondi, G., Papaleo, E., Volpe, A., 2013. Individualization of the FSH starting dose in IVF/ICSI cycles using the antral follicle count. J. Ovarian Res. 6, 11. https://doi.org/10.1186/1757-2215-6-11.

La Marca, A., Sunkara, S.K., 2014. Individualization of controlled ovarian stimulation in IVF using ovarian reserve markers: from theory to practice. Hum. Reprod. Update 20, 124–140. https://doi.org/10.1093/humupd/dmt037.

Mascarenhas, M.N., Flaxman, S.R., Boerma, T., Vanderpoel, S., Stevens, G.A., 2012. National, regional, and global trends in infertility prevalence since 1990: A systematic analysis of 277 health surveys. PLoS Med. 9. https://doi.org/10.1371/journal.pmed.1001356.

Moon, K.Y., Kim, H., Lee, J.Y., Lee, J.R., Jee, B.C., Suh, C.S., Kim, K.C., Lee, W.D., Lim, J.H., Kim, S.H., 2016. Nomogram to predict the number of oocytes retrieved in controlled ovarian stimulation. Clin. Exp. Reprod. Med. 43, 112–118. https://doi.org/10.5653/cerm.2016.43.2.112.

Nisal, A., Diwekar, U., Bhalerao, V., 2019. Personalized medicine for in vitro fertilization procedure using modeling and optimal control. J. Theor. Biol. (accepted).

Randolph, A.D., 1988. Theory of Particulate Processes: Analysis and Techniques of Continuous Crystallization. Academic Press.

Nyboe Andersen, A., Nelson, S.M., Fauser, B.C.J.M., García-Velasco, J.A., Klein, B.M., Arce, J.-C., Tournaye, H., De Sutter, P., Decleer, W., Petracco, A., Borges, E., Barbosa, C.P., Havelock, J., Claman, P., Yuzpe, A., Višnová, H., Ventruba, P., Uher, P., Mrazek, M., Andersen, A.N., Knudsen, U.B., Dewailly, D., Leveque, A.G., La Marca, A., Papaleo, E., Kuczynski, W., Kozioł, K., Anshina, M., Zazerskaya, I., Gzgzyan, A., Bulychova, E., Verdú, V., Barri, P., García-Velasco, J.A., Fernández-Sánchez, M., Martin, F.S., Bosch, E., Serna, J., Castillon, G., Bernabeu, R., Ferrando, M., Lavery, S., Gaudoin, M., Nelson, S.M., Fauser, B.C.J.M., Klein, B.M., Helmgaard, L., Mannaerts, B., Arce, J.-C., 2017. Individualized versus conventional ovarian stimulation for in vitro fertilization: a multicenter, randomized, controlled, assessor-blinded, phase 3 noninferiority trial. Fertil. Steril. 107. https://doi.org/10.1016/j.fertnstert.2016.10.033. 387–396.e4.

Practice Committee of the American Society for Reproductive Medicine, 2018. Comparison of pregnancy rates for poor responders using IVF with mild ovarian stimulation versus conventional IVF: a guideline. Fertil. Steril. 109 (6), 993–999.

Rombauts, L., 2007. Is there a recommended maximum starting dose of FSH in IVF? J. Assist. Reprod. Genet. 24, 343–349. https://doi.org/10.1007/s10815-007-9134-9.

Scoccia, H., 2017. IVF Program Protocols for Assisted Reproductive Technologies. University of Illinois at Chicago.

Sighinolfi, G., Grisendi, V., La Marca, A., 2017. How to personalize ovarian stimulation in clinical practice. J. Turk. Ger. Gynecol. Assoc. 18, 148–153. https://doi.org/10.4274/jtgga.2017.0058.

Simopoulou, M., Sfakianoudis, K., Antoniou, N., Maziotis, E., Rapani, A., Bakas, P., Anifandis, G., Kalampokas, T., Bolaris, S., Pantou, A., Pantos, K., Koutsilieris, M., 2018. Making IVF more effective through the evolution of prediction models: is prognosis the missing piece of the puzzle? Syst. Biol. Reprod. Med. 64, 305–323. https://doi.org/10.1080/19396368.2018.1504347.

Sunderam, S., Kissin, D.M., Crawford, S.B., Folger, S.G., Boulet, S.L., Warner, L., Barfield, W.D., 2018. Assisted reproductive technology surveillance—United States, 2015. MMWR Surveill. Summ. 67, 1–28. https://doi.org/10.15585/mmwr.ss6703a1.

Ulas Acikgoz, S., Diwekar, U.M., 2010. Blood glucose regulation with stochastic optimal control for insulin-dependent diabetic patients. Chem. Eng. Sci. 65, 1227–1236. https://doi.org/10.1016/j.ces.2009.09.077.

van Tilborg, T.C., Torrance, H.L., Oudshoorn, S.C., Eijkemans, M.J.C., Koks, C.A.M., Verhoeve, H.R., Nap, A.W., Scheffer, G.J., Manger, A.P., Schoot, B.C., Sluijmer, A.V., Verhoeff, A., Groen, H., Laven, J.S.E., Mol, B.W.J., Broekmans, F.J.M., 2017. Individualized versus standard FSH dosing in women starting IVF/ICSI: an RCT. Part 1. The predicted poor responder. Hum. Reprod. 32, 2496–2505. https://doi.org/10.1093/humrep/dex318.

Yenkie, K.M., Diwekar, U., 2014. Optimal control for predicting customized drug dosage for superovulation stage of in vitro fertilization. J. Theor. Biol. 355, 219–228. https://doi.org/10.1016/j.jtbi.2014.04.013.

Yenkie, K.M., Diwekar, U., 2012. Stochastic optimal control of seeded batch crystallizer applying the ito process. Ind. Eng. Chem. Res. 52, 108–122.

Yenkie, K.M., Diwekar, U.M., Bhalerao, V., 2013. Modeling the superovulation stage in in vitro fertilization. IEEE Trans. Biomed. Eng. 60, 3003–3008.

Youssef, M.A., van Wely, M., Al-Inany, H., Madani, T., Jahangiri, N., Khodabakhshi, S., Alhalabi, M., Akhondi, M., Ansaripour, S., Tokhmechy, R., Zarandi, L., Rizk, A., El-Mohamedy, M., Shaeer, E., Khattab, M., Mochtar, M.H., van der Veen, F., 2016. A mild ovarian stimulation strategy in women with poor ovarian reserve undergoing IVF: a multicenter randomized non-inferiority trial. Hum. Reprod. https://doi.org/10.1093/humrep/dew282.

Yovich, J.L., Alsbjerg, B., Conceicao, J.L., Hinchliffe, P.M., Keane, K.N., 2016. PIVET rFSH dosing algorithms for individualized controlled ovarian stimulation enables optimized pregnancy productivity rates and avoidance of ovarian hyperstimulation syndrome. Drug Des. Devel. Ther. 10, 2561–2573. https://doi.org/10.2147/DDDT.S104104.

Models based on cellular automata for the analysis of biomedical systems

14

Alejandro Talaminos-Barroso, Javier Reina-Tosina, Laura María Roa-Romero
Biomedical Engineering Group, Universidad de Sevilla, Seville, Spain

1 Introduction

Recent advances in biomedicine have led to the growth of large amounts of biological data in multiple fields of life sciences at different levels of biological organization (Bartocci and Lió, 2016). The exploitation and use of these data requires a multidisciplinary approach and computational techniques in order to generate clinical knowledge (Holt et al., 2018). From this perspective, experimental researchers are responsible for collecting data from laboratory and providing all related information and appropriate documentation to mathematicians, scientists, and engineers, who use specific tools to analyze these data and build computational models (Brodland, 2015). The predictions of the developed models are studied by experimentalists and, after the validation, models are released to the scientific and clinical community to analyze the results of simulations according to the requirements of different experiments.

In the context of biological systems, one of the main problems is the understanding of how various subsystems interact to generate structural and behavioral patterns from a spatial/temporal perspective. Biological models are used to reproduce the dynamics of the interactions between the components under study, including short- and long-range mechanochemical interactions (Deutsch and Dormann, 2005). These behaviors can combine spatial, temporal, and interaction factors, so that the methodology differs according to the objective that the model is intended to achieve.

In a traditional approach, models aimed at describing space-time dynamics are based on nonlinear differential equations in partial derivatives (Murray, 1981). Another approach widely used for biological systems is compartmental modeling, which subdivides the system under study into a limited set of subsystems (compartments) considering the principle of mass conservation. Each compartment represents a population of elements (e.g., concentrations, individuals, or others) that are equivalent with respect to the relationships between compartments. In this sense, each relationship is described by a transfer coefficient, which determines the quantity of mass or energy of a subsystem flowing from one compartment to other. In this sense, these relations between compartments are established by means of differential equations, considering all the elements that are object of study as unique and

Control Applications for Biomedical Engineering Systems. https://doi.org/10.1016/B978-0-12-817461-6.00014-7

indivisible entities, and ignoring spatial aspects that influence the process. This methodological approach is appropriate when the number of elements considered is huge and the objectives of the model can be covered with results on average; however, in certain cases the homogeneity between elements is not clearly defined and it may be interesting to establish individual characteristics, rather than using averages. One way to address this problem and consider individual properties and spatial-temporal characteristics is through the use of models based on cellular automata (CA).

A CA is a discrete model characterized by a matrix of cells (box) that can contain entities with clearly established positions and properties changing over time. The main characteristic of CA-based models is their complex global evolution, based on rules and interconnection patterns between elements. These fundamental characteristics are homogeneity, parallelism, and locality (Lin, 2015). On the one hand, homogeneity refers to the fact that all cells are equivalent and updated by the same set of transition rules throughout the entire cell space. On the other hand, parallelism indicates that the states of the cells are updated simultaneously. Finally, locality is the characteristic that indicates that all transition rules are local by nature.

The inherent complexity of CA has historically hindered their development due to their close dependence on computer science and, for this reason, the applications of CA began to spread with the advent of the digital computer. In the biomedical engineering field, CA have been successfully applied to a wide variety of biological phenomena, such as tumor growth, biological reproductive processes, genetics, epidemiology, heart electrical conduction system, and many others (Ghosh et al., 2018).

This chapter analyzes the use of CA models as a tool for the simulation and reproduction of dynamic behaviors in biological systems with complex local interactions. The chapter is structured as follows: after this introduction, in Section 2, an overview of the basic terms associated with CA modeling is exposed, including concepts such as grid, cell, neighborhood rules, among others. Section 3 presents a brief historical evolution, from a chronological point of view, of the beginnings of CA modeling. These first two sections serve as a first approximation to CA models for undergraduate students who are interested in the study of this computational technique. The methodology followed in these sections will be to introduce the basic concepts with the simplest cellular spaces and examples so that any reader can easily follow them, and the difficulty will gradually increase for those graduate students with a better knowledge. In Section 4, CA models that have emerged in recent decades and that have had a greater impact on biomedical research are analyzed. This section also makes a brief physiological introduction to the systems under analysis by CA for the sake of clarity to readers not knowledgeable on these topics. Next, Section 5 presents some software techniques that are currently applied for the development of CA, particularly involving the use of compute-intensive techniques and GPU (graphics processing unit). The aim of this section is to introduce the graduate student to the latest software techniques for building CA models. Finally, conclusions highlight the historical relevance of CA to provide a better understanding of biological systems.

2 Basic concepts of cellular automata

CA are computational models that try to reproduce the dynamic behaviors of heterogeneous objects that interact with each other through an evolution based on discrete-time steps. These types of models are characterized by having a grid (cellular space) composed of boxes or cells containing objects that make up the CA. The geometrical shape of the cells is a relevant aspect in the construction process of a CA model (Bays, 2009) and the choice of one shape or another will depend on the objectives to be achieved with the model. Fig. 1 shows the most popular types of cells in a cellular space ξ.

The distribution of the cells in space can be one-, two-, or three-dimensional, although two-dimensional grids are usually the most used (Li et al., 2018). Examples of quadrangular one-, two-, and three-dimensional grids are shown in Fig. 2.

To ease the understanding of the concepts discussed later, this chapter focuses on CA with discrete-time evolution as the only independent variable. In addition, cellular spaces of one or two dimensions will be used to simplify the understanding and vision of these concepts.

Fig. 1 Types of cells in a cellular space ξ: triangular, quadrangular, and hexagonal.

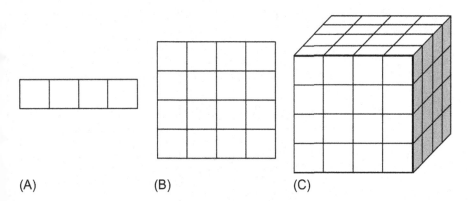

(A) (B) (C)

Fig. 2 Examples of unidimensional grid (A), bidimensional grid, (B) and tridimensional grid (C).

The contents of boxes in a grid have individual characteristics associated and states, which refer to the representation of the dynamics of each object in a specific time instant. This state of the object can be changed over time. For example, in a model for the spread of infectious diseases, the box content would be a person. The internal characteristics could be the immunological conditions and physical status, while the states could be healthy or infected. Depending on the type of infection studied and the health conditions of the person, their status could change from healthy (H) to infected (I), as presented in Fig. 3. This change from one state to another is called state transition.

The number of states of a CA can be defined by a constant denoted as k, S being the set of possible states that an object can acquire. This number of possible states for all the objects of the CA is expressed in Eq. (1).

$$S = \{0, 1, ..., k-1\} \tag{1}$$

In the previous example of the infectious disease model, $k=2$ and $S=\{0,1\}$, where 0 represents the healthy state and 1 is the infected state.

In general, considering a number of boxes (i), each object of the cellular space can transit between a set of states that is defined by S. For example, in a 2×2 space (Fig. 4), each of these four cells can have associated a state that belongs to S for each time instant.

In the previous example, each of the objects A, B, C, and D ($i=4$) represents a person, and healthy and infected conditions are the possible states in which a person can be found. In this way, x can be defined as the set of all possible states to which persons can transit (2).

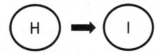

Fig. 3 Example of state transition of a person from the healthy (H) to the infected state (I) in an epidemiology model.

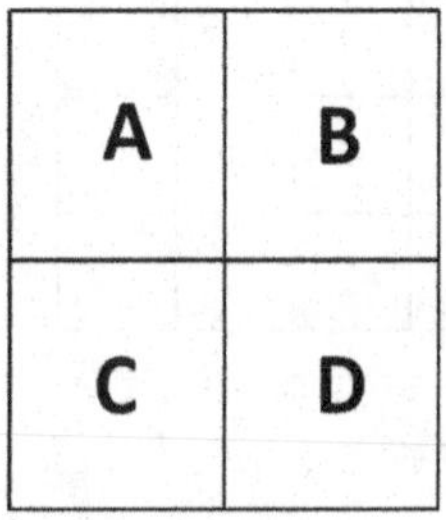

A	B
C	D

Fig. 4 Example of quadrangular CA with dimension 2×2 and four cells: A, B, C, and D.

$$x = \{A_0, A_1, B_0, B_1, C_0, C_1, D_0, D_1\} \tag{2}$$

In general, considering the simultaneity of change in the cells at each discrete-time interval, it can be established that x_i^{t+1} is the state of object i at $t+1$. In the example, x_0^{t+1}, corresponding to person A at $t+1$, is in A_0 state (0). In the same way, x_1^{t+1} applies to person B, which is in the B_1 state (1), and so on as shown in Fig. 5.

In addition to the states and their internal characteristics, transitions from one state to another in a discrete-time interval are also affected by so-called transition rules. These transition rules depend not only on their state, but also on the state of their adjacent objects. In this sense, the concept of neighborhood refers to the influence that the environment has on the dynamics of the evaluated object. Among the most used neighborhood configurations are the von Neumann and Moore neighborhoods.

The von Neumann neighborhood (Fig. 6) is the most commonly used configuration in a two-dimensional automaton with quadrangular boxes. In this type of neighborhood, only contiguous cells to the four sides of the evaluated cell are considered.

On the other hand, the Moore neighborhood (Fig. 7) considers the eight boxes immediately surrounding a box, including diagonal contacts. As it is obvious, boxes on the edges of the grid have fewer neighbor boxes under both von Neumann and Moore neighborhood.

The concept of neighborhood is also related to the radius (r), understood as the spatial domain surrounding a given box. The radius can be of one, two, or more dimensions, although the multidimensional case increases the complexity of the neighborhood analysis, as well as the computational cost. For example, considering

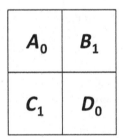

Fig. 5 Example of the person states for a specific instant of time.

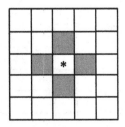

Fig. 6 Example of von Neumann neighborhood for the object represented with an asterisk.

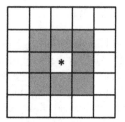

Fig. 7 Example of Moore neighborhood for the object represented with an asterisk.

a two-dimensional CA with von Neumann configuration, the number of neighboring boxes (n) affecting a given object (n_i) is represented by Eq. (3):

$$n_i = 2r(r+1) \tag{3}$$

Thus, in the previous example of the von Neumann neighborhood represented in Fig. 6 with $r = 1$, four boxes in the neighborhood of object i are taken in a two-dimensional CA. These four boxes are considered in the analysis of the state transition of the object evaluated.

In the Moore configuration, the number of neighboring boxes in relation to the radius is presented in Eq. (4a):

$$n_i = (2r + 1)^2 - 1 \tag{4a}$$

In the CA shown in Fig. 7, $r = 1$, and therefore, the number of neighboring boxes is eight in a two-dimensional CA.

Fig. 8 represents both von Neumann's and Moore's neighborhood configurations for $r = 2$.

In a context of three-dimensional CA, the von Neumann and Moore configurations can be extended to consider more complex neighborhoods. For example, Fig. 9 presents the neighborhood region in a three-dimensional CA with von Neumann (A) and Moore (B) configurations and $r = 1$.

Other neighborhood relationships also used in domain-specific modeling problems are the Brickwall (Akdur, 2011) and Morgolus (Toffoli and Margolus, 1987) neighborhoods. The neighborhood conditions for hexagonal grids are extended to a

Fig. 8 The von Neumann (A) and Moore (B) neighborhood for $r = 2$.

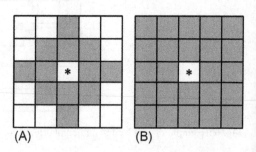

(A)   (B)

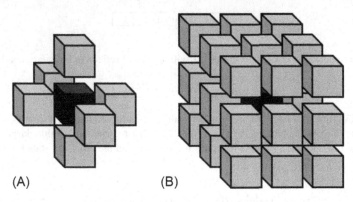

Fig. 9 The von Neumann (A) and Moore (B) neighborhoods for a three-dimensional CA with $r=1$.

hexagonal neighborhood (Siap et al., 2011), while a configuration of Moore and von Neumann is usually chosen for triangular grids (Warne and Hayward, 2015). Each of these neighborhood settings has advantages or disadvantages (Li et al., 2018). For example, the triangular configuration contains a small number of neighboring boxes, and for this reason the complexity of neighborhood can be simplified. On the other hand, quadrangular geometry is simple and intuitive, and it is also especially suitable for matrix-form computation. Finally, hexagonal configuration is computationally more complex to describe, but it has historically been used for the analysis of fluid and gas dynamics (Wolfram, 1986). The choice of the neighborhood setting depends on the type of application under study and the neighborhood relationships needed.

As seen earlier, in the context of the neighborhood configuration, not all boxes have the same neighboring boxes because the edges of the grid have a limited neighborhood. Strategies for modeling the contours of a CA include null, periodic, reflective, and fixed boundaries (Bhattacharjee et al., 2018). The different boundaries are described later, and an example is presented of each one considering a one-dimensional CA with neighborhood radius equal to one and two states: 0 and 1 (referring to the previous example, a person can be healthy or infected). These one-dimensional CA with two states are also referred to as elementary cellular automaton.

- Null boundary condition (Fig. 10): edge boxes have no interaction beyond the cellular space in which they are found, while the rest of the boxes have a neighborhood to the left and right.
- Periodic boundary condition (Fig. 11): edge boxes interact with the boxes of the opposite edge and consider them in their neighborhood region.

$$\boxed{1\mid 0\mid 1\mid 0}\cdots\boxed{0\mid 1\mid 1\mid 0}$$

Fig. 10 Example of a one-dimensional CA with $r=1$ and null boundary condition.

$$\boxed{1\mid 0\mid 1\mid 0}\cdots\boxed{0\mid 1\mid 1\mid 1}$$

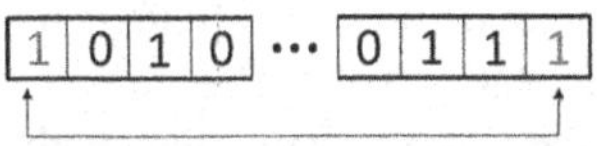

Fig. 11 Example of a one-dimensional CA with $r=1$ and periodic boundary condition.

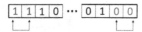

Fig. 12 Example of a one-dimensional CA with $r = 1$ and reflective boundary condition.

$$\boxed{1}\boxed{0}\boxed{1}\boxed{0} \cdots \boxed{0}\boxed{1}\boxed{1}\boxed{0}$$

Fig. 13 Example of a one-dimensional CA with $r = 1$ and fixed boundary condition, where the first and last boxes of the CA have a fixed state that remains unalterable during the simulation time. In this case, one has state 1 (first cell) and the other has state 0 (last cell).

- Reflective boundary condition (Fig. 12): the boxes found on the edges are reflected, maintaining their state.
- Fixed boundary condition (Fig. 13): edge boxes have a fixed state that is not altered during the simulation of the CA. Fig. 13 shows how the states assigned to the edges of the one-dimensional CA have a fixed state.

In general, the transition rules for each of the objects that make up the CA are evaluated at each discrete time interval, where x_i^t is the state of object i at instant t. In addition, f is the function representing the transition rule that establishes the state of object x_i in the next time instant (x_i^{t+1}), as expressed in Eq. (4b):

$$x_i^t \xrightarrow{f} x_i^{t+1} \tag{4b}$$

Among the most commonly used transition rules are those in which each cell is assigned an integer associated with its state. In these cases, the transition function is the sum of the states associated with the objects in the neighborhood zone and using a normalization algorithm (e.g., the modulo operation). For example, considering a one-dimensional CA with neighborhood radius equal to one ($r = 1$) and two possible states (0 or 1), the transition function for this case can be expressed from a function (f) applied to the sum of the current value of the left adjacent cell (x_{i-1}^t), the current value of the object evaluated (x_i^t), and the current value of the right adjacent cell (x_{i+1}^t), as expressed in Eq. (5):

$$x_i^{t+1} = f\left[x_{i-1}^t + x_i^t + x_{i+1}^t\right] \tag{5}$$

The transition function (f) must limit the state of the evaluated object at the next time instant (x_i^{t+1}), so that its value is 0 or 1. A possible solution is the use of the modulo operation, presented in Eq. (6):

$$x_i^{t+1} = \left[x_{i-1}^t + x_i^t + x_{i+1}^t\right] \bmod(k) \tag{6}$$

where $\bmod(k)$ is the remainder of the sum divided by the number of states (k) to limit the result to two possible states: 0 or 1. The transition function can be represented in a table (Table 1) that indicates the possible values that an object x_i can take at the next instant (x_i^{t+1}).

Table 1 Possible states of an object at the next time instant considering a one-dimensional CA with two possible states and the transition function (6)

x_{i-1}^t	x_i^t	x_{i+1}^t	x_i^{t+1}
0	0	0	0
0	0	1	1
0	1	0	1
0	1	1	0
1	0	0	1
1	0	1	0
1	1	0	0
1	1	1	1

These types of CA are called totalistic and can be classified into two categories: purely totalistic or totalistic, as in the earlier example, and outer-totalistic or semi-totalistic. In the case of the outer-totalistic CA, the state of the evaluated object and the sum of the neighboring object are normalized and the result is assigned to the state of the evaluated object using the same normalization technique as in the previous step. Thus, the state of the evaluated object at the next time instant (x_i^{t+1}) can be formulated by considering a transition function, as expressed in Eq. (7):

$$x_i^{t+1} = f\left(x_{i-1}^t + x_{i+1}^t, x_i^t\right) \tag{7}$$

A possible example of transition function (f) is presented in Eq. (8), where the normalization technique, as in the previous example, is the modulo operation.

$$x_i^{t+1} = \left[\left(\left[x_{i-1}^t + x_{i+1}^t\right] \bmod(k)\right) + x_i^t\right] \bmod(k) \tag{8}$$

Table 2 summarizes the possible states of an object at the next time instant in an outer-totalistic CA considering the transition function expressed in Eq. (8).

Table 2 Possible states of an object in the next time instant considering a one-dimensional outer-totalistic CA with two possible states and the transition function (8)

x_{i-1}^t	x_{i+1}^t	$\delta = [x_{i-1}^t + x_{i+1}^t]\bmod(2)$	x_i^t	$[\delta + x_i^t]\bmod(2)$
0	0	0	0	0
1	0	1	0	1
0	1	1	0	1
1	1	0	0	0
0	0	0	1	1
1	0	1	1	0
0	1	1	1	0
1	1	0	1	1

A more complex example of outer-totalistic CA is the famous Conway's Game of Life (Adamatzky, 2010), which considers a two-dimensional cell space with Moore neighborhood configuration and two possible states: alive (1) and dead (0) states. In this case, the sum (ω) of all neighboring object values is expressed in Eq. (9):

$$\omega = x_{i-1,j-1}^t + x_{i,j-1}^t + x_{i+1j-1}^t + x_{i-1,j}^t + x_{i+1,j}^t + x_{i-1,j+1}^t + x_{i,j+1}^t + x_{i+1,j+1}^t \qquad (9)$$

where i and j correspond to the value of the column and row, respectively. The transition function indicates that a dead object will become alive in the next iteration ($t + 1$) if it has exactly three neighbors alive. On the other hand, an alive object will die in the next iteration except if it has exactly two or three neighbors alive. Table 3 summarizes the combination of different values of ω against the two possible states in which the evaluated object ($x_{i,j}^t$) can be found. In the upper row of the table, different values of the sum of the states of the neighboring objects are presented and the first column contains the two possible states of the evaluated object at generation t. Table entries correspond to object state for generation $t + 1$.

So far the transition function has been regarded deterministic. However, neighborhood relations can also be expressed as a probability associated with space-time considerations, the state, or characteristics of the object itself. In this sense, probability values are in the range from 0 to 1. If probability is 0, the state of the evaluated object remains unchanged in the next iteration. If probability is 1, then the state of the object will change in the next iteration. For intermediate values of probability, the transition function will be affected accordingly.

An example of probabilistic CA-based model is the analysis of the spread of an infectious disease where each individual can be in the H or I state, assuming the von Neumann neighborhood and a neighborhood radius equal to one. The transition function establishes that the evaluated object will move to the infected state if it has at least one infected cell in its neighborhood. This transition function, due to different considerations associated with the type of disease and the intrinsic characteristics of the individual, may be assigned a probability of occurrence. Fig. 14 shows the behavior with different probabilities assigned to the transition function: in case (A), the state of the evaluated object remains unchanged due to the low probability (0.25); in case (B), there is the same probability (0.5) of staying in the current state or changing to the state I; in the last case (C), the state of the evaluated object will change to the state I due to the high probability (0.75).

Spatial conditions can also be introduced into a CA, allowing object not only to change the state, but also to shift their position. In this way, objects can move horizontally, vertically, or even diagonally to a region in their neighborhood that is not

Table 3 Combination of the different values of ω versus the two possible states of the object $x_{i,j}^t$ that are being evaluated

ω	0	1	2	3	4	5	6	7	8
$x_{i,j}^t$ 0	0	0	0	1	0	0	0	0	0
1	0	0	1	1	0	0	0	0	0

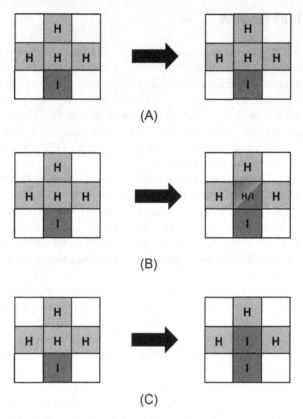

Fig. 14 Probabilistic model of infectious diseases considering different probabilities of occurrence. (A) probability: 0.25, (B) probability: 0.5, and (C) probability: 0.75.

occupied by another object. These spatial considerations must be included within the transition function and be evaluated at each instant of time. However, due to the fact that object mobility is based on probabilities, the behavior of the CA is stochastic and the final result varies with each simulation. In general, two transition rules associated with object mobility can be considered. On the one hand, the degree to which two adjacent objects A and B tend to stay together is called the probability of rupture PB(A,B), and on the other hand, the tendency to move in cell space is referred to as probability of free movement PM(A). Occasionally, for certain applications, the force of gravity can be included to increase the probability of movement towards lower zones of the cellular space.

All these possibilities of neighborhood configurations, boundary conditions, intrinsic properties that characterize an object, mobility and transition rules, provide a wide variety of dynamics with local interactions. These characteristics allow to simulate complex behavior patterns, including interactions with neighboring elements, evolution in time, and random events. In summary, CA represent a powerful tool to address the modeling of diverse and complex systems in different fields of science, including biomedicine.

3   Historical review

The concept of CA was introduced by John von Neumann, who in the early 1940s defined a machine capable of building another identical machine. This machine was called universal constructor (Beuchat and Haenni, 2000) and von Neumann's main motivation was to try to mimic the self-reproducing capacity of living organisms (Deutsch and Dormann, 2005). The replicas obtained from the universal constructor were generated from the instructions of the universal constructor itself. This idea of von Neumann's self-replication shares similarities with the DNA structure discovered a few years later by Watson and Crick (1953).

Ulam et al. (1991), who worked with von Neumann, suggested the development of the self-replicating machine using a grid space formed by a set of boxes that could contain objects called cells; hence, the generalization of cellular spaces into automata. Cells switch between a finite number of states through discrete time intervals, where the transition from one state to another depends on the current state of the cell and the state of the adjacent cells, including, on a first approximation, the nearest four (North, South, East, and West).

After von Neumann, his work was continued by Arthur Burks (Banks, 1970), who proposed a Turing machine embedded in a two-dimensional cellular space with 29 states for each cell and 5 neighborhood levels. However, it was never implemented (Schiff, 2011). Later, Codd (1968) extended von Neumann's work by showing that a set of eight states was sufficient, but his idea was not developed either, mainly due to its huge size (Hutton, 2010).

Along with the development of modern computers, it became evident that these abstract ideas could be applied in a useful way for the analysis of physical and biological phenomena that occur in nature, with achievements and illustrative results (Vichniac, 1984). The use of CA models began to become popular in the 1970s, among other reasons, due to the introduction of John Conway's Game of Life (Gardner, 1970), an example of simulation based on life self-organization (Sipper, 2017). This study proposed a two-dimensional CA with square boxes, each box surrounded by four neighboring boxes, plus four diagonally adjacent boxes. The state of each cell was binary, either 0 (dead) or 1 (alive). A set of rules were evaluated over time and the state of the cells could change at each time step by considering the number of alive and dead cells adjacent to each other. The result of several iterations could yield different configurations depending on the initial spatial distribution of the alive cells: (1) cells that do not change their shape from one iteration to the next (still life); (2) structures that recover their initial conditions in one or more iterations (oscillator); (3) patterns that recover their initial conditions, but with a displacement (intermittent); and (4) producer of new structures (generator). Fig. 15 shows examples of these four typical patterns of the Game of Life. In case (A), there is no change in the transition of the alive cells; in case (B), an oscillatory behavior is represented, where a new pattern is created in the first transition and in the following one it returns to the initial configuration (oscillation of period two); in case (C), the initial distribution returns after four iterations, with a shift to the right and down in the cell space; in the last case (D), the initial distribution of alive cells and the iteration of new cells are shown, after 50 iterations.

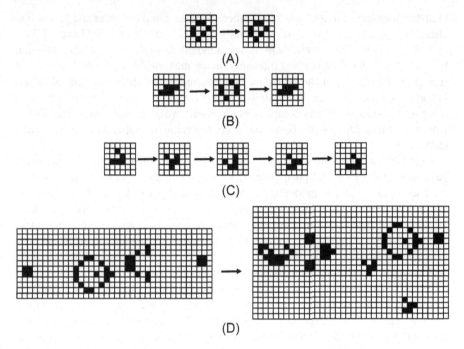

Fig. 15 Typical patterns of the Game of Life: (A) still life, (B) oscillator, (C) intermittent, and (D) generator (after 50 iterations).

Years later, Wolfram suggested his own description of the CA, proposing that deterministic rules were not absolutely necessary and probability was useful in a broad variety of complex physical and chemical phenomena (Kauffman, 1984). Also, Wolfram developed a classification for CA based on the evolutionary characteristics for this type of model, concluding that there were four classes of CA. In particular, the first two classes are characterized by predictable behavior following a set of stable or periodic structures. The other two classes exhibit chaotic behavior, sensitivity to disturbances, and finally, complex transformations over time. Subsequently, this work served as a basis for later studies with other types of classifications according to different criteria (Gutowitz, 1991; Martinez et al., 2013; Wuensche, 1997; Zenil, 2013). Wolfram concentrated his research in the 1990s on the study of one-dimensional CA through the analysis of related works, emphasizing the adaptation ability as one aspect of autonomy to generate different behaviors that reproduced a broad range of phenomena in nature.

At the same time, Langton (1990) studied the dynamic properties of CA by explaining why several rules resulted in periodic, random, or highly structured and complex behaviors. Langton described the capabilities of a CA using three properties (memory, transmission, and information processing) and proposed rules for a CA space considering the transition region between the predictable and the chaotic behaviors. The results provided by a CA simulation were categorized by Langtom using a lambda parameter ranging from 0 to 1, where 0 meant complete order and 1, absolute chaos. For example, the value of lambda is 0.273 in Conway's Game of Life (Ventrella, 2006). However, there is no standard formula for lambda, although Langton found that memory and transmission are maximized near the edge of chaos (Langton, 1990), i.e., at the boundary between order and chaos. In general, it has historically been considered that biological, economic, or social systems operate in a region between order and complete randomness, where complexity is maximum. However, these ideas have been found controversial by other authors (Mitchell et al., 1993).

In the following decades and up to the present day, mathematicians and engineers have applied CA to model phenomena that occur in nature as an alternative to other methodologies used for modeling a physical system. Models based on CA have demonstrated throughout history their potential to simulate systems dynamics, such as biomedical systems, with an huge amount of local interactions, including behaviors that are difficult to detect and complex to be described mathematically using conventional methods (Kier et al., 2005).

The scope of application of CA is very broad and heterogeneous in different fields of science. A review on CA models was recently published (Yanhua et al., 2017). In this sense, the first work appeared in 1965 and from then on, over 10,000 papers have been presented, with an exponential increase over the years. Among the physical phenomena and systems most modeled throughout history are traffic flow, solidification, microstructures, or urban growth. Additionally, new technologies and methodologies, such as genetic algorithms, geographic information systems, numerical simulation, parallel computing, and sensorization, have also increased the application of CA to other fields of knowledge.

4 Applications of cellular automata

In the biomedical field, CA models have also been used for different applications. The following are some of the most relevant models that have had a considerable impact on the literature in the fields of epidemiology, oncology, and cardiac electrical conduction.

4.1 Epidemiology

The incidence of contagious diseases has a great importance in health policy (van Panhuis et al., 2013) and mathematical models in epidemiology have been used to analyze, predict, and control the outbreak of infectious epidemics and propose possible control strategies.

Generally, epidemiological models can be classified into two types of approaches: on the one hand, those focused on the propagation and infection dynamics in a population clearly delimited at a geographical level and, on the other hand, a biological study on the mechanisms involved in the expansion of a virus at the cellular level within an organism.

With respect to the first group of works, classic models in epidemiology describe the different states in which population groups can be found considering the transmission of an infectious disease, including aspects such as the nonhomogeneous distribution of the population, migratory movements, and interactions in a local context. In addition, CA allow the modeling of the physiological characteristics of individuals susceptible to contracting a disease or being infected with a virus. Each individual of the population considered is an object contained in a box in the CA and the transition of the different states depends on the health condition of the individual, in addition to the conditions of the surrounding individuals (Sharma and Gupta, 2017). In general, the simplest models are those that consider two groups of populations: susceptible (S) and infected (I), called SI models. Other approaches include three subpopulations: susceptible-infectious-susceptible (SIS models), where infection does not confer immunity and infected individuals return to the susceptible state. There are other models that consider a new population group called exposed (E), formed by people who are not infected but who do not present symptoms, and depending on the disease involved, can be infected or not. These models are called SEIS models and describe the flows of people between susceptible-exposed-infectious-susceptible. Another type of models of four populations are those that change from the susceptible to recovered group (R), which are those individuals who become immunized once they have overcome an infection. These types of models with four population groups (susceptible-exposed-infectious-recovered) are called SEIR models. Finally, it is also possible to consider models with five groups of populations to consider an infection that does not let any immunity and recovered individuals return to being susceptible again (SEIRS models). Several examples of models based on the classic approaches presented earlier are discussed now.

Situngkir (2004) explores the epidemiological impact of avian influenza disease in Indonesia using a SIR model. The model uses a two-dimensional cellular space with a von Neumann neighborhood and results showed that the economic factors and control policies of Southeast Asian countries have a strong impact on the spread of this disease. On the other hand, the geographical characteristic formed by a conglomerate of islands reduces the severity of the spread. Another similar model for the same disease is presented in Pfeifer et al. (2008), although the model described can be adapted and applied to other epidemiological processes. In this case, five possible states are considered (latent, infectious, recovered, incubating, and symptomatic) assuming the Moore neighborhood and including the possibility of classifying the population by age groups. In particular, this study focuses on the region of the State of Tyrol (Austria), where different population densities, due to orography and large valleys, hinder the spread of the disease. Finally, a recent work based on CA (Yang et al., 2017) studied the infectious spread of HIV in Chongqing (China) by applying a two-dimensional cell space of more than 1000 boxes (corresponding to an area with a radius of 10 km and four divisions with respect to economic and geographical levels) and a fixed boundary condition. The results obtained with the model were consistent with available evidence.

On the other hand, epidemiological models based on CA that study the spread of a virus at cellular level within an organism are more numerous and diverse.

For example, the outbreak of Ebola in 2014–15 led to the emergence of models, such as the one presented in Burkhead and Hawkins (2015). This study proposed a two-dimensional CA considering four possible states: healthy cell (H), infected cell (I1), delayed infected cell (I2), and dead cell (D). The transition rules consider that the two types of infected cells are also infectious and when the cell reaches the state D, a new healthy cell replaces it in the next iteration. The work presents simulations based on different conditions considering the state of the immune system: fully functional, with delayed and with compromised response. In the proposed scenario with a fully functional immune system, state I2 is never reached because a cell in state I1 passes quickly to state D in the next iteration. In this case, the rules of state transition are as follows:

- A cell in state H changes to state I1 in the next iteration if there is at least one cell with state I1 around it.
- A cell in state I1 changes to state D in the next iteration.
- A cell in state D changes to state H in the next iteration.

Fig. 16 shows the example of a scenario described with fully functional immune system, limited to nine cells to simplify the problem and considering initial conditions with two cells in state I1.

In the other two scenarios presented, the immune system with delayed response gives rise to the appearance of cells with state I2 after previously passing through state I1. On the other hand, cells are not regenerated and die at some point in the simulation when the immune system is compromised, reaching the irreversible state D, assuming at least one cell in the state I1 or I2 at the initial instant. Fig. 17 shows the simulation for three iterations considering the immune system with delayed response (A) and compromised response (B), and the same initial conditions as in the previous case.

The infectious processes induced by slow virus disease usually require a greater number of states. For example, in another work by the same authors, but focused on HIV (Burkhead et al., 2009), a two-dimensional CA with seven states is presented to model the spread of infection in the lymph node, considering different levels of infection. Another example simplified to four states (healthy, infected-A1, infected-A2, and dead) was presented in Zorzenon dos Santos and Coutinho (2001), describing the importance of the lymph node, spatial location, and local interactions in HIV infection dynamics, as suggested in the literature. In this line, the work in Jafelice et al. (2009) presents a two-dimensional CA based on quadrangular boxes in a 31×31 space for the study of HIV through the circulatory system, where infected or uninfected cells are considered, as well as the blood and antibodies. The aim of this

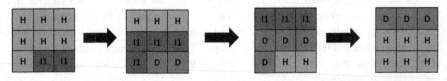

Fig. 16 Simulation of three iterations for the example presented in Burkhead and Hawkins (2015) with fully functional immune system.

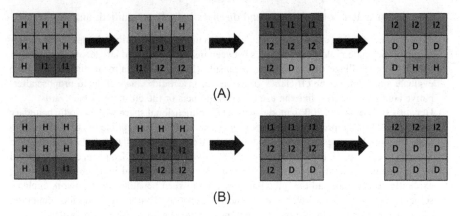

Fig. 17 Simulation of three iterations for the example presented in Burkhead and Hawkins (2015) with delayed response (A) and compromised immune system (B).

work was to reproduce the evolution of HIV in the circulatory system, considering the effects of antiretroviral therapy. In general, HIV has been one of the most studied viruses from the point of view of CA modeling and there are an important number of published works.

Another virus that has been extensively studied from an automaton-based modeling perspective is the avian influenza virus and its different subtypes. For example, (Beauchemin et al., 2005) presents a model where the transmission of the avian influenza virus in an organism is analyzed considering two types of cells: epithelial cells and immune cells. On the one hand, epithelial cells can be found in five states: healthy, infected, expressing, infectious, and dead. The following are the main transitions between states:

- Healthy cells are transformed into infected cells assigning a probability associated with the number of neighboring infectious cells.
- The expressing state represents the infected cells that begin to release the viral peptide and become infectious cells after a period of time.
- All cells die after a certain time associated to a life expectancy parameter.
- Expressing and infectious cells die when they are recognized by an immune cell.
- The regeneration of dead cells is included in the model.

On the other hand, immune cells can be found in two states: virgin or mature. The virus is not considered as a type of cell in the automaton, but the infection is determined by direct propagation from one epithelial cell to another. Dimensionally, this model is characterized by a two-dimensional space with a quadrangular grid, where each epithelial cell is located in a fixed position throughout the simulation period, while immune cells move randomly and interact under certain circumstances with epithelial cells. The results obtained showed that the model can be used as a testing tool to investigate various theoretical aspects of viral infections, particularly in the context of avian influenza disease. Other models related to this type of virus have been presented in recent years (Guan et al., 2016; Situngkir, 2004; Zhang, 2017).

To a lesser extent, other viruses and diseases have been studied, among them:

- Chagas disease (Slimi et al., 2009), studied with a model that describes the space-time inter-action of the epidemiological dynamics between insects (adults and larvae) at the geographical level of a village in the Yucatan peninsula (Mexico). The aim of this work was to evaluate and improve the efficiency of the use of insecticides, as well as to propose alternative control strategies, like the use of mosquito nets or the cleaning of backyards.
- Hepatitis B (Xiao et al., 2006), modeled by a two-dimensional space with periodic boundary and two cell types representing hepatocytes: resistant (R) or susceptible (S) to viral replication. The possible states considered in this model were: healthy cell type R, healthy cell type S, infected cell type R, infected cell type S, defectively infected cell, and dead cell (removed by the immune system or after their lifetime). Viral particles and immune cells were not explicitly considered, and the infection is spread between hepatocytes. The work explains some dynamics associated with transmission of hepatitis B virus, allowing the adaptation of the model to be applied to other persistent infections with replicating parasites.
- Zika virus (Alvarado et al., 2018), studied from a model-based perspective at the cellular level on the basis of experimental data observed in laboratory. For this purpose, a two-dimensional CA model of 400 × 400 boxes was proposed, considering circular boxes to better mimic the shape in a cell culture. Biologically, Zika virus-infected cells take time to release viral particles, increasing the likelihood of infecting neighboring cells until the resources are consumed. In that moment, a new phase leads to cell death. In the proposed model, four possible states were considered: healthy (H), infected (I), condensed (C), which are those cells that have been infected for a period of time, and dead (D). Finally, the model can be adapted to other viral cultures.
- Lyme disease is a type of bacterial infection transmitted to humans by the bite of an infected tick. The work in Li et al. (2012) presented a model for the transmission of this bacterium in an environment with Ixodidae ticks. The cellular space considered in the model was quadrangular, with a dimension of 50 × 50 boxes and two types of objects in the automaton: ticks and hosts. There are three stages associated with the development of the tick: larva, nymph, and adult; and hosts can be of two types: on the one hand, reservoir hosts, including vertebrates such as rodents, insects, and bird species that contribute to the maintenance of transmission; on the other hand, reproduction hosts such as fallow deer and cows, which provide blood to a significant number of adult ticks. The cellular space is divided into different habitats where tick-host interaction is analyzed. With respect to the habitat, three possible types of the zones are considered: woodland, grassland, and nonvegetated. The host has the ability to move through the cellular space, considering different movement patterns that differ from the type of habitat in which they are found. On the other hand, transition rules for tick development and pathogen transmission may vary depending on the habitat and the development stage of the tick. Different types of simulation were performed with habitat blocks of 1 × 1, 2 × 2, 5 × 5, and 10 × 10 boxes, where each block represents a habitat type. The results showed that habitat fragmentation is a risk for the acquisition of Lyme disease, as supported by other authors.
- Hantavirus is a group of viruses that are transmitted to humans through contact with urine, saliva, or feces from infected rodents. A model based on CA was presented (Abdul Karim et al., 2009) for describing the spread of this virus between rodents. For this purpose, a cell space was proposed with Moore neighborhood, toroidal boundary conditions, and two states: infected and uninfected. Two characteristics of hantavirus infection are included in the model. On the one hand, the complete disappearance of the infection in a rodent population when certain climatic conditions associated with seasonality (temporal characteristic). On

the other hand, the population of rodents can increase or reduce taking into account the availability of food resources (spatial characteristic). The results showed that susceptible and infected populations reach a stable value after a series of time intervals, depending on temporal and spatial characteristics as well as initial conditions.

Three-dimensional cellular spaces have also been used to model the propagation of infectious diseases. For example, (Khabouze et al., 2015) presents a CA model considering a 3D cellular space to describe the interactions between cells and the hepatitis B virus, both at the surface level and within the liver. The model is based on the von Neumann neighborhood with four possible states: healthy hepatocytes R (resistant to viral replication), healthy hepatocytes S (susceptible to viral replication), infected hepatocytes I and dead hepatocytes D. Each simulation performed in the work initializes the healthy hepatocytes R and S with different percentages. The results showed that increasing the percentage of hepatocytes S impact on viral load and decrease the number of healthy cells, which explains why a child with a smaller liver is more susceptible to developing the virus than an adult.

4.2 Oncology

The study of tumors using CA has also received attention in the biomedicine field. Cancer is usually caused by mutation or irregular activation of genes that control cell growth and mitosis (Hall, 2015). This genetic alteration causes a normal cell to become a cancer cell. Cancer cells acquire functions such as uncontrolled growth and division beyond the normal limits of the affected organ, as well as the ability to invade nearby organs (metastasis) by spreading through the lymphatic or circulatory system. The mass of cancer cells is called tumor. Fig. 18 summarizes the reproductive behavior of normal cells and cancer cells with damaged DNA.

There are two types of tumors: benign and malignant. A benign tumor has a slow growth and does not spread to other organs, while abnormal cells that make up a malignant tumor are characterized by a fast uncontrolled growth, invading other tissues. However, the growth rate and aggressiveness of the tumor mass and the probability of spreading are very variable and depend on the type of tumor and where it is located. Also, individual tumors respond differently to treatments. Some cases are treated with surgery, others with drugs or chemotherapy, and sometimes two or more treatments are alternated at the same time to provide better results. An example is presented in Fig. 19, which shows statistics from the National Cancer Database for the treatment of the non-Hodgkin lymphoma (Miller et al., 2016), a type of blood cancer.

In any case, untreated malignant tumors yield a progressive deterioration and death within a variable period of time. Cancer cells consume huge amounts of nutrients and energy required by other tissues and their uncontrolled growth increase this demand. To achieve this goal, some types of cancer produce angiogenic factors (TAF) that form new blood vessels into the tumor to provide nutrients to cancer cells, competing with normal cells for space and energy. As a consequence, the normal tissues near the tumor gradually suffer nutritive death due to indefinite proliferation of cancer cells

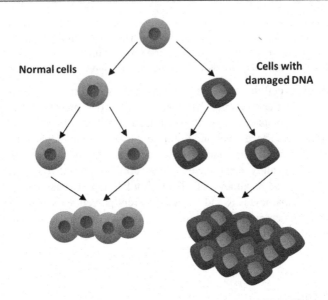

Fig. 18 Reproduction of normal cells and cancer cells with damaged DNA.

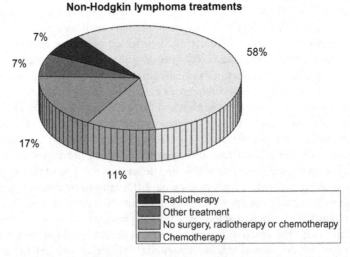

Fig. 19 National Cancer Database statistics for the treatment of the non-Hodgkin's lymphoma (Miller et al., 2016).

(Hall, 2015). Fig. 20 illustrates how angiogenic factors induce the development of new blood vessels, creating channels for nutrient acquisition and metastasis spread.

From the modeling point of view, one of the first CA-based models emerged in 1985 (Düchting and Vogelsaenger, 1985). In this work, the authors highlighted that CA modeling could help clinicians to better understand the dynamics of tumor growth.

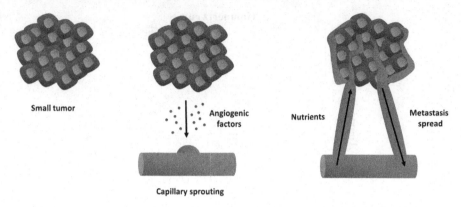

Fig. 20 Angiogenic factors inducing the development of new blood vessels.

The model considered a three-dimensional cell space and a von Neumann neighborhood, which allowed the simulation of tumor growth with different kinds of treatment in a vascularized tissue. The authors showed that modeling and simulation were a powerful tool for analyzing and optimizing tumor treatment. The computational limitations were also highlighted with respect to the representation of a three-dimensional structure of the tumor evolution, as well as the difficulties for validation of the results due to the lack of knowledge about some values of the parameters associated with the phases of the cell cycle.

In the same line, Qi et al. (1993) proposed a CA that described the interaction between the immune system and cancer. The model was based on a Gompertz growth curve (Fig. 21) and analyzed the rate of cancer cell proliferation, cytotoxic rate, and other relevant factors affecting cancer growth curve.

In this study of tumor growth, a two-dimensional cellular space was proposed, using quadrangular boxes and a von Neumann neighborhood. Each box contains one cell of the following types: cancer cells (C), effector or cytotoxic cells (E_0), complexes produced in the cytotoxic process (E), and dead cells (D). The biochemical processes of the cells are represented as chemical reactions. In Eq. (10), tumor cell proliferation describes how a cell C divides into two cells at a division rate (k_1). An enzymatic reaction is represented by Eq. (11), where cell C reacts with effector E_0, producing the complex E at a rate k_2 and its dissolution at rate k_3 into E_0 and dead cells; finally, reaction (12) presents the elimination of dead cells at a rate k_4.

$$C \xrightarrow{k_1} 2C \tag{10}$$

$$C + E_0 \xrightarrow{k_2} E \xrightarrow{k_3} E_0 + D \tag{11}$$

$$D \xrightarrow{k_4} \tag{12}$$

Gompertz curve

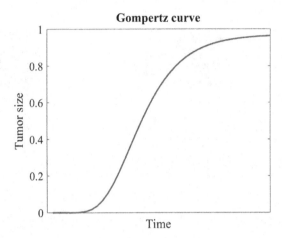

Fig. 21 Typical Gompertz curve to fit data of tumor growth, where the size of the tumor is scaled between 0 (no tumor cells) and 1 (maximum tumor cell population).

Results obtained by this model were in agreement with data observed by other authors. However, in the Qi model (Qi et al., 1993), cells could only divide when a neighboring box was empty; hence, the model was not realistic enough. Other related work (Smolle and Stettner, 1993) demonstrated the importance of the intrinsic and microenvironmental properties of cells in their division, migration, and death in tumor tissue. A two-dimensional cell space with 100×300 boxes and a Moore neighborhood was proposed, considering that a cell can perform three types of actions: division, migration, and death. Each cell has the same probability of being selected for a particular action, with probabilities represented by $p(div)$, $p(mot)$, $p(del)$, respectively, and Eq. (13) is met. The direction in which an action will be executed is randomly selected, with each of the eight possible directions having the same probability. In case of migration, the displacement distance for each iteration was randomly selected within a range.

$$p(div) + p(mot) + p(del) = 1 \tag{13}$$

All the models mentioned so far used a grid with square boxes and, therefore, a similar distance from one cell to another. This approach was not completely satisfactory and some models (Kansal et al., 2000) began to use other distributions based on Voronoi's diagrams (Voronoi, 1908). Fig. 22 presents this type of representation, considering a set of polygons in the plane and randomly distributed points in such a way that the perimeters of the polygons are equidistant from the nearest neighboring points. This type of polygonal distributions are currently widely used in computational geometry and applied to the modeling of brain tumor growth (Rejniak and Anderson, 2011).

One of the first models to incorporate the phenomenon of angiogenesis was published in the late 1990s (Anderson and Chaplain, 1998). A more recent model of this type (Gönczy et al., 2018) describes the basic behavior of the tumor, including proliferation

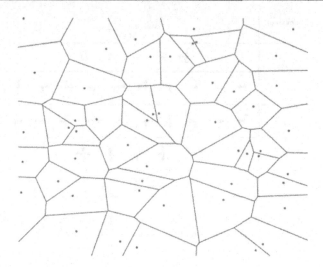

Fig. 22 Partitioning of a plane into regions considering a Voronoi diagram.

and necrosis, different vascularization mechanisms, and the diffusion of TAF. This work proposes a CA with a quadrangular box space and a von Neumann configuration, where each box of the automaton corresponds to a tissue area of 20 μm × 20 μm. The model considers the following cell types:

- Extracellular matrix cells, which may change to tumor cells or capillaries cells (blood vessels) depending on the status of their neighboring cells, the amount of nutrients and TAF in the cell being evaluated.
- Living tumor cells: a grid cell can contain living tumor cells or necrotized tumor cells. Cells that are on the periphery of the tumor are considered living.
- Necrotized tumor cells: tumor cells, surrounded by other tumor cells absorbing nutrients, become part of the necrotic core of the tumor.
- Capillaries cells: represent the blood vessels and provide nutrients to neighboring tumor cells.

Lattice gas automata (LGCA) are another approach to tumor modeling (Yuce, 2018). These models were introduced in 1976 (Hardy et al., 1976) to describe the molecular dynamics of gas movement. An LGCA employs a finite cell space and also includes a finite set of particle states, a neighborhood of interaction, and a set of local rules that determine the movement of particles (cells) and their transitions between states. The difference of an LGCA with respect to a traditional CA is the incorporation of a finite number of directions and velocity channels. Thus, in a two-dimensional LGCA model, it is possible to consider five direction channels: standing, left, right, up, and down. Similarly, in an LGCA three-dimensional model, two more direction channels are included: front and back. From these directional channels, different velocity channels can be formed considering different types of directions. Fig. 23 presents five types of velocities (Shrestha et al., 2014), with different types of directions.

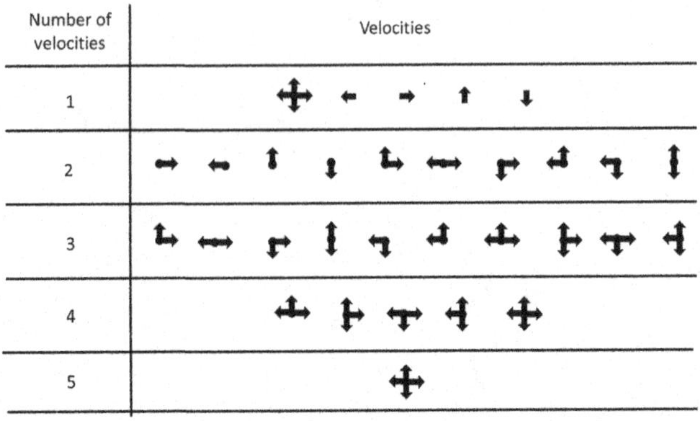

Number of velocities	Velocities
1	✛ ← → ↑ ↓
2	→ ← ↑ ↓ ↳ ↔ ↱ ↰ ↴ ↕
3	↳ ↔ ↱ ↓ ↴ ↰ ↲ ↦ ↧ ↫
4	↲ ↦ ↧ ↫ ✛
5	✛

Fig. 23 Different types of velocity considering a lattice gas automata (LGCA) of two dimensions. The number of possible directions increases with velocity.

The resting condition is considered one of the states and as the number of velocities increases, so does the variability of cell movement in multiple directions. In general, each cell has one and only one velocity channel associated with it. Also, two transition rules are proposed. The first one is associated to the change of each cell state, considering transitions to reproduction, necrosis or cellular elimination. The second rule is associated to cell movement, considering the five proposed velocity channels (Fig. 23).

A recent work (Interian et al., 2017) presents some general assumptions that should be taken into account in the development of this type of CA: spherical tumor and symmetry are maintained forever; there is an adhesion of cancer cells at the boundary of tumor that maintains a solid form and a balance with the expansive forces generated by internal cells; the tumor is in a continuous state of expansion; or the existence of the increased concentration of nutrients at the boundary of the tumor.

Finally, as stated in previous sections, given the underlying complexity, computational performance in the context of CA applied to the development of tumor growth is important. For this reason, some works (Bowles and Silvina, 2016; Poleszczuk and Enderling, 2014) have provided tools and methods to optimize the development of CA models applied to tumor growth.

4.3 Heart electrical conduction system

CA have been extensively applied to the modeling of heart electrical conduction. At physiological level, heart is made up of two pressure and suction pumps (Hall, 2015): a right compartment that pumps blood to the lungs and a left compartment that pumps blood to other tissues and organs. Each of these two compartments has a double-chamber made up of an atrium and a ventricle. Fig. 24 shows a schematic

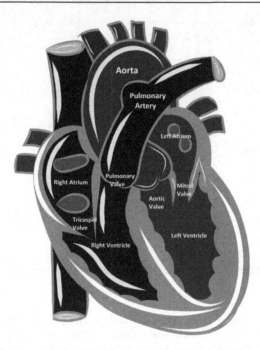

Fig. 24 Schematic representation of the four chambers of the heart, valves, arteries, and veins. Credit: Pixabay License (free right to use).

representation of the four chambers of the heart: left atrium, right atrium, left ventricle, and right ventricle, as well as the different valves that separate each of the atria with their respective ventricles (tricuspid and mitral valves) and the valves that connect the ventricles to the pulmonary and aortic arteries (pulmonary and aortic valves).

Muscle fibers of atria and ventricles contract similar to skeletal muscle, but for much longer. The coordinated activation and stimulation of these heart muscles is controlled by muscle fibers that make up the electrical conduction system of the heart. When a cell is electrically excited, the distribution of its electrical charge is modified, producing a set of actions associated with the entry and exit of cations and anions, triggering an action potential that propagates the electrical stimulation to nearby cells. In this way, the electrical signal travels from one cell to another until it passes through all the cells of the heart.

The cardiac muscle is characterized by having a potential action different from that of skeletal muscle, with a voltage that is around -85 mV at rest and approximately 20 mV in excitable condition for ventricular myocytes and Purkinje cells, which are considered the reference model to understand the cardiac action potential. In this sense, the action potential has five phases (Fig. 25) which are:

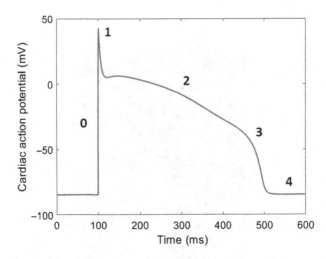

Fig. 25 Different phases of cardiac action potential in ventricular myocytes and Purkinje fibers.

- In phase 0, a fast depolarization occurs due to the opening of the fast Na^+ channels, which generates an increase in the conductance of the membrane, allowing the entry of these ions from the extracellular space into the cell. This causes a change in the action potential, from -85 mV at rest to 30 mV, approximately.
- In phase 1, a small repolarization takes place due to the inactivation of the fast Na^+ channels and the activation of the K^+ channels, allowing these ions to pass from the inside of the cell to the extracellular space.
- Phase 2 is characterized by a plateau lasting about 0.2 s in the atrial muscle and 0.3 s in the ventricular muscle. During this phase, the action potential is maintained in an equilibrium condition due to the movement of Ca^{2+} inwards through the ion channels for calcium and the outward movement of K^+ through the slow potassium channels. Currents produced by the Na-Ca and Na-K pumps play a minor role during this phase too. This plateau is the reason why the contraction of the cardiac muscle is much greater than that of the skeletal.
- During phase 3, called fast repolarization, calcium channels are closed, while potassium channels remain open. The positive net current to the outside causes cellular repolarization and the reduction of the action potential to its resting value. Once that value is reached, potassium channels close.
- The resting potential of the membrane is reached in phase 4. The cell remains in this phase until it is electrically activated again by an adjacent cell.

During phases 0, 1, 2, and partially 3, the cell is refractory to the initiation of the action potential and cannot initiate a new action potential. This is known as the absolute refractory period and it is a protection mechanism to limit the frequency of action potentials. After the end of this period, the relative refractory period begins, in which a depolarization above a certain threshold is necessary to trigger a new action potential.

The electrical impulse or pacemaker that initiates heartbeat is originated in the sinoatrial node (SA), located in the upper lateral wall of the right atrium. From this

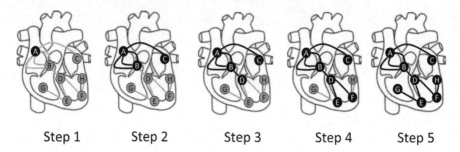

Step 1   Step 2   Step 3   Step 4   Step 5

Fig. 26 Propagation of the electrical impulse across the different cardiac tissues.

nodule, the electrical impulse is transmitted to the atria through the internodal pathways and later reaches the atrioventricular nodule (AV). After a brief time lapse, the electrical impulse moves through the His bundle and then divides into two branches for each of the ventricles. Finally, the ventricular contraction is triggered after the distribution of the electrical impulse in the ventricles through the Purkinje fibers. Fig. 26 shows the propagation of the electric impulse in five steps, considering eight nodes: (A) sinoatrial node, (B) atrioventricular node, (C) muscle fibers of the left atrium, (D) His bundle, (E) right ventricular branch, (F) left ventricular branch, (G) right Purkinje fibers, and (H) left Purkinje fibers.

The propagation velocity of the action potential changes in the different types of cardiac tissues (Cuenca, 2006; Méry and Singh, 2012). Table 4 shows the different propagation velocities and activation times for each of the constituent tissues of the heart's electrical conduction system.

Heart electrical activity is clinically measured by an electrocardiograph, which records this activity through electrodes placed on different body positions. This activity is represented in the electrocardiogram (ECG) (Fig. 27), characterized by the

Table 4 Electrical propagation velocities in each of the cardiac tissues that form the heart electrical conduction system

Cardiac tissue	Electric propagation velocity (cm/s)
SA node	30–50
Auricular myocardial cells	30–50
AV node	100–200
His bundle	100–200
Purkinje fibers (left)	300–400
Purkinje fibers (right)	300–400

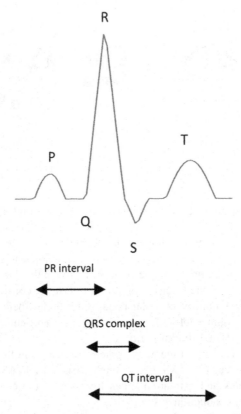

Fig. 27 Electrocardiogram with P, Q, R, S, and T waves that correspond to each of the cardiac action potential phases.

presence of P, Q, R, S, and T waves. In this sense, P wave represents atrial depolarization (phase 0), QRS complex characterizes the ventricular depolarization, and finally T wave represents ventricle repolarization. Mechanical systole or ventricular contraction starts at the beginning of the QRS complex and finishes before the T wave ends. Diastole, or ventricular relaxation, starts after the P wave begins and corresponds to atrial contraction.

One of the most common rhythmic alterations in the heart electrical conduction process is arrhythmias, which can be both ventricular and atrial. They usually appear because the electrical impulse is not properly generated or originates in the wrong place, or also due to the alteration of the electric conduction paths (such as the atrial re-entrances).

From a modeling point of view, the first work emerged in the 40s (Wiener and Rosenblueth, 1946), where the problem of impulse conduction was modeled through a network connected with excitable elements that was assimilated to the concept of cellular automaton. This model formally introduced the concept of excitable medium as a dynamic system capable of being activated from a disturbance that exceeded a certain electrical threshold.

Later, one of the classic works of CA applied to the study of heart electrical propagation modeling was published by Moe Moe et al. (1964). This work presented a computational model formed by a cellular space with hexagonal grid and a neighborhood configuration that affected six neighboring cells, with an electric current moving with a propagation speed of 80 cm/s. Each iteration of the automaton supposed a time interval of 5 ms, associated with a box diameter of 4 mm. On the other hand, the excitability of the cells was modeled by five states. The first state corresponds to the relative refractory period. States two, three, and four resemble the absolute refractory period, with a gradual increase in the propagation velocity from ¼ to ½ with respect to the normal driving velocity. Finally, state five returns to normal driving velocity. The results of Moe's work reproduced behaviors associated with the uncoordinated beats under atrial fibrillation, one of the most frequent arrhythmias in the clinical practice.

Shortly after, Krinsky (1966) made a modification of the original model of Wiener and Rosenblueth (1946) to explain at modeling level some mechanisms associated with atrial fibrillation such as re-entry, multiple waves, and other physiological aspects of arrhythmias. The work also presented a state of the art of mathematical models related to the fundamental mechanisms of cardiac arrhythmias.

In the 70s, Reshod'ko et al. (1977) showed the usefulness of considering cellular biological systems as CA, for the analysis of some biological phenomena difficult to study through experimentation. The work analyzed the excitatory propagation through the smooth muscle cells, which have similarities in many properties to a myocardial cell. For this study, a CA was proposed considering a Moore neighborhood configuration and a grid of 7200 boxes (60 × 120).

Years later, Mitchell et al. (1993) presented a theoretical model-based analysis of electrical instability (fibrillation) in the myocardium. The model allows to simulate some behaviors associated with the generation of arrhythmias to reproduce different varieties of ventricular arrhythmias. To this purpose, a quadrangular cell space of 2500 boxes (50 × 50) was considered, with a Moore neighborhood and four possible cellular states: inactive (at rest), excited (depolarized), absolute refractory (depolarized and nonexcitable), and relative refractory (recovery of excitability). The transition of states in the cells is associated with the value of the action potential at the neighboring cells. Thus, a cell will remain in the resting state until it receives an excitation from at least one of its eight neighboring cells. As shown in Fig. 28, the greater the duration of the relative refractory period, the fewer number of neighboring excited cells necessary for the evaluated cell to reach the stimulation.

However, these classical models omitted important aspects associated with the propagation of the electrical signal, such as the effects of the shape of the electrical impulse, as well as the minimum velocity to guarantee a stable propagation that avoids conduction block (Cabo et al., 1994). A simple representation of the waveform using a

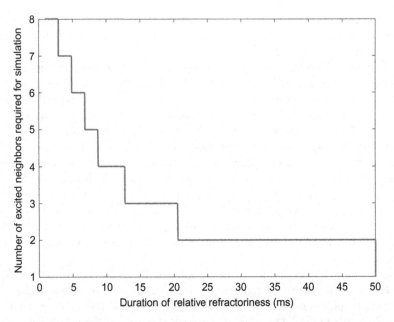

Fig. 28 Number of neighboring excitable cells versus the duration of the refractory period.

piecewise linear function was proposed by Siregar et al. (1996). This work presented a CA of 25,000 square boxes with cells in four states: rest, depolarization, absolute refractory, and relative refractory. Another work from the same period (Feldman et al., 1999) introduced complex propagation regimes with different signal rates and provided physiological parameters to simulate the effects of a specific disease or pharmacological intervention.

Parallel to these works, the first three-dimensional models were emerging. For example, Wei et al. (1990) proposed a complete 3-D heart model, including atria, ventricles, and electrical conduction system. The geometric shape of the automaton was an inclined parallelepiped with 50 planes in the frontal axis, 50 in the sagittal, and 81 in the transversal. Different types of cells were distributed throughout the cellular space, including normal myocardial cells, abnormal or malfunctioning cells, and special electrical conduction systems such as the bundle of Kent, which allowed to simulate supraventricular tachycardias of re-entry and alterations of the ECG caused by Wolff-Parkinson-White syndrome (Fig. 29). Another example of a 3-D model was presented in Harrild and Henriquez (2000), which describes atrial activation, including the two atria and the main atrial muscles. The results of the model demonstrated the normal sinus rhythm of the heartbeat and allowed the extraction of interatrial septum activation patterns, although it included assumptions and simplifications in electrical transmission to reduce computational complexity.

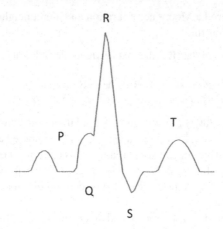

Fig. 29 Alteration of the ECG associated with Wolf-Parkinson-White syndrome (QR interval).

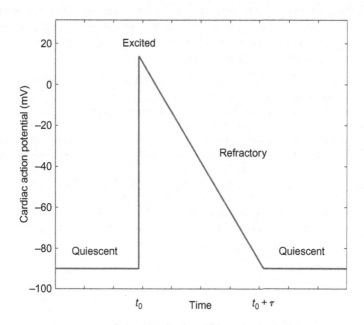

Fig. 30 Evolution of the transmembrane potential for a cell during the idle-excitation-refractory-inactivity cycle in the Barbosa model (Barbosa, 2003).

A more recent and realistic model was presented in 2003 (Barbosa, 2003), where a detailed description of a CA-based model for the propagation of the action potential in a cardiac tissue was proposed, considering three possible states for each cell depending on the transmembrane potential that each cell can take: inactive (I), excited (E), or refractory (R). The evolution of the transmembrane potential for a cell is shown in Fig. 30.

This work considered a Moore configuration and a unit neighborhood radius, with the following transition rules:

- A cell in state I moves to state E in the next iteration when it has at least one neighbor cell in state E.
- A cell in state E changes to the state R in the next iteration.
- A cell in state R returns to state I after several iterations due to the existing time delay.

This model (Barbosa, 2003) introduced rectangular cardiac cells, due to the fact that cells of the real heart do not have a square shape, and considered a brickwall pattern or brick-like cellular distribution, closer to reality. This pattern encompassed a neighborhood with ten cells, including different driving velocities depending on the propagation direction. Fig. 31 shows the brickwall-type cellular distribution with ten affected cells in the neighborhood zone.

A CA that considered a probabilistic and deterministic combination was presented soon after (Atienza et al., 2005). In this work, each cell can be found in three possible states: rest (relaxed or excitable), refractory 1 (excited and able to excite the neighboring cells), or refractory 2 (excited, but unable to excite the neighbors). Refractory period 1 is kept for a fraction of time of about 10% of the total action potential duration. During the rest of the time, the cell remains in the refractory period 2. Transitions between states are governed by three laws:

- Partial repolarization: transition from the refractory state 1 to the refractory state 2, which occurs deterministically after a specific period.
- Total repolarization: transition from the refractory state to the resting state, which, like the previous transition, also occurs deterministically.
- Depolarization: transition from state of rest to refractory state 1, dependent on probabilistic terms that are based on two factors: cell excitability (E), which increases with the amount of time the cell remains at rest, and the magnitude of excitability around the cell (Q). Thus, the greater the amount of excitation, the greater the probability that the cell will be excited. Eq. (14) presents the probability that a cell j is excited by the set of neighboring cells i, where d_{ij} is the distance between cell j and the rest of cells and A_i is the excitation state of cell i, which is modeled as a binary quantity: 1 if cell i is in the refractory period and 0 otherwise).

$$P_j^{exc} = E \cdot Q = E \cdot \sum_{i \neq j} \frac{A_i}{d_{ij}^2} \tag{14}$$

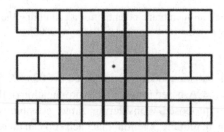

Fig. 31 Cell distribution with brickwall pattern used in the work of Barbosa (2003).

Later on, a very complete model of heart electrical conduction system was published (Méry and Singh, 2012). The work considered electrical conduction in different cardiac tissues to reproduce the propagation from sinoauricular nodule to Purkinje fibers. The model used a CA with quadrangular boxes and a von Neumann neighborhood configuration, three possible states (active, passive, and refractory) and different propagation velocities in each of the tissues, as it occurs in reality.

Finally, recent advances in computing, as well as the use of multicore processors (Mata et al., 2018) and the emergence of new computer tools for 3-D modeling such as Comsol (Cocherová et al., 2017), have led to the emergence of even more sophisticated models focused on specific aspects of electric ventricular conduction (Cocherová et al., 2019), atrial (Avdeev and Bogatov, 2014; Lin et al., 2017) or ventricular fibrillation analysis (Zheng et al., 2015). Other relevant aspects that are also being investigated are electromechanical interaction associated with the contraction of the ventricles and atria muscle cells (Campos et al., 2015), analysis of arrhythmias through simulations of the electrochemical processes that alter the generation of the cardiac action potential (Avdeev and Bogatov, 2014), or sudden death (Sabzpoushan and Pourhasanzade, 2011).

5 Software techniques

The computational techniques for the modeling of CA are based on three fundamental principles: simplicity, parallelism, and ubiquity (Sipper, 2017). Simplicity refers to the elementary characteristics of a single cell, without the capacity to perform any complex action. Parallelism is associated with the simultaneity of actions and interactions of multiple cells at the same time. Finally, ubiquity highlights the importance of the location of a cell and how it interacts with its environment considering the neighborhood radius, but at the same time ignores the characteristics of the entire cell space and the effects are restricted to specific areas. These three principles have implications for both the design and implementation of a cellular automaton model. In general, the simulation of a CA can represent a great computational challenge, especially when the number of cells is high and the transition rules are complex (Marzolla, 2018). However, the evaluations and the cell transitions to new states are independent tasks and can therefore be computed in parallel, at least for a particular iteration. Today, any modern computer provides parallel computing resources due to the existence of multiple processing units (CPUs) with multicore architecture. However, the use of multicore CPU is not the best solution to execute a huge amount of parallel tasks (as happens in the case of CA), although performance is higher when compared with single-core processors.

In recent years, the use of graphical processing units (GPUs) for data-intensive computing has become more relevant. Unlike CPUs, GPUs are made up of a large number of processing units that exploit massive parallelism by executing many concurrent threads. A last-generation graphics card can now reach 4000 cores (Wynters, 2018), while CPU has usually a few tens.

In terms of disadvantages, GPU has a low data transfer rate because it does not have direct access to the random access memory (RAM) and GPU video memory is slower due to the peripheral component interconnect express (PCIe) (Kauffmann and Piché, 2010). Another major disadvantage is the restriction on the types of data that can be managed, because a GPU core is different from a CPU core in its design and simplicity. This limitation may have an impact in high-precision floating-point arithmetic operations. Finally, the use of parallelization techniques requires the learning of complex computer science concepts and highly specialized knowledge (Marzolla, 2018).

In general, the main software techniques to execute applications on GPUs, and more particularly in the simulation of CA, are the following (Gimenes et al., 2018): Open Computing Language (OpenCL), Open Accelerators (OpenACC), Message Passing Interface (MPI), Open Multi-Processing (OpenMP), Single Instruction, Multiple Data (SIMD), and Compute Unified Device Architecture (CUDA). OpenMP and OpenACC are implemented using compiler directives for C, C++, or FORTRAN (although there are tools for other programming languages) and developers have no control about low-level operations. In contrast, OpenCL and CUDA are implemented as libraries for C and C++ and provide the programmer high-to-very-low-level details. With respect to parallelism, all technologies support data parallelism and asynchronous tasks (Memeti et al., 2017), included at GPU level. In the case of OpenMP, although this is a relatively old technology focused particularly on the use of CPU, version 4.0 of the application programming interface (API) included the possibility of using GPU acceleration (Martineau et al., 2016). On the other hand, MPI cannot employ GPU on its own, but can be combined with CUDA to provide it. Table 5 presents a summary of some features for these technologies discussed earlier. A more comprehensive description of the complex functionalities associated with task parallelization can be found in the literature (Memeti et al., 2017).

There is a significant number of studies in the literature that analyze the advantages and disadvantages of each technique discussed (Memeti et al., 2017; Mishra et al., 2017; Thouti and Sathe, 2012; Wang et al., 2017). From another perspective, different works have focused on evaluating the performance of these technologies through the use of CA (Marzolla, 2018; Millán et al., 2017). However, GPU programming requires specific programming skills and detailed knowledge on certain hardware aspects about how a computer works. For this reason, some papers have presented middleware that reduces the complexity, providing intuitive interfaces to simplify the implementation without requiring knowledge of the specific details about the computer architecture. A recent example of this type of tool is ParaCells (Song et al., 2018), a GPU simulation architecture that uses CUDA to simplify the development of cell-centered models in computational biology.

Table 5 Summary of some features of different software interfaces for the implementation of applications with parallelization of tasks

	Stable release	Written in	Architectures	Operating systems	License
OpenCL	2.2–7 (2018)	C y C++	x86-64 y ARM, Cell, IA-32 and POWER	Android, Unix-based, macOS y Windows	OpenCL specification license
Open MPI (MPI)	4.0.0 (2018)	C	x86-64	Unix-based and macOS	New BSD License
GCC (OpenMP, OpenACC and SIMD)	8.2 (2018)	C and C++	Multiple	Unix-based, macOS and Windows	General Public License (GNU)
CUDA	10.0 (2018)	C y C++	Supported GPUs	Unix-based, macOS and Windows	Freeware

6 Conclusions

CA have historically been employed in many knowledge fields since the mid-60s. In the context of biomedical systems, infectious disease epidemiology, oncology, and heart electrical conduction system have been some of the application targets of CA. With the gradual increase of the computational capacity, intensive computing techniques, parallel computing and cloud computing, everything seems to indicate that the use of CA still has a wide range of applications of very diverse nature.

This chapter presents a general vision about CA, including a historical perspective and a brief introduction to the theoretical concepts. Applications of computational automata in biomedicine have been exposed in a descriptive way, to make the contents more accessible to graduate students interested in this type of modeling techniques.

At the end of the chapter, a section devoted to current software techniques and practices most used in the implementation of CA has been presented, including a bibliographic review for those graduate students or readers who have an interest in the implementation of this type of modeling techniques in the context of GPU programming.

Acknowledgments

This work was supported in part by "Fondo de Investigaciones Sanitarias" (Instituto de Salud Carlos III, Spain) under Grants PI15/00306 and DTS15/00195, in part by "Fundación Progreso y Salud" (Government of Andalucía, Spain) under Grants PI-0010-2013, PI-0041-2014 and PIN-0394-2017, and in part by "Fundación Mutua Madrileña" under Grant VÍA-RENAL.

References

Abdul Karim, M.F., Md Ismail, A.I., Ching, H.B., 2009. Cellular automata modelling of hantarvirus infection. Chaos Solitons Fract. 41, 2847–2853.

Adamatzky, A. (Ed.), 2010. Game of Life Cellular Automata. Springer-Verlag, London.

Akdur, G., 2011. The use of biological cellular automaton models in medical, health and biological studies. In: Procedia—Social and Behavioral Sciences, World Conference on Educational Technology Researches—2011, 28, pp. 825–831.

Alvarado, A., Corrales, R., Leal, M.J., la Ossa, A.D., Mora, R., Arroyo, M., Gomez, A., Calderon, A., Arias-Arias, J.L., 2018. Cellular-level characterization of dengue and zika virus infection using multiagent simulation. In: 2018 IEEE International Work Conference on Bioinspired Intelligence (IWOBI). Presented at the 2018 IEEE International Work Conference on Bioinspired Intelligence (IWOBI), pp. 1–6.

Anderson, A.R.A., Chaplain, M.A.J., 1998. Continuous and discrete mathematical models of tumor-induced angiogenesis. Bull. Math. Biol. 60, 857–899.

Atienza, F.A., Carrión, J.R., Alberola, A.G., Álvarez, J.L.R., Muñoz, J.J.S., Sánchez, J.M., Chávarri, M.V., Atienza, F.A., Carrión, J.R., Alberola, A.G., Álvarez, J.L.R., Muñoz, J.J.S., Sánchez, J.M., Chávarri, M.V., 2005. A probabilistic model of cardiac electrical activity based on a cellular automata system. Rev. Esp. Cardiol. 58, 41–47.

Avdeev, S.A., Bogatov, N.M., 2014. Simulation of electrochemical processes in cardiac tissue based on cellular automaton. IOP Conf. Ser.: Mater. Sci. Eng. 66, 012019.

Banks, E.R., 1970. Universality in cellular automata. In: 11th Annual Symposium on Switching and Automata Theory (Swat 1970) (FOCS), pp. 194–215.

Barbosa, C.R.H., 2003. Simulation of a plane wavefront propagating in cardiac tissue using a cellular automata model. Phys. Med. Biol. 48, 4151.

Bartocci, E., Lió, P., 2016. Computational modeling, formal analysis, and tools for systems biology. PLOS Comput. Biol. 12, e1004591.

Bays, C., 2009. Cellular automata in triangular, pentagonal and hexagonal tessellations. In: Meyers, R.A. (Ed.), Encyclopedia of Complexity and Systems Science. Springer New York, New York, NY, pp. 892–900.

Beauchemin, C., Samuel, J., Tuszynski, J., 2005. A simple cellular automaton model for influenza A viral infections. J. Theor. Biol. 232, 223–234.

Beuchat, J.L., Haenni, J.O., 2000. Von Neumann's 29-state cellular automaton: a hardware implementation. IEEE Trans. Educ. 43, 300–308.

Bhattacharjee, K., Naskar, N., Roy, S., Das, S., 2018. A survey of cellular automata: types, dynamics, non-uniformity and applications. Nat. Comput. 1–29.

Bowles, J.K.F., Silvina, A., 2016. Model checking cancer automata. In: 2016 IEEE-EMBS International Conference on Biomedical and Health Informatics (BHI). Presented at the 2016 IEEE-EMBS International Conference on Biomedical and Health Informatics (BHI), IEEE, Las Vegas, NV, USA, pp. 376–379.

Brodland, G.W., 2015. How computational models can help unlock biological systems. Semin. Cell Dev. Biol. 47–48, 62–73 Coding and non-coding RNAs & Mammalian development.

Burkhead, E., Hawkins, J., 2015. A cellular automata model of Ebola virus dynamics. Phys. A: Stat. Mech. Appl. 438, 424–435.

Burkhead, E.G., Hawkins, J.M., Molinek, D.K., 2009. A dynamical study of a cellular automata model of the spread of HIV in a lymph node. Bull. Math. Biol. 71, 25–74.

Cabo, C., Pertsov, A.M., Baxter, W.T., Davidenko, J.M., Gray, R.A., Jalife, J., 1994. Wavefront curvature as a cause of slow conduction and block in isolated cardiac muscle. Circ. Res. 75, 1014–1028.

Campos, R.S., Rocha, B.M., da Barra, L.P.S., Lobosco, M., dos Santos, R.W., 2015. A parallel genetic algorithm to adjust a cardiac model based on cellular automaton and mass-spring systems. In: Parallel Computing Technologies, Lecture Notes in Computer Science. Presented at the International Conference on Parallel Computing Technologies. Springer, Cham, pp. 149–163.

Cocherová, E., Svehlikova, J., Tysler, M., 2019. Activation propagation in cardiac ventricles using the model with the conducting system. In: Lhotska, L., Sukupova, L., Lacković, I., Ibbott, G.S. (Eds.), World Congress on Medical Physics and Biomedical Engineering 2018, IFMBE Proceedings. Springer Singapore, pp. 799–802.

Cocherová, E., Svehlíková, J., Zelinka, J., Tysler, M., 2017. Activation propagation in cardiac ventricles using homogeneous monodomain model and model based on cellular automaton. In: 2017 11th International Conference on Measurement. Presented at the 2017 11th International Conference on Measurement, pp. 217–220.

Codd, E.F., 1968. Cellular Automata, ACM Monograph Series. Academic Press, New York.

Cuenca, E.M., 2006. Fundamentos de fisiología. Editorial Paraninfo.

Deutsch, A., Dormann, S., 2005. Cellular Automaton Modeling of Biological Pattern Formation: Characterization, Applications, and Analysis, Modeling and Simulation in Science, Engineering and Technology. Birkhäuser, Basel.

Düchting, W., Vogelsaenger, T., 1985. Recent progress in modelling and simulation of three-dimensional tumor growth and treatment. Biosystems 18, 79–91.

Feldman, A.B., Chernyak, Y.B., Cohen, R.J., 1999. Cellular automata model of cardiac excitation waves. Herzschr. Elektrophys. 10, 92–104.

Gardner, M., 1970. The fantastic combinations of John Conway's new solitaire game "life" Sci. Am. 223, 120–123.

Ghosh, M., Kumar, R., Saha, M., Sikdar, B.K., 2018. Cellular automata and its applications. In: 2018 IEEE International Conference on Automatic Control and Intelligent Systems (I2CACIS). Presented at the 2018 IEEE International Conference on Automatic Control and Intelligent Systems (I2CACIS), pp. 52–56.

Gimenes, T.L., Pisani, F., Borin, E., 2018. Evaluating the performance and cost of accelerating seismic processing with CUDA, OpenCL, OpenACC, and OpenMP. In: 2018 IEEE International Parallel and Distributed Processing Symposium (IPDPS). Presented at the 2018 IEEE International Parallel and Distributed Processing Symposium (IPDPS)pp. 399–408.

Gönczy, T., Csercsik, D., Kovács, L., 2018. A hybrid cellular automaton model of tumor-induced angiogenesis. In: 2018 IEEE International Conference on Automation, Quality and Testing, Robotics (AQTR). Presented at the 2018 IEEE International Conference on Automation, Quality and Testing, Robotics (AQTR), pp. 1–6.

Guan, P., Bi, X., Fei, L., Huang, D., Liu, L., 2016. Simulation of the epidemic of influenza A (H1N1) in a university using cel-lular automata model. Chinese J. Infect. Control 15, 79–82.

Gutowitz, H., 1991. Cellular Automata: Theory and Experiment. MIT Press.

Hall, J.E., 2015. Guyton and Hall Textbook of Medical Physiology, 13 ed. Saunders, Philadelphia, PA.

Hardy, J., de Pazzis, O., Pomeau, Y., 1976. Molecular dynamics of a classical lattice gas: transport properties and time correlation functions. Phys. Rev. A 13, 1949–1961.

Harrild, D., Henriquez, C., 2000. A computer model of normal conduction in the human atria. Circ. Res. 87, E25–E36.

Holt, W.V., Cummins, J.M., Soler, C., 2018. Computer-assisted sperm analysis and reproductive science; a gift for understanding gamete biology from multidisciplinary perspectives. Reprod. Fertil. Dev. 30, iii–v.

Hutton, T.J., 2010. Codd's self-replicating computer. Artif. Life 16, 99–117.

Interian, R., Rodríguez-Ramos, R., Valdés-Ravelo, F., Ramírez-Torres, A., Ribeiro, C., Conci, A., 2017. Tumor growth modelling by cellular automata. Math. Mech. Complex Syst. 5, 239–259.

Jafelice, R.M., Bechara, B.F.Z., Barros, L.C., Bassanezi, R.C., Gomide, F., 2009. Cellular automata with fuzzy parameters in microscopic study of positive HIV individuals. Math. Comput. Model. 50, 32–44.

Kansal, A.R., Torquato, S., Harsh IV, G.R., Chiocca, E.A., Deisboeck, T.S., 2000. Cellular automaton of idealized brain tumor growth dynamics. Biosystems 55, 119–127.

Kauffman, S.A., 1984. Emergent properties in random complex automata. Phys. D: Nonlinear Phenom. 10, 145–156.

Kauffmann, C., Piché, N., 2010. Seeded ND medical image segmentation by cellular automaton on GPU. Int. J. CARS 5, 251–262.

Khabouze, M., Hattaf, K., Yousfi, N., 2015. Three-dimensional cellular automaton for modeling the hepatitis B virus infection. Int. J. Comput. Biol. (IJCB) 4, 13–20.

Kier, L.B., Seybold, P.G., Cheng, C.-K., 2005. Modeling Chemical Systems using Cellular Automata. Springer Netherlands.

Krinsky, V.I., 1966. Excitation propagation in heterogeneous medium (modes similar to cardiac fibrillation). Bioifzika 11, 676–683.

Langton, C.G., 1990. Computation at the edge of chaos: phase transitions and emergent computation. Phys. D: Nonlinear Phenom. 42, 12–37.

Li, S., Hartemink, N., Speybroeck, N., Vanwambeke, S.O., 2012. Consequences of landscape fragmentation on lyme disease risk: a cellular automata approach. PLoS One 7, e39612.

Li, X., Wu, J., Li, X., 2018. Theory of Practical Cellular Automaton. Springer Singapore.

Lin, Y., 2015. Unstructured Cellular Automata in Ecohydraulics Modelling, first ed. CRC Press.

Lin, Y.T., Chang, E.T.Y., Eatock, J., Galla, T., Clayton, R.H., 2017. Mechanisms of stochastic onset and termination of atrial fibrillation studied with a cellular automaton model. J. Roy. Soc. Interface. 14, 20160968.

Martineau, M., McIntosh-Smith, S., Gaudin, W., 2016. Evaluating OpenMP 4.0's effectiveness as a heterogeneous parallel programming model. In: 2016 IEEE International Parallel and Distributed Processing Symposium Workshops (IPDPSW). Presented at the 2016 IEEE International Parallel and Distributed Processing Symposium Workshops (IPDPSW), pp. 338–347.

Martinez, G.J., Seck-Tuoh-Mora, J.C., Zenil, H., 2013. computation and universality: class IV versus class III cellular automata. J. Cell. Automata 7, 393–430.

Marzolla, M., 2018. Parallel Implementations of Cellular Automata for Traffic Models. arXiv:1804.07981 [cs].

Mata, A.N., Abrego, N.P.C., Alonso, G.R., García, M.A.C., Garza, G.L., Fernández, J.R.G., 2018. Parallel simulation of sinoatrial node cells synchronization. In: 2018 26th Euromicro International Conference on Parallel, Distributed and Network-Based Processing (PDP). Presented at the 2018 26th Euromicro International Conference on Parallel, Distributed and Network-Based Processing (PDP), pp. 126–133.

Memeti, S., Li, L., Pllana, S., Ko\lodziej, J., Kessler, C., 2017. Benchmarking OpenCL, OpenACC, OpenMP, and CUDA: programming productivity, performance, and energy consumption. In: Proceedings of the 2017 Workshop on Adaptive Resource Management and Scheduling for Cloud Computing, ARMS-CC'17. ACM, New York, NY, USA, pp. 1–6.

Méry, D., Singh, N.K., 2012. Formalization of heart models based on the conduction of electrical impulses and cellular automata. In: Liu, Z., Wassyng, A. (Eds.), Foundations of Health Informatics Engineering and Systems, Lecture Notes in Computer Science. Springer, Berlin Heidelberg, pp. 140–159.

Millán, E.N., Wolovick, N., Piccoli, M.F., Garino, C.G., Bringa, E.M., 2017. Performance analysis and comparison of cellular automata GPU implementations. Cluster Comput. 20, 2763–2777.

Miller, K.D., Siegel, R.L., Lin, C.C., Mariotto, A.B., Kramer, J.L., Rowland, J.H., Stein, K.D., Alteri, R., Jemal, A., 2016. Cancer treatment and survivorship statistics, 2016. CA: Cancer J. Clin. 66, 271–289.

Mishra, A., Li, L., Kong, M., Finkel, H., Chapman, B., 2017. Benchmarking and evaluating unified memory for OpenMP GPU offloading. In: Proceedings of the Fourth Workshop on the LLVM Compiler Infrastructure in HPC, LLVM-HPC'17. ACM, New York, NY, USA. 6, pp. 1–6:10.

Mitchell, M., Hraber, P.T., Crutchfield, J.P., 1993. Revisiting the edge of chaos: evolving cellular automata to perform computations. Complex Syst. 7, 89–130.

Moe, G.K., Rheinboldt, W.C., Abildskov, J.A., 1964. A computer model of atrial fibrillation. Am. Heart J. 67, 200–220.

Murray, J.D., 1981. A pre-pattern formation mechanism for animal coat markings. J. Theor. Biol. 88, 161–199.

Pfeifer, B., Kugler, K., Tejada, M.M., Baumgartner, C., Seger, M., Osl, M., Netzer, M., Handler, M., Dander, A., Wurz, M., Graber, A., Tilg, B., 2008. A cellular automaton framework for infectious disease spread simulation. Open Med. Inform. J. 2, 70–81.

Poleszczuk, J., Enderling, H., 2014. A high-performance cellular automaton model of tumor growth with dynamically growing domains. Appl. Math. (Irvine) 5, 144–152.

Qi, A.S., Zheng, X., Du, C.Y., An, B.S., 1993. A cellular automaton model of cancerous growth. J. Theor. Biol. 161, 1–12.

Rejniak, K.A., Anderson, A.R.A., 2011. Hybrid models of tumor growth. Wiley Interdisc. Rev.: Syst. Biol. Med. 3, 115–125.

Reshod'ko, L., Kolarzh, P., Grim, I., 1977. Cellular automata and cellular biological systems. Cybernetics 13, 454–465.

Sabzpoushan, S.H., Pourhasanzade, F., 2011. A cellular automata-based model for simulating restitution property in a single heart cell. J. Med. Signals Sens. 1, 19–23.

Schiff, J.L., 2011. Cellular Automata: A Discrete View of the World. John Wiley & Sons.

Sharma, N., Gupta, A.K., 2017. Impact of time delay on the dynamics of SEIR epidemic model using cellular automata. Phys. A: Stat. Mech. Appl. 471, 114–125.

Shrestha, S.M.B., Joldes, G., Wittek, A., Miller, K., 2014. Modeling three-dimensional avascular tumor growth using lattice gas cellular automata. In: Doyle, B., Miller, K., Wittek, A., Nielsen, P.M.F. (Eds.), Computational Biomechanics for Medicine. Springer New York, pp. 15–26.

Siap, I., Akin, H., Uğuz, S., 2011. Structure and reversibility of 2D hexagonal cellular automata. Comput. Math. Appl. 62, 4161–4169.

Sipper, M., 2017. Paul Rendell: turing machine universality of the Game of Life. Genet. Progr. Evol. Mach. 18, 115–117.

Siregar, P., Sinteff, J.P., Chahine, M., Lebeux, P., 1996. A cellular automata model of the heart and its coupling with a qualitative model. Comput. Biomed. Res. 29, 222–246.

Situngkir, H., 2004. Epidemiology Through Cellular Automata: Case of Study Avian Influenza in Indonesia. Working Paper WPF2004, Bandung Fe Institute. arXiv:nlin/0403035.

Slimi, R., El Yacoubi, S., Dumonteil, E., Gourbière, S., 2009. A cellular automata model for Chagas disease. Appl. Math. Model. 33, 1072–1085.

Smolle, J., Stettner, H., 1993. Computer simulation of tumour cell invasion by a stochastic growth model. J. Theor. Biol. 160, 63–72.

Song, Y., Yang, S., Lei, J., 2018. ParaCells: A GPU Architecture for Cell-Centered Models in Computational Biology. IEEE/ACM Transactions on Computational Biology and Bioinformatics, p. 1.

Thouti, K., Sathe, S.R., 2012. Comparison of OpenMP & OpenCL parallel processing technologies. Int. J. Adv. Comput. Sci. Appl. 3, 56–61.

Toffoli, T., Margolus, N., 1987. Cellular Automata Machines: A New Environment for Modeling. MIT Press, Cambridge, MA, USA.

Ulam, S.M., Ulam, F., Mycielski, J., Hirsch, D., Matthews, W.G., 1991. Adventures of a Mathematician, Edición: Reprint. University of California Press, Berkeley.

van Panhuis, W.G., Grefenstette, J., Jung, S.Y., Chok, N.S., Cross, A., Eng, H., Lee, B.Y., Zadorozhny, V., Brown, S., Cummings, D., Burke, D.S., 2013. Contagious diseases in the United States from 1888 to the present. N. Engl. J. Med. 369, 2152–2158.

Ventrella, J., 2006. A particle swarm selects for evolution of gliders in non-uniform 2d cellular automata. In: In Artificial Life X: Proceedings of the 10th International Conference on the Simulation and Synthesis of Living Systems, pp. 386–392.

Vichniac, G.Y., 1984. Simulating physics with cellular automata. Phys. D: Nonlinear Phenom. 10, 96–116.

Voronoi, G., 1908. Nouvelles applications des paramètres continus à la théorie des formes quadratiques. Deuxième mémoire. Recherches sur les paralléllloèdres primitifs. J. Reine Angew. Math. (134), 198–287.

Wang, Q., Xu, P., Zhang, Y., Chu, X., 2017. EPPMiner: an extended benchmark suite for energy, power and performance characterization of heterogeneous architecture. Proceedings of the Eighth International Conference on Future Energy Systems, E-Energy'17. ACM, New York, NY, USA, pp. 23–33.

Warne, D.J., Hayward, R.F., 2015. The dynamics of cellular automata on 2-manifolds is affected by topology. J. Cell. Automata 10, 319–339.

Watson, J.D., Crick, F.H.C., 1953. Molecular structure of nucleic acids: a structure for deoxyribose nucleic acid. Nature 171, 737–738.

Wei, D., Yamada, G., Musha, T., Tsunakawa, H., Tsutsumi, T., Harumi, K., 1990. Computer simulation of supraventricular tachycardia with the Wolff-Parkinson-White syndrome using three-dimensional heart models. J. Electrocardiol. 23, 261–273.

Wiener, N., Rosenblueth, A., 1946. The mathematical formulation of the problem of conduction of impulses in a network of connected excitable elements, specifically in cardiac muscle. Arch. Inst. Cardiol. Mex. 16, 205–265.

Wolfram, S., 1986. Cellular automaton fluids 1: basic theory. J. Stat. Phys. 45, 471–526.

Wuensche, A., 1997. Attractor Basins of Discrete Networks: Implications on Self-Organisation and Memory (PhD). University of Sussex.

Wynters, E., 2018. C++ Amp makes it easy to explore parallel processing on GPUs in a college course or research project. J. Comput. Sci. Coll. 33, 197–199.

Xiao, X., Shao, S.-H., Chou, K.-C., 2006. A probability cellular automaton model for hepatitis B viral infections. Biochem. Biophys. Res. Commun. 342, 605–610.

Yang, S., He, D., Luo, J., Chen, W., Yang, X., Wei, M., Kong, X., Li, Y., Feng, X., Zeng, Z., 2017. Simulation of HIV/AIDS distribution using GIS based cellular automata model. Biomed. Res. 28, 4053–4057.

Yanhua, Z., Wan-Yi, L., Huanhuan, W., Song, H., Eaijun, W., 2017. A bibliographic review of cellular automaton publications in the last 50 years. J. Cell. Automata 12, 475–492.

Yuce, G., 2018. Lattice-Gas Cellular Automata in Modeling Biological Pattern Formation. Illinois State University.

Zenil, H., 2013. Asymptotic behaviour and ratios of complexity in cellular automata. Int. J. Bifurcat. Chaos. 231350159.

Zhang, P., 2017. Modeling the avian influenza H5N1 virus infection in human and analyzing its evolution. In: Yuan, H., Geng, J., Bian, F. (Eds.), Geo-Spatial Knowledge and Intelligence, Communications in Computer and Information Science. Springer Singapore, pp. 339–352.

Zheng, Y., Wei, D., Zhu, X., Chen, W., Fukuda, K., Shimokawa, H., 2015. Ventricular fibrillation mechanisms and cardiac restitutions: an investigation by simulation study on whole-heart model. Comput. Biol. Med. 63, 261–268.

Zorzenon dos Santos, R.M., Coutinho, S., 2001. Dynamics of HIV infection: a cellular automata approach. Phys. Rev. Lett. 87, 168102.

Index

Note: Page numbers followed by *f* indicate figures and *t* indicate tables.

Printed in the United States
By Bookmasters